Microbial Nexus for Sustainable Wastewater Treatment

Microbial ecology is pivotal in wastewater treatment, where microorganisms play a vital role in breaking down organic matter and ultimately reduce the levels of contaminants in treated water, making it safe for reuse in agriculture, industry, and other applications. The book, *Microbial Nexus for Sustainable Wastewater Treatment: Resources, Efficiency, and Reuse*, ventures into the dynamic world of microbial ecosystems, unveiling their pivotal role in reshaping wastewater treatment technologies. This book addresses novel microbial techniques related to sustainable, efficient technologies of wastewater treatment and wastewater reuse as well as obtaining high-quality effluents from treatment plants.

Features:

- Unveils the potential of high-throughput microbial biotechnology for transforming wastewater management.
- Describes the microbial nexus involved in the biodegradation of pharmaceutical micropollutants.
- Highlights the valuable materials recoverable from wastewater, associated challenges, and diverse opportunities arising from effective wastewater management.
- Covers advanced bioremediation technologies designed to handle emerging pollutants. Demonstrates the integration of nanotechnology with bioaugmentation, exploring potential advantages and disadvantages that shape the future of wastewater treatment.
- Provides insights into adopting a circular economy model aligning with sustainable development goals for resource extraction.

This book is tailored for graduate students and researchers in wastewater treatment, waste valorization, environmental engineering, and hazardous waste management.

Environmental Nexus in Waste Management

Series Editors: Vineet Kumar, *Department of Microbiology, School of Life Sciences at the Central University of Rajasthan, Rajasthan, India* and **Sunil Kumar**, *Researcher in Environmental Engineering and Science*

The book series on "Environmental Nexus for Waste Management" addresses novel approaches and techniques linked to sustainable development paths with an interdisciplinary focus on waste resources monitoring, assessment, management, water/wastewater reclamation and recycling as well as discussing solutions for more sustainable use of natural resources. It is a broad compendium of waste management related to research and development and discussions on the most up-to-date tackling strategies. The series of books constitutes an information source and facilitator for the transfer of knowledge, focusing on practical solutions, and better understanding towards achieving sustainable development.

Solid Waste Treatment Technologies: Challenges and Perspectives
Edited by Pratibha Gautam, Vineet Kumar, and Sunil Kumar

Microbial Nexus for Sustainable Wastewater Treatment: Resources, Efficiency, and Reuse
Edited by Vineet Kumar, Sunil Kumar, Pradeep Verma, and Sartaj Ahmad Bhat

For more information about this series, please visit: www.routledge.com/Environmental-Nexus-in-Waste-Management/book-series/CRCENWM

Microbial Nexus for Sustainable Wastewater Treatment

Resources, Efficiency, and Reuse

Edited by
Vineet Kumar, Sunil Kumar, Pradeep Verma,
and Sartaj Ahmad Bhat

CRC Press
Taylor & Francis Group
Boca Raton London New York

CRC Press is an imprint of the
Taylor & Francis Group, an informa business

Designed cover image: shutterstock

First edition published 2025
by CRC Press
2385 NW Executive Center Drive, Suite 320, Boca Raton FL 33431

and by CRC Press
4 Park Square, Milton Park, Abingdon, Oxon, OX14 4RN

CRC Press is an imprint of Taylor & Francis Group, LLC

ISBN: 9781032528595 (hbk)
ISBN: 9781032578040 (pbk)
ISBN: 9781003441069 (ebk)

DOI: 10.1201/9781003441069

Typeset in Times
by Newgen Publishing UK

Dedication

This book is dedicated to my family for their unwavering support, understanding, love, encouragement, and education to date, without them it would not have been possible. ***Vineet Kumar***

Dedicated to our students, teachers, and mentors, from whom I continue to learn, and to my family for their support, blessings, motivation, and love. ***Sunil Kumar***

This book is dedicated to my children and my wife. Their patience and understanding have given me the time and inspiration to research and edit this project. ***Pradeep Verma***

Dedicated to my family, especially my wife without whose support this book would not have been possible. ***Sartaj Ahmad Bhat***

Contents

Preface

Rapidly depleting freshwater supplies and increasing demand for wastewater reclamation or reuse are important aspects of the current scenario. A massive amount of wastewater discharged from various sectors such as domestic, industry, and municipal poses serious environmental pollution and human health problems as the discharged wastewater contains refractory organic and inorganic compounds. However, wastewater also contains valuable resources that can be recovered and reused, making wastewater treatment an important component of the circular economy. Resource recovery from wastewater facilities in the form of energy, reusable water, biosolids, and other resources, such as nutrients, represents an economic and financial benefit that contributes to the sustainability of the water supply. The discharged wastewater also contains energy that can be recovered through processes such as anaerobic digestion, which converts organic matter into biogas that can be used as a renewable energy source. Thus, efficient wastewater treatment processes can reduce the energy and chemical inputs required for treatment, reducing the overall environmental impact of the process that ensures the safe disposal of wastewater into the environment and promotes the circular economy by recovering valuable resources. After effective treatment, the treated wastewater can be reused for irrigation, industry, and other applications, reducing the demand for freshwater resources. Therefore, resource recovery, efficient treatment processes, and water reuse are critical components of sustainable wastewater treatment. This demands emerging wastewater treatment approaches that ensure the safe disposal of wastewater into the environment enabling the sustainability of water and an environmentally beneficial ecosystem.

In recent years, microbial ecology has become a crucial aspect of wastewater treatment, as microorganisms play a vital role in the breakdown of organic matter. The microbial community in wastewater treatment systems is diverse, with a wide range of bacterial, fungal, and yeast species present, which are involved in various metabolic processes that facilitate the breakdown of organic matter and ultimately, reduce the levels of contaminants in treated water, making it safe for reuse in agriculture, industry, and other applications. The microbial nexus refers to the complex interactions among microorganisms, such as bacteria, archaea, fungi, and protozoa in wastewater treatment systems. These interactions can significantly impact the efficiency of the treatment process. Each microorganism has a unique role in separating and removing different contaminants from wastewater. For example, some bacteria are responsible for breaking down organic matter, while others remove nitrogen and phosphorus compounds. The selection of appropriate microbial species and their interactions can enhance the breakdown of organic matter and the removal of contaminants. The complex interactions among microorganisms (microbial nexus) in wastewater treatment can improve treatment efficiency by optimizing the microbial community structure and function, and facilitate water reuse by improving treated water quality. This can reduce the energy and chemical inputs required for wastewater treatment, reducing the overall environmental impact of the process. Overall, the microbial nexus in wastewater treatment plays a critical role in the efficiency and effectiveness of the treatment process.

In view of the above, the book, *Microbial Nexus for Sustainable Wastewater Treatment: Resources, Efficiency, and Reuse*, addresses novel microbial techniques related to sustainable, efficient technologies of wastewater treatment and wastewater reuse as well as obtaining high-quality effluents from treatment plants. The book is spread over 17 chapters, with contributions by leading workers in their fields and drawn from the world over, providing the latest research and development in different aspects of wastewater treatment, resource recovery, and reuse.

Chapter 1 explores omics' approaches to uncover the concealed diversity within microbial groups, both culturable and unculturable. Despite their low relative abundance, these groups play vital roles in significant metabolic pathways. The chapter discusses how these omics' approaches aid in identifying new functions and metabolic potentials. It also provides insights into successfully co-enriching various classified and unclassified microbial strains without the risk of antagonistic effects. In Chapter 2, the focus shifts to the role of microbes in membrane technologies for wastewater reclamation. The chapter delves into the challenges posed by microbes to membranes and membrane-based technologies. It also highlights successful processes that integrate microbes effectively into wastewater treatment. Chapter 3 provides a general overview of anaerobic digestion and its application in removing emerging pollutants. Recent advances in the process are examined, with a focus on key parameters, operation conditions, and applications that enhance microbial communities at each stage of anaerobic digestion. The chapter also presents approaches to fine-tuning microbial communities within anaerobic digestion. Chapter 4 contributes a significant perspective on the nexus of energy and water, water and waste, and waste and energy. Essential insights into the energy recovery perspective for sustainable development are outlined. In Chapter 5, a comprehensive review of different wastes is provided, followed by an examination of traditional waste treatment methods. The chapter categorizes waste materials and explains advances in both conventional and modern waste treatment methods. It emphasizes novel combinations of these approaches and includes a thorough risk analysis of waste treatment technologies. Chapter 6 delves into the microbial nexus involved in the biodegradation of pharmaceutical micropollutants and elucidates their mode of action. Chapter 7 highlights recent progress in the water supply and sanitation sectors, focusing on the circular economy concept. It discusses value-added materials recoverable from wastewater, associated challenges, and diverse opportunities stemming from effective wastewater management. Chapter 8 updates the status of microbial enzymes associated with wastewater management. Chapter 9 concentrates on the microbial nexus in effluent treatment plants. Chapter 10 revisits recent advances in wastewater treatment through electrochemical systems, evaluating their potential applications in future effluent treatment facilities. Chapter 11 discusses microbial technologies and their practical applications, emphasizing economic feasibility as a potential solution to long-term wastewater treatment challenges. Chapter 12 presents insights into valuable resources recoverable from wastewater using a circular economy model, aligning with sustainable development goals. Chapter 13 highlights recent progress in the management, treatment, and recycling of wastewater. It explores advanced physicochemical, biological, and other methods for wastewater treatment and reuse, addressing their environmental sustainability challenges. Chapter 14 evaluates the behavior of organic and inorganic compounds in the dairy industry's effluent, proposing their use as a culture medium for fermenting microorganisms capable of generating microbial cellulose. Chapter 15 unveils the prospects of high-throughput microbial biotechnology for effective wastewater management. Chapter 16 discusses bioremediation technologies for handling emerging pollutants, integrating nanotechnology with bioaugmentation, and exploring potential advantages and disadvantages. Chapter 17 addresses the advantages of using microbial niches in high Andean areas for wastewater treatment. It proposes new strategies to study and recover native microbes with high genetic value for application in different pollutant treatment systems.

Thus, in our opinion, this book is extremely useful for scientists, environmentalists, and ecotoxicologists, working in the field of microbial degradation and bioremediation and is explicitly targeted as teaching material for undergraduate, postgraduate, and more seasoned researchers, and recommended reading for everyone interested in cleaning the polluted environment and making the depleted or degraded water bodies fertile and rejuvenated to maintain sustainability.

Vineet Kumar
Rajasthan, India
Sunil Kumar
Maharashtra, India
Pradeep Verma
Rajasthan, India
Sartaj Ahmad Bhat
Gifu, Japan

Acknowledgments

Writing a book is never a solitary endeavor, and this book would not have been possible without the help, support, and encouragement of many people. We would like to take this opportunity to express our sincere gratitude to all the individuals and organizations who have contributed to the development and publication of this book, *Microbial Nexus for Sustainable Wastewater Treatment: Resources, Efficiency, and Reuse*. Their dedication, expertise, and support have been instrumental in making this project a success.

First and foremost, we would like to thank the contributing authors for their insightful contributions, expertise, and dedication to the subject matter. Their contributions have helped to create a comprehensive and informative resource for researchers, practitioners, and students in the field of wastewater treatment and resource recovery.

We would also like to thank the reviewers for their time, effort, and commitment to ensuring a high-quality manuscript for this book. Their constructive feedback and suggestions have been invaluable in improving the content, structure, and clarity of the book.

We would also like to express our gratitude to our colleagues, collaborators, and mentors who have shared their expertise, insights, and perspectives with us. Their guidance, feedback, and constructive criticism have been invaluable in shaping this book. Thank you all for your contributions, support, and encouragement.

We are deeply grateful to CRC Press (Taylor & Francis Group) for the opportunity to undertake this project and share the knowledge with the academic and scientific fraternity. We are particularly indebted to Dr. Gagandeep Singh, Senior Publisher (Engineering) at CRC Press (Taylor & Francis Group India Pvt. Ltd.), for the execution of the publishing agreement and valuable suggestions. We are humbled by your unconditional support and encouraged by your commitment for timely delivery of the book manuscript for publication.

We want to acknowledge Ms. Jahnavi Vaid, Editorial Assistant at CRC Press (Taylor & Francis Group), India, for her constructive criticism, sound advice, and invaluable support throughout the project.

We are thankful to Anitha AL, project manager at Newgen Knowledge Works P Ltd., Chennai, India, for the skillful organization and management of the entire book project in many ways (illustrations, graphics/book cover designing, copyediting, typesetting, proofing, etc.).

We would also like to acknowledge the staff of the publisher, CRC Press, Taylor & Francis Group, Florida, who have worked tirelessly to bring this book to fruition. Their professionalism, attention to detail, and dedication to quality have been crucial in ensuring that this book meets the highest standards of excellence.

We hope that this book will make a meaningful contribution to the field of wastewater treatment and the broader goal of achieving sustainable and environmentally friendly solutions for managing our precious water resources

Finally, we would like to express our gratitude to the readers of this book, who we hope will find the content informative, engaging, and useful in their respective fields. We hope that this book will inspire and inform your thinking about sustainable wastewater treatment, resource recovery, efficient treatment processes, and water reuse, and inspire further research and innovation in this important area. We welcome your feedback and comments, and we look forward to continuing the conversation about these important issues.

Vineet Kumar
Sunil Kumar
Pradeep Verma
Sartaj Ahmad Bhat

Editor Biographies

Vineet Kumar is currently a National Postdoctoral Fellow in the Department of Microbiology, School of Life Sciences at the Central University of Rajasthan, Rajasthan, India. He got his M.Sc. (2008) and M.Phil. (2012) in Microbiology at Ch. Charan Singh University, Meerut, India. Subsequently, he earned his Ph.D. (2018) in Environmental Microbiology from Babasaheb Bhimrao Ambedkar (A Central) University, Lucknow, India, focusing on the treatment of distillery effluent in a two-step treatment system toward environmental sustainability. Dr. Kumar's research work has mainly been dedicated to wastewater treatment and solid waste management, focusing on the valorization of agro-industrial wastes for bioenergy production and the development of sustainable approaches to mitigate the environmental impact of refractory pollutants of discharge in wastewater and sludge. As a postdoctoral fellow, he works on the role of biofilm-forming microbial communities in situ and ex situ remediation of pulp-paper mill waste. In recent years, his research has been conducted in collaboration with some of the leading international environmental microbiology and biotechnology research groups across the world (Brazil, Mexico, China, USA, among others) with diverse expertise. He has published more than 55 articles in peer-reviewed international journals of repute, 24 books, and 52 book chapters, on various aspects of science and engineering, with more than 2800 citations, and a h-index of 31 (Google Scholar). Dr. Kumar has been serving as a guest editor and reviewer for more than 65 prestigious international journals. He has served on the editorial board of various reputed journals. Dr. Kumar is a recipient of several prestigious fellowships such as the Rajiv Gandhi National Fellowship by the University Grants Commission, India, and the National Postdoctoral Fellowship by the Science and Engineering Research Board (SERB), Department of Science and Technology, Government of India. He has presented several papers relevant to his research areas at national and international conferences. He is an active member of numerous scientific societies including the Microbiology Society (UK), the Indian Science Congress Association (India), the Association of Microbiologists of India (India), etc. He is serving as the President of the Society for Green Environment, India (website: www.sgeindia.org). He can be reached at drvineet.micro@gmail.com; vineet.way18@gmail.com.

Sunil Kumar is a well-rounded researcher with more than 24 years of experience in leading, supervising, and undertaking research in the broader field of environmental engineering and science with a focus on solid and hazardous waste management. Dr. Kumar is a graduate in Environmental Engineering and Management from the Indian Institute of Technology, Kharagpur, India. He completed his Ph. D. in Environmental Engineering from Jadavpur University, Kolkata, India. His primary area of expertise is solid waste management (municipal solid waste, electronic waste etc.) over a wide range of environmental topics including contaminated sites, EIA, and wastewater treatment. His contributions in these fields have led to 20000 citations, a h-index of 69, and an i10-index of 285 (Google Scholar). His contributions since inception at the CSIR-National Environmental Engineering Research Institute (NEERI), India in 2000 include 261 refereed publications, five books, 40 book chapters, 10 edited volumes, and numerous project reports to various governmental bodies and private, local, and international academic/research bodies. He is the Associate Editor of peer-reviewed journals of international repute, i.e., Environmental Chemistry Letter, International Journal of Environmental Science & Technology, ASCE Journal of Hazardous, Toxic and Radioactive Waste. He also served on the Editorial Board of Bioresource Technology, Elsevier. He has completed many research projects as principal investigator with 18 (13 awarded) Ph. D. and 20 M. Phil/M. Tech theses/dissertations. Dr. Kumar was awarded the most prestigious award, the Alexander von Humboldt-Stiftung Jean-Paul-Str.12 D-53173 Bonn, Germany as a senior researcher for developing a global network and excellence for more advanced research and technology innovation.

Pradeep Verma is working as Professor in the Department of Microbiology, School of Life Sciences at Central University of Rajasthan, Rajasthan, India. He is a well-rounded researcher with more than 21 years of experience in leading, supervising, and undertaking research in the broader field of bioprocess and bioenergy production from lignocellulosic waste with a focus on waste management. He earned his PhD in Microbiology from Sardar Patel University Gujarat, India in 2002. His research area of expertise involves microbial diversity, bioremediation, bioprocess development, lignocellulosic and algal biomass based biorefinery. He has more than 75 research articles in peer-reviewed international journals and has contributed to 50 book chapters in different edited books with more than 6500 citations, and a h-index of 42. He has also edited four books by international publishers such as Springer, Taylor and Francis, CRC Press, and Elsevier. He holds 12 international patents in the field of microwave assisted biomass pretreatment and bio-butanol production. He is a guest editor to several journals such as Biomass Conversion and Biorefinery (Springer), Frontier in Nanotechnology (Frontiers), and International Journal of Environmental Research and Public Health (MDPI). He is also an editorial board member for the journal Current Nanomedicine (Bentham Sciences). He is acting as reviewer for more than 40 journals in different publication houses such as Springer, Elsevier, RSC, ACS, Nature, Frontiers, MDPI, etc. He is also a recipient of various prestigious fellowships and awards, such as the JSPS Postdoctoral Fellowship and Ron-Cockcroft award by Swedish society, UNESCO Fellow ASCR Prague. He has been awarded with a Fellow of Mycological Society of India (MSI-2020), Prof. P.C. Jain Memorial Award, Mycological Society of India 2020, and Fellow of Biotech Research Society, India (2021). He is a member of various national and international societies/academies.

Sartaj Ahmad Bhat is working as a JSPS postdoctoral researcher at the River Basin Research Center, Gifu University, Japan. He received his Ph.D. in Environmental Sciences from Guru Nanak Dev University, Amritsar, India in 2017. His research interests focus on vermicomposting treatment of various solid wastes, especially investigations on the fate and behavior of emerging pollutants during biological treatment of organic wastes. He has published more than 60 papers in peer-reviewed journals and edited over 13 books published by Elsevier Science, CRC Press, and Springer Nature. Dr. Bhat is serving as an Associate/Academic Editor and Editorial Board Member/Advisory Board Member of more than 15 journals published by Frontiers, Springer, Elsevier, PLOS, Hindawi, and De Gruyter. Dr. Bhat is a recipient of several prestigious awards such as the JSPS Postdoctoral Fellowship to pursue research at the River Basin Research Center, Gifu University, Japan, the Basic Scientific Research Fellowship (BSR JRF, SRF) by the University Grants Commission (UGC) India, the DST-SERB National Postdoctoral Fellowship at CSIR-NEERI, Nagpur, India, and Swachhta Saarthi Fellowship by the Govt. of India. Dr. Bhat has also received the 2020 Outstanding Reviewer Award by the International Journal of Environmental Research and Public Health, MDPI, and Top Peer Reviewer 2019 Award in Environment and Ecology by Web of Science and has more than 750 verified reviews and 65 editor records to his credit.

Contributors

Alnaqbi, Mariam
Department of Chemical and Biological Engineering
American University of Sharjah
Sharjah, United Arab Emirates

Al-Othman, Amani
Department of Chemical and Biological Engineering
American University of Sharjah
Sharjah, United Arab Emirates

Abdulla, Hesham
Lecturer of Microbiology
Botany and Microbiology Department
Faculty of Science
Suez Canal University
Egypt

Ayeleru, Olusola Olaitan
Centre for Nanoengineering and Tribocorrosion
University of Johannesburg
Johannesburg, South Africa

Acosta, Richard Andi Solorzano
Dirección de Supervisión y Monitoreo en las Estaciones Experimentales Agrarias, Instituto Nacional de Innovación Agraria (INIA), Lima, Peru

Behera, Surendra
Department of Botany
Fakir Mohan University
Balasore, Odisha, India

Cesca, Karina
Department of Chemical Engineering and Food Engineering
Federal University of Santa Catarina
Florianópolis, Brazil

Carciofi, Bruno Augusto Mattar
Department of Chemical Engineering and Food Engineering
Federal University of Santa Catarina
Florianópolis, Brazil

Cancino, Milagros Estefani Alfaro
Faculty of Pharmacy and Biochemistry
Biotechnology and Omics in Life Sciences Research Group
Universidad Nacional Mayor de San Marcos
Lima, Peru

Corona, Francisco
CARTIF Technology Center
Circular Economy Area
Valladolid, Spain

de Andrade, Cristiano José
Department of Chemical Engineering and Food Engineering
Federal University of Santa Catarina, Florianópolis
Brazil

Dwivedi, Shruti
Department of Biotechnology
Deen Dayal Upadhyay Gorakhpur University
Gorakhpur, Uttar Pradesh, India

Dlova, Sisanda
Centre for Nanoengineering and Tribocorrosion
University of Johannesburg
Johannesburg, South Africa

Eltaher, Nour
Department of Chemical and Biological Engineering
American University of Sharjah
Sharjah, United Arab Emirates

Gutkoski, Júlia Pedó
Department of Chemical Engineering and Food Engineering
Federal University of Santa Catarina
Florianópolis, Brazil

Gupta, Supriya
Department of Biotechnology
Deen Dayal Upadhyay Gorakhpur University
Gorakhpur, Uttar Pradesh, India

Giyahchi, Minoo
Department of Microbiology
School of Biology, College of Science
University of Tehran
Tehran, Iran

Harirchi, Sharareh
Swedish Centre for Resource Recovery
University of Borås
Borås, Sweden

Hidalgo, Dolores
CARTIF Technology Center
Circular Economy Area
Valladolid, Spain

Hualpa-Cutipa, Edwin
Faculty of Pharmacy and Biochemistry
Biotechnology and Omics in Life Sciences Research Group
Universidad Nacional Mayor de San Marcos
Lima, Peru

Hossain, Anamica
Department of Microbiology
University of Dhaka
Dhaka, Bangladesh

Hossain, M. Anwar
Department of Microbiology
University of Dhaka
Dhaka, Bangladesh

Ismail, Engy
Department of Chemical and Biological Engineering
American University of Sharjah
Sharjah, United Arab Emirates

Kaushik, Garima
Department of Environmental Science
School of Earth Sciences
Central University of Rajasthan
Bandar Sindri, Ajmer, Rajasthan, India

Kelbert, Maikon
Department of Chemical Engineering and Food Engineering
Federal University of Santa Catarina
Florianópolis, Brazil

Kumar, Anil
School of Chemical and Metallurgical Engineering
University of the Witwatersrand
Johannesburg, South Africa;
DSI/NRF SARChI: Hydrometallurgy and Sustainable Development
University of the Witwatersrand
Johannesburg, South Africa

Lak, Mandana
Department of Microbiology
Faculty of Biological Sciences
Alzahra University
Tehran, Iran

Langbehn, Rayane Kunert
Department of Chemical Engineering and Food Engineering
Federal University of Santa Catarina
Florianópolis, Brazil

Luna-Maldonado, Alejandro Isabel
Universidad Autónoma de Nuevo León
Facultad de Agronomía
Laboratorio de Ingeniería. Av. Francisco Villa S/N
Colonia Ex Hacienda El Canadá
General Mariano Escobedo
México

Luis-Alaya, Bernabe
Departamento de Microbiología y Parasitología, Facultad de Ciencias Biológicas, Universidad Nacional Mayor de San Marcos, Lima-Perú

Mannrich, Adrielle Helena
Department of Chemical Engineering and Food Engineering
Federal University of Santa Catarina
Florianópolis, Brazil

Michels, Camila
Department of Chemical Engineering and Food Engineering
Federal University of Santa Catarina
Florianópolis, Brazil

Mirshafiei, Mojdeh
Department of Biotechnology
School of Chemical Engineering
College of Engineering
University of Tehran
Tehran, Iran

Mekuto, Lukhanyo
Department of Chemical Engineering
University of Johannesburg
Doornfontein Campus
Johannesburg, South Africa

Martín-Marroquín, Jesús M.
CARTIF Technology Center
Circular Economy Area
Valladolid, Spain

Modekwe, Helen Uchenna
Renewable Energy and Biomass Research Group
Department of Chemical Engineering
University of Johannesburg
Johannesburg, South Africa

Moghimi, Hamid
Department of Microbiology
School of Biology, College of Science
University of Tehran
Tehran, Iran

Márquez-Reyes, Julia Mariana
Universidad Autónoma de Nuevo León
Facultad de Agronomía
Laboratorio de Biotecnología Microbiana. Av. Francisco Villa S/N
Colonia Ex Hacienda El Canadá
General Mariano Escobedo
México.

Méndez-Zamora, Gerardo
Universidad Autónoma de Nuevo León
Facultad de Agronomía, Av. Francisco Villa S/N
Colonia Ex Hacienda El Canadá
General Mariano Escobedo
México

Nojoumi, Seyed Ali
Microbiology Research Center
Pasteur Institute of Iran
Tehran, Iran;
Department of Mycobacteriology and Pulmonary Research
Pasteur Institute of Iran
Tehran, Iran

Ndlovu, Sehliselo
School of Chemical and Metallurgical Engineering
University of the Witwatersrand
Johannesburg, South Africa;
DSI/NRF SARChI: Hydrometallurgy and Sustainable Development

University of the Witwatersrand
Johannesburg, South Africa

Nizhum, Aksa Hossain
Department of Microbiology
University of Dhaka
Dhaka, Bangladesh

Olubambi, Peter Apata
Centre for Nanoengineering and Tribocorrosion
University of Johannesburg
Johannesburg, South Africa

Ouda, Amira
Botany and Microbiology Department
Faculty of Science
Suez Canal University
Egypt

Poddar, Kasturi
Department of Biotechnology and Medical Engineering
National Institute of Technology Rourkela
Odisha, India

Palsania, Preksha
Department of Environmental Science
School of Earth Sciences
Central University of Rajasthan
BandarSindri, Ajmer, Rajasthan, India

Rezaei, Zeinab
Department of Microbiology
School of Biology, College of Science
University of Tehran
Tehran, Iran

Ratnasari, Anisa
Department of Environmental Engineering
Institut Teknologi Sepuluh Nopember
Surabaya, Indonesia

Ramezani, Mohadasseh
Microorganisms Bank
Iranian Biological Resource Centre (IBRC)
ACECR, Tehran, Iran

Rasekh, Behnam
Environment and Biotechnology Division
West Blvd. of Azadi Sport Complex
Research Institute of Petroleum Industry (RIPI)
Tehran, Iran

Ramatsa, Ishmael Matala
Renewable Energy and Biomass Research Group
Department of Chemical Engineering
University of Johannesburg
Johannesburg, South Africa

Rodríguez-Romero, Beatriz Adriana
Universidad Autónoma de Nuevo León
Facultad de Agronomía, Av. Francisco Villa S/N
Colonia Ex Hacienda El Canadá
General Mariano Escobedo
México

Shailesh, Shiromi
Department of Environmental Science
School of Earth Sciences
Central University of Rajasthan
BandarSindri, Ajmer, Rajasthan, India

Sultana, Munawar
Department of Microbiology
University of Dhaka
Dhaka, Bangladesh

Sharma, Sakshi
Department of Environmental Science
School of Earth Sciences
Central University of Rajasthan
BandarSindri, Ajmer, Rajasthan, India

Shyu, Douglas J. H.
Department of Biological Science and Technology
National Pingtung University of Science and Technology
Pingtung, Taiwan

Salazar-Manzanares, Martha Irene
Universidad Autónoma de Nuevo León
Facultad de Agronomía
Laboratorio de Biotecnología Microbiana. Av. Francisco Villa S/N
Colonia Ex Hacienda El Canadá
General Mariano Escobedo, México

Sahu, Jyotsna Rani
Department of Botany and Biotechnology
Ravenshaw University
Cuttack, Odisha, India

Sayed, Rania
Materials Testing and Surface Chemical Analysis Lab
National Institute of Standards
Egypt

Singh, Shubhra
Department of Tropical Agriculture and International Cooperation,
National Pingtung University of Science and Technology
Pingtung, Taiwan;
Department of Biological Science and Technology
National Pingtung University of Science and Technology
Pingtung, Taiwan.

Sarkar, Debapriya
Department of Biotechnology and Medical Engineering
National Institute of Technology Rourkela
Odisha, India

Sarkar, Angana
Department of Biotechnology and Medical Engineering
National Institute of Technology Rourkela
Odisha, India

Shahin, Moh'd Basel
Department of Chemical and Biological Engineering
American University of Sharjah
Sharjah, United Arab Emirates

Soares, Hugo Moreira
Department of Chemical Engineering and Food Engineering
Federal University of Santa Catarina
Florianópolis, Brazil

Tawalbeh, Muhammad
Sustainable and Renewable Energy Engineering Department
University of Sharjah, Sharjah, United Arab Emirates;
Sustainable Energy & Power Systems Research Centre
RISE, University of Sharjah
Sharjah, United Arab Emirates

Tanveer, Aiman
Department of Biotechnology
Deen Dayal Upadhyay Gorakhpur University
Gorakhpur, Uttar Pradesh, India

Taher, Heba Saad
Lecturer of Microbiology
Botany and Microbiology Department
Faculty of Science
Suez Canal University
Egypt

Wahdan, Sara Fareed Mohamed
Lecturer of Microbiology
Botany and Microbiology Department

Faculty of Science
Suez Canal University
Egypt

Yazdani, Ramin
Department of Life Science Engineering
Faculty of New Sciences and Technologies
University of Tehran
Tehran, Iran

Yazdian, Fatemeh
Department of Life Science Engineering
Faculty of New Sciences and Technologies
University of Tehran
Tehran, Iran

Yadav, Kanchan
Department of Biotechnology
Deen Dayal Upadhyay Gorakhpur University
Gorakhpur, Uttar Pradesh, India

Yadav, Dinesh
Department of Biotechnology
Deen Dayal Upadhyay Gorakhpur University
Gorakhpur, Uttar Pradesh, India

1 Enhancing Microbial Cooperation within the Microbial Nexus-based System

Zeinab Rezaei, Minoo Giyahchi, and Hamid Moghimi

1.1 FUNCTIONAL MICROBIAL COMMUNITIES IN WASTEWATER TREATMENT: ECOLOGICAL AND BIOGEOCHEMICAL PERSPECTIVES

Biological wastewater treatment plants (WWTPs) are one of the most critical applications in biotechnology and play key roles in wastewater treatment (WWT). These plants' stable and effective functions depend on microbial communities' structures, diversity, and richness in activated sludge or biofilm structures. Given the importance of microbial metabolic activities in wastewater treatments, understanding the composition of their community, how they interact with each other, and environmental factors is crucial (Zhang et al., 2019; Wani et al., 2022). Biological WWT can be carried out aerobically or anaerobically depending on the source and concentration of the pollutants, the treatment's objectives, and the waste's nature. It has been believed that the main population of microbial cells in aerobic plants are bacteria and protozoa. Fungi and rotifers were assumed to be present in lower abundance; there was no information about archaea and their participation in the treatment process, and the diversity of the microbial population was believed to be low (Sperling & De Lemos Chernicharo, 2005). Recently, new insights about microbial diversity in natural and artificial environments have been revealed thanks to molecular biology. It is now proved that activated sludge, as an artificial ecosystem that is the most applied biological treatment community, harbors thousands of operational taxonomic units and more than 700 known genera of microorganisms including bacteria, fungi, yeasts, and protozoa. Previous studies have shown that a high diversity of microbial communities will keep the system more stable (Chen et al., 2017; Kumar & Chandra, 2020). Bacteria, the most abundant group of microorganisms in WWT, are the key players in biological oxygen demand decrements, implemented by heterotrophic ones and determinant in shaping the structure of flocs, biofilms, and granules (Sperling

DOI: 10.1201/9781003441069-1

& De Lemos Chernicharo, 2005) that are the heart of biological treatment plants. In fact, the physicochemical properties of activated sludge are defined by its microbial composition and richness. Filamentous bacteria, the most important group in flocs' stability and integrity, are present in a low abundance in WWTPs. The overgrowth of these groups is the leading cause of sludge bulking, one of the main operating bottlenecks of WWTPs (Cydzik-Kwiatkowska & Zielińska, 2016). Protozoa, as one of the key players in the microbial loop, besides playing a role in organic matter removal and participating in shaping flocs, could control the population of bacterial cells in the community structure. Their contribution is crucial to the deterioration of suspended solids and pathogens, resulting in higher effluent quality (Sperling & De Lemos Chernicharo, 2005). Viruses (phages) are the most prevalent biological forms in almost every natural and artificial habitat, including WWTPs, playing the most crucial role in maintaining microbial diversity, biological elemental cycle performance, microbial evolution, horizontal gene transfer, and in some cases, harboring auxiliary metabolic genes (Chevallereau et al., 2022) that could help the host degrade xenobiotics and emerging compounds. A recent study revealed that some phages encode antibiotic synthesis genes that could give their host a survival advantage over their competitors (Fan et al., 2023).

From an ecological perspective, each habitat is occupied by three main groups of microorganisms: generalists, transients, and specialists, based on their presence frequency. Generalists are frequent residents of an environment and those who can adapt to a wide range of habitats. Transient populations are the type of species that appear in the presence of certain circumstances in the niche and disappear if the environmental factors receive signals from surrounding changes. Specialists are the species adapted to the conditions mainly unique to that niche and not found in every analogous environment. Specialists are the group that could make the system more functional and heterogeneous (Yu et al., 2021; Lynch & Neufeld, 2015) and may play an important role in the removal of emerging compounds due to their adaptation capability. The mechanism and the reason why a microorganism becomes generalist, transient, or specialist is not yet fully understood, but it has been revealed that specialists are more competitive than the generalist population and may have specific metabolic requirements or interactions (Xu et al., 2021; Yu et al., 2021). Therefore, interspecies interactions are also key factors in determining community composition by influencing species fitness in their habitat. New information about amplicon sequence variants of different full-scale aerobic and anaerobic WWTPs around the world has shown that the generalist group in activated sludge systems includes *Bacteroidetes*, *Planctomycetes*, *Proteobacteria*, and *Chloroflexi* phyla. The abundance of the transient and specialist groups was estimated to be almost 67% and 27%, respectively. In anaerobic digesters (ADs), the generalists include *Proteobacteria*, *Bacteroidetes*, and *Euryarchaeota* but in various abundances in different samples with nearly 68% and 19% transient and specialist taxa, respectively. In the same report, the presence of aromatic and ammonia-based compounds in AD was shown to be related to increases in the abundance of *Planctomycetes* as the generalists that harbor nitrogen cycle genes (Yu et al., 2021).

From the biogeochemical perspective, nitrogen, phosphorus, and carbon cycles are the primary elemental cycles in WWTPs. Reactive nitrogen compounds are crucial for living organisms, but in excess amounts can cause serious harm to all trophic levels. Nitrification and denitrification have long been known and applied in wastewater treatment for reactive nitrogen removal of ammonia-oxidizing, nitrite-oxidizing bacteria and heterotrophic denitrificant ability. *Nitrosomonas* and *Nitrosospira* are two well-known genera in ammonium oxidation that carry out the first subreaction of nitrification. *Nitrobacter* and *Nitrosospira* are genera known for their nitrite oxidation ability. Denitrification is the second reaction in reactive nitrogen removal and is known to be carried out by heterotrophic denitrificants, mainly *Pseudomonas* and *Bacillus* strains. It has long been believed that nitrification–denitrification is the only pathway to the complete removal of ammonium, and the groups mentioned above have been considered key players in this process. Oxygen supply and organic substances are the two main requirements of nitrification and denitrification, respectively; both making conventional nitrogen removal a highly energy-consuming process. In recent years, with the advance of molecular biology approaches that eliminate the necessity of culturing for the identification of microorganisms and their metabolic potentials, versatile new pathways in elemental cycles, including nitrogen have been discovered in which microorganisms use alternative electron donors and acceptors:

1. Anammox, or anaerobic ammonium oxidation is a new N-cycling pathway, discovered and applied in WWTPs. In this pathway, ammonium is oxidized to N_2 by reducing nitrite as the electron acceptor. This process has been identified in five *Candidatus* bacteria with unique physiological properties. These bacteria harbor a cellular compartment called an "anammoxosome," which contains a specific membrane lipid compound called "ladderane" lipids. The anammox is carried out in this compartment. There are several advantages of anammox over conventional denitrification, the most important being energy saving as it is a set of reactions run by autotrophic bacteria without requiring organic carbon. Acquiring more energy through methane production in ADs from excess organic materials is another perspective that could be considered when using anammox in nitrogen removal. A higher affinity to nitrite is a further property of these bacteria compared to heterotrophic denitrificants, making this group more nitrite tolerant. These features make anammox bacteria good candidates for large-scale biological nitrogen removal (Ren et al., 2020; Peeters & Niftrik, 2019), but there are still several limitations that make using this process hard to scale up: 1) the slow growth rate of anammox bacteria; and 2) the nitrite requirement in influent (Zhu et al., 2022; Zhu et al., 2016).
2. Comammox, or complete ammonium oxidation has recently been discovered in some of the *Candidatus Nitrospira* species and can carry out the nitrification process in a single reaction. These bacteria show better growth and performance in nutrients and copper deficiency (which suppresses ammonium monooxygenase (AMO)) compared to ammonium oxidizing bacteria (AOB)

because of the higher affinity to ammonium and a more diverse substrate range. Furthermore, some reports indicate that comammox *Nitrospira* can tolerate lower dissolved oxygen and hence, outcompete AOB in the conditions mentioned earlier and make them dominant nitrifiers. Some studies have also shown comammox bacteria to be the most abundant group in nitrogen removal at a high retention time and attached growth-based plants (Vilardi et al., 2022; Roots et al., 2019).

3. Feammox, or anaerobic ammonium oxidation coupled to the reduction of Fe(III) (Zhu et al., 2022) is another chemolithotrophic pathway of ammonium oxidation, using ferric iron as an electron acceptor. This is a unique metabolic form, first found in the *Acidimicrobiaceae bacterium* A6. Without the need for an oxygen supply and reducing N_2O evolution, feammox is less energy consuming and more efficient in ADs. Furthermore, the potential of cometabolism in feammox bacteria to degrade chlorinated compounds alongside nitrogen removal brings hope that these newly investigated metabolism pathways can also remediate new organic contaminants. However, more investigations are needed to evaluate the effect of Fe(III) on anaerobic nitrogen removal and enrichment of feammox bacteria in WWTPs, as this method is applied in wetlands and not yet in full-scale WWTPs (Ren et al., 2020).
4. Denitrifying anaerobic methane oxidation (DAMO), in which ammonium oxidation occurs through transferring electrons to methane as the final electron acceptor, is found in both bacterial and archaeal species. These species show a higher affinity to nitrite than canonical denitrificants, and by consuming methane as a greenhouse gas, they have the advantage of being an environmentally friendly nitrogen denitrificant. DAMO strains can be enriched in bioreactors that are connected to AD output to supply methane as their electron acceptor. One of the limitations facing the application of this group is their long doubling time, making them still kinetically far from being applied in full-scale nitrogen removal plants (Fu et al., 2017; Luesken et al., 2011; Shi et al., 2013; Ren et al., 2020).
5. Dirammox, or direct ammonium oxidation, was first found in *Alcaligenes ammonioxydans* HO-1 and has now been discovered in some *Alcaligenes* species. This pathway is different from denitrification and anammox in the way that it occurs aerobically. N_2 is produced directly from ammonium through hydroxylamine (NH_2OH), making this pathway the most direct ammonium oxidation reaction among the known pathways and the least in terms of electron demand. As a consequence, this pathway requires less organic matter supply. Furthermore, the high tolerance to ammonium concentration makes this group of bacteria good candidates in high reactive nitrogen concentration wastewater removal (Hou et al., 2022; Pous et al., 2023).

Phosphorus (P) is an element mainly sourced from anthropogenic activities that contributes to aquatic eutrophication and whose efficient removal from wastewater is essential. However, eco-friendly and cost-effective biological removal of this element is challenging due to its operator dependence and sensitivity to fluctuation processes. Biological P removal (BPR) is mainly carried out by polyphosphate accumulating organisms (PAOs), which, due to this potential, have the advantage over other heterotrophs and are suitable for application in the enhanced biological phosphorus removal (EBPR) process. EBPR can be applied in membrane bioreactors, granular sludge reactors, and sequencing batch film bioreactors. *Accumulibacter* is the most common and dominant PAO genera, but other heterotrophic genera, including *Pseudomonas* and *Enterobacter*, have also been detected in EPBR systems. It has been proved that the number of PAOs directly affects the phosphorus removal efficiency. PAOs grow aerobically and accumulate phosphate in the aerobic phase of treatment. Their metabolic activity depletes the oxygen and organic supplies, causing the formation of an anaerobic zone in which accumulated polyphosphates are fermented to volatile fatty acids, which can be consumed as their energy source. The anaerobic phase is essential in P removal as in the presence of oxygen or nitrate, the accumulated phosphate will release and render the treatment process (M.P.C.A, 2006; Bunce et al., 2018). Although it has not yet been widely scaled-up, microalgal strains can accumulate polyphosphates in the absence of inorganic phosphorus. *Scenedesmus* sp. and *Chlorella* sp. are two examples of microalgae that can be used in P removal systems (Azad & Borchardt, 1970), especially in the form of a biofilm in which their tolerance to fluctuations in environmental factors increase the rate of the removal (up to 98%). Despite their efficiency, algal system-based WWTPs are expensive due to the sloughing effect and, therefore, hard to scale up (Bunce et al., 2018).

The sulfur cycle is another process that could occur in sulfate-containing WWT. Sulfate-reducing bacteria (SRB), by reducing sulfate to sulfite and finally sulfide, are key players in sulfate-containing waste treatment. These bacteria are heterotrophic and can consume organic carbon as the electron donor in this process. Therefore, SRB could interlink the sulfur and carbon cycles in anaerobic plants. Sulfammox or sulfate-reducing ammonium oxidation is a pathway in which ammonium can reduce to nitrite through passing electrons to sulfate as the final electron acceptor. This reaction exemplifies the nexus system in which two elemental cycles can be interconnected (Fig. 1.1). Sulfammox can be efficient in some industrial wastewater treatments with a high load of ammonium and sulfate, which are characteristic of seafood, paper, and sugar production, etc. (Rikmann et al., 2014). This reaction is sulfide-free and is a toxic intermediate. Furthermore, the elemental sulfur (S^0) that is produced in this process is a valuable product that could be recovered from the system and used as the electron donor in other systems that use autotrophic denitrificants for nitrogen removal (Toro & Cervantes, 2019). Sulfammox can be co-employed with anammox in a single plant to treat N and S removal simultaneously, which is sufficient for high nitrogen and sulfate-containing wastes (Dominika et al., 2021). These potentials, although they seem interesting, are still far from being applied in full-scale WWTPs as complete information about their operation has not been revealed.

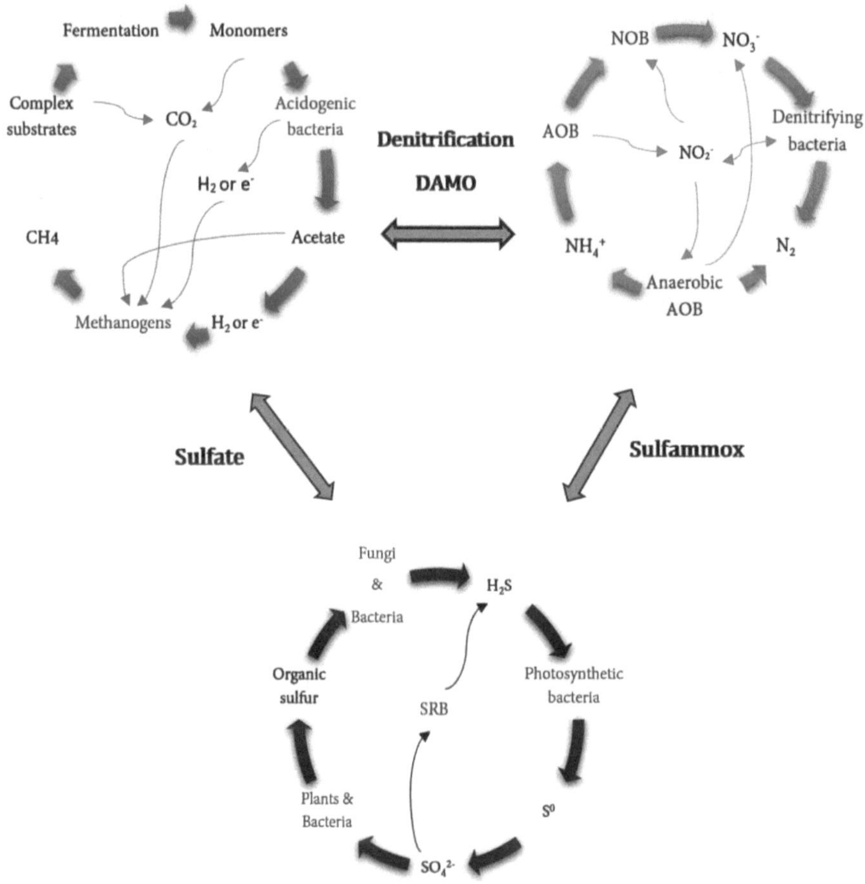

FIGURE 1.1 Interconnection of biogeochemical cycles.

1.2 MICROBIAL NEXUS-BASED SYSTEMS IN WASTEWATER TREATMENT: IMPORTANCE, FUNCTIONS, AND INTERACTIONS

Despite discovering new metabolic pathways and potentials that provide hope of overcoming the limitations in WWTPs' operation during past years, these systems are still facing challenges due to the presence of emerging compounds that threaten the biota's health and decrease discharged water quality (Kumar et al., 2022a, 2022b). As a new concept in sustainable development, the nexus refers to the interlinking and connection of water, food resources, energy, and ecosystems, and applying this concept to WWT to access sustainable treatment is one of the new approaches in this field (Wu & Yin, 2020). Consequently, a microbial nexus means tuning microbial activity and bioremediating potential through optimizing this network. It is proven that the microbial community structure, richness, abundance, and activity, which depend on their adaptation and selection in certain conditions, are directly

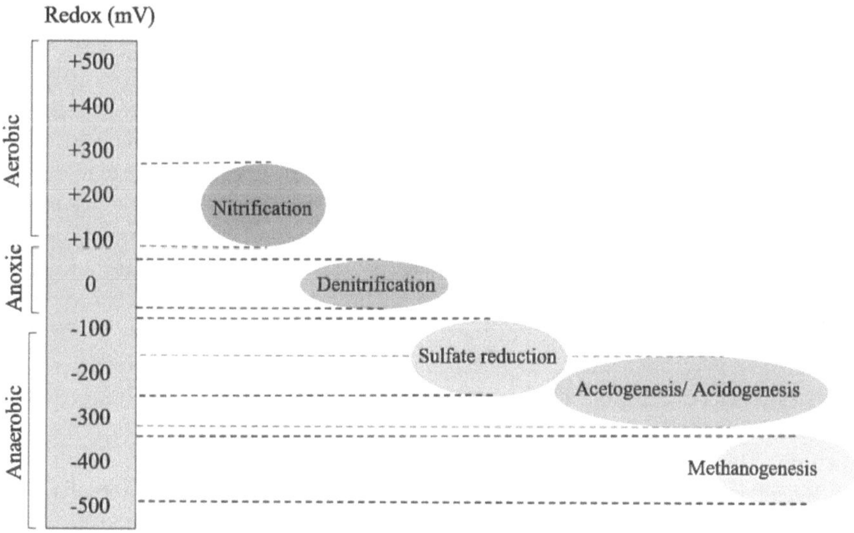

FIGURE 1.2 The range of redox potential of nitrogen, sulfur, and carbon cycles' pathways.

determined and affected by the interaction of microbial strains with either the environmental conditions or other microorganisms. To optimize and enhance this interconnection, complete information about the composition and metabolic activity of the microbial community is required. New molecular biology approaches have helped to access the more extensive library of this information. However, many questions remain unanswered as this subject is so complicated and newly discovered. It has been reported that many of the families and genera of the AS and AD systems are among the unclassified taxa, which means their metabolic potential is unrevealed (Yu et al., 2021).

Conventional wastewater treatment is mainly applied for carbon, nitrogen, and phosphorus removal in the range of the determined standards, each driven by different microbial groups and in different abiotic conditions at certain potential value ranges (Fig. 1.2) that should be considered in designing and implementing WWTPs to achieve an acceptable removal rate and quality. This objective can be achieved through the niche-based design of microbial nexus systems.

According to the microbial nexus concept, biogeochemical cycles could also be interconnected. The connection between denitrification and carbon, sulfur, and phosphorus cycles are interlinked, and denitrificants, with their diverse metabolic pathways, could act as the interlinker of the nitrogen cycle to the carbon, sulfur, and phosphorus cycles.

1.2.1 Microbial Cooperation in WWTPs

The presence of diverse microbial species in the WWTPs' microbiome leads to the formation of a network of interactions in both spatial and temporal scales (Sidhu et al., 2017; Hibbing et al., 2010). Microbial interactions are mediated by different metabolites like exoenzymes, exopolysaccharides (in biofilm structures), signaling molecules, siderophores, etc. (Cremer et al., 2019).

The carbon cycle in ADs is an example of cooperating bioremediation of wastewater. In these systems, organic carbon turns into intermediate metabolites, which are taken up by acidogenic microorganisms. This group converts these metabolites to hydrogen and acetate, the main substrates for methanogenic bacteria. Methanogens are an anaerobic group of microorganisms that produce methane from CO_2 and hydrogen or CO_2 and acetate as carbon and energy sources, respectively. The syntrophic interaction between methanogenic and acetogenic microorganisms, which is based on the hydrogen and format interchange of these species, has long been known as the only mechanism of syntrophic methanogenesis until recently when direct interspecies electron transfer was discovered. This process is an electron carrier-free transfer system in which evolved electrons are received directly through cytochromes or conductive pilli structures. This shortcut pathway decreases energy consumption during methanogenesis. Recent studies have revealed the potential of anammox and comammox bacteria to cooperate in lab-scale experiments. This interaction could be formed based on different metabolic trade-offs between these species, including nitrite transfer from comammox to anammox bacteria or passing nitrate to denitrifying bacteria and then transferring evolved nitrite to anammox bacteria. Both cooperative metabolisms are advantageous over conventional nitrification–denitrification pathways in which N_2O, as a hazardous compound, is released. In addition, with a lower affinity for ammonia, comammox bacteria could expel the available nitrite from nitrite-oxidizing bacteria (NOB), thus protecting anammox bacteria from wash-out from the system (Kits et al., 2017, 2019; Sakoula et al., 2021).

1.3 FACTORS DRIVING MICROBIAL COMMUNITY STRUCTURE AND INTERACTIONS

The assembly of microbial consortia is closely related to microbial function. Theoretically, the microbial community is shaped by two types of microbial assembly processes: stochastic and deterministic (Jiang et al., 2021; Yuan et al., 2019). In deterministic processes, community assembly is thought to be governed by both abiotic (environmental filtering: changes in temperature, pH, salinity, etc.) and biotic (competition, facilitation, mutualism, predation, etc.) factors (Trego et al., 2021; Yuan et al., 2019). Many studies have demonstrated the links between microbial community structures of activated sludge (AS) and environmental factors. Influent loadings such as chemical oxygen demand (COD) and total nitrogen (TN), bioreactor pH, temperatures, dissolved oxygen concentrations (DO), geographical locations, hydraulic retention time, treatment processes, and membrane filtration have been shown to exert significant impacts on AS microbial community structures. Furthermore, network analysis has revealed that biological interactions play a dominant role in determining microbial community dynamics (Xia et al., 2018). On the other hand, stochasticity in community assembly was established under neutral theory, and includes random processes such as microorganism drift, diversification, and random dispersal events

(Yu et al., 2021). Previous investigations have focused on the effects of stochastic processes during microbial succession in different ecosystems. Those studies consistently indicated that microbial community succession is initially determined by stochastic processes and progressively driven by environmental factors. Thus, it is reasonable to hypothesize that deterministic and stochastic processes can interact in shaping the community structure (Yuan et al., 2019).

As mentioned above, environmental conditions, influent composition, and operational parameters are the main factors that affect microbial consortia structure. Based on the rules of these factors, in this part, we will introduce some strategies that can be applied to regulate the microbial community structure.

1.4 STRATEGIES AND APPROACHES FOR MICROBIAL COMMUNITY STRUCTURE REGULATION

1.4.1 Bioaugmentation (BA)

One of the most valuable methods to improve biological systems for wastewater treatment is bioaugmentation (BA). This technique adds a pre-adapted pure bacterial strain or pre-adapted consortia into the wastewater treatment system, resulting in a change in microbial population (Herrero & Stuckey, 2015; Chen et al., 2017). Due to the high development ability, BA has mainly been applied to a non-adapted or stressed system to enhance bioactivity, increase the degradation and removal of undesired contaminants, and recover deteriorated systems caused by transient toxicants (Zhang et al., 2017). The success of BA depends on the selection of an appropriate organism that is able to tolerate the environmental challenges during the treatment process (Raper et al., 2018; Herrero & Stuckey, 2015). This strain can be selected from the indigenous population or an alternative site. Isolation of a strain from the aboriginal population is appropriate since the strain is already adapted to the chosen ecosystem. When a compound cannot be eliminated by organisms already present at the site, selecting one strain from another site may be the best option. Genetic manipulation offers another opportunity to remove compounds for which there is no pollutant-degrading natural strain. Genetically engineered microorganisms can overexpress degradation-associated genes or show more chance of survival (Raper et al., 2018). Moreover, the other factors that influence BA are operational regulation factors such as dosing technique, dosing location, dosing size and regime, and BA times and frequency (Zhang et al., 2017; Raper et al., 2018).

Several studies have applied BA to increase the efficiency of conventional biological wastewater systems. For instance, Chen et al. (2022) demonstrated that adding *Pseudomonas* sp. JMSTP to AS influenced the variety of microbial communities. This research suggested that BA could be a promising strategy for enhancing TN removal in municipal sewage treatment plants (Chen et al., 2022). In another study, Cheng et al. (2021) used novel functional microorganisms in toxic processes to eliminate organic contaminants in tannery wastewater treatment. During the 328-day operation, the average COD removal reached 95.2% (Cheng et al., 2021).

Overall, BA is a helpful technique to enhance the efficiency of wastewater treatment systems. Besides strain selection, the introduction and maintenance of selected microbes and their metabolites into the community and the scaling and design of the bioreactor are also critical in this method. However, the capacity of this approach can be extended to take advantage of significant developments in the fields of microbial ecology, molecular biology, immobilization methods, and the design of advanced bioreactors (Herrero & Stuckey, 2015).

1.4.2 Operational Parameter Optimization and Wastewater Treatment Process Adjustment

Operational parameter optimization and wastewater treatment process adjustment are other important strategies for effectively removing pollutants and resource recycling (Chen et al., 2017). However, there needs to be more research on how to regulate microbial consortia and maintain stable process operations under distinct circumstances (Zhou et al., 2018). For example, in one study, Roots et al. (2019) demonstrated that under an aerobic, low ammonia environment and long sludge retention period, comammox *Nitrospira* comprised 94% of the microbial population, and the rate of ammonium removal was 58.6 mg N/L/d. Thus, these variables are important for the proliferation of comammox *Nitrospira*. Moreover, microelements such as copper are very limited in real wastewater compared to synthetic wastewater. Such copper deficiency may compromise the conventional nitrification process and inhibit the function of AMO for AOB. However, comammox *Nitrospira* could survive in the copper-limited environment because of the potential to use urea for ammonium production (Ren et al., 2020).

Another study was conducted by Yang et al. (2018), who considered the addition of ferrihydrite into an AD. Their results showed that the number of feammox bacteria increased, and anaerobic ammonium oxidation was also induced. Ruiz-Urigüen et al. (2018) suggested a new way to replace anodes with ferric iron to enhance the effectiveness of feammox, and found that in this condition, *Acidimicrobiaceae* bacterium A6 can transfer electrons onto electrodes, and ammonia degradation could be increased (Ren et al., 2020).

Furthermore, most organic micropollutants cannot be permanently removed effectively using conventional methods based on AS. Thus, various alternative technologies should be applied (Grandclément et al., 2017). Different investigations have indicated that a hybrid sludge/biofilm system (the modified process of the conventional AS) is an effective replacement for biological wastewater management. This technology combines the benefits of suspended AS with the attached biofilm and thus permanently removes contaminants, especially under low temperatures. The MBBR system has also been used for carbon removal, nitrification, and denitrification. Moreover, BAF, an attached growth treatment system, has an excellent capacity for organic matter and ammonia (Zhou et al., 2018; Grandclément et al., 2017).

In addition, chemical phosphorus precipitation and the use of new technologies (membrane bioreactors (MBRs), aerobic granular sludge, partial nitritation (PN), partial nitritation/anammox (PN/A), SND, etc.) are other approaches for developing and maintaining the performance of urban wastewater treatment systems (Zhou et al., 2018).

1.4.3 The r/K Selection Theory

Although several species' characteristics used in the biodegradation of various contaminants have been identified, selecting suitable organisms and regulating microbial consortia remains a critical problem. To resolve this, functional theories are essential (Yin et al., 2022). The r/K selection theory was proposed to explain the growth of two common types of microorganisms. Evolutionarily, two kinds of populations could be selected in each ecosystem, including r- and K-strategists (Wu & Yin, 2020). The r-strategist microorganisms have a rapid growth rate in an uncrowded and resource-rich environment, while the K-strategist microorganisms grow slowly and are able to adapt to a crowded and resource-limited environment (Fig. 1.3). Both types of organisms have their specific niches in each ecosystem (Guo et al., 2022; Yin et al., 2022).

Organisms involved in environmental bioprocesses, including nitrification, anammox, and methanogenesis, show different r-/K-properties. Therefore, identifying appropriate r-/K-strategists and the selective enrichment of desired microorganisms by adjusting environmental factors could be a potential approach to improve new bioprocesses (Yin et al., 2022). The r-strategists may dominate the front part of a plug flow reactor and be responsible for high contaminant elimination under high nutrient concentrations. At the same time, K-strategists may be enriched in the latter part of the reactor, where oligotrophic conditions dominate. Besides the potential to maintain under oligotrophic circumstances, K-strategist organisms may have a strong affinity for micropollutants if they have related genes and proteins (Wu & Yin, 2020).

Based on this theory, systems with high substrate gradients can achieve an effective removal rate by offering suitable niches for both types of organisms. Additionally, various microbes can increase the system's stability and resilience under different conditions (Yin et al., 2022).

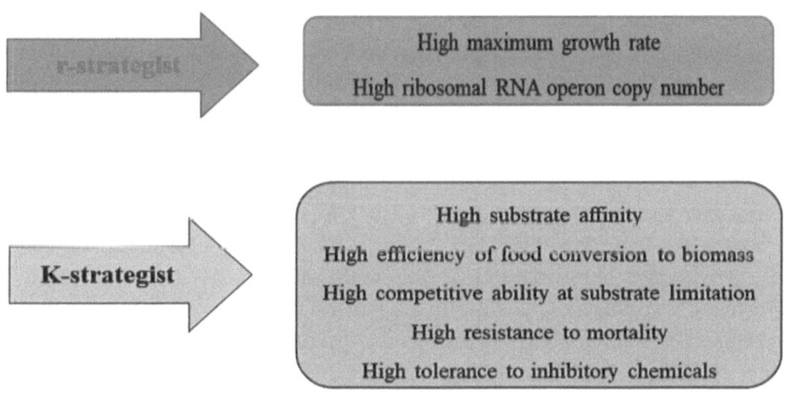

FIGURE 1.3 The characteristics of r- and K-strategists.

Some crucial variables can affect this theory. Since growth rate and substrate affinity determine the selection of r-/K-strategists, the food-to-microorganism (F/M) ratio, which reveals the association between substrate and biomass concentrations, is an important factor influencing the enrichment of r-/K-strategists. The F/M ratio can be determined by regulating substrate or biomass levels and feeding times. It should be noted that mass transfer is also essential because it can change F/M ratios in the system. The other significant factor is the operating mode of the reactor. Sequencing batch reactors (SBRs) and plug flow systems can enrich r-strategists, while systems such as MBRs provide ideal conditions for selecting K-strategists. In addition, the r-/K-properties are affected by cell size, shape, and structure. To fully apply the r/K selection theory in bioprocesses, these factors should be considered. Generally, using this theory is the primary step in bioprocesses, but r/K selection-based technologies and concepts enable biologists to exploit the abilities of r-/K-strategists (Yin et al., 2022).

1.5 TOOLS AND METHODS TO IMPROVE MICROBIAL CONSORTIA AND METABOLIC COOPERATION

Diverse techniques have been utilized to investigate microbial ecology (including genetics, variety, metabolic pathways, and microbial cooperation) in biological wastewater treatment systems. Common techniques like cultivation-dependent methods cannot be applied to explore the ecology of complex microbial consortia because only 1% of microbial populations in wastewater can be artificially cultured. Therefore, it is hard to comprehensively identify the microbial communications and metabolic processes, while novel molecular and bioinformatics techniques can identify uncultured microbes and their functions (G. Chen et al., 2022; Wu & Yin 2020).

Integrated omics' analysis (metagenomics, metatranscriptomics, metaproteomics, and metabolomics) are gaining more attention as they offer a better understanding of microbial communities' structure, function, and dynamics (Narayanasamy et al., 2015).

Metagenomics is the analysis of total microbial genetic material collected directly from environmental samples to find information about the state of microbes in a particular environment. Metagenomics is divided into function and sequence-based, and both are applied to obtain metagenomes from environmental samples (Chen et al., 2022; Kumar et al., 2022c). Functional metagenomics involves forming a metagenomic library by cloning the population DNA in a particular host, followed by screening the library for activities related to microorganisms and metabolic processes. Therefore, it is mainly used to investigate functional genes and enzymes, such as genes associated with nitrogen and phosphorus metabolism and contaminant removal. Contrarily, sequence-based metagenomics utilizes next-generation sequencing (NGS) technology to sequence large-scale DNA segments and identify functional genes using bioinformatics' approaches. It can provide genetic data for a whole population. Thus, it can help construct metabolic pathways and predict the function of potential genes (Chen et al., 2022; Charles et al., 2017). From 2011 to 2021, several studies focused on applying metagenomics for the biological elimination of phosphorus, nitrogen, and other contaminants in wastewater treatment (Table 1.1). Metatranscriptomic methods

TABLE 1.1
Examples of studies in which metagenomic analysis was used in biological wastewater treatment systems (from 2015 to 2021)

Methods	Results	References
Metagenomic analysis of the genome of comammox *Nitrospira*	Understanding a novel metabolic pathway of comammox	Daims et al. (2015)
Metagenomic binning technology for investigating seven clades of *Ca. Accumulibacter*	Studying the variety of *Ca. Accumulibacter*	Skennerton et al. (2015)
Metagenomic and metaproteomic analysis for investigating the proteomic diversities between two biofilm communities in EBPR reactors	Annotating proteins and studying the mechanism of change in the metabolic process of phosphorus removal from flocculation to granules	Barr et al. (2016)
Metagenomic analysis of the genomes of denitrification microorganisms	Providing information about denitrification microorganisms' diversity	Speth et al. (2016)
Metagenomic analysis for determining the microbial consortia in anammox samples	Indicating that anammox-associated *Planctomycetes* was the main phylum	Aktan et al. (2018)
Metagenomic analysis for studying the vanadate reduction pathway in a membrane biofilm reactor	Showing the link between secretion of extracellular polymeric substances and the microbial metabolism related to vanadate reduction	Lai et al. (2018)
Metagenomic analysis	Indicating the presence of *Halomonas* sp. which can remove nitrate and chromate in denitrifying granular sludge	Reddy and Nancharaiah, (2017)
Nanopore metagenomic sequencing	Showing mobile antibiotic resistance genes in wastewater	Che et al. (2019)
Metagenomic analysis for vanadate reduction in a methane-based membrane biofilm batch reactor	Showing the enrichment of genes associated with methane oxidation (methane-oxidizing microbes provide electron donors for other microbes to reduce vanadate)	Wang and Lai (2019)
Metagenomic and metatranscriptomic analysis for understanding the process of substrate competition between *Ca. Jettenia* and *Ca. Brocadia*	Finding that *Ca. Brocadia* has more functional genes for nitrogen metabolism and chemotaxis in comparison to *Ca. Jettenia*	Zhao et al. (2019)

(*continued*)

TABLE 1.1 (Continued)
Examples of studies in which metagenomic analysis was used in biological wastewater treatment systems (from 2015 to 2021)

Methods	Results	References
Metagenomic analysis for determining the impact of particle size on the performance of granule based PN/A	Indicating that large particles harbor high microbial diversity and support more diverse functions	Chen et al. (2020)
Metagenomic analysis for characterizing the microbial consortia and functional features in anammox aggregates	Providing new insight into the stable anammox process	Jia et al. (2021)
Metagenomic analysis for analyzing a sequencing batch bioreactor for Cu(II) containing wastewater treatment	Finding that 5 mg/L Cu(II) inhibited the degradation of organic matter	Zhao et al. (2021)

are also utilized to obtain information about the activities of environmental microbial consortia. Metatranscriptomics refers to the analysis of metagenomic mRNA for understanding complicated microbial populations' regulation and expression profiles. Due to problems related to the processing of environmental RNA samples, in situ metatranscriptomic investigations are limited, but the issues can be resolved by the use of direct cDNA sequencing through NGS-based techniques (Suneja & Sharma, 2020; Bharagava et al., 2018). Furthermore, metaproteomics is defined as the characterization of all the proteins recovered from environmental samples. It involves protein extraction and purification, separating peptides, mass spectrometry, and computational analysis (Bharagava et al., 2018; Suneja & Sharma, 2020). The first metaproteomic study in wastewater treatment focused on the proteins expressed by the genes involved in EBPR by using an activated sludge-based SBR. Optimization of this process helped scientists to identify important microorganisms and biochemical pathways in diverse treatment processes. However, there are some challenges related to metaproteomic analyses, such as unbalanced species distribution, the wide range of protein expression levels, and the significant genetic heterogeneity in microbial consortia (Suneja & Sharma, 2020).

Another method is metabolomics, which includes the comprehensive analysis of the metabolites synthesized by the microbial population of a specific ecosystem. Along with other omics' methods, it determines microbial cooperation and their interactions with the surrounding environment (Suneja & Sharma, 2020; Bharagava et al., 2018). For instance, Tang et al. (2018) applied both metabolomic and metagenomic tools in a study of anaerobic ammonium oxidation. They indicated regulatory processes for acyl homoserine lactones that influence the growth of the anammox. Overall, integrated omics will become a powerful approach to analyzing microbial communities (Suneja & Sharma, 2020).

1.6 FUTURE PERSPECTIVES

Natural microbial consortia have higher metabolic abilities than monocultures and show more stability in environmental conditions. They also exhibit reduced metabolic burdens because of a division of labor (DOL) and resource exchange. These unique characteristics of microbial communities make them an interesting platform for scientists to change microbes to perform more complex tasks in both closed and defined bioreactors and open and natural environments. Thus, the focus of synthetic biology is changing from the engineering of single microorganisms to engineering multiple species within dynamic communities (McCarty & Ledesma-Amaro, 2019; Amor & Bello, 2019). In addition, engineering a single microbe has some limitations, such as inefficient delivery of large amounts of DNA into the target microbe, the absence of cofactors and precursors in the host cell, etc. These problems may be solved by creating synthetic microbial consortia consisting of two or more genetically engineered species/strains (McCarty & Ledesma-Amaro, 2019).

It is essential to utilize natural microbial consortia to construct an artificial microbial community. In a natural consortium, it is challenging to understand which organisms are participating in bioremediation. Therefore, a synthetic microbial consortium is a novel approach for designing an artificial microbial population with function-specific microbes (Jaiswal & Shukla, 2020). Synthetic biologists have applied engineering methods such as computer models and modular DNA fragments to make rational changes in living organisms. Through the development of genetic engineering and rapid DNA assembly techniques, modular components can be interlinked to construct metabolic processes and biological networks to control cellular behavior (McCarty & Ledesma-Amaro, 2019).

For engineering microbial communities toward bioremediation, two major strategies can be used: (1) top down and (2) bottom up (Che & Men, 2019; Li et al., 2021).

Top-down engineering is an effective strategy that uses a natural microbial population to accomplish desired functions by changing physiochemical conditions and abiotic and biotic processes. For instance, research has indicated that adding electron donors facilitates microbial communication and enhances the elimination of toxic chlorinated pollution. However, top-down engineering overlooks the complicated interconnected metabolism and interactions in microbial populations, thus restricting system optimization (Li et al., 2021).

On the other hand, bottom-up engineering is related to species relationships and metabolism. With the advance of omics' technology, scientists have suggested different ways to manipulate metabolic networks and microbial cooperation. In this approach, a microbial community with defined tasks can be built and optimized, and this method can be used to remove complex contaminants. However, species with inaccurate or incomplete data about metabolic pathways, genes, or proteins with unknown functions present a central challenge in engineering microbial communities (Li et al., 2021).

As the DOL in a population is necessary for the degradation of persistent contaminants, which typically needs several stages, artificial microbial communities may play significant roles in bioremediation (Che & Men, 2019). Furthermore, synthetic microbial communities are better than complex consortia for bioremediation because substrate competition may be limited by creating microbial communities that only include the contributing species (Che & Men, 2019).

Although synthetic microbial consortia already provide opportunities for improvements in biotechnology applications, some challenges remain (McCarty & Ledesma-Amaro, 2019). The development of enhanced synthetic biology tools and the high-throughput analysis of natural microbial ecosystems will influence future progress in the engineering of microbial consortia. Some methods like directed evolution, genomic and metabolic engineering, and artificial cell-to-cell communications can be used to optimize this system (Che & Men, 2019).

REFERENCES

Aktan, Cigdem Kalkan, Ayse Ekin Uzunhasanoglu, and Kozet Yapsakli. 2018. "Speciation of Nickel and Zinc, Its Short-Term Inhibitory Effect on Anammox, and the Associated Microbial Community Composition." *Bioresource Technology*. https://doi.org/10.1016/j.biortech.2018.08.011

Amor, Daniel Rodríguez, and Martina Dal Bello. 2019. "Bottom-up Approaches to Synthetic Cooperation in Microbial Communities." *Life* 9 (1). https://doi.org/10.3390/life9010022

Azad, Hardam S, and Jack A Borchardt. 1970. "Variations in Phosphorus Uptake by Algae." *Environmental Science & Technology* 4 (9): 737–43. https://doi.org/10.1021/es60044a008

Barr, Jeremy J, Bas E Dutilh, Connor T Skennerton, Toshikazu Fukushima, Marcus L Hastie, Jeffrey J Gorman, Gene W Tyson, and Philip L Bond. 2016. "Metagenomic and Metaproteomic Analyses of Accumulibacter Phosphatis-Enriched Floccular and Granular Biofilm." 18: 273–87. https://doi.org/10.1111/1462-2920.13019

Bharagava, Ram N, Diane Purchase, Gaurav Saxena, and Sikandar I Mulla. 2018. *Applications of Metagenomics in Microbial Bioremediation of Pollutants: From Genomics to Environmental Cleanup. From Genomics to Environmental Cleanup. Microbial Diversity in the Genomic Era*. Elsevier Inc. https://doi.org/10.1016/B978-0-12-814849-5.00026-5

Bunce, Joshua T, Edmond Ndam, Irina D Ofiteru, Andrew Moore, and David W Graham. 2018. "A Review of Phosphorus Removal Technologies and Their Applicability to Small-Scale Domestic Wastewater Treatment Systems." *Frontiers in Environmental Science*. https://doi.org/10.3389/fenvs.2018.00008

Charles, Trevor C, Liles, Mark R, Sessitsch, Angela. 2017. Functional Metagenomics: Tools and Applications. 1–253.

Che, Shun, and Yujie Men. 2019. "Synthetic Microbial Consortia for Biosynthesis and Biodegradation: Promises and Challenges." *Journal of Industrial Microbiology and Biotechnology* 46 (9–10): 1343–58. https://doi.org/10.1007/s10295-019-02211-4

Che, You, Yu Xia, Lei Liu, An-dong Li, Yu Yang, and Tong Zhang. 2019. "Mobile Antibiotic Resistome in Wastewater Treatment Plants Revealed by Nanopore Metagenomic Sequencing." *Microbiome* 7(44): 1–13. https://doi.org/10.1186/s40168-019-0663-0

Chen, Geng, Rui Bai, Yiqing Zhang, Biyi Zhao, and Yong Xiao. 2022. "Application of Metagenomics to Biological Wastewater Treatment." *Science of the Total Environment* 807. https://doi.org/10.1016/j.scitotenv.2021.150737

Chen, Hui, Tao Liu, Jie Li, Likai Mao, Jun Ye, Xiaoyu Han, Mike S M Jetten, and Jianhua Guo. 2020. "Larger Anammox Granules Not Only Harbor Higher Species Diversity but Also Support More Functional Diversity." *Environmental Science and Technology* 2020 (54): 14664–14673. https://doi.org/10.1021/acs.est.0c02609

Chen, Tianyuan, Xiaoyong Yang, Qian Sun, Anyi Hu, Dan Qin, Jiangwei Li, Yinhan Wang, and Chang Ping Yu. 2022. "Changes in Wastewater Treatment Performance and the Microbial Community during the Bioaugmentation of a Denitrifying Pseudomonas Strain in the Low Carbon–Nitrogen Ratio Sequencing Batch Reactor." *Water (Switzerland)* 14 (4). https://doi.org/10.3390/w14040540

Chen, Yangwu, Shuhuan Lan, Longhui Wang, Shiyang Dong, Houzhen Zhou, Zhouliang Tan, and Xudong Li. 2017. "A Review: Driving Factors and Regulation Strategies of Microbial Community Structure and Dynamics in Wastewater Treatment Systems." *Chemosphere* 174: 173–82. https://doi.org/10.1016/j.chemosphere.2017.01.129

Cheng, Yu, Kangmin Chon, Xianghao Ren, Yingying Kou, Moon Hyun Hwang, and Kyu Jung Chae. 2021. "Bioaugmentation Treatment of a Novel Microbial Consortium for Degradation of Organic Pollutants in Tannery Wastewater under a Full-Scale Oxic Process." *Biochemical Engineering Journal* 175 (July): 108131. https://doi.org/10.1016/j.bej.2021.108131

Chevallereau, Anne, Benoît J Pons, Stineke van Houte, and Edze R Westra. 2022. "Interactions between Bacterial and Phage Communities in Natural Environments." *Nature Reviews Microbiology* 20 (1): 49–62. https://doi.org/10.1038/s41579-021-00602-y

Cremer, Jonas, Anna, Melbinger, Karl, Wienand, Tania, Henriquez, Heinrich, Jung, and Erwin, Frey. 2019. "Cooperation in Microbial Populations: Theory and Experimental Model Systems." *Journal of Molecular Biology* 431 (23): 4599–4644. https://doi.org/10.1016/j.jmb.2019.09.023

Cydzik-Kwiatkowska, Agnieszka, and Magdalena Zielińska. 2016. "Bacterial Communities in Full-Scale Wastewater Treatment Systems." *World Journal of Microbiology and Biotechnology* 32 (4): 1–8. https://doi.org/10.1007/s11274-016-2012-9

Daims, Holger, Elena V Lebedeva, Petra Pjevac, Ping Han, Craig Herbold, Mads Albertsen, Nico Jehmlich, et al. 2015. "Complete Nitrification by Nitrospira Bacteria." *Nature*. https://doi.org/10.1038/nature16461

Dominika, Grubba, Majtacz Joanna, and Mąkinia Jacek. 2021. "Environmental Technology & Innovation Sulfate Reducing Ammonium Oxidation (SULFAMMOX) Process under Anaerobic Conditions." *Environmental Technology & Innovation* 22: 101416. https://doi.org/10.1016/j.eti.2021.101416

Fan, Xiangyu, Mengzhi Ji, Kaili Sun, and Qiang Li. 2023. "Microbial and Phage Communities as Well as Their Interaction in PO Saponi Fi Cation Wastewater Treatment Systems". *Water Sci. Technol.* 87 (2): 354–65. https://doi.org/10.2166/wst.2022.422

Fu, Liang, Jing Ding, Yong-Ze Lu, Zhao-Wei Ding, Ya-Nan Bai, and Raymond J Zeng. 2017. "Hollow Fiber Membrane Bioreactor Affects Microbial Community and Morphology of the DAMO and Anammox Co-Culture System." *Bioresource Technology* 232 (May): 247–53. https://doi.org/10.1016/j.biortech.2017.02.048

Grandclément, Camille, Isabelle Seyssiecq, Anne Piram, Pascal Wong-Wah-Chung, Guillaume Vanot, Nicolas Tiliacos, Nicolas Roche, and Pierre Doumenq. 2017. "From the Conventional Biological Wastewater Treatment to Hybrid Processes, the Evaluation of Organic Micropollutant Removal: A Review." *Water Research* 111: 297–317. https://doi.org/10.1016/j.watres.2017.01.005

Guo, Qiannan, Qidong Yin, Jin Du, Jiane Zuo, and Guangxue Wu. 2022. "New Insights into the r/K Selection Theory Achieved in Methanogenic Systems through Continuous-Flow and Sequencing Batch Operational Modes." *Science of the Total Environment* 807: 150732. https://doi.org/10.1016/j.scitotenv.2021.150732

Herrero, Mónica, and David C, Stuckey. 2015. "Bioaugmentation and Its Application in Wastewater Treatment: A Review." *Chemosphere* 140: 119–28. https://doi.org/10.1016/j.chemosphere.2014.10.033

Hibbing, Michael E, Clay Fuqua, Matthew R, Parsek, and Brook S, Peterson. 2010. "Bacterial Competition: Surviving and Thriving in the Microbial Jungle." *Nature Reviews Microbiology* 8 (1): 15–25. https://doi.org/10.1038/nrmicro2259

Hou, Ting Ting, Li Li Miao, Ji Sen Peng, Lan Ma, Qiang Huang, Ying Liu, Meng Ru Wu, Guo Min Ai, Shuang Jiang Liu, and Zhi Pei Liu. 2022. "Dirammox Is Widely Distributed and Dependently Evolved in Alcaligenes and Is Important to Nitrogen Cycle." *Frontiers in Microbiology* 13 (2): 1–12. https://doi.org/10.3389/fmicb.2022.864053

Jaiswal, Shweta, and Pratyoosh Shukla. 2020. "Alternative Strategies for Microbial Remediation of Pollutants via Synthetic Biology." *Frontiers in Microbiology* 11 (May): 1–14. https://doi.org/10.3389/fmicb.2020.00808

Jia, Fangxu, Yongzhen Peng, Jianwei Li, Xiyao Li, and Hong Yao. 2021. "Metagenomic Prediction Analysis of Microbial Aggregation in Anammox- Dominated Community." *Water Environment Research* 1–10. https://doi.org/10.1002/wer.1529

Jiang, Yiming, Haiying Huang, Yanrong Tian, Xuan Yu, and Xiangkai Li. 2021. "Stochasticity versus Determinism: Microbial Community Assembly Patterns under Specific Conditions in Petrochemical Activated Sludge." *Journal of Hazardous Materials* 407: 124372. https://doi.org/10.1016/j.jhazmat.2020.124372

Kits, Dimitri K, Man-Young Jung, Julia Vierheilig, Petra Pjevac, Christopher J Sedlacek, Shurong Liu, Craig Herbold, et al. 2019. "Low Yield and Abiotic Origin of N(2)O Formed by the Complete Nitrifier Nitrospira Inopinata." *Nature Communications* 10 (1): 1836. https://doi.org/10.1038/s41467-019-09790-x

Kits, Dimitri K, Christopher J Sedlacek, Elena V Lebedeva, Ping Han, Alexandr Bulaev, Petra Pjevac, Anne Daebeler, et al. 2017. "Kinetic Analysis of a Complete Nitrifier Reveals an Oligotrophic Lifestyle." *Nature* 549 (7671): 269–72. https://doi.org/10.1038/nature23679

Kumar, Vineet, Ram Chandra, 2020. Metagenomics Analysis of Rhizospheric Bacterial Communities of *Saccharum arundinaceum* Growing on Organometallic Sludge of Sugarcane Molasses-based Distillery. *3 Biotech* 10 (7): 316. https://doi.org/10.1007/s13205-020-02310-5

Kumar, Vineet, Muhammad Bilal, Luiz Fernando Romanholo Ferreira, 2022a. Editorial: Recent Trends in Integrated Wastewater Treatment for Sustainable Development. *Frontiers in Microbiology* 13, 846503. https://doi.org/10.3389/fmicb.2022.846503

Kumar, Vineet, Fuad Ameen, M Amirul Islam, Sakshi Agrawal, Ankit Motghare, Abhijit Dey, Maulin P Shah, Juliana Heloisa Pine Américo-Pinheiro, Simranjeet Singh, Praveen C Ramamurthy, 2022b. Evaluation of Cytotoxicity and Genotoxicity Effects of Refractory Pollutants of Untreated and Biomethanated Distillery Effluent Using *Allium cepa. Environmental Pollution* 300, 118975. https://doi.org/10.1016/j.envpol.2022.118975

Kumar, Vineet, Bilal, Muhammad, Shahi, Sushil Kumar, & Garg, Vinod Kumar (Eds.). (2022c). *Metagenomics to Bioremediation: Applications, Cutting Edge Tools, and Future Outlook*. Academic Press.

Lai, Chun-yu, Qiu-yi Dong, Jiaxian Chen, Quan-song Zhu, Xin Yang, Wen-da Chen, He-ping Zhao, and Liang Zhu. 2018. "Role of Extracellular Polymeric Substances in a Methane

Based Membrane Biofilm Reactor Reducing Vanadate." Environmental Science and Technology 52 (18): 10680–10688. https://doi.org/10.1021/acs.est.8b02374

Li, Xianglong, Shanghua Wu, Yuzhu Dong, Haonan Fan, Zhihui Bai, and Xuliang Zhuang. 2021. "Engineering Microbial Consortia towards Bioremediation." *Water (Switzerland)* 13 (20): 1–10. https://doi.org/10.3390/w13202928

Luesken, Francisca A, Jaime Sánchez, Theo A van Alen, Janeth Sanabria, Huub J M Op den Camp, Mike S M Jetten, and Boran Kartal. 2011. "Simultaneous Nitrite-Dependent Anaerobic Methane and Ammonium Oxidation Processes." *Applied and Environmental Microbiology* 77 (19): 6802–7. https://doi.org/10.1128/AEM.05539-11

Lynch, Michael D J, and Josh D Neufeld. 2015. "Ecology and Exploration of the Rare Biosphere." *Nature Publishing Group* 13 (April). https://doi.org/10.1038/nrmicro3400

M.P.C.A. 2006. "Phosphorus Treatment and Removal Technologies – View-Document.Html." *Minnesota Pollution Control Agency*, no. June: 1–5.

McCarty, Nicholas S, and Rodrigo Ledesma-Amaro. 2019. "Synthetic Biology Tools to Engineer Microbial Communities for Biotechnology." *Trends in Biotechnology* 37 (2): 181–97. https://doi.org/10.1016/j.tibtech.2018.11.002

Narayanasamy, Shaman, Emilie E.L. Muller, Abdul R Sheik, and Paul Wilmes. 2015. "Integrated Omics for the Identification of Key Functionalities in Biological Wastewater Treatment Microbial Communities." *Microbial Biotechnology* 8 (3): 363–68. https://doi.org/10.1111/1751-7915.12255

Peeters, Stijn H, and Laura Van Niftrik. 2019. "ScienceDirect Trending Topics and Open Questions in Anaerobic Ammonium Oxidation." *Current Opinion in Chemical Biology* 49: 45–52. https://doi.org/10.1016/j.cbpa.2018.09.022

Pous, Narcís, Lluis Bañeras, Philippe F.-X. Corvini, Shuang-Jiang Liu, and Sebastià Puig. 2023. "Direct Ammonium Oxidation to Nitrogen Gas (Dirammox) in Alcaligenes Strain HO-1: The Electrode Role." *Environmental Science and Ecotechnology* 15: 100253. https://doi.org/https://doi.org/10.1016/j.ese.2023.100253

Raper, Eleanor Claire, Tom, Stephenson, David R, Anderson, Raymond, Fisher, and Ana, Soares. 2018. "Industrial Wastewater Treatment through Bioaugmentation." *Process Safety and Environmental Protection* 118: 178–87. https://doi.org/10.1016/j.psep.2018.06.035

Reddy, Kiran Kumar G, and Y V Nancharaiah. 2017. "Sustainable Bioreduction of Toxic Levels of Chromate in a Denitrifying Granular Sludge Reactor." no. Atsdr 2014.

Ren, Yi, Huu Hao Ngo, Wenshan Guo, Dongbo Wang, Lai Peng, Bing Jie Ni, Wei Wei, and Yiwen Liu. 2020. "New Perspectives on Microbial Communities and Biological Nitrogen Removal Processes in Wastewater Treatment Systems." *Bioresource Technology* 297: 122491. https://doi.org/10.1016/j.biortech.2019.122491

Rikmann, Ergo, Ivar Zekker, Martin Tomingas, Priit Vabamäe, Kristel Kroon, Alar Saluste, Taavo Tenno, et al. 2014. "Comparison of Sulfate-Reducing and Conventional Anammox Upflow Anaerobic Sludge Blanket Reactors." *Journal of Bioscience and Bioengineering* 118 (4): 426–33. https://doi.org/10.1016/j.jbiosc.2014.03.012

Roots, Paul, Yubo Wang, Alex F Rosenthal, James S Grif, Fabrizio Sabba, Morgan Petrovich, Fenghua Yang, Joseph A Kozak, Heng Zhang, and George F Wells. 2019. "Comammox Nitrospira Are the Dominant Ammonia Oxidizers in a Mainstream Low Dissolved Oxygen Nitri Fi Cation Reactor." *Water Research* 157 (2): 396–405. https://doi.org/10.1016/j.watres.2019.03.060

Ruiz-Urigüen, M., Shuai, W., Jaffé, P., R., 2018. "Feammox Acidimicrobiaceae sp. A6, a lithoautotrophic electrode-colonizing bacterium." *Applied and Environmenatl Microbiology AEM*. 02029–02018.

Sakoula, Dimitra, Hanna Koch, Jeroen Frank, Mike S M Jetten, Maartje A H J van Kessel, and Sebastian Lücker. 2021. "Enrichment and Physiological Characterization of a Novel

Comammox Nitrospira Indicates Ammonium Inhibition of Complete Nitrification." *The ISME Journal* 15 (4): 1010–24. https://doi.org/10.1038/s41396-020-00827-4

Shi, Ying, Shihu Hu, Juqing Lou, Peili Lu, Jurg Keller, and Zhiguo Yuan. 2013. "Nitrogen Removal from Wastewater by Coupling Anammox and Methane-Dependent Denitrification in a Membrane Biofilm Reactor." *Environmental Science & Technology* 47 (20): 11577–83. https://doi.org/10.1021/es402775z

Sidhu, Chandni, Surendra Vikram, and Anil Kumar Pinnaka. 2017. "Unraveling the Microbial Interactions and Metabolic Potentials in Pre- and Post-Treated Sludge from a Wastewater Treatment Plant Using Metagenomic Studies." *Frontiers in Microbiology* 8 (July): 1–10. https://doi.org/10.3389/fmicb.2017.01382

Skennerton, Connor T, Jeremy J Barr, Frances R Slater, Philip L Bond, and Gene W Tyson. 2015. "Expanding Our View of Genomic Diversity in Candidatus Accumulibacter Clades." *Environmental Microbiology* 17: 1574–85. https://doi.org/10.1111/1462-2920.12582

Sperling, Marcos Von, and Carl Augusto De Lemos Chernicharo. 2005. Biological Wastewater Treatment in Warm Climate Regions. IWA Publishing. Vol. 1.

Speth, Daan R, Simon Guerrero-cruz, Bas E Dutilh, and Mike S M Jetten. 2016. Genome-based Microbial Ecology of Anammox Granules in a Full-scale Wastewater Treatment System. *Nature Communications* 7: 1172. https://doi.org/10.1038/ncomms11172

Suneja, Garima, and Rakesh Sharma. 2020. *Insights into the Microbial Processes Prevalent in Wastewater Treatment Systems Using Omics Approaches. Removal of Toxic Pollutants through Microbiological and Tertiary Treatment: New Perspectives.* Elsevier B.V. https://doi.org/10.1016/B978-0-12-821014-7.00016-2

Tang, Xi, Yongzhao Guo, Shanshan Wu, Liming Chen, Huchun Tao, and Sitong Liu. 2018. Metabolomics Uncovers the Regulatory Pathway of Acyl-homoserine Lactones based Quorum Sensing in Anammox Consortia. *Environmental Science & Technology*, 52: 2206–2216. https://doi. org/10.1021/acs.est.7b05699.

Toro, Emilia Rios-del E, and Francisco J Cervantes. 2019. "Anaerobic Ammonium Oxidation in Marine Environments: Contribution to Biogeochemical Cycles and Biotechnological Developments for Wastewater Treatment." *Reviews in Environmental Science and Bio/ Technology* 3. https://doi.org/10.1007/s11157-018-09489-3

Trego, Anna Christine, Paul G McAteer, Corine Nzeteu, Therese Mahony, Florence Abram, Umer Zeeshan Ijaz, and Vincent O'Flaherty. 2021. "Combined Stochastic and Deterministic Processes Drive Community Assembly of Anaerobic Microbiomes During Granule Flotation." *Frontiers in Microbiology* 12 (May): 1–14. https://doi.org/ 10.3389/fmicb.2021.666584

Vilardi, Katherine, Irmarie Cotto, Megan Bachmann, Mike Parsons, Stephanie Klaus, Christopher Wilson, Charles Bott, Kelsey Pieper, Ameet Pinto, and Ameet Pinto Katherine Vilardi, Irmarie Cotto, Megan Bachmann, Mike Parsons, Stephanie Klaus, Christopher Wilson, Charles Bott, Kelsey Pieper. 2022. "Co-Occurrence and Cooperation between Comammox and Anammox Bacteria in a Full-Scale Attached Growth Municipal Wastewater Treatment Process." *BioRxiv*, January, 2022.12.05.519185. https://doi.org/ 10.1101/2022.12.05.519185

Wang, Zhen, and Ling-dong Shi Chun-yu Lai. 2019. "Methane Oxidation Coupled to Vanadate Reduction in a Membrane Biofilm Batch Reactor under Hypoxic Condition." *Biodegradation* 6. https://doi.org/10.1007/s10532-019-09887-6

Wani, Atif Khurshid, Akhtar, Nahid, Naqash, Nafiaah, Chopra, Chirag, Singh, Reena, Kumar, Vineet, Kumar, Sunil, Mulla, Sikandar I., Américo-Pinheiro, Juliana Heloisa Pinê, 2022. Bioprospecting Culturable and Unculturable Microbial Consortia through Metagenomics for Bioremediation. *Cleaner Chemical Engineering* 2, 100017. https:// doi.org/10.1016/j.clce.2022.100017

Wu, Guangxue, and Qidong Yin. 2020. "Microbial Niche Nexus Sustaining Biological Wastewater Treatment." *Npj Clean Water* 3 (1). https://doi.org/10.1038/s41545-020-00080-4

Xia, Yu, Xianghua Wen, Bing Zhang, and Yunfeng Yang. 2018. "Diversity and Assembly Patterns of Activated Sludge Microbial Communities: A Review." *Biotechnology Advances* 36 (4): 1038–47. https://doi.org/10.1016/j.biotechadv.2018.03.005

Xu, Qicheng, Gongwen Luo, Junjie Guo, and Fengge Zhang. 2021. "Microbial Generalist or Specialist: Intraspecific Variation and Dormancy Potential Matter," no. October. https://doi.org/10.1111/mec.16217

Yang, Yafei, Zhen Jin, Xie Quan, Yaobin Zhang. 2018. "Transformation of Nitrogen and Iron Species during Nitrogen Removal from Wastewater via Feammox by adding Ferrihydrite." *ACS Sustainable Chemistry and Engineering* 6 (11): 14394–14402.

Yin, Qidong, Yuepeng Sun, Bo Li, Zhaolu Feng, and Guangxue Wu. 2022. "The r/K Selection Theory and Its Application in Biological Wastewater Treatment Processes." *Science of the Total Environment* 824 (February). https://doi.org/10.1016/j.scitotenv.2022.153836

Yu, Jinjin, Siang Nee Tang, and Patrick K.H. Lee. 2021. "Microbial Communities in Full-Scale Wastewater Treatment Systems Exhibit Deterministic Assembly Processes and Functional Dependency over Time." *Environmental Science and Technology* 55 (8): 5312–23. https://doi.org/10.1021/acs.est.0c06732

Yuan, Heyang, Ran Mei, Junhui Liao, and Wen Tso Liu. 2019. "Nexus of Stochastic and Deterministic Processes on Microbial Community Assembly in Biological Systems." *Frontiers in Microbiology* 10 (July): 1–12. https://doi.org/10.3389/fmicb.2019.01536

Zhang, Lei, Zhen Shen, Wangkai Fang, and Guang Gao. 2019. "Composition of Bacterial Communities in Municipal Wastewater Treatment Plant." *Science of the Total Environment* 689: 1181–91. https://doi.org/10.1016/j.scitotenv.2019.06.432

Zhang, Qian Qian, Guang Feng Yang, Lei Zhang, Zheng Zhe Zhang, Guang Ming Tian, and Ren Cun Jin. 2017. "Bioaugmentation as a Useful Strategy for Performance Enhancement in Biological Wastewater Treatment Undergoing Different Stresses: Application and Mechanisms." *Critical Reviews in Environmental Science and Technology* 47 (19): 1877–99. https://doi.org/10.1080/10643389.2017.1400851

Zhao, Jianguo, Yu Li, Yahe Li, Ke Zhang, Hongzhong Zhang, and Yanfei Li. 2021. "Effects of Humic Acid on Sludge Performance, Antibiotics Resistance Genes Propagation and Functional Genes Expression during Cu (II) -Containing Wastewater Treatment via Metagenomics Analysis." *Bioresource Technology* 23: 124575. https://doi.org/10.1016/j.biortech.2020.124575

Zhao, Yunpeng, Ying Feng, Liming Chen, Zhao Niu, and Sitong Liu. 2019. "Genome-Centered Omics Insight into the Competition and Niche Differentiation of Ca . Jettenia and Ca . Brocadia Affiliated to Anammox Bacteria." *Applied Microbiology and Biotechnology* 103 (19): 8191–8202. https://doi.org/10.1007/s00253-019-10040-9

Zhou, Hexi, Xin Li, Guoren Xu, and Huarong Yu. 2018. "Overview of Strategies for Enhanced Treatment of Municipal/Domestic Wastewater at Low Temperature." *Science of the Total Environment* 643: 225–37. https://doi.org/10.1016/j.scitotenv.2018.06.100

Zhu, Ting-ting, Wen-xia Lai, Yao-bin Zhang, and Yi-wen Liu. 2022. "Feammox Process Driven Anaerobic Ammonium Removal of Wastewater Treatment under Supplementing Fe (III) Compounds." *Science of the Total Environment* 804: 149965. https://doi.org/10.1016/j.scitotenv.2021.149965

Zhu, Tingting, Yaobin Zhang, Guanhong Bu, Xie Quan, and Yiwen Liu. 2016. "Producing Nitrite from Anodic Ammonia Oxidation to Accelerate Anammox in a Bioelectrochemical System with a given Anode Potential." *Chemical Engineering Journal* 291: 184–91. https://doi.org/https://doi.org/10.1016/j.cej.2016.01.099

2 Microbes in Membrane Technologies

Prospects for Sustainable Wastewater Treatment

Moh'd Basel Shahin, Engy Ismail, Nour Eltaher, Mariam Alnaqbi, Amani Al-Othman, and Muhammad Tawalbeh

2.1 INTRODUCTION

Wastewater treatment is critical for protecting human health and conserving the environment by removing dangerous chemicals and pathogens from wastewater before it is discharged into the environment. These pollutants include suspended particles and organic debris, as well as antibiotics and pesticides, which are predominantly released by household waste, crop production, and industrial effluents. Untreated wastewater has the potential to pollute water bodies, disrupt aquatic ecosystems, and infect humans and animals with waterborne diseases. Endocrine disrupting compounds found in pharmaceutical effluents, for example, can impact the immune system in both humans and animals and cause reproductive and developmental disorders (Sacdal et al., 2020; Al Sharabati et al., 2021). Therefore, wastewater treatment is a crucial process to minimize the adverse impacts of wastewater on public health and the environment.

Traditional wastewater treatment methods include primary and secondary approaches. There are three categories of wastewater treatment techniques: biological, chemical, and physical (Tawalbeh et al., 2023). These include adsorption, advanced oxidation, chemical coagulation, electrocoagulation, and membrane separation (Al-Bsoul et al., 2020; Bani-Melhem et al., 2023; Darra et al., 2023; Halalsheh et al., 2022; Kazim et al., 2023; Shamaei et al., 2018). Physical methods, such as sedimentation and screening are used in the initial phase to remove largely sized solids and suspended particles from wastewater. Secondary treatment is the process of removing organic substances and nutrients from wastewater using biological processes. Traditional technologies, while effective at removing some pollutants from wastewater, have several drawbacks that hinder their progress. Conventional wastewater treatment processes have limitations in terms of pollutant removal. Primary treatment removes large solids and suspended particles but not pollutants such as dissolved organic compounds. In addition, novel pollutants, such as viruses are

DOI: 10.1201/9781003441069-2

detected more frequently and in higher quantities in traditional wastewater treatment systems than in current ones (Schmitz et al., 2016). The amount of energy required by standard techniques is another problem. Traditional systems demand a significant amount of energy to run, particularly for secondary treatment aeration (Macintosh et al., 2019). This has the potential to increase wastewater treatment costs while also contributing to greenhouse gas emissions. Finally, conventional wastewater treatment has an adverse environmental impact, particularly in terms of sludge generation and disposal. Sludge can pose environmental and health problems, as well as increase the total operating costs of wastewater treatment plants if it not disposed of appropriately (Englande et al., 2015).

With many benefits over conventional techniques, membrane technologies are developing as a promising solution for wastewater treatment. Membrane technologies utilize semi-permeable membranes to extract pollutants from wastewater. Microfiltration (MF), ultrafiltration (UF), nanofiltration (NF), and reverse osmosis (RO) are membrane separation technologies (Tawalbeh et al., 2018). The ability of membrane technology to effectively eliminate contaminants from wastewater is one benefit. Membrane pores have a wide size range for filtration, which allows them to eliminate the smallest nanosized particles, such as the viruses detected in conventional treatment (Schmitz et al., 2016). Additionally, membrane filtration can generate highly treated water that satisfies water quality requirements. The fact that membrane technologies use less energy than conventional techniques is another benefit. Membrane filtration is more energy efficient and economical than aeration for secondary treatment since it uses less energy (Macintosh et al., 2019). Since they provide a concentrated waste stream that can be treated separately, membrane technologies can help limit the production and disposal of sludge. By doing so, the hazards to the ecosystem and people's health posed by the disposal of sludge can be reduced (Englande et al., 2015). With overall low energy consumption, lack of chemical use, and reasonable operating conditions, membrane filtration can be a more sustainable solution in comparison to conventional wastewater treatment methods (Al Aani et al., 2020). This chapter will begin with a general overview of various kinds of membrane methods used in wastewater treatment. Then, the chapter will delve into the challenges that microbes present to membranes and membrane-based technologies and highlight the processes that have successfully integrated microbes in wastewater treatment.

2.2 ROLE OF MICROBES IN MEMBRANE TECHNOLOGIES

Although membrane technologies have shown significant potential in treating wastewater, the presence of microorganisms in wastewater may negatively impact their overall performance. A major challenge in membrane process operations that is caused by microbes is microbial fouling. Microbial fouling is triggered by the attachment of microbes on the membrane surface, which could reduce the efficiency of membrane-based processes. This microbial-related challenge limits the utilization of membranes in wastewater treatment. On the other hand, researchers have recognized the potential of combining microbes with membrane technologies to create new and promising wastewater treatment methods. Membrane bioreactors (MBRs) are an example of this, where the biological treatment of contaminants is combined with membrane filtration to efficiently treat wastewater. Another example of the successful integration of microbes and membranes for wastewater treatment is microbial fuel cells (MFCs) that incorporate microbes into electrochemical cells to produce electricity. As they can oxidize both organic and inorganic materials found in water, MFCs are useful in the treatment of wastewater (Gude, 2016). Ultimately, the role of microbes in membrane and wastewater treatment technologies is complex, with the potential for both positive and negative impacts. Therefore, researchers can strive toward a more sustainable future for wastewater treatment by investigating the potential of microbes in membrane technologies and developing effective strategies to overcome their limitations.

2.3 MEMBRANE TECHNOLOGIES IN WASTEWATER TREATMENT

Membrane processes are widely employed in wastewater treatment, with four main types, including MF, UF, NF, and RO. The major distinction between each membrane process is its pore size, which determines the types of contaminants that can be removed. Figure 2.1 provides a helpful overview of the pore sizes associated with each membrane process and the typical contaminants that can be removed by each (Anis et al., 2019b). By understanding the differences between each type of membrane process, it becomes possible to select the most suitable one for a particular application.

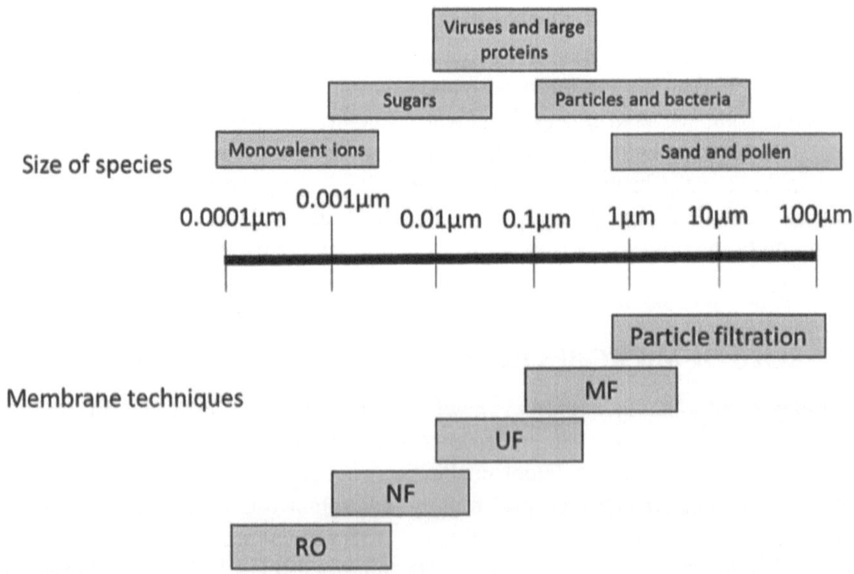

FIGURE 2.1 Overview of pore size and materials removed by different membrane processes.

Source: Anis et al. (2019b).

2.3.1 Microfiltration

Among the earliest pressure-driven membrane processes is MF. It can prevent a variety of matter from passing through, including bacteria, proteins, and other micrometer-sized materials (Anis et al., 2019a). The typical size of the pores of a microfilter membrane ranges from 0.1 to 10 micrometers (Singh & Purkait, 2019). Because of this wide range of pore sizes, MF can be applied to a variety of processes, including the removal of pharmaceuticals and personal care products (Wang et al., 2015), the processing of dairy products (France et al., 2021), the field of biotechnology (Brown et al., 2009), and most importantly, the treatment of wastewater (Manni et al., 2020).

In water treatment plants and other industrial processes, MF membranes are often arranged into equipment, which are known as modules, to facilitate the membrane separation process. There are various categories of modules used for membrane filtration, including the plate and frame, spiral wound, tubular, and hollow fiber membranes (Singh & Purkait, 2019). In MF, hollow fiber membranes are the modules that are used most frequently. They are composed of long, hollow fibers that form thin tubes. Moreover, they can be designed in a manner that allows water to flow from the outside to the inside of the hollow fiber via a vacuum or from the inside to the outside through pressure (Tawalbeh et al., 2021). Hollow fiber membranes are favorable modules because they allow for backwashing, provide a high surface area for separation, and can be fabricated at a low to moderate cost (American Water Works Association, 2016; Hosseini et al., 2022).

The effectiveness of membranes can be affected by the materials used to construct them. MF membranes were commonly made from polymers due to their ease of production and low cost. However, the main disadvantage of polymeric membranes is that they are prone to fouling, resulting in a limited lifespan (Singh & Purkait, 2019). As a result, ceramic membranes prepared from inorganic materials, such as alumina, titania, and zeolites, have become the preferred choice for industrial applications. In particular, zeolites have received great attention due to their numerous advantages, such as perfect shape selectivity and thermal, chemical, and structural stabilities (Tawalbeh et al., 2012; Tawalbeh et al., 2013). The high cost of ceramic membranes has hindered their advancement in the industry; however, researchers are actively experimenting with a range of cost-effective ceramic membrane materials to tackle this issue (Manni et al., 2020).

2.3.2 Ultrafiltration

UF is the mechanical separation of suspended particles, macromolecular species, and organic matter from a liquid using driven pressure or a concentration gradient across a semi-permeable membrane. Using porous membranes with pore sizes of 2–100 nm, this separation process targets nanometer-sized particles with higher molecular weights (Hampu et al., 2020). Because of the vast variety of particle sizes this technology can remove; its applications vary across a multitude of industries. These applications range from dairy production and medical use to wastewater treatment and the filtration of bacteria and viruses (Al Aani et al., 2020). UF is a more appealing substitute to conventional purification processes due to its minimal energy consumption, lack of chemical use, and moderate operating conditions. These traits allow for simultaneous expense reduction and increased sustainability (Al Aani et al., 2020).

UF membranes, like NF membranes, use a spiral wound membrane (SWM) configuration. The wide use of this configuration results from its high surface area, which allows for efficient particle separation. This structure also allows for enhanced membrane stability at high-operating pressures because of the spacers and the packed module design (Kammakakam & Lai, 2023). However, it is also important to consider flow configuration for the effective application of UF processes. Crossflow and dead-end filtration are the two primary flow configurations employed in UF, as illustrated in Figure 2.2. Greater water recovery is achieved with dead-end filtration because the feed flow crosses the membrane perpendicularly (Kammakakam & Lai, 2023). Contrarily, crossflow filtration involves a parallel flow of liquid and particles across the membrane, therefore leading to improved stability and a longer membrane lifetime (Kammakakam & Lai, 2023).

Membrane material is an important factor for UF applicability and efficiency. Using inorganic ceramic membranes for protein separation in the dairy industry is one example of this. Compared to other membrane materials, inorganic ceramics have better thermal and mechanical characteristics, which enhances the performance of UF membranes in this specific application (Al Aani et al., 2020). Furthermore, ceramic membranes have a lifetime exceeding 10 years and can be used in both dead-end and crossflow configurations (Igunnu & Chen, 2014). On the contrary, more than 90% of UF membranes are constructed using polymer materials, such as polyacrylonitrile and polyvinylidene (Igunnu & Chen, 2014; Kammakakam & Lai, 2023). These materials have revolutionized UF membranes due to the creation of asymmetric membranes, which possess exceptional permeation qualities and high sustainability (Kammakakam & Lai, 2023). Therefore, membrane material should be selected according to the application it will serve.

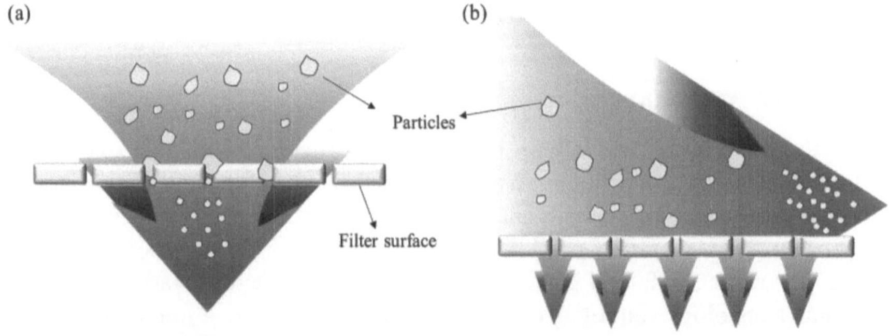

FIGURE 2.2 Comparison of (a) dead-end and (b) crossflow filtration.

Source: Kammakakam & Lai (2023).

2.3.3 Nanofiltration

NF is a pressure-based membrane filtration process that utilizes a filter with a pore size between 0.5–7 nanometers (Madhavi & Ramesh, 2023). The nano filter rejects any particles that have a size larger than its pore diameter. Most NF membranes are thin-film composites (TFC) that are rolled in a spiral, cylindrical shape referred to as a SWM (Chen et al., 2020). The feed to the membrane enters from one side of the SWM and the permeate leaves through a center tube that collects the purified water. The remaining water exits the membrane as a concentrated stream. Figure 2.3 shows the SWM build model (Chen et al., 2020).

For the feed water to pass through the membrane, pressure must be applied to drive the filtration process. This pressure ranges from 4 to 40 bar, which is sufficient for removing contaminants, such as microorganisms, droplets, dust particles, and dyes to name a few (García Doménech et al., 2020). Operating an NF process under elevated pressure generates high-quality water and high contaminant removal efficiency, which makes the process cost and energy effective compared to other pressure-driven membrane filtrations (Mohammad et al., 2015). Additionally, NF membranes exhibit better mechanical and thermal stability, and their scalability is more flexible compared to UF, MF, and RO membrane processes (Madhavi & Ramesh, 2023).

The NF membrane is often synthesized from polymeric and ceramic materials. However, polymeric membranes are favored due to mechanical resistance, flexibility, and forming properties. Among these polymers are cellulose acetate (CA), polyimides (PI), and polyether sulfones (PES), which are widely used (Madhavi & Ramesh, 2023). Furthermore, several factors limit the operation of NF membrane

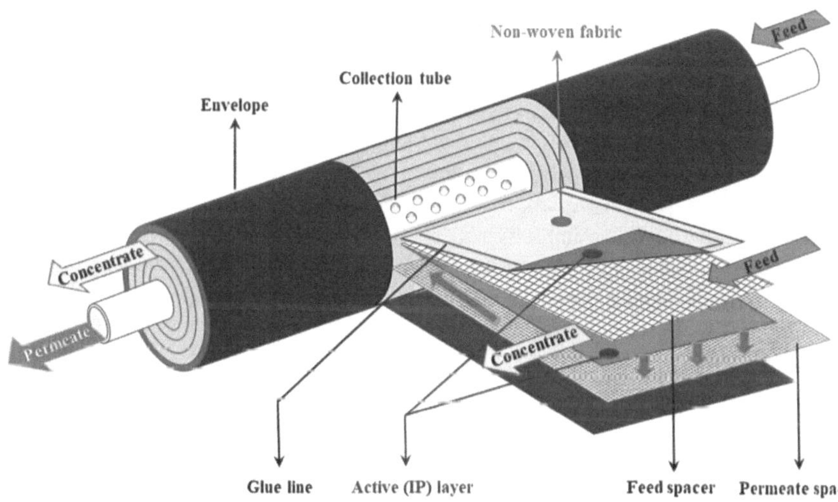

FIGURE 2.3 SWM build model for TFC-NF.

Source: Chen et al. (2020).

processes. These factors are molecular weight cut-off, maximum operating temperature, pH range, and stabilized salt rejection (Mohammad et al., 2015).

2.3.4 Reverse Osmosis

Reverse osmosis (RO) is a water filtration technique that removes ions, molecules, bacteria, and larger particles from drinkable water using a semi-permeable membrane and applied pressure. In RO, osmotic pressure, a cooperative characteristic caused by chemical potential differences of the solvent, a thermodynamic parameter, is overcome by the applied pressure (Ahuchaogu et al., 2018). The RO removes bacteria and dissolved particles from the water to produce potable water. The membrane rejects the large molecules and the small molecules, such as water, can pass through it (Warsinger et al., 2016). RO is best known for its use in the purifying of potable water from seawater by desalination and the removal of other waste products from water. Bacteria, chlorides, heavy metals, and dissolved compounds like nitrate, sodium, cryptosporidium, fluorine, sulfides, giardia, arsenic, mercury, uranium, and lead can be removed using this method (Ahuchaogu et al., 2018).

Normally, osmosis involves molecules that move from a region with high water potential (low solute concentration) to low water potential (high solute concentration). In RO, as the name suggests, the natural flow of molecules is reversed and they move from a region of high solute concentration to low solute concentration, due to an external pressure applied.

Since RO membranes are practically nonporous, they can successfully filter out particles as well as a variety of low-molar-mass species, such as salt ions, organics, and other molecules. The RO membranes are usually comprised of CA or polysulfone coated with aromatic polyamides. The most common module for RO is what is called the spiral wound module and it features a perforated permeate collecting tube covered in a flat sheet membrane (Baker, 2004). The feed flows into the membrane and on the other side of the membrane, the permeate is gathered and spirals inward toward the central collecting tube (Sagle & Freeman, n.d.).

This technique benefits from a membrane-based process that achieves concentration and separation without a state change, the use of chemicals, or thermal energy. As a result, the process is energy efficient and ideal for recovery applications (Tawalbeh et al., 2023). There are various applications for RO systems, including the treatment of wastewater in the beverage industry and distillery spent wash as well as the recovery of phenol compounds and groundwater treatment (Puri et al., 2022), Therefore, it can be concluded that RO membranes are effective and have a wide range of applicability (Garud et al., 2011).

2.4 MICROBES IN MEMBRANE TECHNOLOGIES

2.4.1 Membrane Bioreactors (MBRs)

Microbes can be used in conjunction with membrane technologies to create an effective and promising system for treating wastewater. MBRs are considered innovative and advanced wastewater treatment methods that integrate biological and membrane technologies into one unit (Asante-Sackey et al., 2022). It is a process that exploits the benefits of microbes and membrane filtration processes in wastewater treatment to produce high-quality water.

MBRs make use of the process in which organic matter is consumed by microorganisms. It involves the consumption of these pollutants by microbes within an aerobic (oxygen-rich) environment (Bilad, 2017). The aerobic environment in the MBR is provided by an aeration tube placed inside the bioreactor vessel (Sengupta et al., 2022). Through this process, microorganisms efficiently convert organic pollutants into harmless byproducts, mainly carbon dioxide, water, and biomass. The resulting microorganisms and biomass form flocs that are subsequently separated using a membrane (Bilad, 2017). Various types of microorganisms can grow inside of the bioreactor by allowing the biomass to remain within the tank for a sufficient time. Examples of such microbes include *Rhodococcus rhodochrous*, *Bacillus subtilis*, and *Acinetobacter* (Femina Carolin et al., 2021).

MBRs were initially built with the membrane placed on the outside of the aerobic bioreactor, as in the side stream MBR system (Femina Carolin et al., 2021). However, the current configuration of the MBR involves placing the membrane inside the bioreactor, as in the submerged MBR system (Sengupta et al., 2022). In most MBR systems, the preferred types of membranes utilized are MF or UF membranes (Asante-Sackey et al., 2022) and they are usually placed in either tubular modules, flat sheet modules, or hollow fiber modules (Femina Carolin et al., 2021; Sengupta et al., 2022). Smaller pore-sized membranes, like NF or RO membranes, are utilized for the removal of specific types of contaminants, such as pharmaceutical compounds, for which MF and UF membranes are ineffective due to their pore size (Femina Carolin et al., 2021). In submerged MBRs, hollow fiber and flat sheet modules (1–3 mm in diameter) are typically used, whereas in side stream MBRs (> 5 mm in diameter), tubular modules are suitable to enable flow from inside the membrane to the outside (Femina Carolin et al., 2021). Figure 2.4 shows the two main types of MBR configurations: side stream MBR and submerged MBR (Ng & Kim, 2007).

The main difference between submerged MBRs and side stream MBRs is the location of the membrane. In submerged MBRs, the MF or UF membrane is placed inside the bioreactor and the driving force for filtration is either pressure or suction

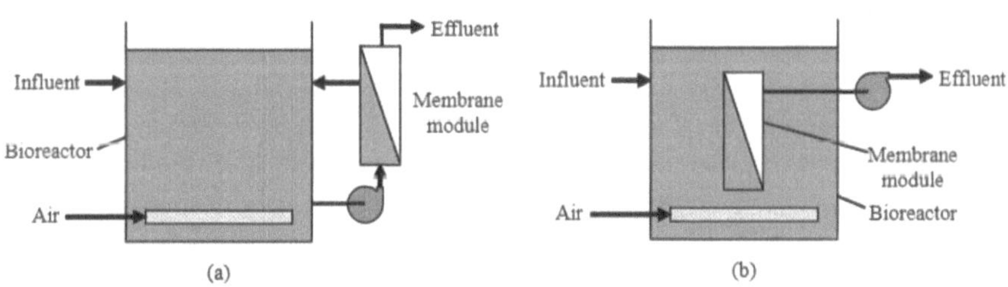

FIGURE 2.4 Illustration of MBR configurations: (a) side stream MBR and (b) submerged MBR.

Source: Ng & Kim (2007).

(Mutamim et al., 2012). Following separation, the filtered components of the water, known as sludge, are recycled into the bioreactor, while the treated water is pumped out of the tank as effluent. On the other hand, the membrane in side stream MBRs is placed outside of the bioreactor tank. A portion of the biologically treated water is removed from the bioreactor and pumped through the membrane module in a different location. In both configurations, air is supplied to the bioreactor tank to create an aerobic environment for microorganisms to grow (Mutamim et al., 2012). These air bubbles are essential for submerged MBRs because they act as a source of agitation that minimizes membrane fouling. However, increased bubbling of air is required for side stream MBRs to reduce membrane fouling (Sengupta et al., 2022). In comparison to other MBR configurations, submerged MBRs have been proven to be more energy and cost efficient (Judd & Turan, 2018).

The MBR operation can also occur in anaerobic conditions where the biological treatment takes place in the absence of oxygen. Anaerobic MBRs can be used to recover energy because they lead to the production of methane and could also result in a reduced footprint when compared to aerobic MBRs (Sengupta et al., 2022). However, anaerobic MBR operation has a long startup time and is more suitable for high-strength water treatment (Mutamim et al., 2012). Moreover, anaerobic operation is more prone to membrane fouling compared to aerobic operation.

Overall, the implementation of MBRs in wastewater treatment showcases the great potential of integrating microorganisms with membrane processes, resulting in a highly effective technological solution. This is supported by research, which has demonstrated that MBRs produce high-quality treated water with minimal organic and inorganic pollutants (Asante-Sackey et al., 2022). Furthermore, the use of MBRs eliminates the need for conventional treatment methods, such as UV light disinfection and tertiary membrane filtration, which can decrease the overall footprint of the wastewater treatment plant (Asante-Sackey et al., 2022). MBRs are promising alternatives to conventional treatment processes, however, their operation is limited by several factors that necessitate further research. These factors include membrane fouling and high cost and energy requirements.

2.4.2 Microbial Fuel Cells (MFCs)

MFCs are another good representation of the beneficial integration of microbes and membranes within a single process, offering sustainable technology that can be used for wastewater treatment. They are a new bioelectrochemical method that uses electrons derived from biological reactions catalyzed by bacteria to generate electricity. MFC wastewater systems produce electricity by the oxidation of organic or inorganic compounds present in urban sewage, agricultural, dairy, food, and industrial wastewater. Electrons generated by bacteria during substrate oxidation in the anode compartment (the negative terminal) are transmitted to the cathode compartment (the positive terminal) via an electron conductive medium (e.g., a wire or a load) in MFCs (Al-Murisi et al., 2022). Electrons react with oxygen in the cathode, while protons diffuse across a proton exchange membrane (PEM). During operation, MFCs necessitate constant electron release in the anode and constant electron consumption in

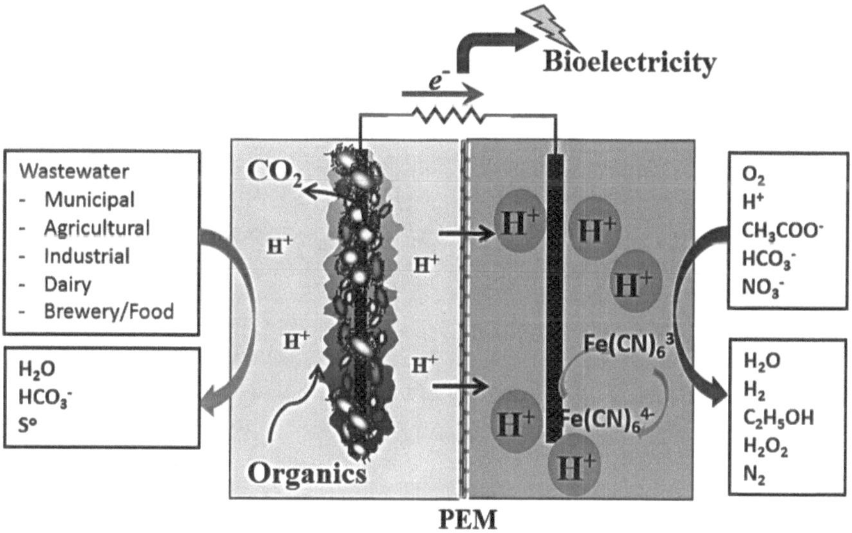

FIGURE 2.5 Illustration of the MFC and its application in wastewater treatment.

Source: Gude, (2016).

the cathode (Harnisch et al., 2011). Figure 2.5 shows a schematic of an MFC and its application in wastewater treatment (Gude, 2016).

Traditional aerobic treatment of low-strength wastewater requires significant capital as well as operational and energy consumption costs. For example, the aeration of sewage needs around 0.5 kWh/m^3, leading to an annual energy usage of approximately 30 kWh per person (Malik et al., 2023). Wastewater contains three forms of energy, organic matter, nutritional elements, such as nitrogen and phosphorus, and thermal energy. Domestic wastewater contains both chemical energy (26%) and thermal energy (24%) (Gude, 2016). Chemical energy in the form of carbon can be measured using chemical oxygen demand (COD) and it is also present in the form of nutritional molecules, such as nitrogen and phosphorus. Although thermal energy accounts for most of the energy produced by MFCs, it cannot be efficiently retrieved unless a heat pump is used. On the other hand, chemical energy can be easily recovered. Due to the recovery of this energy, wastewater treatment can be considered an energy-producing process, rather than an energy-consuming one, while reducing environmental pollution. MFCs can greatly help solve the challenges that most wastewater treatment processes encounter. They have been proven to be a reliable technology that produces low amounts of sludge and is energy efficient (Malik et al., 2023). MFC power outputs are growing today, and significant obstacles have been overcome due to the implementation of both microbial and electrochemical elements.

2.4.2.1 MFC Configurations

Double-chamber MFC, single-chamber MFC, up-flow MFC, and stacked MFC are the fundamental designs utilized in laboratories for MFC applications. The double-chamber MFC is the most basic MFC design and can be shaped like a bottle or a cube (Kumar et al., 2016). Typically, one bottle can come in different designs and serves as an anode and the other as a cathode, separated by a PEM. A particular medium (or substrate) in the anode and a defined catholyte solution are frequently used to generate energy in a two-chamber MFC (Kumar et al., 2017). A single-chamber MFC consists of a single chamber that houses both the anode and the cathode. The anode is either far or close to the cathode, separated by a PEM. In comparison to the double-chamber MFC, the single-chamber MFC is simple and affordable, and also produces a considerable amount of power (Ringeisen et al., 2006). The up-flow MFC has a cylinder form; the cathode chamber is at the top of the MFC, and the anode is at the bottom. There are no separate anolytes or catholytes in this design nor is there a physical separation (Kumar et al., 2017). A stacked MFC is a collection of MFCs that are either coupled in series or parallel to increase power output. By connecting some MFCs and multiplying individual power output or current output, the output of the MFC can be increased.

2.4.2.2 Applications of MFCs

MFCs can be used for a variety of purposes, including wastewater treatment, environmental monitoring, and renewable energy production. They may use a variety of organic substrates as fuel sources, including sewage, agricultural waste, and food waste, making them viable technology for long-term waste management and energy generation. Furthermore, MFCs have the potential to be integrated into a variety of systems, such as microbial desalination cells (MDCs) for creating freshwater from seawater or self-powered sensors for remote environmental monitoring (Boas et al., 2022).

MFCs can be used in wastewater treatment plants to process organic waste while simultaneously producing electricity. This is achieved as the bacteria break down organic molecules in wastewater and generate energy as a byproduct. As a result, MFCs can be regarded as a feasible and eco-friendly wastewater treatment alternative since they not only purify the waste but also generate clean energy (Alhajar et al., 2022).

MFC technology is a feasible sustainable energy option for tackling issues such as nonrenewable energy use and pollution. In environmental monitoring, MFCs are used to sense and detect contaminants in water or soil. MFC microorganisms can be tailored to respond particularly to specific contaminants and their activity can be evaluated through changes in electricity output (Sonawane et al., 2020). As a result, MFCs are a promising technology for monitoring environmental pollutants in real time.

MFCs can be transformed into MDCs by modifying their primary operating principle through adding an extra chamber between the anodic and cathodic chambers on a standard MFC. An anion exchange membrane and a cation exchange membrane divide the chambers (Tawalbeh et al., 2020). MDCs are capable of performing desalination while also treating wastewater and producing energy.

Microbial activity in MFCs provides electrical signals that can be associated with analyte concentration, resulting in a simple and cost-effective biosensing approach (Ivars-Barceló et al., 2018). The capacity to generate electrical current in response to biological behavior is extremely tempting for biosensor applications. MFCs have been utilized as biosensors for water quality (pH, temperature, and heavy metal concentrations), identifying specific chemicals (mostly acetate, propionate, and butyrate), and monitoring biochemical oxygen demand (Ivars-Barceló et al., 2018).

2.4.2.3 Challenges of MFCs

While MFCs are known to be a sustainable and environmentally friendly technology, the high capital cost associated with factors such as manufacturing and expensive electrodes remains their greatest limitation (Yaqoob et al., 2020). To decide on a cost-effective and efficient electrode, some factors such as conductivity, stability, surface area, porosity, cost-effectiveness, and corrosion resistance must be considered. The high cost of electrodes is a significant contributor to the overall expense of MFC technology, which has hindered its widespread adoption in the industry. Although carbon felt is acknowledged as a low-cost electrode material for use in MFCs, the mass transfer of various products and substrates onto the inner surface of carbon felt limits the bacterial growth rate (Wei et al., 2011). As a result, research is being conducted in the hopes of discovering a low-cost substitute for carbon-felt electrodes that has no such limitations. One such alteration is substituting it with graphitic mesoporous carbon, which is a relatively cheaper, more conductive, and biocompatible electrode ideal for MFC applications (Mahmoud & El-Khatib, 2020). The transfer of electrons generated by bacteria in an anaerobic anodic chamber system toward the cathode is critical to the operation of an MFC. As a result, the greater the conductivity of the electrodes the greater the power generation. To ensure the long-term operation of MFCs, the chemical and mechanical stability of electrodes is an essential factor to address. If not accounted for, the electrode can undergo swelling, breakage, and cracking which disrupts the electrode's physical stability (Ying et al., 2018). To increase the MFC's power production, the surface area must be increased, which simultaneously lowers the internal resistance (Li et al., 2017). In addition, the greater the surface area, the larger the site available for bacterial reaction. As for porosity, this can influence the electrode's performance negatively if not selected based on the desired percentage. An electrode with high porosity can decrease electrical conductivity and an electrode with low porosity can cause clogging in the system (Neethu et al., 2022). Since sensitive microorganisms are involved, each component of a bioelectrochemical system must be biocompatible. The anode is always close to the living creature, and in some cases, such as microbial carbon-capture cells, the cathode is also near the microalgae (Neethu et al., 2022). As a result, for an MFC to function properly, the electrode must be biocompatible.

Dependence on microorganisms is another challenge. Microorganisms utilized in MFC technology are highly susceptible to pH fluctuations; therefore, it is essential to carefully monitor and regulate pH as a critical parameter. The best level of performance is obtained in a neutral environment; thus, a highly acidic or basic environment is not favorable. Another study that adjusted the anolyte pH revealed that the MFC performed best when operated at a pH of 7.3 for wastewater in the anodic chamber (Jia et al., 2014). Yet, in another study on the influence of pH on anodic biofilm, a pH of up to 9.5 was shown not to inhibit anodic biofilm activity (Neethu et al., 2022). Furthermore, the efficient utilization of PEM between the anodic and cathodic chambers is critical for maintaining the proper pH in the chambers (Neethu et al., 2022).

2.4.3 Microbial Fouling

Over the years, membrane technologies have gained popularity over conventional wastewater treatment methods due to several advantages they offer, including their capability to easily scale up, occupy less space, and selectively remove contaminants. Despite these benefits, the application of membrane technologies suffers from a major challenge: a gradual reduction in permeate flux during their operation due to membrane fouling (Guo et al., 2012). Membrane technologies are affected by various types of foulants and fouling mechanisms, each of which necessitates different control strategies. There are four main types of membrane foulants: particulates, which are organic or inorganic particles that cause the formation of a cake layer or clog the porous surface of the membrane. Organic foulants bind to the surface of the membrane by adsorption; inorganic foulants, which form on the surface of the membrane due to oxidation or variation in pH; and finally microbial foulants, which attach to the membrane causing the formation of a biofilm on the membrane surface (Guo et al., 2012). Microbial fouling or biofouling is a major operational challenge in membrane applications, which has led to an increasing focus on developing effective solutions. Therefore, in this section, a comprehensive analysis of microbial fouling will be presented, along with a review of current literature solutions aimed at addressing its effects on membrane technologies.

2.4.3.1 Microbial Fouling in Membrane Technologies

Microbial fouling is the unwanted buildup of microbiological organisms on the membranes, which takes place through the deposition, metabolism, and growth of bacterial cells (Guo et al., 2012). It is considered a significant vulnerability in the application of membrane processes for water treatment. This is because, even after eliminating the majority of microorganisms from the water, a small number of remaining microorganisms can continue to grow by utilizing biodegradable substances present in the water (Nguyen et al., 2012). Microbial fouling is considered a leading factor in over 45% of all instances of membrane fouling (Komlenic, 2010). Moreover, microbial fouling affects all types of membranes and membrane technologies, such as UF, NF, MF, RO, and MBR, however, it has been outlined as a major challenge in NF, RO, and MBR filtration. The effects of microbial fouling on membrane processes can be significant and may include the following (Cui et al., 2021; Nguyen et al., 2012):

- Reduction of membrane flux due to biofilm formation on the surface of the membrane.
- Higher energy consumption due to a higher pressure required for membrane cleaning and counteracting the reduction in permeate flux.

- Biodegradation of some membrane materials due to the formation of acidic byproducts. For instance, CA-based membranes are prone to this type of biodegradation.
- Decline in product water quality due to the increase in salt passage. This occurs due to the buildup of dissolved ions within the biofilm.

Therefore, to mitigate the impact of microbial fouling on the efficiency of membrane processes, it is necessary to understand the mechanism of microbial attachment to membranes, the monitoring methods used to detect the problem, and the strategies available for its control.

2.4.3.2 Process of Microbial Attachment to the Membrane Surface: Biofilm Formation

In membrane processes, the biofilm formation process by microorganisms typically takes place in four phases. The first phase is the adsorption of organic and inorganic compounds, such as proteins and macromolecules on the membrane surface (Cui et al., 2021; Guo et al., 2012). This causes the formation of a conditioning film, which encourages microorganisms to attach to the membrane.

The second phase involves the transportation and adhesion of microbes to the surface of the membrane (Cui et al., 2021). The primary method of microbial transportation to the membrane surface occurs via fluid dynamic forces and, after the successful transportation of microbes to the surface, the initial adhesion occurs via hydrophobic and electrokinetic interactions with the membrane to form the biofilm layer (Nguyen et al., 2012). Various factors affect microbial attachment to the surface of the membrane, including the membrane's roughness, material of construction, and porosity, the feed water's pH level and temperature, and the microorganism's species and population density.

In the third phase of the process, microbes begin to grow and multiply on the surface of the membrane by utilizing the adsorbed organic matter on the membrane or the nutrients present in the feed water (Nguyen et al., 2012). Moreover, during their proliferation process, microbes begin to secrete different types of organic materials, such as lipids, proteins, and polysaccharides, which are known as extracellular polymeric substances (EPS) (Cui et al., 2021; Nguyen et al., 2012). The EPS that dissolve in the bordering solution or weakly bind to microbial cells are called soluble microbial products (SMP). The EPS and SMP bind the microbes together and contribute to the structural integrity of the biofilm formed on the membrane surface (Nguyen et al., 2012). In addition, they add to the stability of the biofilm layer and help in the organization of microbes at the colonized surface. The reduction in membrane permeability can mainly be attributed to the accumulation of EPS on its surface, rather than the accumulation of microorganisms. Because of their ability to accumulate at the surface or infiltrate the pores, SMP were found to be among the most commonly occurring organic foulants in UF, NF, and RO.

Finally, the fourth phase involves the detachment of the mature cells from the biofilm layer where the growth begins to be limited by the shear forces of the fluid (Cui et al., 2021; Nguyen et al., 2012). This results in microbial fouling being in a steady state at the membrane surface.

2.4.3.3 Microbial Fouling Monitoring Techniques

Being able to accurately identify and characterize the biofilm layer on the surface of the membrane is essential for developing effective microbial fouling prevention strategies. Moreover, the membrane surface needs to be continuously analyzed continuously without damaging the membrane material.

A common monitoring technique is to assess the microbial fouling potential of the water being fed to the membrane process. By taking samples of the feed water, important biological parameters can be tested. Some of these parameters include the adenosine triphosphate (ATP) content, which is present in every bacterial cell, and the total direct cell counts, which reveal the concentration of microbes in the water (Nguyen et al., 2012). It should be noted however that these parameters cannot be continuously tested as they require sampling of the feed water. Hence, this method is mainly useful as a preliminary screening technique for the detection of microbial fouling in membrane processes.

An alternative monitoring approach is to make use of developed monitoring systems, such as the membrane fouling simulator (MFS) or the Silent Alarm™ system (Nguyen et al., 2012). The MFS is a monitoring system originally designed for the early detection of microbial fouling in RO and NF processes with spiral wound modules. It monitors the pressure drop in the feed water channel through control devices such as flow controllers and pressure drop transmitters. The MFS is also considered an early screening technique for the detection of microbial fouling. It is a well-suited method for assessing the possibility of biofilm formation in high-operating pressure processes, such as RO (Nguyen et al., 2012). The Silent Alarm™ was also specifically designed for RO and NF membranes, however, it monitors the process in real time. It can detect the initial phases of microbial fouling and utilizes a scaling factor known as the fouling monitor (FM) (Saad, 2004). When the FM reaches 5%, it indicates that fouling has commenced. However, if the FM reading exceeds 20%, it indicates that the membrane has already reached a state of irreversible fouling and would probably need to be replaced (Saad, 2004).

Other monitoring strategies found successful in laboratory experiments include biosensors (or nanosensors) and ultrasonic time-domain reflectometry (UTDR), which utilizes sound waves to identify the location of the biofilm layer as well as its characteristics (Wang et al., 2018; Xu & Lee, 2020).

2.4.3.4 Control of Microbial Fouling

Different methods can be employed to prevent or control microbial fouling in membrane processes. These methods can be broadly categorized into three main strategies: physical, chemical, and biological. Each strategy has its advantages and limitations, and as a result, research is continuously being done to find the most viable method for the prevention and control of microbial fouling.

Microbial fouling can be mitigated by the physical cleaning of the membrane when a drop in permeating flux is noticed in the process. Physical cleaning aims to remove the microbial foulants on the surface of the membrane or to reduce their interactions within the biofilm layer. It can be done hydraulically, which involves consistent backwashing of the membrane to ensure that the microbes detach from the surface (Ruigómez et al., 2022). It can also be done pneumatically, which includes air bubbling, scouring, or sparging (Nguyen et al., 2012). The most widespread control method for microbial fouling is hydraulic backwashing. In MBR systems and SWM modules, backwashing (hydraulic) combined with air sparging (pneumatic) is the most commonly employed technique (Nguyen et al., 2012). Additionally, mechanical cleaning is another form of physical cleaning, which can entail using a sponge ball to scrub the surface of the membrane.

Chemical membrane cleaning is another method for dealing with microbial foulants. The chemical agents used are mostly oxidants but can involve other types, such as acids, bases, and enzymes (Cui et al., 2021; Nguyen et al., 2012). The main limitation of chemical cleaning is the production of toxic byproducts. To improve the effectiveness of microbial fouling control, chemical cleaning is combined with physical cleaning in a process called chemical-enhanced backwashing (Ruigómez et al., 2022).

Antimicrobial chemical agents or biocides can also be added to the water to attack and kill the microorganisms (Da-Silva-Correa et al., 2022). This can be done before the membrane process as a disinfection stage. Chlorine is an example of an oxidizing biocide that is used for this purpose. However, chlorine was found to affect polymeric-based membranes, such as polyamide membranes negatively. Therefore, alternative biocides for disinfection that are relatively less toxic include chloramines and chlorine dioxide, as well as non-oxidizing biocides, such as formaldehyde and glutaraldehyde (Nguyen et al., 2012). Ozonation and UV light radiation are other types of biocide disinfection treatments that destroy the presence of microbes in water, which can be useful to control and prevent microbial fouling from occurring in membrane processes (Kim et al., 2009).

The use of biological strategies to control microbial fouling is a relatively new concept, and the majority of research in this field has been done on a laboratory scale (Cui et al., 2021). A biological treatment technique called quorum quenching (QQ) relies on the idea that microorganisms communicate with one another to carry out important biofouling processes, including the creation of the biofilm. This communication happens via a quorum sensing (QS) mechanism where the bacteria send signaling molecules known as auto-inducers (Millanar-Marfa et al., 2020). QQ works by the obstruction of the QS in the microbial community through mechanisms such as enzymatic treatment and utilizing chemical reaction approaches. This method minimizes microbial fouling by blocking the auto-inducer production in the microbial community or hindering the ability of the signaling molecule to attach to receptors. A study done on RO membranes that contained foulants showed that 60% of microbial species generated signaling molecules (Nguyen et al., 2012). This emphasizes the great potential of using QQ to control microbial fouling in membrane processes.

An alternative biological-based strategy is to use bacteriophages, which are viruses that can infect and reproduce inside microbial cells to kill them. This method was found to be effective in MBRs by minimizing the microbial attachment to the membrane, and because some bacteriophages are larger than the pores of NF, UF, and RO, they remain attached to the surface and continue to infect the bacterial cells coming from the feed water. However, this approach has two major drawbacks: it may result in an elimination of beneficial bacterial cells required for wastewater processing, and it can cause the microbial cells to slowly develop an immune system, which causes the bacteriophages to become inactive (Cui et al., 2021).

2.4.3.5 Microbial Fouling in MFCs

MFCs have also been found to suffer from microbial fouling. This is related to the PEM that is used in MFCs to separate the cathode and the anode compartments of the cell (Flimban et al., 2020). The membrane should possess important characteristics, including high proton conductivity, great mechanical strength, and excellent chemical resistance (Javed et al., 2022). Additionally, the PEM, in general, can operate at low temperatures (below 80 °C) or high temperatures (between 120–200 °C) using proper PEM materials or high-temperature hydrocarbons, respectively (Shi et al., 2015). Unfortunately, this membrane is susceptible to microbial fouling, which jeopardizes the efficiency of the fuel cell and reduces energy generation (Flimban et al., 2020). The presence of foulant results in the partial or complete blockage of the membrane, which prevents the migration of protons from the anode toward the cathode compartments. Furthermore, metal ions present in the cell often bind to the negatively charged sulfonate groups in the PEM in place of protons. This binding creates an intermolecular bridging between the foulants and the sulfonate groups leading to the formation of biofilms (Xu et al., 2012). For the cell to continue producing electricity, the PEM must be changed or recovered, which is very costly. Therefore, microbial fouling is one of the main challenges hindering the scalability of MFCs (Noori et al., 2019).

The electricity generated from MFCs is rarely utilized due to several losses (Liu et al., 2018). These losses include activation loss, ohmic loss, and concentration loss (Noori et al., 2019). Furthermore, solutions proposed to minimize such losses include modification of the MFC design. For example, a possible modification is to reduce the distance between the electrodes in the cell assembly. Another example would be to use a highly electrochemical catalyst, such as gold, silver, or platinum (Noori et al., 2019). Moreover, the rapid growth of biofilms in MFCs significantly reduces their power output. During the operation of the MFC, the biofouling layer thickness was found to reach 14.7, 165.1, and 250.1 μm after two, four, and six months of operation, respectively (Miskan et al., 2016). More importantly, the MFC showed a 55% power density reduction in the six-month period from 1 W/m^2 in the second month to 0.45 W/m^2 by the end of the sixth month (Miskan et al., 2016).

2.4.3.6 Promising Solutions to Fouling and Loss of Power in MFCs

The antifouling properties of PEM are improved through doping with different materials. For example, a composite membrane composed of sulfonated graphene oxide and silicon dioxide (SGO@SiO$_2$) had higher ion exchange capacity and proton conductivity compared to Nafion-117 membranes (Xu et al., 2019). However, such modifications are still emerging, and their implementation is yet to dominate.

Improving MFC power output can be achieved through MBR–MFC coupling, which is used to overcome the limitations of both technologies. The main objective of this coupling is to reduce membrane fouling in the MBR and utilize the energy produced by the MFC (Liu et al., 2018). For example, a hollow fiber MBR–MFC system improved the removal efficiency of COD, ammonia-nitrogen, and total nitrogen by 4.4%, 1.2%, and 10.3%, respectively (Tian et al., 2015). Additionally, the maximum power density for the MFC was 2.18 W/m^2 and the average voltage produced by the cell was 0.15 V. This minute electricity is useful in mitigating microbial fouling in MBRs (Tian et al., 2015). The bioelectric potential offered by the microbial cell redirects the negatively charged foulants away from the membrane surface, thus, reducing the foulant accumulation rate.

2.5 CONCLUSIONS

In conclusion, microbial technologies offer a wide range of applications in wastewater treatment and reuse. MBRs combine membranes and microbial technologies where microorganisms break down the organic matter present in wastewater while the membrane separates the liquid from the solids in the reactor. Unfortunately, the presence of microbes in conventional membrane processes, such as MF, UF, NF, RO, and MBRs presents a serious challenge regarding microbial fouling. Microbial fouling is especially detrimental for NF, RO, and MBR where the performance of the membrane is significantly hindered. The drawbacks include the reduction of membrane flux, increased energy consumption, biodegradation of membrane material, and decline in water quality. To overcome these drawbacks, several techniques have been developed to predict the potential of biofouling by sampling the membrane feed water. These samples are tested for ATP content and total direct cell count, which reveal the concentration of microbes in the sample. Another example of a monitoring technique mainly used in NF and RO is the MFS. The simulator detects any pressure drop in the feed water due to microbial fouling and allows for early mitigation of this fouling. To control and counteract existing microbial fouling, physical, chemical, and biological strategies have been developed. Physical methods include hydraulic backwashing and other mechanical methods, such as scrubbing. Chemical cleaning is done by reacting the membrane with reagents like oxidants, acids, bases, or enzymes. Lastly, the main biological strategy is QQ, which works on the microorganism communication principle. Microorganisms can minimize microbial fouling by blocking receptors from signaling molecules. Another promising application utilizing both microbes and membranes is MFCs. These cells break down organic matter to produce electricity. The design of MFCs could be a single chamber, double chamber, up-flow, and stacked. The current applications for MFCs are still laboratory scale and they include wastewater treatment, environmental monitoring, desalination, and biosensors. Unfortunately, MFCs suffer from several disadvantages that prevent their large-scale implementation, which are high cost, dependence on microorganisms, inability to collect the generated power, and microbial fouling. Microbial fouling in MFCs occurs in the PEM, which reduces the efficiency and power output of the process. To counteract this fouling, a promising antifouling composite membrane improves the ion exchange capacity and proton conductivity. Additionally, coupling MBRs and MFCs helps reduce membrane fouling in MBRs by utilizing the bioelectric field generated by the MFC's electric potential. Overall, effective integration of microbes in membrane technologies requires further research to overcome microbial fouling and create sustainable water treatment processes.

REFERENCES

Ahuchaogu, Ahamefula A., Okonkwo Joseph Chukwu, A. I. Obike, Chitua E. Igara, Innocent Chidi Nnorom, and John Bull Onyekachi Echeme. 2018. "Reverse Osmosis Technology, Its Applications and Nano-Enabled Membrane." *International Journal of Advanced Research in Chemical Science (IJARCS)* 5(2):20–26. doi: 10.20431/2349-0403.0502005

Al Aani, Saif, Tameem N. Mustafa, and Nidal Hilal. 2020. "Ultrafiltration Membranes for Wastewater and Water Process Engineering: A Comprehensive Statistical Review over the Past Decade." *Journal of Water Process Engineering* 35:101241. doi: 10.1016/J.JWPE.2020.101241

Al-Bsoul, Abeer, Mohammad Al-Shannag, Muhammad Tawalbeh, Ahmed A. Al-Taani, Walid K. Lafi, Amani Al-Othman, and Mohammad Alsheyab. 2020. "Optimal Conditions for Olive Mill Wastewater Treatment Using Ultrasound and Advanced Oxidation Processes." *Science of The Total Environment* 700:134576. doi: 10.1016/j.scitotenv.2019.134576

Al-Murisi, Mohammed, Dana Al-Muqbel, Amani Al-Othman, and Muhammad Tawalbeh. 2022. "Integrated Microbial Desalination Cell and Microbial Electrolysis Cell for Wastewater Treatment, Bioelectricity Generation, and Biofuel Production: Success, Experience, Challenges, and Future Prospects." pp. 145–66 in *Integrated Environmental Technologies for Wastewater Treatment and Sustainable Development*. Elsevier.

Al Sharabati, Miral, Raed Abokwiek, Amani Al-Othman, Muhammad Tawalbeh, Ceren Karaman, Yasin Orooji, and Fatemeh Karimi. 2021. "Biodegradable Polymers and Their Nano-Composites for the Removal of Endocrine-Disrupting Chemicals (EDCs) from Wastewater: A Review." *Environmental Research* 202:111694. doi: https://doi.org/10.1016/j.envres.2021.111694

Alhajar, Abdallah, Muhammad Tawalbeh, Dana Arjomand, Nooruddin Abdel Rahman, Hassan Khan, and Amani Al-Othman. 2022. "Chapter 14 – Integrating Forward Osmosis into Microbial Fuel Cells for Wastewater Treatment." pp. 321–36 in *Integrated Environmental Technologies for Wastewater Treatment and Sustainable Development*, edited by V. Kumar and M. B. T.-I. E. T. for W. T. and S. D. Kumar. Elsevier.

American Water Works Association. 2016. "Introduction to Low-Pressure Membrane Processes." pp. 1–7 in *M53 Microfiltration and Ultrafiltration Membranes for Drinking Water*.

Anis, Shaheen Fatima, Raed Hashaikeh, and Nidal Hilal. 2019a. "Microfiltration Membrane Processes: A Review of Research Trends over the Past Decade." *Journal of Water Process Engineering* 32:100941. doi: 10.1016/J.JWPE.2019.100941

Anis, Shaheen Fatima, Raed Hashaikeh, and Nidal Hilal. 2019b. "Reverse Osmosis Pretreatment Technologies and Future Trends: A Comprehensive Review." *Desalination* 452:159–95.

Asante-Sackey, Dennis, Sudesh Rathilal, Emmanuel Kweinor Tetteh, and Edward Kwaku Armah. 2022. "Membrane Bioreactors for Produced Water Treatment: A Mini-Review." *Membranes* 12(3):275. doi: 10.3390/MEMBRANES12030275

Baker, Richard W. 2004. "Overview of Membrane Science and Technology." pp. 1–14 in *Membrane Technology and Applications*. John Wiley & Sons, Ltd.

Bani-Melhem, Khalid, Muhammad Rasool Al-Kilani, and Muhammad Tawalbeh. 2023. "Evaluation of Scrap Metallic Waste Electrode Materials for the Application in Electrocoagulation Treatment of Wastewater." *Chemosphere* 310:136668. doi: 10.1016/j.chemosphere.2022.136668

Bilad, Muhammad Roil. 2017. "Membrane Bioreactor for Domestic Wastewater Treatment: Principles, Challanges and Future Research Directions." *Indonesian Journal of Science & Technology* 2(1):97–123. doi: 10.17509/ijost.v2i1

Boas, Joana Vilas, Vânia B. Oliveira, Manuel Simões, and Alexandra M. F. R. Pinto. 2022. "Review on Microbial Fuel Cells Applications, Developments and Costs." *Journal of Environmental Management* 307:114525. doi: 10.1016/J.JENVMAN.2022.114525

Brown, A. I., P. Levison, N. J. Titchener-Hooker, and G. J. Lye. 2009. "Membrane Pleating Effects in 0.2 Mm Rated Microfiltration Cartridges." *Journal of Membrane Science* 341(1–2):76–83. doi: 10.1016/J.MEMSCI.2009.05.044

Chen, Bo Zhi, Xiaohui Ju, Ning Liu, Chang Hui Chu, Jin Peng Lu, Chen Wang, and Shi Peng Sun. 2020. "Pilot-Scale Fabrication of Nanofiltration Membranes and Spiral-Wound Modules." *Chemical Engineering Research and Design* 160:395–404. doi: 10.1016/J.CHERD.2020.06.011

Cui, Yin, Huan Gao, Ran Yu, Lei Gao, and Manjun Zhan. 2021. "Biological-Based Control Strategies for MBR Membrane Biofouling: A Review." *Water Science and Technology* 83(11):2597–2614. doi: 10.2166/WST.2021.168

Darra, Rasha, Maryam Bin Hammad, Fatma Alshamsi, Shatha Alhammadi, Waad Al-Ali, Ahmed Aidan, Muhammad Tawalbeh, Neda Halalsheh, and Amani Al-Othman. 2023. "Wastewater Treatment Processes and Microbial Community." pp. 329–55 in *Metagenomics to Bioremediation*. Elsevier.

Da-Silva-Correa, Luiz H., Hayley Smith, Matthew C. Thibodeau, Bethany Welsh, and Heather L. Buckley. 2022. "The Application of Non-Oxidizing Biocides to Prevent Biofouling in Reverse Osmosis Polyamide Membrane Systems: A Review." *Aqua Water Infrastructure, Ecosystems and Society* 71(2):261–92. doi: 10.2166/aqua.2022.118

Englande, A. J., Peter Krenkel, and J. Shamas. 2015. "Wastewater Treatment &Water Reclamation." in *Earth Systems and Environmental Sciences*. Elsevier.

Femina Carolin, C., Senthil P. Kumar, Janet G. Joshiba, and Vinoth V. Kumar. 2021. "Analysis and Removal of Pharmaceutical Residues from Wastewater Using Membrane Bioreactors: A Review." *Environmental Chemistry Letters* 19(1):329–43. doi: 10.1007/s10311-020-01068-9

Flimban, Sami G. A., Sedky H. A. Hassan, Md Mukhlesur Rahman, and Sang Eun Oh. 2020. "The Effect of Nafion Membrane Fouling on the Power Generation of a Microbial Fuel Cell." *International Journal of Hydrogen Energy* 45(25):13643–51. doi: 10.1016/J.IJHYDENE.2018.02.097

France, Thomas C., Alan L. Kelly, Shane V. Crowley, and James A. O'mahony. 2021. "Cold Microfiltration as an Enabler of Sustainable Dairy Protein Ingredient Innovation." *Foods* 10(9):2091. doi: 10.3390/FOODS10092091

García Doménech, Natalia, Finn Purcell-Milton, and Yurii K. Gun'ko. 2020. "Recent Progress and Future Prospects in Development of Advanced Materials for Nanofiltration." *Materials Today Communications* 23:100888. doi: 10.1016/J.MTCOMM.2019.100888

Garud, R. M., S. V. Kore, V. S. Kore, and G. S. Kulkarni. 2011. "A Short Review on Process and Applications of Reverse Osmosis." *Universal Journal of Environmental Research and Technology* 1(3):233–38.

Gude, Veera Gnaneswar. 2016. "Wastewater Treatment in Microbial Fuel Cells – An Overview." *Journal of Cleaner Production* 122:287–307. doi: 10.1016/j.jclepro.2016.02.022

Guo, Wenshan, Huu Hao Ngo, and Jianxin Li. 2012. "A Mini-Review on Membrane Fouling." *Bioresource Technology* 122:27–34. doi: 10.1016/j.biortech.2012.04.089

Halalsheh, Neda, Odey Alshboul, Ali Shehadeh, Rabia Emhamed Al Mamlook, Amani Al-Othman, Muhammad Tawalbeh, Ali Saeed Almuflih, and Charalambos Papelis. 2022. "Breakthrough Curves Prediction of Selenite Adsorption on Chemically Modified Zeolite Using Boosted Decision Tree Algorithms for Water Treatment Applications." *Water* 14(16):2519. doi: 10.3390/w14162519

Hampu, Nicholas, Jay R. Werber, Wui Yarn Chan, Elizabeth C. Feinberg, and Marc A. Hillmyer. 2020. "Next-Generation Ultrafiltration Membranes Enabled by Block Polymers." *ACS Nano* 14(12):16446–71. doi: 10.1021/ACSNANO.0C07883/SUPPL_FILE/NN0C07883_SI_001.PDF

Harnisch, F., F. Aulenta, and U. Schröder. 2011. "Microbial Fuel Cells and Bioelectrochemical Systems: Industrial and Environmental Biotechnologies Based on Extracellular Electron Transfer." *Comprehensive Biotechnology, Second Edition* 6:643–59. doi: 10.1016/B978-0-08-088504-9.00462-1

Hosseini, Mahsa Keyvan, Lei Liu, Parisa Keyvan Hosseini, Anisha Bhattacharyya, Kenneth Lee, Jiahe Miao, and Bing Chen. 2022. "Review of Hollow Fiber (HF) Membrane Filtration Technology for the Treatment of Oily Wastewater: Applications and Challenges." *Journal of Marine Science and Engineering* 10(9):1313. doi: 10.3390/JMSE10091313

Igunnu, Ebenezer T., and George Z. Chen. 2014. "Produced Water Treatment Technologies." *International Journal of Low-Carbon Technologies* 9(3):157–77. doi: 10.1093/IJLCT/CTS049

Ivars-Barceló, Francisco, Alessio Zuliani, Marjan Fallah, Mehrdad Mashkour, Mostafa Rahimnejad, and Rafael Luque. 2018. "Novel Applications of Microbial Fuel Cells in Sensors and Biosensors." *Applied Sciences*, 8(7):1184. doi: 10.3390/APP8071184

Javed Nauman, Rana Muhammad, Amani Al-Othman, Paul Nancarrow, and Muhammad Tawalbeh. 2022. "Zirconium Silicate-Ionic Liquid Membranes for High-Temperature Hydrogen PEM Fuel Cells." *International Journal of Hydrogen Energy*. doi: https://doi.org/10.1016/j.ijhydene.2022.05.009

Jia, Qibo, Liling Wei, Hongliang Han, and Jianquan Shen. 2014. "Factors That Influence the Performance of Two-Chamber Microbial Fuel Cell." *International Journal of Hydrogen Energy* 39(25):13687–93. doi: 10.1016/J.IJHYDENE.2014.04.023

Judd, Simon, and Farid Turan. 2018. "Sidestream vs Immersed Membrane Bioreactors: A Cost Analysis." *Proceed. Water Environ. Fed.* 10:3722–33. doi: 10.2175/193864718825136008

Kammakakam, Irshad, and Zhiping Lai. 2023. "Next-Generation Ultrafiltration Membranes: A Review of Material Design, Properties, Recent Progress, and Challenges." *Chemosphere* 316:137669. doi: 10.1016/J.CHEMOSPHERE.2022.137669

Kazim, Hisham, Moin Sabri, Amani Al-Othman, and Muhammad Tawalbeh. 2023. "Artificial Intelligence Application in Membrane Processes and Prediction of Fouling for Better Resource Recovery." *Journal of Resource Recovery* 1:1008. doi: 10.52547/jrr.2303.1008

Kim, Dooil, Seunghoon Jung, Jinsik Sohn, Hyungsoo Kim, and Seockheon Lee. 2009. "Biocide Application for Controlling Biofouling of SWRO Membranes – an Overview." *Desalination* 238(1–3):43–52. doi: 10.1016/j.desal.2008.01.034

Komlenic, Rodney. 2010. "Rethinking the Causes of Membrane Biofouling." *Filtration & Separation* 47(5):26–28. doi: 10.1016/S0015-1882(10)70211-1

Kumar, Ravinder, Lakhveer Singh, and A. W. Zularisam. 2016. "Exoelectrogens: Recent Advances in Molecular Drivers Involved in Extracellular Electron Transfer and Strategies Used to Improve It for Microbial Fuel Cell Applications." *Renewable and Sustainable Energy Reviews* 56:1322–36. doi: 10.1016/j.rser.2015.12.029

Kumar, Ravinder, Lakhveer Singh, and A. W. Zularisam. 2017. "Microbial Fuel Cells: Types and Applications." pp. 367–84 in *Waste Biomass Management – A Holistic Approach*. Springer International Publishing.

Li, Shuang, Chong Cheng, and Arne Thomas. 2017. "Carbon-Based Microbial-Fuel-Cell Electrodes: From Conductive Supports to Active Catalysts." *Advanced Materials* 29(8):1602547. doi: 10.1002/ADMA.201602547

Liu, Wenbin, Hui Jia, Jie Wang, Hongwei Zhang, Changchun Xin, and Yingjie Zhang. 2018. "Microbial Fuel Cell and Membrane Bioreactor Coupling System: Recent Trends." *Environmental Science and Pollution Research* 25(24):23631–44. doi: 10.1007/S11356-018-2656-0/FIGURES/7

Macintosh, C., S. Astals, C. Sembera, A. Ertl, J. E. Drewes, P. D. Jensen, and K. Koch. 2019. "Successful Strategies for Increasing Energy Self-Sufficiency at Grüneck Wastewater Treatment Plant in Germany by Food Waste Co-Digestion and Improved Aeration." *Applied Energy* 242:797–808. doi: 10.1016/J.APENERGY.2019.03.126

Madhavi, Vemula, and Thotakura Ramesh. 2023. "Introduction and Basic Principle of Nanofiltration Membrane Process." pp. 1–15 in *Nanofiltration Membrane for Water Purification*. Springer, Singapore.

Mahmoud, Mohamed, and K. M. El-Khatib. 2020. "Three-Dimensional Graphitic Mesoporous Carbon-Doped Carbon Felt Bioanodes Enables High Electric Current Production in Microbial Fuel Cells." *International Journal of Hydrogen Energy* 45(56):32413–22. doi: 10.1016/J.IJHYDENE.2020.08.207

Malik, Sumira, Shristi Kishore, Archna Dhasmana, Preeti Kumari, Tamoghni Mitra, Vishal Chaudhary, Ritu Kumari, Jutishna Bora, Anuj Ranjan, Tatiana Minkina, and Vishnu D. Rajput. 2023. "A Perspective Review on Microbial Fuel Cells in Treatment and Product Recovery from Wastewater." *Water* 15(2):316. doi: 10.3390/W15020316

Manni, A., B. Achiou, A. Karim, A. Harrati, C. Sadik, M. Ouammou, S. Alami Younssi, and A. El Bouari. 2020. "New Low-Cost Ceramic Microfiltration Membrane Made from Natural Magnesite for Industrial Wastewater Treatment." *Journal of Environmental Chemical Engineering* 8(4):103906. doi: 10.1016/J.JECE.2020.103906

Millanar-Marfa, Jessa Marie J., Laura Borea, Shadi W. Hasan, Mark Daniel G. de Luna, Vincenzo Belgiorno, and Vincenzo Naddeo. 2020. "6 – Advanced Membrane Bioreactors for Emerging Contaminant Removal and Quorum Sensing Control." pp. 117–47 in *Current Developments in Biotechnology and Bioengineering*. Elsevier B.V.

Miskan, Madihah, Manal Ismail, Mostafa Ghasemi, Jamaliah Md Jahim, Darman Nordin, and Mimi Hani Abu Bakar. 2016. "Characterization of Membrane Biofouling and Its Effect on the Performance of Microbial Fuel Cell." *International Journal of Hydrogen Energy* 41(1):543–52. doi: 10.1016/J.IJHYDENE.2015.09.037

Mohammad, A. W., Y. H. Teow, W. L. Ang, Y. T. Chung, D. L. Oatley-Radcliffe, and N. Hilal. 2015. "Nanofiltration Membranes Review: Recent Advances and Future Prospects." *Desalination* 356:226–54. doi: 10.1016/J.DESAL.2014.10.043

Mutamim, Noor Sabrina Ahmad, Zainura Zainon Noor, Mohd Ariffin Abu Hassan, and Gustaf Olsson. 2012. "Application of Membrane Bioreactor Technology in Treating High Strength Industrial Wastewater: A Performance Review." *Desalination* 305:1–11. doi: 10.1016/J.DESAL.2012.07.033

Neethu, B., Amitap Khandelwal, M. M. Ghangrekar, K. Ihjas, and Jaichander Swaminathan. 2022. "Microbial Fuel Cells—Challenges for Commercialization and How They Can Be Addressed." pp. 393–418 in *Scaling Up of Microbial Electrochemical Systems*. Elsevier.

Ng, Aileen N. L., and Albert S. Kim. 2007. "A Mini-Review of Modeling Studies on Membrane Bioreactor (MBR) Treatment for Municipal Wastewaters." *Desalination* 212(1–3):261–81. doi: 10.1016/J.DESAL.2006.10.013

Nguyen, Thang, Felicity A. Roddick, and Linhua Fan. 2012. "Biofouling of Water Treatment Membranes: A Review of the Underlying Causes, Monitoring Techniques and Control Measures." *Membranes* 2(4):804–40.

Noori, Md T., M. M. Ghangrekar, C. K. Mukherjee, and Booki Min. 2019. "Biofouling Effects on the Performance of Microbial Fuel Cells and Recent Advances in Biotechnological and Chemical Strategies for Mitigation." *Biotechnology Advances* 37(8):107420. doi: 10.1016/J.BIOTECHADV.2019.107420, www.sciencedirect.com/science/article/pii/S073497501930120X

Puri, Saurabhi, Reshma Devkate, and Pragati Jadhav. 2022. "A Complete over Review on World of Water Purification of Reverse Osmosis." *International Journal of Creative Research Thoughts (IJCRT)* 10(2):d606–9.

Ringeisen, Bradley R., Emily Henderson, Peter K. Wu, Jeremy Pietron, Ricky Ray, Brenda Little, Justin C. Biffinger, and Joanne M. Jones-Meehan. 2006. "High Power Density from a Miniature Microbial Fuel Cell Using Shewanella Oneidensis DSP10." *Environmental Science and Technology* 40(8):2629–34. doi: 10.1021/ES052254W

Ruigómez, Ignacio, Enrique González, Luis Rodríguez-Gómez, and Luisa Vera. 2022. "Fouling Control Strategies for Direct Membrane Ultrafiltration: Physical Cleanings Assisted by Membrane Rotational Movement." *Chemical Engineering Journal* 436:135161. doi: 10.1016/j.cej.2022.135161

Saad, Mohamad Amin. 2004. "Early Discovery of RO Membrane Fouling and Real-Time Monitoring of Plant Performance for Optimizing Cost of Water." *Desalination* 165:183–91. doi: 10.1016/j.desal.2004.06.021

Sacdal, Rosselle, Jonalyn Madriaga, and Maria Pythias Espino. 2020. "Overview of the Analysis, Occurrence and Ecological Effects of Hormones in Lake Waters in Asia." *Environmental Research* 182:109091. doi: 10.1016/J.ENVRES.2019.109091

Sagle, Alyson, and Benny Freeman. 2004. "Fundamentals of Membranes for Water Treatment." The Future of Desalination in Texas 2.

Schmitz, Bradley W., Masaaki Kitajima, Maria E. Campillo, Charles P. Gerba, and Ian L. Pepper. 2016. "Virus Reduction during Advanced Bardenpho and Conventional Wastewater Treatment Processes." *Environmental Science and Technology* 50(17):9524–32. doi: 10.1021/ACS.EST.6B01384/SUPPL_FILE/ES6B01384_SI_001.PDF

Sengupta, Arijit, Mahmood Jebur, Mohanad Kamaz, and S. Ranil Wickramasinghe. 2022. "Removal of Emerging Contaminants from Wastewater Streams Using Membrane Bioreactors: A Review." *Membranes* 12(1):60. doi: 10.3390/membranes12010060

Shamaei, Laleh, Behnam Khorshidi, Basil Perdicakis, and Mohtada Sadrzadeh. 2018. "Treatment of Oil Sands Produced Water Using Combined Electrocoagulation and Chemical Coagulation Techniques." *Science of The Total Environment* 645:560–72. doi: 10.1016/j.scitotenv.2018.06.387

Shi, Donglu, Zizheng Guo, and Nicholas Bedford. 2015. "Nanoenergy Materials." pp. 255–91 in *Nanomaterials and Devices*. William Andrew Publishing.

Singh, Randeep, and Mihir Kumar Purkait. 2019. "Microfiltration Membranes." pp. 111–46 in *Membrane Separation Principles and Applications*. Elsevier.

Sonawane, Jayesh M., Chizoba I. Ezugwu, and Prakash C. Ghosh. 2020. "Microbial Fuel Cell-Based Biological Oxygen Demand Sensors for Monitoring Wastewater: State-of-the-Art and Practical Applications." *ACS Sensors* 5(8):2297–2316. doi: 10.1021/ACSSENSORS.0C01299/ASSET/IMAGES/MEDIUM/SE0C01299_0010.GIF

Tawalbeh, Muhammad, Fatma Handel, Tezel, Boguslaw, Kruczek, Sadok, Letaief, and Christian, Detellier. 2013. "Synthesis and Characterization of Silicalite-1 Membrane

Prepared on a Novel Support by the Pore Plugging Method." *Journal of Porous Materials* 20(6). doi: 10.1007/s10934-013-9726-y

Tawalbeh, Muhammad , Fatma Handel, Tezel, Sadok, Letaief, Christian, Detellier, and Boguslaw, Kruczek. 2012. "Separation of CO2 and N2 on Zeolite Silicalate-1 Membrane Synthesized on Novel Support." *Separation Science and Technology* 47(11):1606–16. doi: 10.1080/01496395.2012.655836

Tawalbeh, Muhammad, Amani Al-Othman, Noun Abdelwahab, Abdul Hai Alami, and Abdul Ghani Olabi. 2021. "Recent Developments in Pressure Retarded Osmosis for Desalination and Power Generation." *Renewable and Sustainable Energy Reviews* 138:110492. doi: 10.1016/j.rser.2020.110492

Tawalbeh, Muhammad, Amani Al-Othman, Karnail Singh, Ikram Douba, Dania Kabakebji, and Malek Alkasrawi. 2020. "Microbial Desalination Cells for Water Purification and Power Generation: A Critical Review." *Energy* 209:118493. doi: 10.1016/j.energy.2020.118493

Tawalbeh, Muhammad, Shima Mohammed, Amani Al-Othman, Mohammad Yusuf, M. Mofijur, and Hesam Kamyab. 2023. "MXenes and MXene-Based Materials for Removal of Pharmaceutical Compounds from Wastewater: Critical Review." *Environmental Research* 228:115919. doi: 10.1016/j.envres.2023.115919

Tawalbeh, Muhammad, Abdullah Al Mojjly, Amani Al-Othman, and Nidal Hilal. 2018. "Membrane Separation as a Pre-Treatment Process for Oily Saline Water." *Desalination* 447:182–202. doi: 10.1016/J.DESAL.2018.07.029

Tian, Yu, Hui Li, Lipin Li, Xinying Su, Yaobin Lu, Wei Zuo, and Jun Zhang. 2015. "In-Situ Integration of Microbial Fuel Cell with Hollow-Fiber Membrane Bioreactor for Wastewater Treatment and Membrane Fouling Mitigation." *Biosensors and Bioelectronics* 64:189–95. doi: 10.1016/J.BIOS.2014.08.070

Wang, Jinfeng, Hongqiang Ren, Xianhui Li, Jianxin Li, Lili Ding, Jinju Geng, Ke Xu, Hui Huang, and Haidong Hu. 2018. "In Situ Monitoring of Wastewater Biofilm Formation Process via Ultrasonic Time Domain Reflectometry (UTDR)." *Chemical Engineering Journal* 334:2134–41. doi: 10.1016/j.cej.2017.11.043

Wang, Yifei, Jiaxin Zhu, Haiou Huang, and Hyun Hee Cho. 2015. "Carbon Nanotube Composite Membranes for Microfiltration of Pharmaceuticals and Personal Care Products: Capabilities and Potential Mechanisms." *Journal of Membrane Science* 479:165–74. doi: 10.1016/J.MEMSCI.2015.01.034

Warsinger, David M., Emily W. Tow, Kishor G. Nayar, Laith A. Maswadeh, and John H. Lienhard V. 2016. "Energy Efficiency of Batch and Semi-Batch (CCRO) Reverse Osmosis Desalination." *Water Research* 106:272–82. doi: 10.1016/J.WATRES.2016.09.029

Wei, Jincheng, Peng Liang, and Xia Huang. 2011. "Recent Progress in Electrodes for Microbial Fuel Cells." *Bioresource Technology* 102(20):9335–44. doi: 10.1016/J.BIORTECH.2011.07.019

Xu, Jian, and Hyowon Lee. 2020. "Anti-Biofouling Strategies for Long-Term Continuous Use of Implantable Biosensors." *Chemosensors* 8(3):66. doi: 10.3390/chemosensors8030066

Xu, Juan, Guo Ping Sheng, Hong Wei Luo, Wen Wei Li, Long Fei Wang, and Han Qing Yu. 2012. "Fouling of Proton Exchange Membrane (PEM) Deteriorates the Performance of Microbial Fuel Cell." *Water Research* 46(6):1817–24. doi: 10.1016/J.WATRES.2011.12.060

Xu, Qibin, Lei Wang, Chen Li, Xudong Wang, Cuicui Li, and Yatian Geng. 2019. "Study on Improvement of the Proton Conductivity and Anti-Fouling of Proton Exchange Membrane by Doping SGO@SiO2 in Microbial Fuel Cell Applications." *International Journal of Hydrogen Energy* 44(29):15322–32. doi: 10.1016/J.IJHYDENE.2019.03.238

Yaqoob, Asim Ali, Mohamad Nasir Mohamad Ibrahim, and Susana Rodríguez-Couto. 2020. "Development and Modification of Materials to Build Cost-Effective Anodes for Microbial Fuel Cells (MFCs): An Overview." *Biochemical Engineering Journal* 164:107779. doi: https://doi.org/10.1016/j.bej.2020.107779

Ying, Xianbin, Dongsheng Shen, Meizhen Wang, Huajun Feng, Yuan Gu, and Wenting Chen. 2018. "Titanium Dioxide Thin Film-Modified Stainless Steel Mesh for Enhanced Current-Generation in Microbial Fuel Cells." *Chemical Engineering Journal* 333:260–67. doi: 10.1016/J.CEJ.2017.09.132

3 Strategies for Tuning the Microbial Community from Anaerobic Digestion to remove Emerging Pollutants

Maikon Kelbert, Rayane Kunert Langbehn, Adrielle Helena Mannrich, Júlia Pedó Gutkoski, Karina Cesca, Camila Michels, Bruno Augusto Mattar Carciofi, Hugo Moreira Soares, and Cristiano José de Andrade

3.1 INTRODUCTION

Anthropogenic activities significantly impact the quality of water. In addition, social and economic development has increased the production and demand for several chemicals. These chemicals include a large number of substances known as emerging pollutants (EPs) (NORMAN; www.norman-network.com/?q=node/19). They can enter the environment through several pathways, presenting potential risks to the environment and human health (Zhang et al., 2013). The risks that EPs possess were first reported in the 1980s (Ebele et al., 2017). In this period, intensive research began concerning the presence of chemicals used by humans and animals (e.g., personal hygiene products, pharmaceuticals, cleaning products, agricultural compounds, endocrine disruptor compounds, and microplastics) (Luo et al., 2014; Wilkinson et al., 2017; Kumar et al., 2022). Despite being a known issue, the challenge of preventing their release into the environment persists (Taheran et al., 2018; Gogoi et al., 2018). Moreover, this chapter will not focus on microplastics as they require a different approach to their occurrence and treatment.

The main sources of EPs can be household waste, industrial effluents, and untreated domestic and industrial wastewater, as shown in Figure 3.1 (Pereira et al., 2020; Singh et al., 2022). Domestic wastewater is considered the main route to release EPs into the environment because conventional processes are not designed to mitigate or monitor their presence in wastewater (Luo et al., 2014). EPs are usually detected in wastewater treatment plants (WWTPs) at concentrations raging

DOI: 10.1201/9781003441069-3

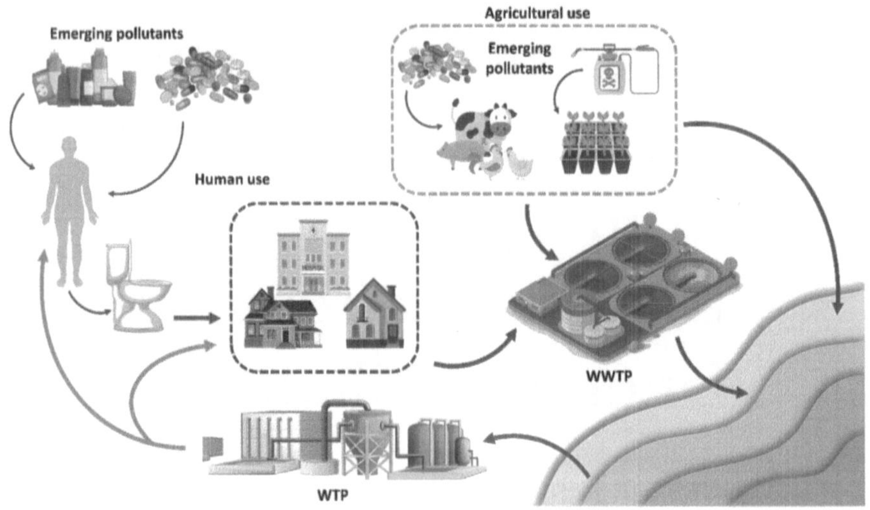

FIGURE 3.1 The main source of emerging pollutants and the pathway to their release to the environment.

between ng.L^{-1} and μg.L^{-1}, and although the impact on the ecosystem is increasingly studied, the long-term effects remain unknown (Negreira et al., 2014; Coetsier et al., 2009). Furthermore, there is a lack of legislation requiring their proper treatment (Gonzalez-Gil et al., 2020). The regulation of EPs remains an evolving topic and many countries are working to improve laws and policies to address these chemical substances. However, government authorities and regulatory agencies are still struggling to monitor EPs and taking steps to reduce their presence in the environment.

Anaerobic digestion (AD) is a biological process that uses microorganisms to decompose complex organic materials in anaerobic conditions (Cremonez et al., 2021). AD is normally applied in WWTPs to treat waste and concomitantly generate biogas (Panigrahi et al., 2019). This process can remove several organic pollutants from wastewater. Studies suggest that AD can effectively remove certain EPs, such as pharmaceuticals and hormones. However, several EPs are recalcitrant to the process, meaning they remain in the treated effluent and can be released into the environment or sorbed in the anaerobic sludge (Haffiez et al., 2022). Some approaches may help improve the AD process and enhance the effectiveness of removing EPs, such as adjusting operating conditions, adding co-solutes, using pretreatment, using specific anaerobic reactors, and combining them with other treatment processes (Panigrahi et al., 2019; Nascimento et al., 2021; Han et al., 2023; Gonzalez-Salgado et al., 2020). In summary, several strategies can be used to enhance the activity of the microbial community from AD and consequently remove EPs. In this chapter, we will provide some necessary definitions and also show strategies that can be used for tuning the microbial community from AD to remove EPs.

3.2 MICROBIAL COMMUNITY FROM AD

AD can be defined as a sequence of biochemical processes that transform organic matter into biogas. A consortium of microorganisms syntrophically interacts to break complex organic matter (Anukam et al., 2019). Hydrolysis, acidogenesis, acetogenesis, and methanogenesis are necessary for complete AD. Table 3.1 presents the microbial community involved in each step of AD.

The reactions involved are still being studied, mainly concerning factors that negatively influence the process, such as the influence of operation parameters or the presence of recalcitrant organic matter (Panigrahi et al., 2019; Haffiez et al., 2022; Elmoutez et al., 2023; Pasalari et al., 2021). The steps of AD are described in Figure 3.2.

3.2.1 Hydrolysis

Hydrolysis consists of transforming complex substances, generally long-chain polymers (carbohydrates, lipids, and proteins), into their smaller derivatives, such as sugars, fatty acids, and amino acids (Elmoutez et al., 2023; Thanarasu et al., 2022; Li et al., 2019). In this step, the enzymes excreted by hydrolytic fermentative bacteria degrade the macromolecule, as described below:

Carbohydrates or polysaccharides: They are degraded and converted mainly into sugars and acetic acid. This degradation occurs due to anaerobic microorganisms that produce enzymes or multienzyme complexes (Cremonez et al., 2021; Elmoutez et al., 2023).

Lipids: Similarly, lipases are excreted by microorganisms to catalyze the breaking of bonds between water and lipids, generating fatty acids that can be saturated or unsaturated (Thanarasu et al., 2022).

TABLE 3.1
Microbial community in anaerobic digestion

Anaerobic digestion steps	Microorganisms
Hydrolysis and acidogenesis	*Acetivibrio* sp., *Bacillus* sp., *Bifidobacterium* sp., *Butyrivibrio* sp., *Clostridium* sp., *Eubacterium* sp., *Selenomonas* sp., *Lachnospira* sp., *Megasphaera* sp., *Peptococcus* sp., *Staphylococcus* sp., among others.
Acetogenesis	*Acetobacterium* sp., *Clostridium* sp., *Desulfotomaculum* sp., *Desulfovibrio* sp., *Syntrophomonas* sp., among others.
Aceticlastic and hydrogenotrophic methanogenesis	Mainly *Methanosarcina* sp., *Methanosaeta* sp., *Methanobrevibacter* sp., *Methanospirillum* sp., and *Methanobacterium* sp.

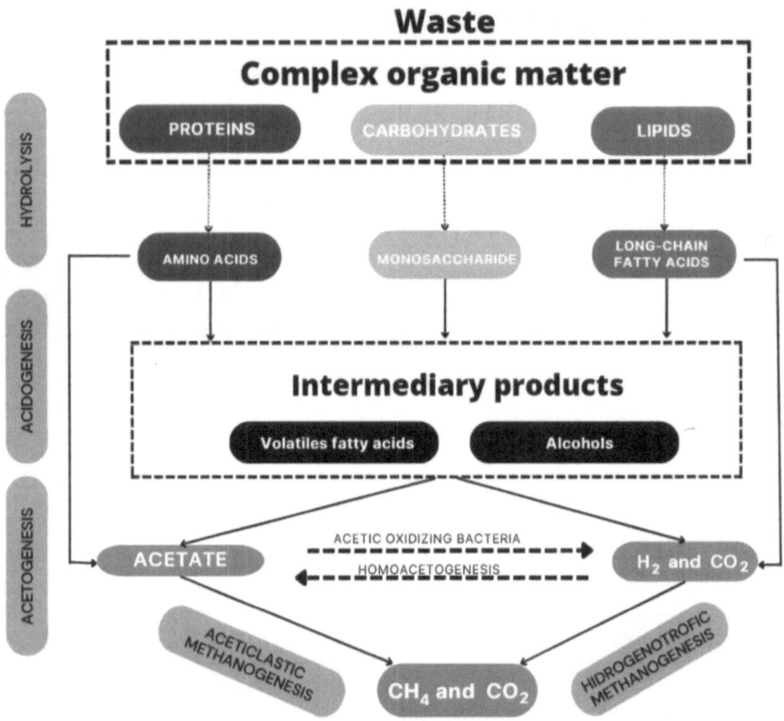

FIGURE 3.2 A representative scheme of anaerobic digestion steps (hydrolysis, acidogenesis, acetogenesis, and methanogenesis).

Proteins: In this process, the enzymes involved are proteases that break proteins into amino acids. Subsequently, they can be transformed into organic acids through oxidation–reduction reactions. It is important to emphasize that ammonia and sulfur in these compounds are converted into ammonia and sulfide, regardless of the degradation route (Li et al., 2019).

In hydrolysis, some factors can affect the degradation behavior, including the pH, size, and nature of the particles in the initial medium, and temperature (Gonzalez-Salgado et al., 2020; Pasalari et al., 2021). The organic load rate and biomass composition also affect the hydrolysis step.

3.2.2 Acidogenesis

A consortium of microorganisms converts molecules from the previous step into substrates for acetogens and methanogens. In this step, intermediate molecules are produced, such as volatile fatty acids (e.g., acetic acid, butyric acid, and propionic acid) (Cremonez et al., 2021). However, acetic acid is most relevant because it is used directly by methanogens to produce methane (Ketheesan et al., 2015). The general chemical reactions involved in acidogenesis are described below:

$$C_6H_{12}O_6 \leftrightarrow 2CH_3CH_2OH + 2CO_2$$

$$C_6H_{12}O_6 + 2H_2 \leftrightarrow 2CH_3CH_2COOH + 2H_2O$$

$$C_6H_{12}O_6 \rightarrow 3CH_3COOH$$

The formation of carbon dioxide and alcohol can also occur. Some microorganisms present in this step may be the same as those found in the hydrolysis phase, such as *Clostridium*, *Flavobacterium*, *Micrococcus*, and *Pseudomonas* (Ketheesan et al., 2015; Xie et al., 2014).

3.2.3 Acetogenesis

Microorganisms can participate in two pathways: i) the conversion of carbon dioxide and hydrogen into acetic acid; or ii) metabolizing the volatile fatty acids produced in acidogenesis and producing acetic acid. The microbial community responsible for consuming carbon dioxide and hydrogen is known as being homoacetogenic (Cremonez et al., 2021; Li et al., 2019; Ketheesan et al., 2015).

The general chemical reactions of this step are represented below:

$$CH_3CH_2COO^- + 3H_2O \leftrightarrow CH_3COO^- + H + HCO_3^- + 3H_2$$

$$C_6H_{12}O_6 + 2H_2O \leftrightarrow 2CH_3COOH + 2CO_2 + 4H_2$$

$$CH_3CH_2OH + 2H_2O \leftrightarrow CH_3COO^- + 3H_2 + H^+$$

The main products of acetogenesis are carbon dioxide, hydrogen, and acetic acid. In addition, nitrates, sulfates, and other protons are formed and are equally important for biogas formation. Some bacteria in this stage are *Sporomusa*, *Clostridium*, and *Acetobacterium* (Elmoutez et al., 2023).

3.2.4 Methanogenesis

Methane production occurs through two groups of archaea, the acetoclastic and hydrogenotrophic methanogens (Cremonez et al., 2021; Yang et al., 2013). However, around 70% of methane is produced by acetoclastic methanogens. *Methanosarcina* and *Methanosaeta* are the main genera and use acetic acid as a substrate to produce methane. The second one uses carbon dioxide and hydrogen to produce methane. *Methanobacterium* is one of the most common genera in hydrogenotrophic methanogenesis (Yang et al., 2014).

$$CH_3COOH \rightarrow CH_4 + CO_2$$

$$CO_2 + 4H_2 \rightarrow CH_4 + 2H_2O$$

$$2CH_3CH_2OH + CO_2 \rightarrow CH_4 + 2CH_3COOH$$

Methanogens are microorganisms that act symbiotically to complete AD and are very sensitive to environmental changes. Temperature variations, the accumulation of volatile fatty acids from previous phases, or the presence of some EPs (e.g., antibiotics) can negatively impact metabolic activity (Shi et al., 2017; Cetecioglu et al., 2012; Aydin et al., 2015; Chakraborty et al., 2002; Zhang et al., 2018). These factors must be constantly monitored, as any change in this step can compromise the entire degradation of the organic load of the process (Aydin et al., 2015).

3.3 PATHWAY TO REMOVE EPS IN AD

Few studies address the biodegradation and biotransformation of EPs, such as pharmaceuticals, by microbial communities in AD systems. Beyond assessing the process by the aqueous phase removal efficiency and by the EPs sorption into sludge, it is necessary to understand the biological mechanisms and factors affecting the removal of EPs under anaerobic conditions (Gonzalez-Gil et al., 2020; Harb et al., 2019). In anaerobic treatments, EPs can undergo biodegradation, biotransformation, and sorption. The nonremoved fraction remains at the aqueous phase as residue. Furthermore, the volatilization is negligible, even for musk fragrances (Gonzalez-Salgado et al., 2020; Azizan et al., 2021; Alvarino et al., 2014; He et al., 2018). Biodegradation is the complete mineralization of the compounds, while biotransformation refers to an EP molecular transformation in its structure through a biological pathway (Lahti & Oikari, 2011). Biotransformation is the main EP removal route in anaerobic systems. However, the presence of substrate and the chemical compatibility of the EPs with the active sites of enzymes produced by microbial communities are essential for biotransformation to occur (Gonzalez-Gil et al., 2020).

To biologically remove EPs, microorganisms must transform them through microbial metabolism or cometabolism. Microbial metabolism occurs when microorganisms use EPs as a substrate, while cometabolism occurs when the EPs are transformed/degraded by enzymes or cofactors produced to degrade another substance (Hazen, 2010; Tran et al., 2013). In the anaerobic reactor, reduction reactions can occur depending on the presence of electron receptors, such as CO_2, NO_3^-, SO_4^{2-}, Fe^{3+}, and Mn^{4+} (Grady et al., 2011). Such receptors can activate the microbial community growth that directly or indirectly metabolizes EPs. Subsequently, the microbial metabolism can then be used as a carbon and energy source. However, this scenario is not feasible since EPs are usually in low concentrations (ng to ug L^{-1}) (Azizan et al., 2021). Cometabolism is the main mechanism responsible for the biotransformation of compounds in anaerobic systems since microorganisms need a high amount of energy to biodegrade recalcitrant molecules such as EPs (Tran et al., 2013). The cometabolic pathway often leads to minor modifications in the EPs' chemical structure. They can generate transformation products (TPs), which are similar to the original molecule. Moreover, they can participate in metabolic reactions and be completely mineralized (Gonzalez-Gil et al., 2020).

The EPs' sorption, i.e., adsorption and absorption, into sludge is another route that can occur in anaerobic systems. Adsorption is ruled by electrostatic interactions with the usually positive functional groups from EPs and the negative biomass surface from microorganisms (Verlicchi et al., 2012). Absorption involves the affinity between the hydrophobic interactions from compounds with aromatic and aliphatic groups and the lipophilic microorganism's cell or the lipid fraction from the suspended solids (Gonzalez-Salgado et al., 2020). The octanol-water partition coefficient ($\log_{Kow}$) is a helpful parameter in the absorption process analysis since it represents the hydrophobic characteristics of the molecule. A lower $\log_{Kow}$ value (> 2.5) implies that the probability of absorption into the lipid fraction will also be lower because of the higher compound hydrophobicity (Carballa et al., 2008). However, for polar compounds with low $\log_{Kow}$ values, there is still relevant sorption into biomass and vice versa. Thus, this value can be applied to lipophilic interactions with nonpolar compounds (Golet et al., 2003). In contrast, adsorption may be the main route for polar compounds as the electrostatic interaction influences the compound hydrophobicity based on the predominant charge, resulting in $\log_{Kow}$ value deviations (Azizan et al., 2021).

The variety of microorganisms involved in the AD of compounds is immense and remains unknown, and different microorganisms can perform the same biological activity through distinct metabolic pathways. Enzymes seem to play an important role in EPs' biotransformation. Furthermore, microorganisms can produce several different enzymes, and several microorganisms may have common enzymatic pathways. However, methanogenic enzymes appear crucial for the biotransformation of compounds, and adequate stimulation of anaerobic microorganisms can allow for the biotransformation of EPs. Applied studies to identify the main microbial communities responsible for the biotransformation of compounds show that different microorganisms may present a similar degree of biotransformation (Gonzalez-Gil et al., 2020; Wolfson et al., 2018).

Some reactions in anaerobic systems involve the cleavage of ether bonds, carboxylation of benzene rings to benzoate, and reductive dehalogenation (Sun et al., 2003; Veetil et al., 2012). Additionally, electron-donating groups in the EPs' structure may influence their fates. Chen et al. (2019) observed that naproxen, which has an electron-donating group ($-OCH_3$), was more efficiently removed (72.1%) in an upflow anaerobic system than carbamazepine (18.5%), ibuprofen (25%), and diclofenac (–50%), which have electron-withdrawing groups ($-CONH_2$ for carbamazepine and -COOH for ibuprofen and diclofenac). However, it is challenging to theoretically predict the biotransformation pathway of complex molecules such as EPs in anaerobic conditions since multiple enzymatic reactions can occur.

Enzymes are selective and efficient catalysts that accelerate the rate of most biochemical reactions by lowering the activation energy without their consumption. Similarly, they catalyze the metabolic or cometabolic transformation of EPs. Therefore, addressing which enzymes are responsible for EPs' biotransformation can provide more accurate information since one microorganism produces many enzymes with multiple enzymatic pathways. Additionally, this can lead to predicting possible TPs and a better understanding of EPs' potential removal and limitations when applied in biological treatments (Gonzalez-Gil et al., 2020). Table 3.2 presents some authors that suggest which enzymes can be linked with EPs' biotransformation and which reaction may have happened, along with their removal efficiencies.

TABLE 3.2
Enzymes involved in the biotransformation of emerging pollutants

EPs	TPs	Reaction	Enzymes	Removal efficiency (%)*	Reference
N-acetyl-sulfamethoxazole	Sulfamethoxazole	Secondary amide hydrolysis; deacetylation	Aryl-acylamidases (EC 3.5.1.13)	83±3	Gonzalez-Gil et al. (2019)
			N-acetyl-phenylethylamine hydrolases (EC 3.5.1.85)		
			Aylamine N-acetyltransferases (EC 2.3.1.5)		
Acetaminophen	n.i.	Secondary amide hydrolysis; deacetylation	Aylamine N-acetyltransferases (EC 2.3.1.5)	92±1	
			Hydrolases of nonpeptide C-N bonds (EC 3.5.1.-)		
Sulfamethoxazole	n.i.	Reduction of N-O bond in isoxazole ring	Ferredoxin hydrogenase (EC 1.12.7.2)	32±4	
			Cytochrome	97±2	Carneiro et al. (2020)
Atenolol	Atenolol acid	Primary amide hydrolysis	Serine proteases (EC 3.4.21)	34±9	Gonzalez-Gil et al. (2019)
Diclofenac	n.i.	Phosphorylation	Acetate kinase (EC 2.7.2.1)	35±7	Carneiro et al. (2020)
		Decarboxylation	Decarboxylases (EC 4.1.1.-)	10±9	Gonzalez-Gil et al. (2019)
				20	Gonzalez-Gil et al. (2017)
Climbazole	Climbazole-alcohol	Ketone reduction	Carbonyl reductase (EC 1.1.1.184)	37±5	Gonzalez-Gil et al. (2019)
Terbutryn	n.i.	S-methyl group	Membrane S-methyltransferases (EC 2.1.1.-)	46±3	

Venlafaxine		O-demethylation	O-methyltransferases (EC 2.1.1.-)	27±6	
Erythromycin		Cleavage of cladinose	Glycosylases (EC 3.2,-)	57±3	Carneiro et al. (2020)
	Erythromycin TP 576	Ester hydrolysis of macrolid group	Esterases (EC 3.1.-)	24±7	Gonzalez-Gil et al. (2019)
			Serine proteases (EC 3.4.21)		
Clarithromycin	Clarithromycin TP 590			34±8	
Ibuprofen	n.i.			26±12	Carneiro et al. (2020)
			Acetate kinase (EC 2.7.2.1)	16	Gonzalez-Gil et al., (2017)
Naproxen				18	
Nonylphenol				24	
Octylphenol				30	
Bisphenol A		Phosphorylation of hydroxyl groups		26±9	Carneiro et al. (2020)
			Acetate kinase (EC 2.7.2.1)	33	Gonzalez-Gil et al. (2017)
Triclosan				14	
	Phenol, catechol, and 2, 4-dichlorophenol	Cleavage of diphenyl ether bond and subsequent reductive dechlorination	Cysteine dioxygenase (EC 1.13.11.20) Methane monooxygenase (EC 1.14.18.3)	87	Veetil et al. (2012)
Galaxolide	n.i.	Reduction of ketone groups	3-hydroxybutyryl-CoA dehydrogenas (EC 1.1.1.157)	50	Gonzalez-Gil et al. (2017)
				96±10	Carneiro et al. (2020)
Tonalide				92±10	
Celestolide				89±10	

TPs: transformation products; n.i.: TP not identified; *Average±standard deviation; The enzymes identified in the EPs without TPs are more uncertain.

Enzymes such as transferases, hydrolases, oxidoreductases, lyases, and kinases are proposed to catalyze EPs' anaerobic biotransformation (Table 3.2). The differences usually observed in EPs' removal efficiency between anaerobic and aerobic systems may be due to the specific enzymes each condition produces, even though the enzymatic pathways are similar (Gonzalez-Gil et al., 2020; Alvarino et al., 2014).

As described above, AD consists of four main steps. In the enzymatic pathway, the first step is conducted by extracellular enzymes, e.g., sulfatases, glucosidases, lipases, phosphatases, and proteases, while in the second step, hexokinases convert monosaccharides into pyruvate and ATP via the glycolytic pathway (Merlin Christy et al., 2014; Zhang et al., 2015). The conversion of intermediate metabolites to acetate occurs through many enzymatic steps by syntrophic acetogens. Via the acetyl Co-A pathway, homoacetongens can also produce acetate by reducing CO_2 with hydrogen (Stams, 1994). *Methanosarcina* and *Methanosaeta* are the main methanogenic bacteria responsible for converting acetate to methane. For *Methanosarcina*, the enzymes involved in the catalytic reaction are acetate kinase and phosphotransacetylase, while for *Methanosaeta*, it is acetyl-CoA synthetase (Merlin Christy et al., 2014; Zhang et al., 2015). Lastly, CO_2 is converted to methane by dehydrogenases, transferases, hydrolases, and reductases, although some authors suggest that methyl-coenzyme M reductase is the critical enzyme in this reaction (Gonzalez-Gil et al., 2020; Zhang et al., 2015).

Wolfson et al. (2018) observed that a specific group of acetogens, acetate oxidizers, and methanogens were responsible for naproxen demethylation. The authors proposed a carbon transfer through the anaerobic food web model that corroborates the significance of linked community interactions in the anaerobic transformation of aromatic O-methyl compounds. Furthermore, the addition of cofactors (e.g., NAD+, NADH, NADP, nucleoside triphosphate ATP, and coenzyme A) positively influenced the removal (> 60%) of four carbamazepine TPs (i.e., 10-OH-CBZ, 10,11-DiOH-CBZ, 2-OH-CBZ, and 3-OH-CBZ), which were not biotransformed by anaerobic microorganisms or by the enzymes alone (Gonzalez-Gil et al., 2019). Therefore, if the anaerobic microorganisms are appropriately stimulated, they could activate enzymes capable of biotransforming EPs that are typically resistant under anaerobic conditions (Gonzalez-Gil et al., 2020).

3.4 TUNING THE MICROBIAL COMMUNITY FROM AD

Some strategies can improve the performance of the microbial community in the AD process and consequently increase the efficiency in removing EPs. Figure 3.3 shows these strategies, which include adjusting operational conditions; anaerobic reactor design; addition of co-solutes or other substances; pretreatment of the substrate; and inoculum selection and acclimation (Panigrahi et al., 2019; Nascimento et al., 2021; Han et al., 2023; Gonzalez-Salgado et al., 2020; Venegas et al., 2021).

Inoculum selection and acclimation is the first step to enhance performance in AD processes for methane production and EPs' removal. The inoculum plays an important role in shaping the microbiome and shortening the adaptation time during the AD start-up stage (Sella et al., 2022; Han et al., 2016). The inoculum source

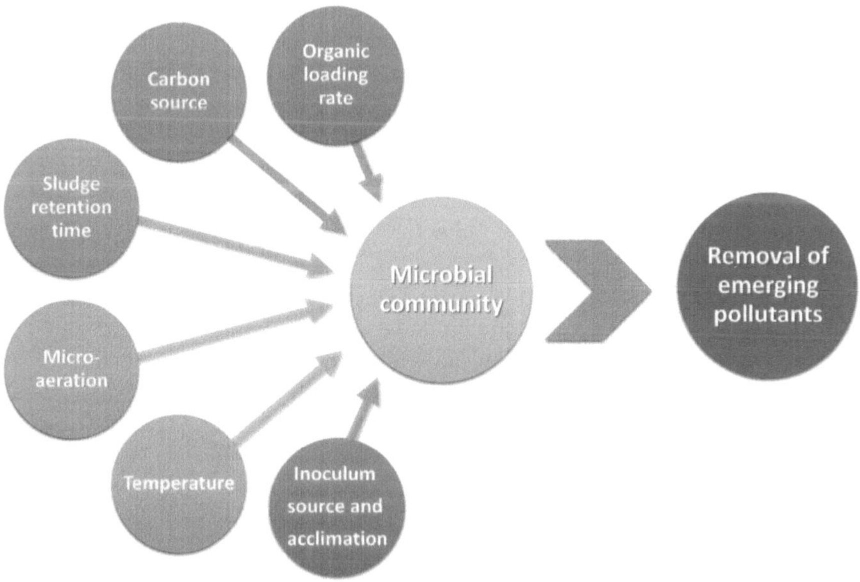

FIGURE 3.3 Approaches to tuning the microbial community in anaerobic digestion to remove emerging pollutants.

effect on removing sulfamethoxazole in an anaerobic structured bed biofilm reactor (ASBBR) using polyurethane foam as support media was investigated, demonstrating that the inoculum source played an important role in antibiotic biodegradation during AD. Poultry slaughterhouse sludge achieved 90 ± 5% of sulfamethoxazole removal, followed by brewery sludge (84 ± 6%) and sewage sludge inoculum (70 ± 5%). However, under the experimental conditions evaluated in this study, the inoculum did not influence AD regarding COD removal and methane production (Sella et al., 2022).

Proper inoculum acclimation brings several benefits to AD, ensuring that the microorganisms are completely adapted to the treatment conditions (Li et al., 2019). Acclimation also makes the microbial communities less susceptible to the toxicity of EPs during the treatment process by developing more resistant microbiota. Moreover, acclimation can reduce the lag phase of biotransformation and further enhance the removal performance in AD (Azizan et al., 2021). The most usual acclimation method for AD is the gradual increase of organic loading rate (OLR) and other operating conditions to the treatment requirement, as well as the use of a microbial consortium already acclimated to the process parameters.

Adjusting the operational conditions of AD is performed by changing variables such as temperature, micro-aeration, pH, solid retention time, and OLR (Haffiez et al., 2022; Gonzalez-Salgado et al., 2020; Fu et al., 2022; Luo et al., 2020). Changing these parameters aims to provide an appropriate environment to prevail over the reactive forms of EPs and to favor alternative metabolic routes through microbial community amendment (Azizan et al., 2021). Therefore, optimizing the process performance and increasing the EPs' removal will be possible.

Temperature is an important parameter for tuning the microbial community in the AD process. It regulates microbial intracellular enzyme activity, growth, and syntrophic interaction between microorganisms (Pasalari et al., 2021). The best AD yields are linked to mesophilic temperatures (30 °C–40 °C, with the optimal between 35 °C–37 °C) and thermophilic temperatures (50 °C–65 °C, with the optimal between 55 °C–62 °C) (Panigrahi et al., 2019). Hence, temperature also influences the removal of EPs in AD by regulating and stimulating the microbiota. Nonetheless, there is no consensus on the ideal temperature range for removing EPs during the AD process. Feng et al. (2017) observed that temperature did not influence the removal of sulfamethoxazole and trimethoprim in lab-scale AD. However, the thermophilic condition improved erythromycin degradation. Similarly, Gonzalez-Gil et al. (2016) concluded that temperature did not exert a significant influence on the biotransformation of several compounds during the operation of a lab-scale AD to treat activated sludge. On the other hand, Zahedi et al. (2021) showed that thermophilic anaerobic treatment was the most suitable process to remove 13 pharmaceuticals from slaughterhouse wastewater, while mesophilic treatment reached the highest methane yield. These conflicting results may be explained by the chemical structure and physicochemical properties of EPs, which make them more or less susceptible to biodegradation under specific conditions.

Micro-aeration involves introducing limited amounts of oxygen into the AD system during pretreatment or directly into the reactor. The oxygen supply changes the microbial community in AD systems without compromising the performance of the methanogen microorganisms. This is possible because facultative anaerobic bacteria quickly consume the oxygen injected into the reactor, protecting methanogens from oxygen stress. Several advantages have been attributed to the micro-aeration in AD: increase of methane yield; stabilization of the AD process; enhanced hydrolysis of the substrate; contribution to hydrogen sulfide stripping; and increase of pollutant degradation (Fu et al., 2022). The removal of EPs is also improved by micro-aeration (e.g., hormones, dyes, pharmaceuticals, and pesticides). Furthermore, the oxygen provided by micro-aeration can initiate degradation by acting as the final electron acceptor in abiotic redox reactions. Additionally, small amounts of oxygen contribute to diversifying the microbial community in AD processes and enabling the growth of hydrolytic and monooxygenase-producing microorganisms (Nascimento et al., 2021; Fu et al., 2022). Nascimento et al. (2021) classified micro-aeration as an effective strategy to improve the removal of seven EPs in a lab-scale UASB fed with synthetic wastewater. This study demonstrated that a flow rate of 4 $mL \cdot min^{-1}$ of synthetic air achieved a removal greater than 80% for all pollutants. Indeed, the removal of EPs under micro-aeration conditions was significantly higher than without micro-aeration. Moreover, Menezes et al. (2019) stated that operating AD under continuous or intermittent micro-aeration resulted in a similar performance for removing tetra-azo dye direct black 22 in lab-scale sequencing batch reactors fed with synthetic wastewater. However, intermittent micro-aeration stood out as a better alternative because it reduced oxygen consumption by 75% compared to continuous micro-aeration conditions and completely removed acute ecotoxicity caused by the pollutant.

The solid retention time (SRT) is referred to as one of the most important parameters for removing recalcitrant pollutants from wastewater through biological processes (Langbehn et al., 2021). It has a positive effect on the biodegradation of EPs, i.e., a longer SRT tends to improve EPs' removal because of the enrichment of the microbial community with slow-growing microorganisms, resulting in a broader variety of microorganisms and likely enzymatic activities (Gonzalez-Gil et al., 2020). Additionally, a long SRT tends to favor the removal pathway by biotransformation rather than sorption (Azizan et al., 2021).

Thus, with the adjustment of operational conditions, it is possible to change the configurations of the used reactors. Some studies suggest that using specific anaerobic reactors (e.g., anaerobic membrane reactors and UASB) may be more effective in removing EPs than conventional anaerobic reactors. The carbon source may affect EPs' removal in the AD process. Oliveira et al. (2019) evaluated the effects of 11 carbon sources on short-term batch assays with synthetic wastewater containing sulfamethazine. Glucose, fructose, sucrose, and meat extract induced antibiotic degradation, with removal efficiencies of 54, 53, 58, and 61%, respectively. This suggests that the results are related to the occurrence of cometabolism, which is associated with the acidogenic stages of AD, as they comprise less specific biochemical reactions. Adding co-solutes, such as short-chain fatty acids, can help improve AD and EPs' removal, as they can increase microbial activity in the process and provide necessary nutrients for the breakdown of chemicals. They can act as a main substrate, and the EPs can be cometabolized in this process (as reported in section 3). Furthermore, physical, chemical, biological, or combined pretreatments can improve AD efficiency and help degrade EPs (Han et al., 2023; Atelge et al., 2020). Pretreatment can make these substances more susceptible to degradation by microorganisms present in AD. Finally, choosing the most appropriate strategy for tuning the microbial community will depend on the specific characteristics of the sewage and the EPs present in the wastewater to be treated.

3.5 RECENT ADVANCES IN AD TO IMPROVE EPS' BIODEGRADATION

Although AD is a well-developed and widely applied WWTP in sludge stabilization and organic matter removal, the performance of this process for EPs' removal remains poorly reported. Furthermore, there are several parameters that EPs can influence during AD. Thus, the scientific community has focused on assessing the damage that EPs can cause in the microbial community with consequent changes in metabolic pathways that lead to changes in process efficiency. The most recent advances for the improvement of the AD process are focused on its main parameters, operating conditions, and applications that allow an improvement in the microbial community for the removal of EPs (Fig. 3.4), as shown by: i) use of differentiated bacterial communities such as purple phototrophic bacteria (de las Heras et al., 2020); ii) different configurations of bioreactors (Panigrahi et al., 2019; Liu et al., 2022); iii) application of pretreatments such as advanced oxidative processes (AOP), hydrothermal treatment, disinfection in an autoclave, ozone and alkaline treatment (Zhang et al.,

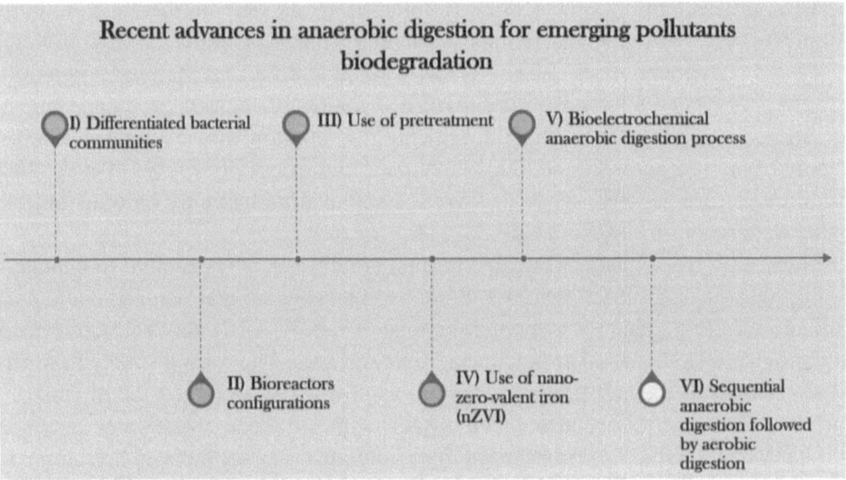

FIGURE 3.4 Recent advances in the anaerobic digestion process.

2019; Wang et al., 2022; Kazimierowicz et al., 2023); iv) use of nano-zero-valent iron (nZVI) (Liu et al., 2022); v) bioelectrochemical systems (Langbehn et al., 2021); and vi) sequential AD followed by aerobic treatment (Tomei et al., 2019).

The use of differentiated bacterial communities (e.g., purple phototrophic bacteria) can enhance EPs' degradation when associated with some improvements in reactors and pretreatments (de las Heras et al., 2020). Changes in the process of photoanaerobic membrane bioreactors (PAnMBR) (de las Heras et al., 2020), mainly in the hydraulic retention time, have been proposed to achieve higher EPs' removal (de las Heras et al., 2017). In the case of using AOP as a pretreatment, the results showed a significant reduction in toxicity, which directly impacted the taxonomic structure of anaerobic bacteria, influencing the efficiency of biogas production and methane content (Kazimierowicz et al., 2023). In addition, when thermal hydrolysis is applied as a pretreatment (high temperature and pressure), it destroys cell walls and may reduce the abundance of antibiotic resistance genes (ARGs) that are susceptible to hydrolytic destruction (Sun et al., 2019; Wang et al., 2022). On the other hand, ozone and alkaline pretreatment have no significant effect on the abundance of ARGs in the pretreatment stage but may increase the removal of ARGs in the subsequent AD process.

The addition of nano-zero-valent iron (nZVI) in an anaerobic membrane bioreactor (AnMBR) can improve the process performance regarding the biodegradation of antibiotics and/or removal of EPs (Liu et al., 2022). Micro-aeration can also improve AD by regulating microbial communities and promoting the growth of facultative microbiota, thus increasing methane production and stabilizing the AD process (Fu et al., 2022).

Bioelectrochemical AD (BEAD) is a promising state-of-the-art technology in wastewater treatment and bioenergy recovery. Although knowledge about the inhibitory effect of EPs on conventional AD processes is increasing, their impact on the

BEAD process remains unknown. Another process that can help in the reduction/degradation of pollutants such as extractable organic halogens, polychlorinated biphenyls, polycyclic aromatic hydrocarbons, di(2-ethylhexyl)phthalate, and alkylphenolethoxylates would be the application of a sequential anaerobic-aerobic digestion process (Tomei et al., 2019).

3.6 CONCLUSIONS

An AD system operating efficiently depends on a perfect balance between the microorganisms involved in each stage. Otherwise, the final effluent will not match the legislation requirements, presenting low organic matter and consequently, low production of biogas. The AD process has the potential to remove several EPs. However, addressing strategies that lead to improvements in the conventional process is necessary. These improvements will directly affect the microbial community involved in the process, improving its ability to remove complex organic matter and EPs. It is important to point out that the adequate removal of EPs will depend on the source of the wastewater, the adjustment of operational parameters, and, mainly, the correct acclimatization of the microbial community. In this way, these parameters need to be adjusted to cause an adequate removal of EPs.

ACKNOWLEDGMENTS

The authors acknowledge Coordination for the Improvement of Higher-Level Personnel (CAPES) and the National Council for Scientific and Technological Development (CNPq) for scholarships and funding sources.

REFERENCES

Alvarino, T.; Suarez, S.; Lema, J. M.; Omil, F. Understanding the Removal Mechanisms of PPCPs and the Influence of Main Technological Parameters in Anaerobic UASB and Aerobic CAS Reactors. *J. Hazard. Mater.*, 2014, 278, 506–513. https://doi.org/10.1016/j.jhazmat.2014.06.031

Anukam, A.; Mohammadi, A.; Naqvi, M.; Granström, K. A Review of the Chemistry of Anaerobic Digestion: Methods of Accelerating and Optimizing Process Efficiency. *Process*, 2019, 7 (8), 504. https://doi.org/10.3390/PR7080504

Atelge, M. R.; Atabani, A. E.; Banu, J. R.; Krisa, D.; Kaya, M.; Eskicioglu, C.; Kumar, G.; Lee, C.; Yildiz, Y.; Unalan, S.; et al. A Critical Review of Pretreatment Technologies to Enhance Anaerobic Digestion and Energy Recovery. *Fuel*, 2020, 270, 117494. https://doi.org/10.1016/j.fuel.2020.117494

Aydin, S.; Cetecioglu, Z.; Arikan, O.; Ince, B.; Ozbayram, E. G.; Ince, O. Inhibitory Effects of Antibiotic Combinations on Syntrophic Bacteria, Homoacetogens and Methanogens. *Chemosphere*, 2015, 120, 515–520. https://doi.org/10.1016/j.chemosphere.2014.09.045

Azizan, N. A. Z.; Yuzir, A.; Abdullah, N. Pharmaceutical Compounds in Anaerobic Digestion: A Review on the Removals and Effect to the Process Performance. *J. Environ. Chem. Eng.*, 2021, 9 (5), 105926. https://doi.org/10.1016/j.jece.2021.105926

Carballa, M.; Fink, G.; Omil, F.; Lema, J. M.; Ternes, T. Determination of the Solid–Water Distribution Coefficient (Kd) for Pharmaceuticals, Estrogens and Musk Fragrances in

Digested Sludge. *Water Res.*, 2008, 42 (1–2), 287–295. https://doi.org/10.1016/J.WATRES.2007.07.012

Carneiro, R. B.; Gonzalez-Gil, L.; Londoño, Y. A.; Zaiat, M.; Carballa, M.; Lema, J. M. Acidogenesis Is a Key Step in the Anaerobic Biotransformation of Organic Micropollutants. *J. Hazard. Mater.*, 2020, 389, 121888. https://doi.org/10.1016/j.jhazmat.2019.121888

Cetecioglu, Z.; Ince, B.; Orhon, D.; Ince, O. Acute Inhibitory Impact of Antimicrobials on Acetoclastic Methanogenic Activity. *Bioresour. Technol.*, 2012, 114, 109–116. https://doi.org/10.1016/j.biortech.2012.03.020

Chakraborty, N.; Sarkar, G. M.; Lahiri, S. C. Effect of Pesticide (Tara-909) on Biomethanation of Sewage Sludge and Isolated Methanogens. *Biomass and Bioenergy*, 2002, 23 (1), 75–80. https://doi.org/10.1016/S0961-9534(02)00021-1

Chen, W. H.; Wong, Y. T.; Huang, T. H.; Chen, W. H.; Lin, J. G. Removals of Pharmaceuticals in Municipal Wastewater Using a Staged Anaerobic Fluidized Membrane Bioreactor. *Int. Biodeterior. Biodegradation*, 2019, 140, 29–36. https://doi.org/10.1016/J.IBIOD.2019.03.008

Coetsier, C. M.; Spinelli, S.; Lin, L.; Roig, B.; Touraud, E. Discharge of Pharmaceutical Products (PPs) through a Conventional Biological Sewage Treatment Plant: MECs vs PECs? *Environ. Int.*, 2009, 35 (5), 787–792. https://doi.org/10.1016/j.envint.2009.01.008

Cremonez, P. A.; Teleken, J. G.; Weiser Meier, T. R.; Alves, H. J. Two-Stage Anaerobic Digestion in Agroindustrial Waste Treatment: A Review. *J. Environ. Manage.*, 2021, *281*, 111854. https://doi.org/10.1016/j.jenvman.2020.111854

de las Heras, I.; Molina, R.; Segura, Y.; Hülsen, T.; Molina, M. C.; Gonzalez-Benítez, N.; Melero, J. A.; Mohedano, A. F.; Martínez, F.; Puyol, D. Contamination of N-Poor Wastewater with Emerging Pollutants Does Not Affect the Performance of Purple Phototrophic Bacteria and the Subsequent Resource Recovery Potential. *J. Hazard. Mater.*, 2020, 385, 121617. https://doi.org/10.1016/j.jhazmat.2019.121617

de las Heras, I.; Padrino, B.; Molina, R.; Segura, Y.; Melero, J. A.; Mohedano, A. F.; Martínez, F.; Puyol, D. Efficient Treatment of Synthetic Wastewater Contaminated with Emerging Pollutants by Anaerobic Purple Phototrophic Bacteria. *Lect. Notes Civ. Eng.*, 2017, 4, 324–330. https://doi.org/10.1007/978-3-319-58421-8_52

Ebele, A. J.; Abou-Elwafa Abdallah, M.; Harrad, S. Pharmaceuticals and Personal Care Products (PPCPs) in the Freshwater Aquatic Environment. *Emerging Contaminants*. Elsevier March 1, 2017, pp. 1–16. https://doi.org/10.1016/j.emcon.2016.12.004

Elmoutez, S.; Abushaban, A.; Necibi, M. C.; Sillanpää, M.; Liu, J.; Dhiba, D.; Chehbouni, A.; Taky, M. Design and Operational Aspects of Anaerobic Membrane Bioreactor for Efficient Wastewater Treatment and Biogas Production. *Environ. Challenges*, 2023, 10, 100671. https://doi.org/10.1016/J.ENVC.2022.100671

Feng, L.; Casas, M. E.; Ottosen, L. D. M., et al. Removal of Antibiotics during the Anaerobic Digestion of Pig Manure. *Sci Total Environ*, 2017, 603–604:219–225. https://doi.org/10.1016/j.scitotenv.2017.05.280

Fu, S.; Lian, S.; Angelidaki, I.; Guo, R. Micro-Aeration. An Attractive Strategy to Facilitate Anaerobic Digestion. *Trends Biotechnol.*, 2022, 1–13. https://doi.org/10.1016/j.tibtech.2022.09.008

Gogoi, A.; Mazumder, P.; Tyagi, V. K.; Tushara Chaminda, G. G.; An, A. K.; Kumar, M. Occurrence and Fate of Emerging Contaminants in Water Environment: A Review. *Groundw. Sustain. Dev.*, 2018, 6, 169–180. https://doi.org/10.1016/j.gsd.2017.12.009

Golet, E. M.; Xifra, I.; Siegrist, H.; Alder, A. C.; Giger, W. Environmental Exposure Assessment of Fluoroquinolone Antibacterial Agents from Sewage to Soil. *Environ. Sci. Technol.*,

2003, 37 (15), 3243–3249. https://doi.org/10.1021/ES0264448/ASSET/IMAGES/LARGE/ES0264448F00004.JPEG

Gonzalez-Gil, L.; Carballa, M.; Lema, J. M. Cometabolic Enzymatic Transformation of Organic Micropollutants under Methanogenic Conditions. *Environ. Sci. Technol.*, 2017, 51 (5), 2963–2971. https://doi.org/10.1021/acs.est.6b05549

Gonzalez-Gil, L.; Carballa, M.; Lema, J. M. Removal of Organic Micro-Pollutants by Anaerobic Microbes and Enzymes. In *Current Developments in Biotechnology and Bioengineering*; Elsevier, 2020; pp. 397–426. https://doi.org/10.1016/b978-0-12-819594-9.00016-4

Gonzalez-Gil, L.; Krah, D.; Ghattas, A. K.; Carballa, M.; Wick, A.; Helmholz, L.; Lema, J. M.; Ternes, T. A. Biotransformation of Organic Micropollutants by Anaerobic Sludge Enzymes. *Water Res.*, 2019, 152, 202–214. https://doi.org/10.1016/j.watres.2018.12.064

Gonzalez-Gil, L.; Papa, M.; Feretti, D.; Ceretti, E.; Mazzoleni, G.; Steimberg, N.; Pedrazzani, R.; Bertanza, G.; Lema, J. M.; Carballa, M. Is Anaerobic Digestion Effective for the Removal of Organic Micropollutants and Biological Activities from Sewage Sludge? *Water Res.,* 2016, 102, 211–220. https://doi.org/10.1016/J.WATRES.2016.06.025

Gonzalez-Salgado, I.; Cavaillé, L.; Dubos, S.; Mengelle, E.; Kim, C.; Bounouba, M.; Paul, E.; Pommier, S.; Bessiere, Y. Combining Thermophilic Aerobic Reactor (TAR) with Mesophilic Anaerobic Digestion (MAD) Improves the Degradation of Pharmaceutical Compounds. *Water Res.*, 2020, 182, 116033. https://doi.org/10.1016/J.WATRES.2020.116033

Grady Jr., C. P. L.; Daigger, G. T.; Love, N. G.; Filipe, C. D. M. Biological Wastewater Treatment. 2011. https://doi.org/10.1201/B13775

Haffiez, N.; Chung, T. H.; Zakaria, B. S.; Shahidi, M.; Mezbahuddin, S.; Hai, F. I.; Dhar, B. R. A Critical Review of Process Parameters Influencing the Fate of Antibiotic Resistance Genes in the Anaerobic Digestion of Organic Waste. *Bioresour. Technol.*, 2022, 354 (April), 127189. https://doi.org/10.1016/j.biortech.2022.127189

Han, S.; Liu, Y.; Zhang, S.; Luo, G. Reactor Performances and Microbial Communities of Biogas Reactors: Effects of Inoculum Sources. *Appl. Microbiol. Biotechnol.*, 2016, 100 (2), 987–995. https://doi.org/10.1007/S00253-015-7062-7/FIGURES/5

Han, Z.; Shao, B.; Lei, L.; Pang, R.; Wu, D.; Tai, J.; Xie, B.; Su, Y. The Role of Pretreatments in Handling Antibiotic Resistance Genes in Anaerobic Sludge Digestion – A Review. *Sci. Total Environ.*, 2023, 869 (November 2022), 161799. https://doi.org/10.1016/j.scitotenv.2023.161799

Harb, M.; Lou, E.; Smith, A. L.; Stadler, L. B. Perspectives on the Fate of Micropollutants in Mainstream Anaerobic Wastewater Treatment. *Curr. Opin. Biotechnol.*, 2019, 57, 94–100. https://doi.org/10.1016/J.COPBIO.2019.02.022

Hazen, T. C. Cometabolic Bioremediation. Handb. Hydrocarb. *Lipid Microbiol.*, 2010, 2505–2514. https://doi.org/10.1007/978-3-540-77587-4_185

He, Y.; Sutton, N. B.; Rijnaarts, H. H. M.; Langenhoff, A. A. M. Pharmaceutical Biodegradation under Three Anaerobic Redox Conditions Evaluated by Chemical and Toxicological Analyses. *Sci. Total Environ.*, 2018, 618, 658–664. https://doi.org/10.1016/J.SCITOTENV.2017.07.219

Kazimierowicz, J.; Dębowski, M.; Zieliński, M. Effect of Pharmaceutical Sludge Pre-Treatment with Fenton/Fenton-like Reagents on Toxicity and Anaerobic Digestion Efficiency. *Int. J. Environ. Res. Public Health*, 2023, 20 (1). https://doi.org/10.3390/ijerph20010271

Ketheesan, B.; Stuckey, D. C. Effects of Hydraulic/Organic Shock/Transient Loads in Anaerobic Wastewater Treatment: A Review. *Crit. Rev. Environ. Sci. Technol.*, 2015, 45 (24), 2693–2727. https://doi.org/10.1080/10643389.2015.1046771

Kumar, V., Agrawal, S., Bhat, S.A., Américo-Pinheiro, J.H.P., Shahi, S.K., Kumar, S., 2022. Environmental Impact, Health Hazards, and Plant-Microbes Synergism in Remediation of Emerging Contaminants. *Cleaner Chemical Engineering* 2, 100030. https://doi.org/10.1016/j.clce.2022.100030

Lahti, M.; Oikari, A. Microbial Transformation of Pharmaceuticals Naproxen, Bisoprolol, and Diclofenac in Aerobic and Anaerobic Environments. *Arch. Environ. Contam. Toxicol.*, 2011, 61 (2), 202–210. https://doi.org/10.1007/S00244-010-9622-2/TABLES/2

Langbehn, R. K.; Michels, C.; Soares, H. M. Antibiotics in Wastewater: From Its Occurrence to the Biological Removal by Environmentally Conscious Technologies. *Environ. Pollut.*, 2021, 275, 116603. https://doi.org/10.1016/j.envpol.2021.116603

Li, Y.; Chen, Y.; Wu, J. Enhancement of Methane Production in Anaerobic Digestion Process: A Review. *Appl. Energy*, 2019, 240, 120–137. https://doi.org/10.1016/j.apenergy.2019.01.243

Liu, W.; Xia, R.; Ding, X.; Cui, W.; Li, T.; Li, G.; Luo, W. Impacts of Nano-Zero-Valent Iron on Antibiotic Removal by Anaerobic Membrane Bioreactor for Swine Wastewater Treatment. *J. Memb. Sci.*, 2022, 659 (May), 120762. https://doi.org/10.1016/j.memsci.2022.120762

Luo, J.; Zhang, Q.; Zhao, J.; Wu, Y.; Wu, L.; Li, H.; Tang, M.; Sun, Y.; Guo, W.; Feng, Q.; et al. Potential Influences of Exogenous Pollutants Occurred in Waste Activated Sludge on Anaerobic Digestion. *A Review. J. Hazard. Mater.*, 2020, 383, 121176. https://doi.org/10.1016/j.jhazmat.2019.121176

Luo, Y.; Guo, W.; Ngo, H. H.; Nghiem, L. D.; Hai, F. I.; Zhang, J.; Liang, S.; Wang, X. C. A Review on the Occurrence of Micropollutants in the Aquatic Environment and Their Fate and Removal during Wastewater Treatment. *Sci. Total Environ.*, 2014, 473–474, 619–641. https://doi.org/10.1016/j.scitotenv.2013.12.065

Menezes, O.; Brito, R.; Hallwass, F.; Florêncio, L.; Kato, M. T.; Gavazza, S. Coupling Intermittent Micro-Aeration to Anaerobic Digestion Improves Tetra-Azo Dye Direct Black 22 Treatment in Sequencing Batch Reactors. *Chem. Eng. Res. Des.*, 2019, 146, 369–378. https://doi.org/10.1016/J.CHERD.2019.04.020

Merlin Christy, P.; Gopinath, L. R.; Divya, D. A Review on Anaerobic Decomposition and Enhancement of Biogas Production through Enzymes and Microorganisms. *Renew. Sustain. Energy Rev.*, 2014, 34, 167–173. https://doi.org/10.1016/J.RSER.2014.03.010

Nascimento, J. G. da S.; Silva, E. V. A.; dos Santos, A. B.; da Silva, M. E. R.; Firmino, P. I. M. Microaeration Improves the Removal/Biotransformation of Organic Micropollutants in Anaerobic Wastewater Treatment Systems. *Environ. Res.*, 2021, 198, 111313. https://doi.org/10.1016/j.envres.2021.111313

Negreira, N.; de Alda, M. L.; Barceló, D. Cytostatic Drugs and Metabolites in Municipal and Hospital Wastewaters in Spain: Filtration, Occurrence, and Environmental Risk. *Sci. Total Environ.*, 2014, 497–498, 68–77. https://doi.org/10.1016/j.scitotenv.2014.07.101

NORMAN. Network of reference laboratories, research centers and related organizations for monitoring of emerging environmental substances www.norman-network.com/?q=node/19

Oliveira, B. M.; Zaiat, M.; Oliveira, G. H. D. The Contribution of Selected Organic Substrates to the Anaerobic Cometabolism of Sulfamethazine. *J. Environ. Sci. Heal. Part B*, 2019, 54 (4), 263–270. https://doi.org/10.1080/03601234.2018.1553909

Panigrahi, S.; Dubey, B. K. A Critical Review on Operating Parameters and Strategies to Improve the Biogas Yield from Anaerobic Digestion of Organic Fraction of Municipal Solid Waste. *Renew. Energy*, 2019, 143, 779–797. https://doi.org/10.1016/j.renene.2019.05.040

Pasalari, H.; Gholami, M.; Rezaee, A.; Esrafili, A.; Farzadkia, M. Perspectives on Microbial Community in Anaerobic Digestion with Emphasis on Environmental Parameters: A Systematic Review. *Chemosphere*, 2021, 270. https://doi.org/10.1016/j.chemosphere.2020.128618

Pereira, C. S.; Kelbert, M.; Daronch, N. A.; Michels, C.; de Oliveira, D.; Soares, H. M. Potential of Enzymatic Process as an Innovative Technology to Remove Anticancer Drugs in Wastewater. *Appl. Microbiol. Biotechnol.*, 2020, 104 (1), 23–31. https://doi.org/10.1007/s00253-019-10229-y

Shi, X.; Leong, K. Y.; Ng, H. Y. Anaerobic Treatment of Pharmaceutical Wastewater: A Critical Review. *Bioresour. Technol.*, 2017, 245, 1238–1244. https://doi.org/10.1016/j.biortech.2017.08.150

Singh, S. P.; Agarwal, A. K.; Gupta, T.; Maliyekkal, S. M. *New Trends in Emerging Environmental Contaminants*; Singh, S. P., Agarwal, A. K., Gupta, T., Maliyekkal, S. M., Eds.; Springer, 2022.

Sella, C. F.; Carneiro, R. B.; Sabatini, C. A.; Sakamoto, I. K.; Zaiat, M. Can Different Inoculum Sources Influence the Biodegradation of Sulfamethoxazole Antibiotic during Anaerobic Digestion? *Brazilian J. Chem. Eng.*, 2022, 39 (1), 35–46. https://doi.org/10.1007/S43153-021-00178-3/TABLES/2

Stams, A. J. M. Metabolic Interactions between Anaerobic Bacteria in Methanogenic Environments. *Antonie Van Leeuwenhoek*, 1994, 66 (1–3), 271–294. https://doi.org/10.1007/BF00871644/METRICS

Sun, C.; Li, W.; Chen, Z.; Qin, W.; Wen, X. Responses of Antibiotics, Antibiotic Resistance Genes, and Mobile Genetic Elements in Sewage Sludge to Thermal Hydrolysis Pre-Treatment and Various Anaerobic Digestion Conditions. *Environ. Int.*, 2019, 133 (October), 105156. https://doi.org/10.1016/j.envint.2019.105156

Sung, Y.; Ritalahti, K. M.; Sanford, R. A.; Urbance, J. W.; Flynn, S. J.; Tiedje, J. M.; Löffler, F. E. Characterization of Two Tetrachloroethene-Reducing, Acetate-Oxidizing Anaerobic Bacteria and Their Description as Desulfuromonas Michiganensis Sp. *Nov. Appl. Environ. Microbiol.*, 2003, 69 (5), 2964–2974. https://doi.org/10.1128/AEM.69.5.2964-2974.2003/ASSET/BE6CB245-0E0C-4056-A411-A2F3F2298B3F/ASSETS/GRAPHIC/AM0531761007.JPEG

Taheran, M.; Naghdi, M.; Brar, S. K.; Verma, M.; Surampalli, R. Y. Emerging Contaminants: Here Today, There Tomorrow! *Environ. Nanotechnology, Monit. Manag.*, 2018, 10, 122–126. https://doi.org/10.1016/j.enmm.2018.05.010

Thanarasu, A.; Periyasamy, K.; Subramanian, S. An Integrated Anaerobic Digestion and Microbial Electrolysis System for the Enhancement of Methane Production from Organic Waste: Fundamentals, Innovative Design and Scale-up Deliberation. *Chemosphere*, 2022, 287, 131886. https://doi.org/10.1016/J.CHEMOSPHERE.2021.131886

Tomei, M. C.; Mosca Angelucci, D.; Mascolo, G.; Kunkel, U. Post-Aerobic Treatment to Enhance the Removal of Conventional and Emerging Micropollutants in the Digestion of Waste Sludge. *Waste Manag.*, 2019, 96, 36–46. https://doi.org/10.1016/j.wasman.2019.07.013

Tran, N. H.; Urase, T.; Ngo, H. H.; Hu, J.; Ong, S. L. Insight into Metabolic and Cometabolic Activities of Autotrophic and Heterotrophic Microorganisms in the Biodegradation of Emerging Trace Organic Contaminants. *Bioresour. Technol.*, 2013, 146, 721–731. https://doi.org/10.1016/J.BIORTECH.2013.07.083

Veetil, P. G. P.; Nadaraja, A. V.; Bhasi, A.; Khan, S.; Bhaskaran, K. Degradation of Triclosan under Aerobic, Anoxic, and Anaerobic Conditions. *Appl. Biochem. Biotechnol.*, 2012, 167 (6), 1603–1612. https://doi.org/10.1007/s12010-012-9573-3

Venegas, M.; Leiva, A. M.; Reyes-Contreras, C.; Neumann, P.; Piña, B.; Vidal, G. Presence and Fate of Micropollutants during Anaerobic Digestion of Sewage and Their Implications for the Circular Economy: *A Short Review. J. Environ. Chem. Eng.*, 2021, 9 (1). https://doi.org/10.1016/j.jece.2020.104931

Verlicchi, P.; Al Aukidy, M.; Zambello, E. Occurrence of Pharmaceutical Compounds in Urban Wastewater: Removal, Mass Load and Environmental Risk after a Secondary Treatment—A Review. *Sci. Total Environ.*, 2012, 429, 123–155. https://doi.org/10.1016/J.SCITOTENV.2012.04.028

Wang, Y.; Zhao, X.; Wang, Y.; Wang, I.; Turap, Y.; Wang, W. Hydrothermal Treatment Enhances the Removal of Antibiotic Resistance Genes, Dewatering, and Biogas Production in Antibiotic Fermentation Residues. *J. Hazard. Mater.*, 2022, 435 (April), 128901. https://doi.org/10.1016/j.jhazmat.2022.128901

Wilkinson, J.; Hooda, P. S.; Barker, J.; Barton, S.; Swinden, J. Occurrence, Fate and Transformation of Emerging Contaminants in Water: An Overarching Review of the Field. *Environ. Pollut.*, 2017, 231, 954–970. https://doi.org/10.1016/j.envpol.2017.08.032

Wolfson, S. J.; Porter, A. W.; Campbell, J. K.; Young, L. Y. Naproxen Is Transformed Via Acetogenesis and Syntrophic Acetate Oxidation by a Methanogenic Wastewater Consortium. *Microb. Ecol.*, 2018, 76 (2), 362–371. https://doi.org/10.1007/S00248-017-1136-2/FIGURES/5

Xie, Z.; Wang, Z.; Wang, Q.; Zhu, C.; Wu, Z. An Anaerobic Dynamic Membrane Bioreactor (AnDMBR) for Landfill Leachate Treatment: Performance and Microbial Community Identification. *Bioresour. Technol.*, 2014, 161, 29–39. https://doi.org/10.1016/J.BIORTECH.2014.03.014

Yang, Y.; Zhang, C.; Hu, Z. Impact of Metallic and Metal Oxide Nanoparticles on Wastewater Treatment and Anaerobic Digestion. *Environ. Sci. Process. Impacts*, 2013, 15, 39–48. https://doi.org/10.1039/c2em30655g

Yang, Y.; Yu, K.; Xia, Y.; Lau, F. T. K.; Tang, D. T. W.; Fung, W. C.; Fang, H. H. P.; Zhang, T. Metagenomic Analysis of Sludge from Full-Scale Anaerobic Digesters Operated in Municipal Wastewater Treatment Plants. *Appl. Microbiol. Biotechnol.*, 2014, 98 (12), 5709–5718. https://doi.org/10.1007/s00253-014-5648-0

Zahedi, S.; Gros, M.; Balcazar, J. L.; Petrovic, M.; Pijuan, M. Assessing the Occurrence of Pharmaceuticals and Antibiotic Resistance Genes during the Anaerobic Treatment of Slaughterhouse Wastewater at Different Temperatures. *Sci. Total Environ.*, 2021, 789, 147910. https://doi.org/10.1016/J.SCITOTENV.2021.147910

Zhang, J.; Chang, V. W. C.; Giannis, A.; Wang, J. Y. Removal of Cytostatic Drugs from Aquatic Environment: A Review. *Sci. Total Environ.*, 2013, 445–446, 281–298. https://doi.org/10.1016/j.scitotenv.2012.12.061

Zhang, J.; Liang, Y.; Yau, P. M.; Pandey, R.; Harpalani, S. A Metaproteomic Approach for Identifying Proteins in Anaerobic Bioreactors Converting Coal to Methane. *Int. J. Coal Geol.*, 2015, 146, 91–103. https://doi.org/10.1016/J.COAL.2015.05.006

Zhang, J.; Lu, T.; Shen, P.; Sui, Q.; Zhong, H.; Liu, J.; Tong, J.; Wei, Y. The Role of Substrate Types and Substrate Microbial Community on the Fate of Antibiotic Resistance Genes during Anaerobic Digestion. *Chemosphere*, 2019, 229, 461–470. https://doi.org/10.1016/j.chemosphere.2019.05.036

Zhang, W.; Dai, K.; Xia, X. Y.; Wang, H. J.; Chen, Y.; Lu, Y. Z.; Zhang, F.; Zeng, R. J. Free Acetic Acid as the Key Factor for the Inhibition of Hydrogenotrophic Methanogenesis in Mesophilic Mixed Culture Fermentation. *Bioresour. Technol.*, 2018, 264, 17–23. https://doi.org/10.1016/j.biortech.2018.05.049

4 Energy–Water–Waste Nexus for Sustainable Wastewater and Development

Anisa Ratnasari

4.1 INTRODUCTION

Water, waste, and wastewater which are utilized and generated by some sectors, such as agricultural sector, hospital sector, and municipal sector have no separable from energy that cover their process (Al-Ghouti et al. 2019; El Messaoudi et al., 2022). Asnoted in the sustainable development goals (SDGs) SDG6, SDG7, SG11, and SDG12, the goals correlated with water, waste, and energy, there is an urgent need to ensure the world's increasingly dwindling water supply is protected and preserved, especially considering, waste found in the environment has increased by 15% over the past 20 years (Yu et al. 2022).

To understand the connection between these things, several techniques are applicable to elaborate the nexus between water, waste, and energy including optimization, life cycle assessment (Friedrich, Poganietz, and Lehn 2020), case study (Vergara-Araya et al. 2020), multivariate analysis/statistic (Dura et al. 2022), integral theory framework (Ahmad et al. 2020), and econometric method (Kiziltan 2021). Generally, water and waste have a negative connection related to pollution in water bodies. Water and energy have both negative and positive connections related to energy generation from water flow and wastewater as positive connection and energy demand from distribution of clean water and wastewater treatment. It is possible for waste and energy to have a positive connection since the conversion of waste provides a promising alternative energy source. These connections show that water, waste, and energy have strong correlation which needs to be better understood.

The nexus of energy, water, and waste has been well studied. Recently, research has been focused on the security nexus of food, water, and energy using microalgae for sustainability in the agricultural sector (Cao et al. 2023), solar evaporation for the nexus between water and energy (Zhang et al. 2023), application of thermochemical technique correlated with energy and food waste (Yadav et al. 2023), photoelectrocatalytic system for the nexus between water and energy (Changotra et al. 2022), and microbial fuel cell for alternative energy generation from wastewater

DOI: 10.1201/9781003441069-4

(Toczyłowska-Mamińska and Mamiński 2022). However, there are limited studies that evaluate the nexus between global water availability, waste generation, and energy production including the role of the microbial nexus of water and waste. As the vast majority of inventions and breakthroughs, this is the reason why it is so important to review these advancements in a detailed manner to enable further improvements.

This chapter aims to provide comprehensive insight into the nexus of energy and water, the nexus of water and waste, as well as the nexus of waste and energy. It also overviews essential insights into the energy recovery perspective for sustainable development.

4.2 GLOBAL WATER AVAILABILITY AND GENERATED WASTE

4.2.1 Global Water Availability

Increases in water demand can influence water availability and water quality (He et al. 2021). The shrinking resources and inadequate water treatment and management are causing water availability to diminish in relation to the growing water demand, highlighting the need for a deeper understanding of water availability in correlation with water demand. As seen in Figure 4.1, global water availability comprises 97% seawater and 3% freshwater (Alhaj, Tahir, and Al-Ghamdi 2022). Only 3% of water in the world can be utilized for human consumption. Moreover, from 3% freshwater, 70% is frozen, 29% is underground water, and 15% is lakes and rivers. This indicates that the possible and reachable water availability is only 15% of the 3% water availability in the world. Consequently, there is only 0.45% water availability to fulfill water demand.

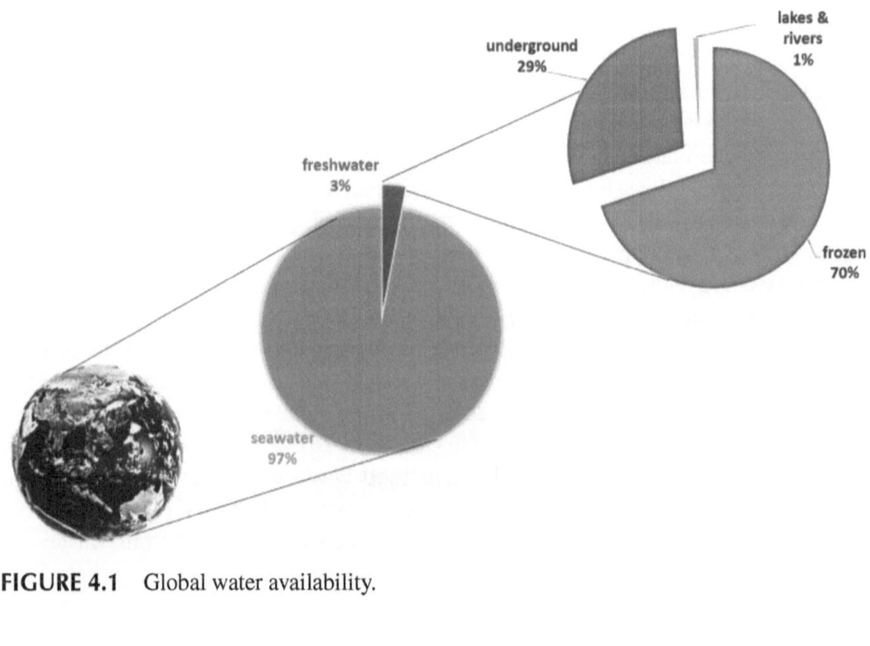

FIGURE 4.1 Global water availability.

It has been noted that there is 97% seawater for providing global water availability, which can be explored using desalination technology to change it into freshwater. Regarding the utilization of desalination technology, less energy required will be more beneficial to collecting freshwater. There are several desalination technologies that can be applied to desalinating seawater into freshwater, such as thermal desalination and membrane desalination. Thermal desalination involves distillation and crystallization techniques. The distillation technique utilizes thermal energy (heat) to extract the water vapor and reassemble it into freshwater (ElHelw et al. 2020). Distillation offers several advantages including low-temperature heat source that can be utilized, simple construction of equipment, extensive use, process reliability, and simple maintenance work. In contrast, crystallization uses thermal energy (freezing) to crystallize water into ice to separate it from saline solution (Chen et al. 2021). Then, the ice is melted and changed into freshwater. For membrane distillation, a non-porous membrane and electrodialysis (ED) membrane is used to convert seawater into freshwater. A non-porous membrane separates saline solution from freshwater depending on pressure, concentration, and electrical potential gradient (Yadav et al. 2021). Nanofiltration membranes and reverse osmosis as non-porous membranes can play a role in converting seawater into freshwater. Meanwhile, ED separates salt ions and charged organic matter using the ion exchange principle (Doornbusch et al. 2020). ED employs two electrodes including a cathode and an anode. An electric potential gradient leads cation to draw near the cathode and go over the cation exchange membranes, which are ensnared by anion exchange membranes. Similarly, anions make a beeline for the anode, travel through the anion exchange membranes, and are withheld by the cation exchange membranes. However, these technologies are purposed to desalinate seawater into freshwater. Assuming the efficiency of these technologies is possible at 10%, freshwater can be achieved up to 9.70% of the water availability in the world (Chen et al. 2021). This indicates that the availability of water that can be consumed by humans reaches approximately 10.15%, consisting of 9.70% from the desalination process of seawater and 0.45% from the availability of freshwater globally. Thus, desalination is a promising technique to meet freshwater needs.

4.2.2 Generated Waste

Along with increasing population in the world, waste generation also increases from year to year (Babayemi et al. 2019; Agrawal et al. 2021). As seen in Figure 4.2, waste generation is estimated to increase from 2030 to 2050. The highest waste generation in 2030 is estimated to be in East Asia and the Pacific with 602 million tonnes per year (MTPY), followed by South Asia (466 MTPY), Europe and Central Asia (440 MTPY), North America (342 MTPY), Latin America and Caribbean (290 MTPY), Sub-saharan Africa (269 MTPY), as well as Middle East and North Africa (177 MTPY) (Sharma and Jain 2020). Line in 2030, in 2050, the highest waste generation is estimated and found in East Asia and Pacific with 714 MTPY, followed by South Asia (661 MTPY), Sub-saharan Africa (516 MTPY), Europe and Central Asia (490 MTPY), North America (396 MTPY), Latin America and Carribean (369 MTPY), as well as Middle East and North Africa (255 MTPY) (Lefeuvre et al. 2019).

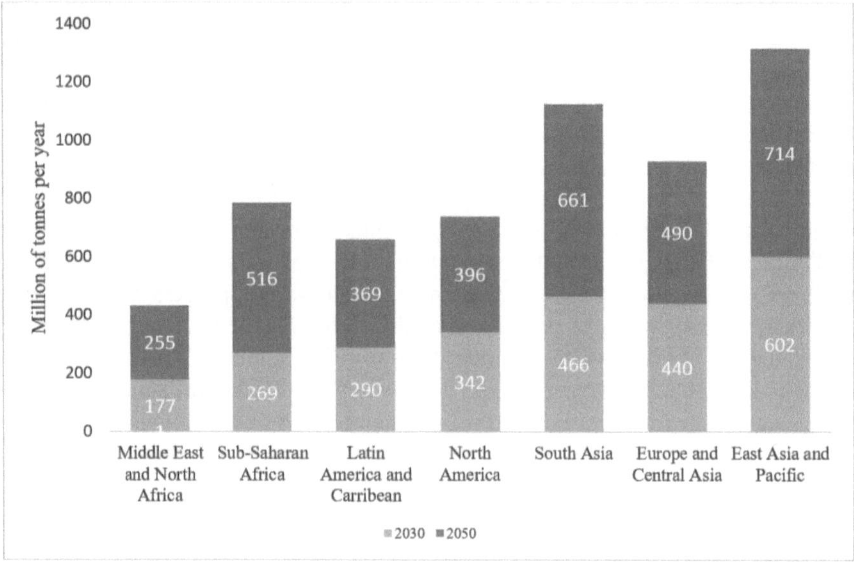

FIGURE 4.2 Global generated waste.

Nonetheless, the highest difference is estimated to be found in Sub-saharan Africa changing from 269 MTPY to 516 MTPY with a difference up to 247 MTPY in 2050, followed by South Asia (195 MTPY), East Asia and Pacific 112 MTPY, Latin America and Carribean (79 MTPY), Middle East and North Africa (78 MTPY), North America 954 MTPY), and Europe and Central Asia (50 MTPY). However, from these groups, it is inferred that Sub-Saharan Africa which is potentially to have a high number of waste generation is highly recommended to give attention to their waste management. It also has been noted that most developing countries need to give attention to their waste management as it was found that Asia, followed by Europe, tends to have a high amount of waste generation.

4.3 THE NEXUS BETWEEN ENERGY, WATER, AND WASTE

Energy, water, and waste are major contributors in the environment and human life for achieving better human health and addressing the issue of water availability, waste generation, and environment contamination (Kumar and Chandra 2017; Kumar et al. 2021; Américo-Pinheiro et al. 2022). The nexus between energy, water, and waste has been previously assessed using optimization, life cycle assessment (Ghani, Silalertruksa, and Gheewala 2019), case study (Liu et al. 2020), multivariate analysis/ statistic (Lazaro et al., 2021), integral theory framework (Fetanat, Tayebi, and Mofid 2021), and econometric methods (Zeeshan et al. 2021). These methods are projected to find the best solution correlated with energy, water, and waste nexus. The nexus thinking across several aspects.

4.3.1 Energy Water Nexus

Energy and water have been mentioned in SDGs SDG6 and SDG7 (Pandey and Asif 2022). As seen in Figure 4.3, it has been shown that energy is closely related to water. To achieve SDG6, it will be essential to ensure the availability of safe and affordable drinking water, sanitation, and hygiene for all by 2030. Besides improving the efficiency of water use across all sectors by 2030, it will ensure sustainable withdrawals of freshwater and supplies to address water scarcity by ensuring sustainable withdrawals and supplies. In terms of SDG7, by 2030, it will be essential to provide universal access to affordable, reliable, and modern energy services to the entire world, as well as dramatically increasing the share of renewable energy in the total global energy mix by 2030. According to SDG6 and SDG7, several sectors have been presented their relation including the treatment of wastewater, taking water to fulfill drinking needs from nature, the distribution of freshwater and drinking water for society, and, for reversed context, the energy power generation requiring water flow in rivers. The economic aspect is also inseparable from the supply chain. As previously mentioned, water distribution, clean water production, and the process until it is accepted by consumers are included in the nexus between energy and water.

Conceptually, the availability of energy and water for fulfilling human needs is still greatly challenging and limiting. As presented in the previous section, water availability is only provided by 10.15% in the world. In addition to water availability, energy availability will also not meet the increasingly diverse human needs. In 2030, it is predicted that energy demand will increase by 50% and water demand will increase by 40%. Moreover, not only human needs are increasing, but also the number of human populations is increasing. Thus, the demand for water and energy will increase as well. The previous study assessed that from 1999–2019 energy and water use increased, while energy and water intensity decreased (Ke et al. 2022).

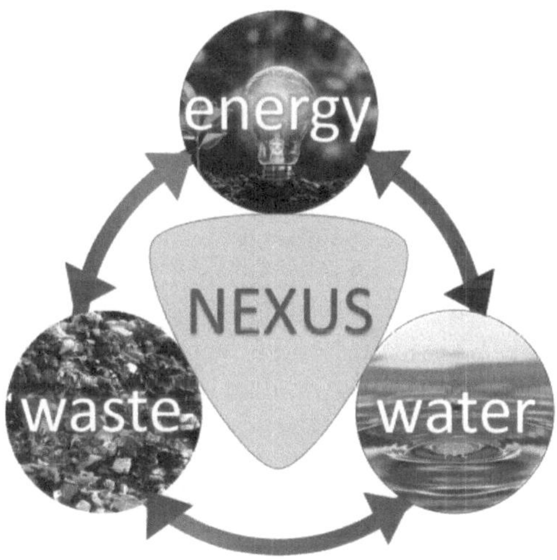

FIGURE 4.3 Energy-water-waste nexus.

Water and energy ratios also decreased, which is inversely proportional to increased economic growth. Pan et al. (2020) evaluated the use of electrical energy to convert brackish water into freshwater for supporting water availability. In general, for the desalination process, membrane technology is applied since it has simple operation, almost no sludge, and has easy fabrication. Nonetheless, the application of membrane technology requires high energy since it involves driving force and pressure. However, an energy recovery device (ERD) has been found that can save 2–3 times more energy compared to non-ERD applications. Therefore, the existence of ERD can help save energy in the desalination process for increasing water availability and dealing with water crises in the future.

4.3.2 Water Waste Nexus

SDG6, SDG11, and SD12 mention water and waste as they relate to sustainable development (Lodhia et al. 2022). As shown in Figure 4.3, it was determined that water has the closest relationship to waste. To achieve SDG6 by 2030, pollution will have to be reduced and dumping eliminated dumping in addition to minimizing the release of contaminants like hazardous chemicals and materials into the environment. Half the proportion of untreated wastewater and recycling and safe reuse will also need increased worldwide by 2030. It is also anticipated that by 2030, relating to SDG11, the adverse per capita impact of cities on the environment will be substantially reduced, through proper attention to air quality, management of municipal waste, and other forms of waste management. Regarding SGD12, it is necessary to ensure that, by 2030, it is possible to manage chemicals and all wastes in an environmentally responsible manner, in compliance with international guidelines, as well as significantly reduce their release into the air, water, and soil. This will minimize their negative impacts on human health and the environment as well as significantly reduce the amount of waste generated by preventing, reducing, recycling, and reusing it as much as possible. According to SDG6, SDG11, and SDG12, there have been several sectors whose relationship has been discussed, including agricultural production, industry, municipalities, and hospitals. Agricultural production requires water and fertilizer for crop growth; however, the use of fertilizer can lead to soil and water contamination. Industry use of hazardous compounds possible to contaminate water bodies. Municipalities have an increasing need to generate waste and engage in extra burden on water and waste management. Hospitals mostly use water and possible excess pharmaceutical active compounds in water bodies.

Waste can contaminate water and must be handled properly together to avoid its presence in water bodies. As presented in the previous section, the increasing waste from 2030 to 2050 is up to 259 in the world. Water and waste have opposite relations, meaning when the presence of water availability decreases the presence of waste and wastewater increases. This also correlates with water use. However, almost no studies discuss the nexus between water and waste. Nonetheless, Goswami and Sadhu (2023) compared floating solar photovoltaic (FSPV) and land-based photovoltaic (LBPV) systems to treat the Kolkata wetland (KWL) in India which resulted in the difference costs for FSPV system focusing on floating structures ($427,050) and contingency 1,5% ($223,368) with no land cost and for LBPV system focusing on land cost ($536,250) and contingency 1,5% ($216,962) with no floating structure cost. However, it is clearly known that waste contaminates water.

4.3.3 Waste Energy Nexus

SDG7 and SDG12 are devoted to waste and energy (Puntillo 2023). According to SDG7, by 2030 access to clean energy research and technology should be available including renewable energy, energy efficiency, and advanced and cleaner technology for fossil fuels. Investment in energy infrastructure and clean energy technologies as well as expanding infrastructure and upgrading technology and promoting investment in renewable energy technologies should be promoted. As part of SDG12, waste generation should be significantly reduced through the use of preventative measures, recycling and reuse.

Instantly, the waste can be changed into valuable products, such as compost and generating energy. The nexus between waste and energy is easily found in a wide variety of sectors including municipal liquid and solid waste that can produce energy from the anaerobic digestion process, industrial waste that can produce energy from pyrolysis, as well as agricultural waste and biomass waste that can provide bioenergy. One study highlighted that thermal route and biochemical treatment have different products as fuel. Thermal waste treatment tends to result in oil, char, and syngas, whilst biochemical treatment for waste tends to produce valuable ethanol, biogas, and hydrogen (Sharma et al. 2020). However, it needs to be studied further for providing and supporting global energy availability.

4.4 WATER-WASTE-MICROBES NEXUS FOR ALLEVIATING CONTAMINANTS IN WASTEWATER

Several methods including biological methods have been proposed to eliminate contaminants in wastewater. In biological methods, microbes such as bacteria, fungi, algae, and archaea take part in alleviating contaminants in wastewater. As seen in Figure 4.4, the microbes, especially bacteria, can be differentiated as sulfate-reducing microbes, sulfammox microbes, and denitrification microbes. Sulfate-reducing microbes use carbon (C) as an energy source and convert them into methane (CH_4) and carbon dioxide (CO_2) and also convert sulfate ions (SO_4^{2-}) into sulfide ions (S^{2-}). A study assessed sulfate-reducing microbes for removing ciprofloxacin using *Paraclostridium* sp. that resulted in strong relation between sulfate ion reduction, COD reduction, and growth of microbes with R^2=0.928 (Fang et al. 2021).

Sulfammox microbes play a role in handling nitrogen and sulfur cycles in wastewater. Sulfammox microbes convert nitrate ions (NO_3^-) into nitrogen (N_2) and ammonium ions (NH_4^+) as well as change sulfate ions into sulfide ions. Microbes such as *Nitrospira*, *Nitrosomonas*, and *Candidatus Kuenenia* can remove ammonium and nitrate in more than 88% of removal percentages (Wang, Zheng, et al. 2022). Aniline was also degraded using sulfamoyl microbe, *Desulfomicrobium*, which attained up to 70% of the removal percentage from high-sulfate wastewater (Li et al.

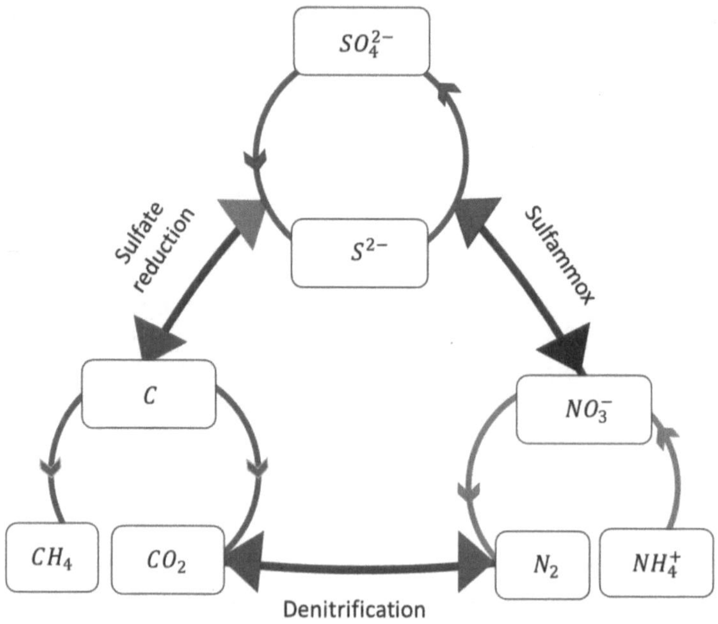

FIGURE 4.4 Role of microbial nexus alleviating contaminants in wastewater.

2020). *Desulfomicrobium* reduced sulfate from high-sulfate wastewater through dissimilatory pathways which transform sulfate into adenosine-5'-phosphosulfate (APS) by sat gene that encodes aprA and aprB, and sulfate adenylyltransferase genes to lead adenylylsulfate reductase subunit A and B alter APS into sulfite. In addition to reducing sulfate, *Desulfomicrobium* also uses aniline which contains nitrate and carbon for their optimal growth and energy source. Thus, aniline is degraded. An earlier study showed that *Bacillus benzoevorans* degraded ammonia and sulfate up to 44.4% and 40.0% (Cai, Jiang, and Zheng 2010). The removal rate for ammonium-nitrogen and sulfate-sulfur was 0.168 mg $(Lh)^{-1}$ (R^2 =0.9802) and 0.191 mg $(Lh)^{-1}$ (R^2 = 0.9525). Brocadia anammoxoglobus sulfate was also confirmed as sulfammox microbes that can be a viable alternative agent to reduce 50% sulfate in wastewater (Liu et al. 2008; Dominika, Joanna, and Jacek 2021).

Denitrification microbe is a key agent in nitrogen and carbon cycles. As seen in Figure 4.4, denitrification microbes convert nitrate ions into nitrogen and ammonium ions. Denitrification microbes use carbon for their energy source and nitrogen for their optimal growth. *Pseudomonas mendocina* S16 and *Enterobacter cloacae* DS'5 were assessed as nitrifying-denitrifying bacteria which have removal efficiencies up to 66.59 %–97.97 % for S16 and 72.27%–96.44 % for DS'5 (Shu et al. 2022). It was also reported that under the carbon source of $Na_2C_2O_4$ nitrogen removal cannot increase, while nitrogen removal achieved up to 87.15%, 70.59%, and 64.16% under the carbon source of $C_6H_5Na_3O_7$, $C_4H_4Na_2O_4$, and $C_2H_3NaO_2$. It strongly indicated that the type of carbon source potentially influences the removal process. A higher result was obtained by employing another nitrifying-denitrifying bacterium, *Rhodococcus* sp. CPZ 24, that had removal efficiencies up to 95.05% and 96.67% to eliminate ammonium and nitrate (Wang, Chen, et al. 2022). As can be seen, water-waste microbes can be used for alleviating contaminants in wastewater.

4.5 ENERGY RECOVERY PERSPECTIVES FOR SUSTAINABLE DEVELOPMENT

It is well-known that energy can be generated by waste using combustion, gasification, esterification, and anerobic digestion. Idowu et al. (2019) assessed livestock waste using a combination treatment of esterification as pre-treatment that was assisted by a microwave technology application to convert animal waste fats into biodiesel. Up to 60% of free fatty acid reduction was achieved in the first 60 min of the reaction. Microwaves were able to fast-track the rate of chemical reaction. Another waste fat which can be potentially converted into alternative energy is waste cooking oils, which are categorized as municipal wastes. Today, waste management of waste cooking oil has not received much attention. The conversion of waste cooking oil into biodiesel can be achieved using a heterogenous catalyst collected from cork biochar (Bhatia et al. 2020). There are two benefits to applying cork biochar to catalyze cooking oil waste converting process into energy from municipalities including the reduction of cork biochar as solid waste and the increasing energy generation from the esterification process of the cooking oil waste. According to fatty acids methyl esters (FAME) analysis, the catalyst from biochar could enhance the FAME percentage which presented maximum waste cooking oil to biodiesel conversion was accomplished up to 98% via transesterification reaction. The catalyst from biochar can be applied for enhancing the waste-to-energy conversion to support energy recovery for sustainable development. In addition, the catalyst can be reused for five cycles.

Aside from animal waste fats and cooking oil waste, agricultural waste such as cocoa pod husk can be converted to alternative energy, for instance, bio-oil, and promoted to increased esterification process of waste by applying its biochar. A study used a thermochemical technique to extract up to 58%wt of bio-oil from cocoa pod husk (Adjin-Tetteh et al. 2018) which confirmed that more than 50%wt of bio-oil, meaning more than 29%wt, contained 9, 12-octadecadienoic acid. In addition, 30%wt. of bio-char and 12%wt. of non-condensable gas were produced as value-added chemical product in the extraction process. However, in addition to catalyst from cork biochar, the catalyst also can be produced from the derivation of cocoa pod husk biochar.

Similarly, cocoa waste including the pericarp, exocarp, mesocarp, and a thin layer of endocarp of the fruit can be converted into valuable alternative energy through an anaerobic digestion process (Acosta et al. 2018). According to lab-scale tests, more than 50% of the chemical oxygen demand that was contained in cocoa waste was able to be recovered as biogas and methane. The results showed that dry anaerobic digestion of cocoa waste had a more stable performance compared to wet anaerobic digestion. In addition, more than 8300 MWh y^{-1} of net electricity was produced during the anaerobic digestion process and was estimated to be able to cover the electricity demand of a small district in Ecuador. Comparable results showed that cocoa waste could produce nearly 139 million MJ to 291 million MJ of electricity for 9 million kg to 19 million kg of dry cocoa waste in Uganda through an anaerobic digestion process (Zhong et al. 2020). It was estimated that nearly 40 GJ of energy can be covered and produced per year from cocoa waste in Uganda. In China, anaerobic digestion-assisted hydrothermal carbonization was also applied for converting sewage sludge and pinewood sawdust into methane (Wang, Lin, et al. 2022). The combined technique recovered the alternative energy up to 91.45%. These studies confirmed that agricultural waste, livestock waste, municipal waste, and wastewater can be used to collect energy through esterification and anaerobic digestion processes.

4.6 CHALLENGES OF ENERGY-WATER-WASTE NEXUS

As discussed, energy, water, and waste have a strong correlation with the environment (Usman et al. 2020). This correlation shows the challenges and limitations required to achieve SDGs and provide for human life as well as the environment. Water management and clean water production for saving water availability, waste and wastewater management for scaling and controlling generated waste, as well as energy production, consumption, and conservation for saving provided energies will need to be explored.

Water management and clean water production need to be taken seriously since the availability of clean water in nature is depleting and water pollution is increasing (Carrard, Foster, and Willetts 2019). Water management strategies such as reducing water use, protecting wetlands, increasing efficiency in the agricultural sector, increasing water storage, and building sustainable clean water treatment systems to produce clean water for providing water availability and avoiding water scarcity need to be emphasizzed. Actions for reducing water use include taking shorter showers, installing low-flow toilets, collecting rainwater for gardening, and reusing greywater (López Zavala, Castillo Vega, and López Miranda 2016). Protecting wetlands is important since it they remove pollutants from water. Increasing efficiency in the agricultural sector can be achieved by simply changing flood irrigation systems to sprinklers and reducing the evaporation process from the soil. Increasing water storage can store flood water, prevent its loss, and save it from scarcity during times of drought. Building sustainable clean water treatment systems ensures that clean water, ultimately freshwater, is provided and easy to access.

Treatment systems are also need for wastewater since wastewater is water that has complex combination contents consisting of 99.9% water and approximately 0.1% waste contents including organic compounds, organic matter, suspended solids (~1000 mg/L), inorganic solids, particulate matter, biological oxygen demand, chemical oxygen demand, microbes (~109 amounts/mL), micro-pollutants heavy metals, and nutrients (Jain et al. 2021). Remediation technology is needed to treat and eliminate this complex content and convert to clean water. Aside from the treatment, waste treatment is also needed for waste management including reducing potential waste generation, reusing refillable items and reusable materials, minimizing the use of paper, recycling cans, glass, cardboard, and plastics, and recovering and converting waste into energy. As mentioned waste, especially organic waste such as animal waste fats, and cooking oil waste, can to be converted into renewable energy (Awasthi et al. 2020). Thus, apart from reducing the amount of waste generation, energy is also produced and captured from waste generation.

Energy production from waste generation is one of the alternative techniques used to conserve and provide energy. Nonetheless, the consumption of energy has to be monitored since generating and collecting energy for fulfilling human needs requires great effort since energy has limited sources. Several actions regarding energy production, consumption, and conservation include turning off lights or appliances when unused, replacing incandescent bulbs or appliances with energy-efficient bulbs or appliances, using natural light from morning until evening, reducing water heating, and enabling computer energy savings systems. These small steps need to be widely applied because they have a big impact on the future and can save energy and water and avoid waste generation. The role of the government, education, and other elements of society should also be considered.

4.7 CONCLUSION

In the history of scientific and artificial processes, energy-water-waste has entered respectable overcritical concentration. Multitudinous treatments were assumed to remove pollutants from water and wastewater as well as to produce energy from the junking process since pollutants have a mischievous jolt on the terrain. There are obvious relations between waste and water due to the presence of pollutants in wastewater. Energy requirements are also a part of waste and wastewater treatment when removing pollutants from water and wastewater. The energy-water-waste nexus can offer the potentiality of saving water, reducing pollutants in water and wastewater, recovering energy from waste and water, and applying energy for sustainable evolution, but more research is needed.

ACKNOWLEDGMENTS

The author thanks the Department of Environmental Engineering in Institut Teknologi Sepuluh Nopember for facilitating this work. A thanks is also extended to the Indonesia Endowment Fund for Education, also known as LPDP (Lembaga Pengelolaan Dana Pendidikan), for supporting this work. A special thanks is also due to Mrs. Suparmi and Mr. Kusnul Yakin for their technical assistance.

REFERENCES

Acosta, Nayaret, Jo De Vrieze, Verónica Sandoval, Danny Sinche, Isabella Wierinck, and Korneel Rabaey. 2018. "Cocoa residues as viable biomass for renewable energy production through anaerobic digestion." *Bioresource Technology* 265:568–572.

Adjin-Tetteh, Maxwell, N. Asiedu, David Dodoo-Arhin, Ayman Karam, and Prince Nana Amaniampong. 2018. "Thermochemical conversion and characterization of cocoa pod husks a potential agricultural waste from Ghana." *Industrial Crops and Products* 119:304–312.

Agrawal, N., Kumar, V., Shahi, S.K., 2021. Biodegradation and detoxification of phenanthrene in in-vitro and in-vivo conditions by a newly isolated ligninolytic fungus *Coriolopsis byrsina* strain APC5 and characterization of their metabolites for environmental safety. *Environmental Science and Pollution Research*. https://doi.org/10.1007/s11356-021-15271-w

Ahmad, Shakeel, Haifeng Jia, Zhengxia Chen, Qian Li, and Changqing Xu. 2020. "Water-energy nexus and energy efficiency: A systematic analysis of urban water systems." *Renewable and Sustainable Energy Reviews* 134:110381.

Al-Ghouti, Mohammad A., Maryam A. Al-Kaabi, Mohammad Y. Ashfaq, and Dana Adel Da'na. 2019. "Produced water characteristics, treatment and reuse: A review." *Journal of Water Process Engineering* 28:222–239.

Alhaj, Mohamed, Furqan Tahir, and Sami G. Al-Ghamdi. 2022. "Life-cycle environmental assessment of solar-driven Multi-Effect Desalination (MED) plant." *Desalination* 524:115451.

Américo-Pinheiro, J.H.P., Paschoa, C.V.M., Salomão, G.R., Cruz, I.A., Isique, W.D., Ferreira, L.F.R., Sher, F., Torres, N.H., Kumar, V., Pinheiro, R.S.B., 2022. Adsorptive remediation of naproxen from water using in-house developed hybrid material functionalized with iron oxide. *Chemosphere* 289, 133222. https://doi.org/10.1016/j.chemosphere.2021.133222

Awasthi, Mukesh Kumar, Surendra Sarsaiya, Anil Patel, Ankita Juneja, Rajendra Prasad Singh, Binghua Yan, Sanjeev Kumar Awasthi, Archana Jain, Tao Liu, Yumin Duan, Ashok Pandey, Zengqiang Zhang, and Mohammad J. Taherzadeh. 2020. "Refining biomass residues for sustainable energy and bio-products: An assessment of technology, its importance, and strategic applications in circular bio-economy." *Renewable and Sustainable Energy Reviews* 127:109876.

Babayemi, Joshua O., Innocent C. Nnorom, Oladele Osibanjo, and Roland Weber. 2019. "Ensuring sustainability in plastics use in Africa: Consumption, waste generation, and projections." *Environmental Sciences Europe* 31 (1):60.

Bhatia, Shashi Kant, Ranjit Gurav, Tae-Rim Choi, Hyun Joong Kim, Soo-Yeon Yang, Hun-Suk Song, Jun Young Park, Ye-Lim Park, Yeong-Hoon Han, Yong-Keun Choi, Sang-Hyoun Kim, Jeong-Jun Yoon, and Yung-Hun Yang. 2020. "Conversion of waste cooking oil into biodiesel using heterogenous catalyst derived from cork biochar." *Bioresource Technology* 302:122872.

Cai, Jing, JianXiang Jiang, and Ping Zheng. 2010. "Isolation and identification of bacteria responsible for simultaneous anaerobic ammonium and sulfate removal." *Science China Chemistry* 53 (3):645–650.

Cao, Thanh Ngoc-Dan, Hussnain Mukhtar, Linh-Thy Le, Duyen Phuc-Hanh Tran, My Thi Tra Ngo, Mai-Duy-Thong Pham, Thanh-Binh Nguyen, Thi-Kim-Quyen Vo, and Xuan-Thanh Bui. 2023. "Roles of microalgae-based biofertilizer in sustainability of green agriculture and food-water-energy security nexus." *Science of The Total Environment* 870:161927.

Carrard, Naomi, Tim Foster, and Juliet Willetts. 2019. "Groundwater as a source of drinking water in Southeast Asia and the Pacific: A multi-country review of current reliance and resource concerns." *Water* 11 (8):1605. https://doi.org/10.3390/w11081605

Changotra, Rahil, Ajay K. Ray, and Quan He. 2022. "Establishing a water-to-energy platform via dual-functional photocatalytic and photoelectrocatalytic systems: A comparative and perspective review." *Advances in Colloid and Interface Science* 309:102793.

Chandra, R., and Kumar, V., 2017. Detection of *Bacillus* and *Stenotrophomonas* species growing in an organic acid and endocrine-disrupting chemicals rich environment of distillery spent wash and its phytotoxicity. *Environmental Monitoring and Assessment* 189, 26. https://doi.org/10.1007/s10661-016-5746-9

Chen, Qian, Faheem Hassan Akhtar, Muhammad Burhan, Kumja M, and Kim Choon Ng. 2021. "A novel zero-liquid discharge desalination system based on the humidification-dehumidification process: A preliminary study." *Water Research* 207:117794.

Chen, Qian, Muhammad Burhan, Muhammad Wakil Shahzad, Doskhan Ybyraiymkul, Faheem Hassan Akhtar, Yong Li, and Kim Choon Ng. 2021. "A zero liquid discharge system integrating multi-effect distillation and evaporative crystallization for desalination brine treatment." *Desalination* 502:114928.

Dominika, Grubba, Majtacz Joanna, and Mąkinia Jacek. 2021. "Sulfate reducing ammonium oxidation (SULFAMMOX) process under anaerobic conditions." *Environmental Technology & Innovation* 22:101416.

Doornbusch, G.J., M. Bel, M. Tedesco, J.W. Post, Z. Borneman, and K. Nijmeijer. 2020. "Effect of membrane area and membrane properties in multistage electrodialysis on seawater desalination performance." *Journal of Membrane Science* 611:118303.

Dura, Codruța Cornelia, Ana Maria Mihaela Iordache, Alexandru Ionescu, Claudia Isac, and Teodora Odett Breaz. 2022. "Analyzing performance in wholesale trade Romanian SMEs: Framing circular economy business scenarios." Sustainability 14 (9):5567. https://doi.org/10.3390/su14095567

ElHelw, Mohamed, Wael M. El-Maghlany, and Walaa M. El-Ashmawy. 2020. "Novel sea water desalination unit utilizing solar energy heating system." *Alexandria Engineering Journal* 59 (2):915–924.

El Messaoudi, N., El Khomri, M., El Mouden, A., Bouich, A., Jada, A., Lacherai, A., Iqbal, H.M.N., Mulla, S.I., Kumar, V., Pinê Américo-Pinheiro, J.H., 2022. Regeneration and reusability of non-conventional low-cost adsorbents to remove dyes from wastewaters in multiple consecutive adsorption-desorption cycles: a review. *Biomass Conversion and Biorefinery*. https://doi.org/10.1007/s13399-022-03604-9

Fang, Heting, Akashdeep Singh Oberoi, Zhiqing He, Samir Kumar Khanal, and Hui Lu. 2021. "Ciprofloxacin-degrading Paraclostridium sp. isolated from sulfate-reducing bacteria-enriched sludge: Optimization and mechanism." *Water Research* 191:116808.

Fetanat, Abdolvahhab, Mohsen Tayebi, and Hossein Mofid. 2021. "Water-energy-food security nexus based selection of energy recovery from wastewater treatment technologies: An extended decision making framework under intuitionistic fuzzy environment." *Sustainable Energy Technologies and Assessments* 43:100937.

Friedrich, Jasmin, Witold-Roger Poganietz, and Helmut Lehn. 2020. "Life-cycle assessment of system alternatives for the Water-Energy-Waste Nexus in the urban building stock." *Resources, Conservation and Recycling* 158:104808.

Ghani, Hafiz Usman, Thapat Silalertruksa, and Shabbir H. Gheewala. 2019. "Water-energy-food nexus of bioethanol in Pakistan: A life cycle approach evaluating footprint indicators and energy performance." *Science of The Total Environment* 687:867–876.

Goswami, Anik, and Pradip Kumar Sadhu. 2023. "Adoption of floating solar photovoltaics on waste water management system: a unique nexus of water-energy utilization, low-cost clean energy generation and water conservation." *Clean Technologies and Environmental Policy* 25 (2):343–368.

Hannan, M.A., R.A. Begum, Ali Q. Al-Shetwi, P.J. Ker, M.A. Al Mamun, Aini Hussain, Hassan Basri, and T.M.I. Mahlia. 2020. "Waste collection route optimisation model for linking cost saving and emission reduction to achieve sustainable development goals." *Sustainable Cities and Society* 62:102393.

He, Chunyang, Zhifeng Liu, Jianguo Wu, Xinhao Pan, Zihang Fang, Jingwei Li, and Brett A. Bryan. 2021. "Future global urban water scarcity and potential solutions." *Nature Communications* 12 (1):4667.

Idowu, Ibijoke, Montserrat Ortoneda Pedrola, Steve Wylie, K.H. Teng, Patryk Kot, David Phipps, and Andy Shaw. 2019. "Improving biodiesel yield of animal waste fats by combination of a pre-treatment technique and microwave technology." *Renewable Energy* 142:535–542.

Jain, Keerti, Anand S. Patel, Vishwas P. Pardhi, and Swaran Jeet Singh Flora. 2021. "Nanotechnology in Wastewater Management: A New Paradigm Towards Wastewater Treatment." 26 (6):1797.

Ke, Jing, Nina Khanna, and Nan Zhou. 2022. "Analysis of water–energy nexus and trends in support of the sustainable development goals: A study using longitudinal water–energy use data." *Journal of Cleaner Production* 371:133448.

Kiziltan, Mustafa. 2021. "Water-energy nexus of Turkey's municipalities: Evidence from spatial panel data analysis." *Energy* 226:120347.

Kumar, V., Shahi, S.K., Ferreira, L.F.R., Bilal, M., Biswas, J.K., Bulgariu, L., 2021. Detection and characterization of refractory organic and inorganic pollutants discharged in biomethanated distillery effluent and their phytotoxicity, cytotoxicity, and genotoxicity assessment using *Phaseolus aureus* L. and *Allium cepa* L. *Environmental Research* 201, 111551. https://doi.org/10.1016/j.envres.2021.111551

Lazaro, Lira Luz Benites, Leandro Luiz Giatti, Celio Bermann, Angelica Giarolla, and Jean Ometto. 2021. "Policy and governance dynamics in the water-energy-food-land nexus of biofuels: Proposing a qualitative analysis model." *Renewable and Sustainable Energy Reviews* 149:111384.

Lefeuvre, Anaële, Sébastien Garnier, Leslie Jacquemin, Baptiste Pillain, and Guido Sonnemann. 2019. "Anticipating in-use stocks of carbon fibre reinforced polymers and related waste generated by the wind power sector until 2050." *Resources, Conservation and Recycling* 141:30–39.

Li, Jun, Ying Liang, Yu Miao, Depeng Wang, Shuyu Jia, and Chang-Hong Liu. 2020. "Metagenomic insights into aniline effects on microbial community and biological sulfate reduction pathways during anaerobic treatment of high-sulfate wastewater." *Science of The Total Environment* 742:140537.

Liu, Sitong, Fenglin Yang, Zheng Gong, Fangang Meng, Huihui Chen, Yuan Xue, and Kenji Furukawa. 2008. "Application of anaerobic ammonium-oxidizing consortium to achieve completely autotrophic ammonium and sulfate removal." *Bioresource Technology* 99 (15):6817–6825.

Liu, Yufei, Yuanchao Hu, Meirong Su, Fanxin Meng, Zhi Dang, and Guining Lu. 2020. "Multiregional input-output analysis for energy-water nexus: A case study of Pearl River Delta urban agglomeration." *Journal of Cleaner Production* 262:121255.

Lodhia, Sumit, Amanpreet Kaur, and Sanjaya Chinthana Kuruppu. 2022. "The disclosure of sustainable development goals (SDGs) by the top 50 Australian companies: Substantive or symbolic legitimation?" *Meditari Accountancy Research* ahead-of-print (ahead-of-print).

López Zavala, Miguel Ángel, Ricardo Castillo Vega, and Rebeca Andrea López Miranda. 2016. "Potential of Rainwater Harvesting and Greywater Reuse for Water Consumption Reduction and Wastewater Minimization." 8 (6):264.

Pan, Shu-Yuan, Andrew Z. Haddad, Arkadeep Kumar, and Sheng-Wei Wang. 2020. "Brackish water desalination using reverse osmosis and capacitive deionization at the water-energy nexus." *Water Research* 183:116064.

Pandey, Asha, and Muhammad Asif. 2022. "Assessment of energy and environmental sustainability in South Asia in the perspective of the Sustainable Development Goals." *Renewable and Sustainable Energy Reviews* 165:112492.

Puntillo, Pina. 2023. "Circular economy business models: Towards achieving sustainable development goals in the waste management sector—Empirical evidence and theoretical implications." *Corporate Social Responsibility and Environmental Management* 30 (2):941–954.

Sharma, Kapil Dev, and Siddharth Jain. 2020. "Municipal solid waste generation, composition, and management: the global scenario." *Social Responsibility Journal* 16 (6):917–948.

Sharma, Surbhi, Soumen Basu, Nagaraj P. Shetti, Mohammadreza Kamali, Pavan Walvekar, and Tejraj M. Aminabhavi. 2020. "Waste-to-energy nexus: A sustainable development." *Environmental Pollution* 267:115501.

Shu, Hu, Huiming Sun, Wen Huang, Yang Zhao, Yonghao Ma, Wei Chen, Yuping Sun, Xiaoying Chen, Ping Zhong, Huirong Yang, Xiaopeng Wu, Minwei Huang, and Sentai Liao. 2022. "Nitrogen removal characteristics and potential application of the heterotrophic nitrifying-aerobic denitrifying bacteria *Pseudomonas mendocina* S16 and Enterobacter cloacae DS'5 isolated from aquaculture wastewater ponds." *Bioresource Technology* 345:126541.

Toczyłowska-Mamińska, Renata, and Mariusz Ł. Mamiński. 2022. "Wastewater as a renewable energy source— utilisation of microbial fuel cell technology." 15 (19):6928.

Usman, Ahmed, Sana Ullah, Ilhan Ozturk, Muhammad Zubair Chishti, and Syeda Maria Zafar. 2020. "Analysis of asymmetries in the nexus among clean energy and environmental quality in Pakistan." *Environmental Science and Pollution Research* 27 (17):20736–20747.

Vergara-Araya, Mónica, Helmut Lehn, and Witold-Roger Poganietz. 2020. "Integrated water, waste and energy management systems – A case study from Curauma, Chile." *Resources, Conservation and Recycling* 156:104725.

Wang, Jingli, Peizhen Chen, Shaopeng Li, Xiangqun Zheng, Chunxue Zhang, and Wenjie Zhao. 2022. "Mutagenesis of high-efficiency heterotrophic nitrifying-aerobic denitrifying bacterium *Rhodococcus* sp. strain CPZ 24." *Bioresource Technology* 361:127692.

Wang, Ruikun, Kai Lin, Daomeng Ren, Pingbo Peng, Zhenghui Zhao, Qianqian Yin, and Peng Gao. 2022. "Energy conversion performance in co-hydrothermal carbonization of sewage sludge and pinewood sawdust coupling with anaerobic digestion of the produced wastewater." *Science of The Total Environment* 803:149964.

Wang, Yanfei, Xiaona Zheng, Guangxue Wu, and Yuntao Guan. 2022. "Removal of ammonium and nitrate through Anammox and FeS-driven autotrophic denitrification." *Frontiers of Environmental Science & Engineering* 17 (6):74.

Yadav, Anshul, Pawan Kumar Labhasetwar, and Vinod Kumar Shahi. 2021. "Fabrication and optimization of tunable pore size poly(ethylene glycol) modified poly(vinylidene-co-hexafluoropropylene) membranes in vacuum membrane distillation for desalination." *Separation and Purification Technology* 271:118840.

Yadav, Sanjeev, Priyanka Katiyar, Mohammed K. Al Mesfer, and Mohd Danish. 2023. "Syngas production from thermochemical conversion of mixed food waste: A review." *Wiley Interdisciplinary Reviews Energy and Environment* n/a (n/a):e468.

Yu, Zhang, Syed Abdul Rehman Khan, Pablo Ponce, Hafiz Muhammad Zia-ul-haq, and Katerine Ponce. 2022. "Exploring essential factors to improve waste-to-resource recovery: A roadmap towards sustainability." *Journal of Cleaner Production* 350:131305.

Zeeshan, Muhammad, Jiabin Han, Alam Rehman, Irfan Ullah, and Fakhr E. Alam Afridi. 2021. "Exploring asymmetric Nexus between CO_2 emissions, Environmental pollution, and household health expenditure in China." *Risk Management and Healthcare Policy* 14:527–539.

Zhang, Xu, Xiaotong Yang, Peixun Guo, Xingjie Yao, Haibing Cong, and Bing Xu. 2023. "Solar interfacial evaporation at the water–energy Nexus: Bottlenecks, approaches, and opportunities." *Solar RRL*,7(9):2201098. https://doi.org/10.1002/solr.202201098

Zhong, Yuan, Rui Chen, Juan-Pablo Rojas-Sossa, Christine Isaguirre, Austin Mashburn, Terence Marsh, Yan Liu, and Wei Liao. 2020. "Anaerobic co-digestion of energy crop and agricultural wastes to prepare uniform-format cellulosic feedstock for biorefining." *Renewable Energy* 147:1358–1370.

5 Recent Advances in Conventional and Modern Wastewater Treatment Approaches

Ramin Yazdani, Sharareh Harirchi, Mandana Lak, Mojdeh Mirshafiei, Seyed Ali Nojoumi, Mohadasseh Ramezani, Behnam Rasekh, and Fatemeh Yazdian

5.1 INTRODUCTION

The uncontrollable production of various wastes in the world is limiting natural and clean resources and causing irreparable changes on the planet. However, the necessity of urbanization and rapid industrialization results in the wide use of different sources, minerals, chemicals, etc., which results in waste generation whether we like it or not. The generated wastes considerably contaminate water, soil, and air. Each living thing relies entirely on natural water supplies; hence, water pollution is today a critical concern for many nations. Every day, millions of tons of pollutants are released into natural water streams, violating legal regulations, and severely damaging the environment (Agrawal et al., 2021; El Messaoudi et al., 2022; Bhat et al., 2023). In defiance of justifiable requirements, large amounts of unsafe substances are discharged into standard water bodies and the climate is being polluted (Yang & Rose, 2003). Due to their detrimental impact on aquatic systems, heavy metals are particularly concerning. The primary source of these nondegradable heavy metals is the effluent discharge of industrial waste (Livingstone, 2001). When hazardous material is improperly buried in the soil, it can cause contamination. Poisonous substances within the waste influence the organic activities of the soil surface, rendering it harmful and sterile. This can disrupt crop production and contaminate groundwater. Crops grown on toxic terrain concentrate toxins and cause humans who consume them to develop potentially fatal diseases (Awasthi et al., 2020).

In addition xylene, acetaldehyde, arsenic, benzene, mercury, styrene, lead, formaldehyde, benzopyrene, 1,3-butadiene, polychlorinated biphenyls (PCBs), dioxins, and polycyclic aromatic hydrocarbons (PAHs) are hazardous wastes that can pollute the air. These contaminants are left behind whenever wastes are disposed of landfilled without sufficient treatment, and under specific circumstances, they start to burn.

DOI: 10.1201/9781003441069-5

Dissolved gases in water are distributed in clouds and also pollute rainwater. In the absence of pyrolysis and combustion, comparable discussed outflow scenarios, such as the volatilization of substances at adequately tall vapor weights and the discharge of methane and other hydrocarbons as a result of anaerobic absorption, can happen. When compounds have moderately high fluid solvency, vapor weight may not precisely anticipate their fugacity from the soil or other unconsolidated media at the transfer location. Here, avoidance may be more exact than Henry's law (Hester & Harrison, 2002: Vallero, 2014).

To overcome and manage waste, it is principally required to reduce waste production from the origin; however, it is essential to create novel and effective methods and upgrade conventional approaches for the treatment of unavoidable generated wastes. In this chapter, firstly, waste definition and classification, associated problems, and different wastewater treatment strategies are given. Recent advances in waste management and treatment approaches are covered and other parts related to integrated technologies and methods of risk analysis are thoroughly discussed.

5.2 CLASSIFICATION OF WASTES

Developing robust and cost-effective waste management strategies requires a clear and in-depth understanding of waste quantities and characteristics. In some more developed countries, waste characterization and quantification are the basis of management and intervention. However, elsewhere little emphasis has been placed on systematic waste surveys; so, characteristics, quantities, seasonal variations, as well as future trends of waste production remain poorly understood. Despite the lack of comprehensive and coherent information, some general trends and common elements are discernible (Wilson et al., 2012; Zhu et al., 2021).

The simplest definition of waste can refer to an unwanted or surplus product or material with little or no value for its owner who wants to throw it away. However, defining a material or product as waste depends on various factors such as time, price, income level, ethical background, religion, personal preference of the owner, geographical location, climate, culture, age or shelf life, and the type and amount of damage. Therefore, it is not easy to make a precise definition for what turns into waste. For example, the European Union provides many regulations and guidelines to define waste due to its imperative subjectivity. Waste can be categorized as solid, gas, or liquid; however, solid wastes have a broader range, including chemicals, materials, and compounds that are not wastewater or flue gases (Christensen, 2011). Generally, compared to developed countries, developing countries produce much less waste per capita. Despite this, managing even small amounts of waste can be challenging. Therefore, a comprehensive understanding of solid waste classification seems necessary. As shown in Figure 5.1, typically, solid wastes can be classified into four main categories based on their types and sources: municipal waste, industrial waste, agricultural waste, and hazardous waste (Wilson et al., 2012; Zhu et al., 2021). Hazardous wastes are often explosive, highly oxidizing, harmful to health, ignitable, radioactive, infectious, irritant, toxic, mutagenic, corrosive, or carcinogenic and can harm humans, plants, animals, and other forms of life in lands, aquatic environments, and even air (Neksumi et al., 2022).

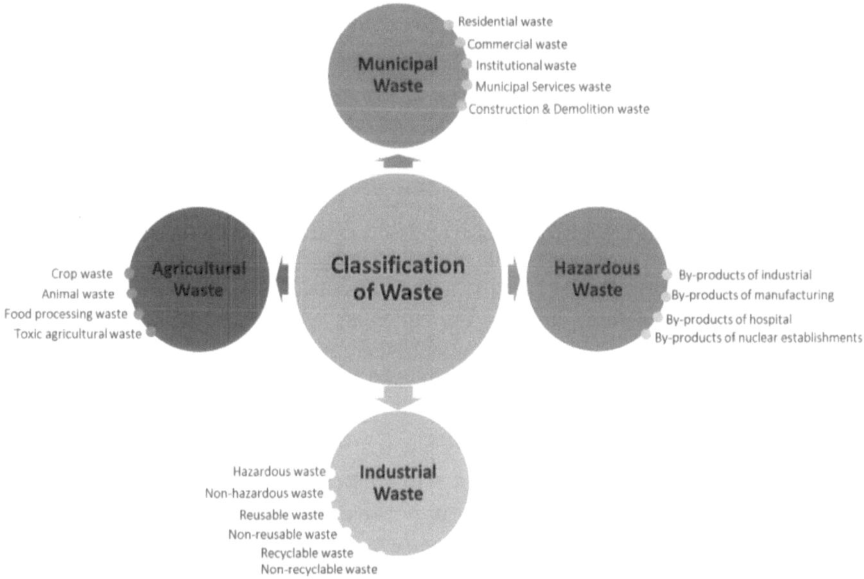

FIGURE 5.1 Classification of wastes.

5.2.1 Municipal Waste

Population growth and the rise of urbanization play a significant role in increasing the amount of municipal solid waste (MSW), which is a byproduct of modern lifestyles. Most MSW is generated by households, hotels, schools, shops, offices, and other establishments. Residential waste such as plastic, food waste, glass, paper, metal, demolition waste, construction debris, and small amounts of hazardous waste like batteries, electric light bulbs, automotive parts, as well as discarded medicines and chemicals comprise MSW. The quantity, composition, nature, and generation rates of MSW correlate with geographical location, season, economic development, activity, culture, etc. Additionally, some municipal authorities characterize and report waste types as part of the municipal waste stream differently. Moreover, urban organic waste is directly influenced by the number of trees and shrubs planted in public places. An appropriate apperceive of the composition and volume of MSW generated results in suitable management practices and provides an approximate idea about energy harvesting from MSW. Recycling or resource recovery is a viable solution that reduces the consumption of raw materials, minimizes greenhouse gas (GHG) emissions, and reduces the amount of waste disposed of, as well as serving as a promising potential energy source. Moreover, waste-to-energy technologies can play a crucial role in the (bio)economy of a region (Vergara & Tchobanoglous, 2012; Shah et al., 2021).

5.2.2 Industrial Waste

Wastes generated during industrial production stepwise within the range of industrial activities are known as industrial solid wastes (ISW), which include hazardous, nonhazardous, recyclable, nonrecyclable, reusable, and nonreusable waste. For example, fly ash, coal gangue, boron sludge, carbide slag, red mud, flue-gas desulfurization gypsum, and aluminum dust are components of ISW (Kumar et al., 2022a, 2022b; Kumar et al., 2021). Depending on raw materials, type of industry, production capacity, and production process technology, the obtained ISW includes different types; therefore, ISW classification is a significant step in identifying and addressing its environmental impacts. Furthermore, ISW can be classified into four types according to nature, pollution characteristics, industrial sectors, and industrial processes. According to the nature of ISW, it is divided into two main groups: organic industrial solid waste (OISW) and inorganic industrial solid waste (IISW). In OISW, waste is either organic or contains organic matter in its chemical composition. Examples of OISW include industries such as water treatment stations, wood manufactories, food preservation, oil extraction, plastic, dying, painting, and tanning. In IISW, the wastes have an inorganic nature or contain inorganic chemical compositions such as ceramic, cement, marble, and granite manufactories. It should be mentioned that IISW exhibits more environmental impact and hazardous effects (Soliman et al., 2019; Zhao et al., 2022).

ISW is classified as hazardous and nonhazardous, based on pollution characteristics. Hazardous industrial solid waste represents waste that exhibits chemical reactivity, toxicity, explosiveness, erosion, or other properties that may have significant effects on both health and/or the environment. In contrast, nonhazardous industrial solid waste refers to waste that does not have hazardous properties and does not require special attention for handling and removal. According to industrial sectors, ISW is classified into various industries; namely mining industry solid waste, metallurgical industry solid waste, chemical industry solid waste, food preservation industry solid waste, and construction material industry solid waste. Finally, corresponding to the industrial process, ISW is classified into fired industrial solid waste which represents all waste produced after the firing step of the industrial procedure; and unfired industrial solid waste which comprises all wastes obtained from the industrial procedure without the firing step. As with MSW, worldwide industrial development and increasing ISW production, and also the inadequacy of existing ISW collection, processing, and disposal systems in many countries pose serious challenges. On the other hand, massive accumulation of ISW not only threatens land resources but also pollutes the air. Therefore, precautions for the management of ISW are necessary and feasible (Soliman et al., 2019; Soliman & Moustafa, 2020).

5.2.3 Agricultural Waste

Developments in agriculture are generally accompanied by residual flows from the cultivation and processing of raw agricultural products. Although agricultural nonproduct outputs may contain beneficial substances for humans, their economic value is lower than the cost of harvesting, processing, and transportation. Agricultural waste includes crop residues, food processing waste, animal waste, and hazardous or toxic agricultural waste (such as pesticides, insecticides, herbicides, etc.) depending on the type of agricultural activities performed. Expanding agricultural production has naturally resulted in increased amounts of animal waste, residues of agricultural products, and agricultural and industrial byproducts. As a result, farmers are often forced to use pesticides to kill insects and prevent epidemic diseases. The packages of these pesticides with the residue of the main contents are usually left in nature; and because of toxic chemicals, they have the potential to cause environmental consequences. In addition, the use of fertilizers to maintain crop quality and productivity results in the uptake of fertilizer compounds such as nitrogen, potassium, and phosphorus in the soil, which also enters ponds, lakes, and/or rivers through surface runoff or irrigation systems and leads to water pollution. Furthermore, livestock wastes such as manure and organic materials in slaughterhouses and also aquaculture wastes are included in this group (Obi et al., 2016).

5.2.4 Hazardous Waste

With continued development in industry, commerce, agriculture, hospitals, and medical facilities, a broad spectrum of toxic chemicals and hazardous wastes are produced. These hazardous wastes are mainly byproducts of industrial, manufacturing, agricultural, hospital, and nuclear establishments. Taking into consideration that the large-scale producers of various hazardous waste sources, such as petrochemicals, chemicals, metal, paper, leather, textile, and power plants produce a large amount of hazardous waste daily, this poses serious threats to the ecosystem. Additionally, medium and small industries such as automotive and equipment workshops, metal plating, textile factories, finishing workshops, hospitals, and health centers play remarkable roles in the production of hazardous waste. However, the types, quantities, and sources of hazardous waste are influenced by the diversity and extent of industrial activities and thus, vary by geographic region (Akpan & Olukanni, 2020; Pappu et al., 2007).

5.3 WASTEWATER TREATMENT STRATEGIES AND APPROACHES

Wastewater treatment strategies and approaches are the most important phase in waste management optimization. However, to remove or significantly reduce the amount of waste before disposal, various unique approaches should be created after considering their effects on the environment. Source reduction, product substitution, and recycling are further categories for these procedures, which are also referred to as waste minimization approaches (Rao et al., 2016). Eliminating waste before it is generated is known as source reduction and is commonly referred to as waste or pollution avoidance. It is an efficient method for reducing waste. The product modification process is the most thought-provoking method because it affects product quality, changes the manufacturing process, machines, and equipment, and increases manufacturing cost (Tomić et al., 2022).

However, altering a product through the configuration, design, or specifications of the final product, crude materials, or basic industrialized process can lead straightforwardly to waste reduction. Recovering nonhazardous waste components and using them as raw materials or feedstock for other industrial processes is called waste recycling. It further stipulates that noncontaminating waste components should be recovered for reuse. Every material has a unique recycling process, mainly used to recover metals from electronic waste (e-waste).

For example, out of 1,000 kilograms (kg) of used mobile phones, 3.5 kg of silver, 0.34 kg of gold, and over 130 kg of copper can be recycled. Therefore, waste recycling must operate carefully (Baskar, 2020; Cheremisinoff, 2003).

By the end of 2020, it was predicted that the worldwide e-waste management business would generate $62.5 billion in income annually through reuse (Rao et al., 2020). Life cycle assessment (LCA) is a popular strategy for reduction. This method considers all resource and energy inputs over the lifetime of a material and helps identify waste generation, pollutants produced, and potential environmental remediation through alternative means. This research will allow us to change processes, replace chemicals with more environmentally friendly ones, and dispense with or indeed avoid the generation of dangerous waste (Hong et al., 2017).

5.3.1 Waste Treatment Processes

Instead of placing waste in landfill, waste treatment involves recycling or destroying it. To find efficient and cost-effective waste management methods, there is a developing process that is rife with experimentation (Kumar & Kumar, 2022). For instance, when treating hazardous waste, physical processes like centrifugation, mechanical collection, coagulation, sedimentation, filtration, leaching, etc., are first used to reduce the volume of the material. In addition, the biological process, which involves microorganisms digesting waste streams in bioreactors, has demonstrated its applicability to treat different waste. Uncontrolled waste incineration for energy recovery is another commonly used waste management technique. However, this practice produces a lot of air pollutants and harms human health and the environment. Therefore, this step should be performed with caution. Hazardous waste disposal can be divided into a number categories: chemical, physical, heat, and biological treatment (VanGuilder, 2018).

5.3.2 Physical Treatment

The process of physically separating hazardous materials from nonhazardous materials is known as physical treatment. The physical steps that make up the process are listed in Table 5.1.

5.3.3 Chemical Treatment

By utilizing particular chemicals such as acids, bases, oxidizing agents, and reducing agents, waste undergoes chemical treatment and becomes less hazardous. Chemical

TABLE 5.1
Different steps of physical waste treatment

Step	Gas cleaning	Liquid-solids separation	Removal of specific materials
1	Mechanical collection	Centrifugation	Adsorption
2	Electrostatic precipitation	Clarification	Crystallization
3	Fabric filter	Coagulation	Dialysis
4	Wet scrubbing	Filtration	Distillation
5	Activated carbon adsorption	Flocculation	Electrodialysis
6	Adsorption	Flotation	Evaporation
7		Foaming	Leaching
8		Sedimentation	Reverse osmosis
9		Thickening	Solvent extraction
10			Stripping

Source: Wang et al. (2019a); Blackman Jr (2016).

treatments also help recover valuable byproducts from toxic waste. As a result, the overall costs of waste disposal can be minimized. Chemical fixation, absorption, oxidation, solidification, precipitation, wet oxidation, ion exchange, neutralization, and dehalogenation are techniques used in chemical processing (Blackman Jr, 2016).

5.3.4 Biological Treatment

The chemical industry typically uses biological treatment to remove organic contaminants from effluent wastewater—the use of an enzyme or microbe to digest the organic waste. Low to moderate concentrations of organic components and less complicated inorganic chemicals are required for this to work. Aerobic systems, activated sludge, anaerobic systems, tricking filters, waste stabilization ponds, spray irrigation, and rotating bio-contactors are examples of biological treatment processes (Blackman Jr, 2016).

5.3.5 Thermal Treatment

High-temperature oxidation is a process that can be used to treat mixtures of organic and inorganic compounds. The process is called combustion and occurs in the presence of an oxidant called an incinerator.

The organic matter in the mix is transformed into nontoxic gases and safely released into the environment through this process, while the remaining inorganic buildup is dumped on land. The high temperature and air cause heavy metal sulfides and cyanides to be desorbed by oxygen and reduced to oxides, which is why this treatment works well with heavy metals. It is an economically advantageous method because it uses the heat generated by burning organic matter as fuel (Blackman Jr, 2016).

FIGURE 5.2 Disposal method. (A) Landfilling, (B) Incineration, (C) Composting.

Source: Wang & Wu (2021).

5.3.6 Advanced Waste Landfilling Processes

Waste landfilling and ocean dumping are two conventional methods for waste disposal following the maximum possible mitigation of hazardous materials in the waste. Landfill is known as a secure method that facilitates the disposal of hazardous waste in which a place isolated from the environment is provided for use for the safe and secure placement of the waste. By using this method, organic solid wastes are combined, compressed, dewatered, and dumped into natural or artificial pits where they are degraded by a variety of microbes (Fig. 5.2A). Landfills' easy operation and management, low costs, and high treatment capacity, as well as generation of landfill gas are advantages of this technology. The limitations including the availability of land for landfill siting, leachate with chemicals and heavy metals, emission of GHGs, and poor sanitation have impacted its widespread use (Wang & Wu, 2021). However, it is essential to design an appropriate site for landfilling to guarantee safe and clean disposal of waste. For this purpose, the selected site should have adequate size and be almost impermeable in order to contain large amounts of waste over a long time without unwanted penetration of the surrounding environment. Moreover, the site should be far from crowded urban areas to decrease public health risks and be near waste generation sites to reduce transportation costs (Qasim & Shareefdeen, 2022).

Ocean dumping is an old method that resembles landfill and has been used for highly toxic and reactive waste such as radioactive waste, heavy oils, and waste containing heavy metals. In this methods, hazardous waste is buried in deep oceans or sea sites at a depth of 1.5–5 km under strict regulations and restrictions (Mead, 2021; Singh & Yadav, 2022).

5.3.7 Advanced Solid Waste Treatment Processes

There has been a rapid increase in urbanization and population growth around the world, leading to large quantities of various solid wastes being produced every day, and the efficient treatment and disposal of these wastes have become major concerns for many countries (Sim & Wu, 2010). Overreliance on landfills and poor waste management have consistently plagued people with economic, safety, and health concerns. Therefore, there is a pressing need for efficient waste treatment approaches and the reduction of adverse environmental effects. In order to reduce solid waste material, limit GHG emissions, and reduce the amount of waste dumped, resource recovery provides a workable solution (Gunarathne et al., 2019). Several conventional methods (Fig. 5.3), which are discussed below, have been widely employed for solid waste treatment. For landfilling, please refer to subsection 5.3.6. Additionally, recent research in the management of solid waste introduced machine learning procedures in which difficult nonlinear processes are modeled to improve solid waste control in a sustainable concept (Xia et al., 2022).

5.3.7.1 Incineration

Incineration has remained an expensive and technically inappropriate solution to facilitate volume reduction and energy recovery. Essentially, it involves the controlled combustion of waste combined with heat recovery to produce steam, which in turn produces electricity (Fig. 5.2B). The net energy yield of this process is determined by composition, percentage of inert materials, moisture content, and waste density. By applying temperatures above 800 ◦C, thermal energy, gas, and ash are released. Byproducts of solid waste combustion, namely ash, could be utilized in construction industries such as cement and building material manufacture. However, the development of incineration plants requires constant monitoring and treatment of exhaust gases, which involves high capital, operation, and maintenance costs; while, air pollution control regulations are increasingly restricting it (Zhang et al., 2021b).

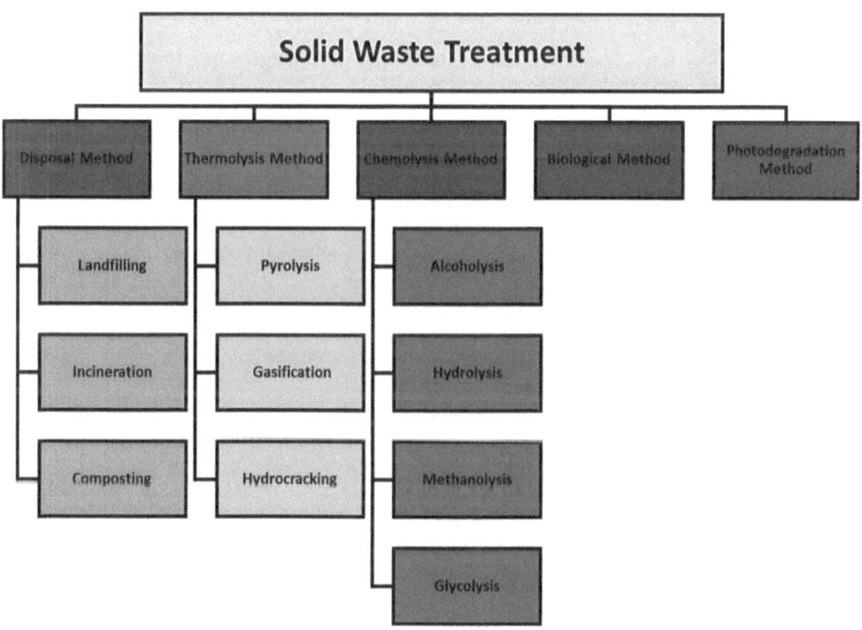

FIGURE 5.3 Classification of advanced processes for solid waste treatment.

5.3.7.2 Composting

Composting of organic waste is a convenient and long-standing biological technology conducted under aerobic conditions with the intention of reducing the quantities of solid waste and creating revenue streams from its sale (Fig. 5.2C). In this regard, organic solid waste transforms into agricultural resources such as fertilizer or soil amendment. Since composting requires high maintenance and operating costs, it is more expensive than commercial fertilizers, while the lack of material separation can cause contamination of the compost with plastic, glass, and toxic materials. In addition, the long processing period, emissions of GHGs and other organic compounds, and loss of nitrogen limit its widespread use. In this circumstance, only a small amount of compost produced is suitable for agricultural application (Wang & Wu, 2021).

5.3.7.3 Pyrolysis

As a thermal waste process treatment, pyrolysis converts long-chain polymers into smaller molecules in the absence of air, typically, at temperatures between 300–900 °C. Three states of gas, liquid oil, and char are formed during pyrolysis. Due to various parameters such as feedstock form, reactor type, composition, retention time, catalysts, temperature, etc., these valuable products are diverse. Moreover, pyrolysis can be divided into thermal pyrolysis and catalytic pyrolysis based on the use of a catalyst. The study of the pyrolysis mechanism is very useful to control pyrolysis conditions in order to achieve efficient resource recovery (Zhang et al., 2021a; Shah et al., 2021). Figure 5.4A illustrates a schematic of a simple pyrolysis procedure.

5.3.7.4 Gasification

Gasification is a process of converting carbon-based feedstock to gaseous products (containing CO_2, CO, H_2, CH_4, and other light hydrocarbons) in the presence of oxygen at high temperatures through a series of processes, including a mixture of combustible gases, small amounts of coal, ash, and synthetic gas. In general, gasification requires higher temperatures than pyrolysis; air gasification is performed at 550–900 °C, while pure oxygen, oxygen-enriched gas is done at 1000–1600 °C. Gasification processes can be an effective and clean strategy to produce hydrogen-enriched gas. Gasification processes are classified according to the type of gasifier and the oxidizing medium (Shah et al., 2021; Zhang et al., 2021a). Figure 5.4B illustrates a schematic of the gasification procedure.

5.3.7.5 Hydrocracking

Besides pyrolysis and gasification, hydrocracking is also a type of thermolysis technology. Hydrocracking is also known as hydrogenation. During this process, long-chain hydrocarbon molecules are broken down into smaller molecules, such as gasoline and kerosene by adding hydrogen under high pressure in a hydrogen atmosphere in the presence of a catalyst. Hydrocracking shows better selectivity and its products have a narrower molecular weight distribution, while it can also increase the efficiency of mass and heat transfer compered to pyrolysis (Zhang et al., 2021a). Figure 5.4C illustrates a schematic of the hydrocracking procedure.

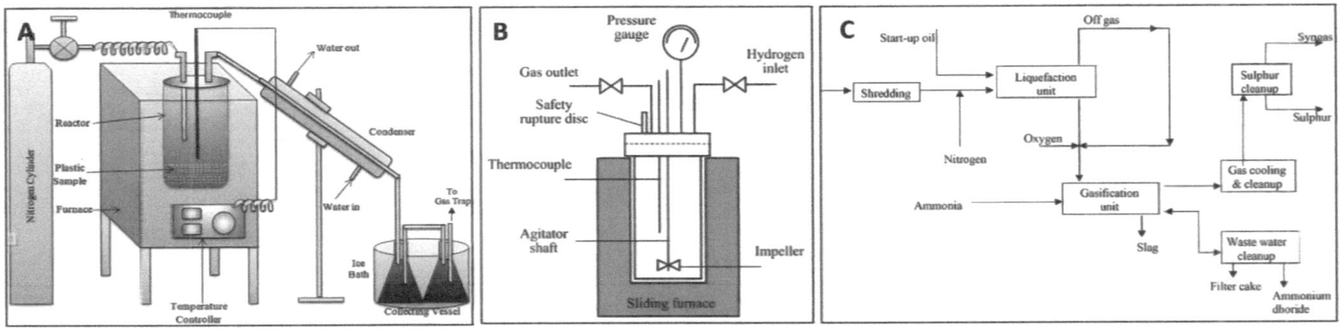

FIGURE 5.4 Schematic process of (A) pyrolysis, (B) gasification, (C) hydrocracking.

Source: Zhang et al. (2021a).

5.3.7.6 Chemolysis

Solid waste can be recycled chemically using a degradation mechanism in the presence of various degradation reagents. Chemolysis is another treatment method that targets unsaturated polyesters and resins, especially those with C-O, C-N, or other hydrophilic groups such as polyurethanes, polyesters, polyethylene terephthalate, polyamides, nylons, etc. Chemolysis is also known as solvolysis or depolymerization. During this process, individual polymers can be chemically converted into monomers at temperatures of 80–280 °C. The reaction pathways of chemolysis mainly include alcoholysis, hydrolysis, methanolysis, and glycolysis (Zhang et al., 2021a; Gadhave et al., 2019).

5.3.7.7 Biological Methods

There are a variety of biological processes available for treating solid waste based on the type of pollutant present. These processes can be categorized as either aerobic or anaerobic, each serving a unique purpose in breaking down and managing different types of pollutants efficiently. Biological treatments are usually combined sequentially with other tertiary treatment processes. As a biological method of treating organic solid wastes, anaerobic digestion has been widely used to simultaneously recycle methane, reduce volume, degrade organics, and inactivate pathogens. Compared to other methods, AD minimizes CO_2 emissions and has a lower impact on air quality. Although organic solid wastes can be used individually or in combination as substrates for AD, some digestates that contain heavy metals, microplastics (MP), and antibiotic resistance genes (ARGs) must be disposed of using another method before being used in agriculture or forestry (Wang & Wu, 2021; Zamri et al., 2021).

5.3.7.8 Photodegradation

Photodegradation is a process in which free radicals are generated by oxidizing and decomposing materials in response to sunlight and ultraviolet radiation. It is also known as photooxidative degradation in the presence of oxygen. Light in the near-ultraviolet region has the potential to cleave C-C bonds in polymer chains due to its high energy. On the other hand, the photodegradation method has an advantage over other degradation processes since it can be spatially and temporally localized and can be controlled in an easy and independent manner (Zhang et al., 2021a; Zhang et al., 2020a).

5.3.8 Advanced Wastewater Treatment Processes

In general, municipal, industrial, and animal wastes along with rain and groundwater generate wastewater or sewage that is a flow of used water. The main weight of wastewater is related to water (99.9%) and the remaining part is dissolved materials or suspended solids. Detergents, excrements, food scraps, salts, grease, oils, plastics, grits, sands, and heavy metals are all mixed in this solid substance. Some examples of wastewater types include municipal wastewater, industrial wastewater, domestic and domestic/industrial wastewater mixtures, and agricultural wastewater. Slaughterhouses, vegetable processing plants, juice and beverage factories, dairy processing plants, meat processing plants, rendering plants, and drainage water irrigation systems are examples of specific agricultural industries (Samer, 2015).

Wastewater contains critical volumes of disintegrated and colloidal particles after essential or actual treatment, which should be wiped out before release. The issue lies in devising a method for transforming dissolved substances or particulate matter into larger particles that the separation processes can eliminate. This can be accomplished with biological treatment or secondary treatment. After removing suspended solids, natural approaches treat wastewater with microorganisms like algae, fungi, or bacteria under aerobic or anaerobic conditions. During this process, organic material in the wastewater is either oxidized or absorbed into cells that can be removed by removal or sedimentation. Biological intervention is referred to as secondary treatment. Some undesirable compounds and heavy metals will react during chemical or chemical-based tertiary treatment, but some contaminating materials will not be affected. Furthermore, this treatment strategy is insufficient due to the cost of chemical-added substances and the genuine issue of managing substantial sums of chemical slime (Manikandan et al., 2022; Samer, 2015).

Instead, biological treatments should be used. This treatment process utilizes commonly occurring microorganisms to change soluble organic matter into a thick biomass that can be isolated from the treated wastewater amid the sedimentation process. The resulting slime will require far less chemical treatment since the microbes utilize the dissolved natural materials as nourishment. As a result, secondary therapy typically entails a biological process combined with a chemical treatment to get rid of harmful chemicals (Samer, 2015).

The phrase "treatment of wastewater by processes including biological treatment with post-purification" can describe post-treatment. In other words, post-processing is a biological process. Precipitated wastewater is fed into carefully designed bioreactors where microorganisms such as bacteria, algae, and fungi digest organic matter under anaerobic or aerobic conditions. The bacteria can reproduce and obtain energy from the dissolved organic matter in the bioreactor under the proper bioenvironmental conditions. As long as the microorganisms are nourished and get oxygen within the settled wastewater, the biological oxidation of dissolved natural matter will proceed. Most of the physical activity is performed by the bacteria that make up the basic trophic level—the place an organism holds in a food chain—of the food chain inside the bioreactor. Total wastewater purification can be achieved through the bioconversion of dissolved organic materials into a thick bacterial biomass. The microbial biomass must, at that point, be isolated by sedimentation from the cleaned-up pro-fluent. The distinction between this auxiliary sedimentation and essential sedimentation is that the sludge within the last mentioned contains bacterial cells contrary fecal solids. The expulsion of natural matter from settled wastewater is carried out by microorganisms, fundamentally heterotrophic microscopic organisms but once in a while, organisms. Microorganisms can break down raw materials through two natural means: natural oxidation and biosynthesis. Some end products of biological oxidation, like minerals, are left in the solution and released with the effluent. Colloidal dissolved organic matter is transformed into new cells during the biosynthetic process, producing a dense biomass that can be removed by sedimentation. On the other hand, in some circumstances, algal photosynthesis is significant (Gray, 2010; Samer, 2015).

5.3.8.1 Petroleum Wastewater Treatment

Traditional methods of oily wastewater treatment are frequently unsatisfactory because of operational challenges, high prices, the need for large amounts of industrial area, the requirement for ex situ treatment, the creation of secondary contaminants, and inadequate treatment efficiency (Adetunji & Olaniran, 2021). In order to optimize separation processes utilizing multidisciplinary approaches, scientists have promoted the use of innovative methods and materials (Abuhasel et al., 2021). For the total removal of contaminants, it may occasionally be required to use a combination of methods (Adetunji & Olaniran, 2021).

The coagulation/flocculation process differs from conventional technology in its simplicity, economy, and use of natural polymeric flocculants in the process to increase sustainability. Zhao et al. (2021) concluded that although this method is a simple method, when eliminating extremely tiny or dissolved oil particles, the desired results are challenging to obtain. In addition, a great advantage with better removal performance can be obtained by combining with the treatment method.

The main goal of Zueva et al. (2020) was to upgrade a refinery wastewater treatment system's technical framework. To increase process effectiveness at a lower cost, they considered replacing expensive filter components in refineries with coagulation. This study proved that $Ca(OH)_2$ and $Al_2(SO_4)_3$ are efficient in treating oily wastewater. $Al_2(SO_4)_3$ dose and pH were both subjected to a central composite design. Under optimal circumstances, turbidity removal reached 100%, total hydrocarbons reached 90%, and COD reached 70%. The amount of total hydrocarbon in the treated wastewater was less than the sewage limit. Charge neutralization caused by the deposition of positively charged aluminum hydroxide on particles with negative charges was the primary mechanism of aggregation. Coagulation substitutes are expensive with an annual cost of €102,600.00 per filtration procedure, resulting in a significant reduction in operating costs of up to €5,436.35/year with the same efficiency.

One of the most popular techniques for eliminating organic chemicals from wastewater is biological treatment. Bacteria are the most common pathogens for petroleum and petroleum product degradation. One of these metabolic byproducts, called a biosurfactant, helps in the breakdown of biological organic compounds by producing micelles, emulsifying substances to make them more soluble, and reducing the interfacial tension of oils (Nejad et al., 2020). With the use of an isolated salt-tolerant strain, *Pseudomonas balearica* strain Z8, Nejad et al. (2020) investigated the biological treatment of oily sludge waste. As a result, the isolated strain was found to be a biosurfactant-producing bacterium. Microorganisms and their metabolites can degrade hydrocarbons into lesser or nontoxic forms through biodegradation. This technique, known as bioremediation, is one of the main ways to eliminate petroleum pollutants from the environment in an easy and affordable manner (Sanghamitra et al., 2021).

Advanced oxidation processes (AOP) are not only effective in inactivating pathogenic microorganisms, but also show great promise because they can treat contaminated sewage with a rapid response speed and without cross-contamination (Elmobarak et al., 2021).

One of the most efficient and promising cutting-edge techniques for the in situ and ex situ transformation of high molecular weight molecules is known as supercritical hydroxylation (SCWO). This technique entails working above the water's critical point (374.3 °C and 22.12 mPa). SCWO is made up of homogenous oxidation reactions with components in solution and hydrothermal combustion that takes place in supercritical water (Krause et al., 2022). The procedure begins by adding oxygen and certain additives and pressurizing the equipment to initiate an oxidation reaction. It is then heated to the critical point temperature. Thermal energy can be recovered following the reaction to create electrical energy for industries. One of the major benefits of this method is how readily harmful substances can be broken down and transformed into safe products like CO_2, water, and N_2. While becoming a remarkably effective method, SCWO has a number of disadvantages, chiefly connected to operation and installation expenses as well as a high risk of equipment and pipe corrosion (Tang et al., 2021). A process's system corrosion is essentially unavoidable because of the circumstances needed for a reaction to occur. After excessive oxygen injection, corrosive anions are formed that accelerate the corrosion rate (Guo et al., 2021). Many SCWO plants are therefore forced to suspend or even stop operation entirely due to severe corrosion of system components. However, several salt clogging and corrosion control methods have been developed to extend equipment life, including coating equipment and pipes with corrosion-resistant materials to pre-neutralize, adsorb, or react in a fluid-solid phase (Medeiros et al., 2022).

Processes for electrochemical catalysis have a number of advantages over other methods for treating water. This is a reliable and simple technique, especially for varying wastewater loads (Muddemann et al., 2019). Graphene oxide-modified Ti/Sb-SnO_2 anodes were employed in an investigation into wastewater treatment from an offshore natural gas extraction platform. These anodes had an energy efficiency of 42.63 g/Kwh and a 58.60% reduction in chemical oxygen demand (Pahlevani et al., 2020). Despite being one of the most effective methods for treating oily water, it is difficult to treat significant amounts of wastewater due to the high price of electrodes, electrochemical catalysts, and certain coatings (Muddemann et al., 2019).

To use sophisticated treatments effectively, a number of aspects must be considered. The effectiveness of photocatalytic oxidation processes depends on the use of affordable and effective catalysts (Ma et al., 2021). The Fenton oxidation technique requires a highly acidic ambient environment with a pH below 3 for optimal process effectiveness. For the best results, an alkaline condition is needed for oxidation by ozonation. Different AOP combinations are necessary, according to Ma et al. (2021) to prevent high running costs and insufficient efficiency. Additionally, advancing the fundamentals of oxidation technology can help to improve AOP.

Biofilter/sieve membrane filters were developed to avoid the shortcomings of traditional processing methods. Research describes the use of natural polymers for their efficiency and sustainable properties (Galdino Jr et al., 2020). Microbial cellulose is one such substance. This compound has a chemically equivalent structure to plant cellulose (Lehtonen et al., 2021). It is produced extracellularly in the form of hydrated films by aerobic bacteria (Wang et al., 2019a,b). An appealing natural biomaterial with desirable qualities for a range of uses is bacterial cellulose (BC) (Hussain et al., 2019). Due to its many applications, recent research indicates that BC is an extremely impressive biopolymer as a replacement for traditional plant cellulose production (de Amorim et al., 2020). BC's hydrophilicity boosts process effectiveness and maintains the final product (water) below environmental disposal guidelines or restrictions for reusing within the same industry (Galdino Jr et al., 2020).

The ability to be washed is BC's key benefit over conventional membranes. It is possible to remove saturated membranes from the filtration system, wash, and reuse them several times without losing filtration effectiveness (Galdino Jr et al., 2020). BC has mechanical qualities that are helpful for the development of designed biomaterials due to its nanostructure and level of polymerization (de Amorim et al., 2020). These characteristics highlight the benefits of employing microbial polymers.

Galdino Jr et al. (2020) showed that BC membranes can remove petroleum from synthetic wastewater almost entirely and can be cleansed and reused over 20 times without losing their structural integrity or filtration capacity. Lehtonen et al. (2021) revealed that controlling pressure and flow parameters is crucial in optimizing the usage of BC for improved utilization as an industrial-scale filtration system. Nanometric fibers are a problem with this kind of processing. Due to this property, the BC membrane is able to completely extract the oil contained in the solution (Galdino Jr et al., 2020). However, there is the issue of modest membrane saturation because of the low porosity of the membrane and its link to the size of the oil molecules to be filtered or adsorbed. Several studies have been done to examine how different production or processing techniques might change the structure of BC (Hou et al., 2019). However, the application of such methods in the industrial sector continues to be restricted by the absence of research on membrane saturation and filtration rate improvement.

5.3.8.2 Hospital Wastewater Treatment

Hospital wastewater is extremely important because it includes high levels of antibiotics, as well as antibiotic resistance genes and bacteria (ARBs and ARGs). The COVID-19 pandemic outbreak prompted a demand for the monitoring and control of viruses and other pathogens in wastewater across the globe (Khan et al., 2021).

For the majority of pharmaceuticals in hospital wastewater (HWW), conventional procedures frequently suffer from a lack of high removal precision. Additionally, because hospital wastewater has a high organic content, AOPs are typically overpriced. As a result, numerous technologies have been investigated to eliminate such micropollutants, including hybrid techniques that combine two or more treatments (Khan et al., 2021; Pariente et al., 2022).

5.4 COMBINED PHYSICAL AND CHEMICAL TREATMENTS

Many organic molecules can be completely mineralized into CO_2, H_2O, and inorganic ions through AOPs. However, the amount of chemicals used in these procedures is typically excessive, which results in high prices. Additionally, these procedures may cause the formation of toxic byproducts. Combination strategies for improving treatment effectiveness have been demonstrated as being practical and useful by numerous investigations (Pariente et al., 2022).

Recently, Della-Flora et al. (2020) investigated the use of adsorption (using activated carbon made from avocado seeds) and the solar photo-Fenton reaction to treat HWW that had been fortified with the anticancer medication flutamide spiked in hospital effluent (C0 = 5 mg/L). Using three administrations of 5 mg/L Fe^{2+} and an initial H_2O_2 concentration of 150 mg/L over the course of two hours, the oxidation procedure enabled a 58% breakdown of the pharmaceutical. Furthermore, in the reactions, up to 13 byproducts were monitored as well.

The electro-Fenton technique was used by Dolatabadi et al. (2020) to study the removal of the nonsteroidal anti-inflammatory drug mefenamic acid that was spiked in HWW (7 mg/L). After a 12-minute reaction period, the elimination efficiency was greater than 95% utilizing 233 mg/L of H_2O_2 and 6.6 mA/cm^2 of current density.

Mahdavi et al. (2020) reported that total dissolved solids, COD, and total coliforms were removed from hemodialysis effluent using a combination of coagulation, flocculation, sedimentation, and ultrafiltration methods. Combining sedimentation and ultrafiltration procedures increased the examined parameters' removal effectiveness to > 99%.

In view of these findings, additional study is required in this area to assess the efficacy of enhanced Fenton technologies for the treatment of HWW, concentrating on the removal of pathogens from the real effluents as well as the elimination of pollutants from spiked-hospital wastewaters (Mackuľak et al., 2019).

5.5 BIOLOGICAL AND PHYSICOCHEMICAL HYBRID TREATMENTS

To reduce the discharge of pharmaceutical compounds (PhCs) into the aquatic environment, hybrid approaches have recently undergone much research and development. The majority of hybrid systems start with biologically based processes and then move on to physical or chemical ones. The most popular method for combining with biological treatments, such as membrane bioreactors or traditional activated sludge treatment, is ozonation (Pirsaheb et al., 2020).

Using MBR systems and the ozonation process, Vo et al. (2019) investigated the efficacy of a sponge-MBR combined with an ozonation process for extracting antibiotics from hospital wastewater. Norfloxacin, erythromycin, tetracycline, and trimethoprim were effectively reduced (more than 90%) among the seven medications that were studied and ciprofloxacin and ofloxacin were highly removed (over 80%). However, only 66% of sulfamethoxazole was effectively eliminated. According to the authors, the presence of sponges greatly improved the removal effectiveness of various medicines, and when compared to other biological wastewater treatment procedures, sponge-MBR and ozonation treatment may be a viable approach for eliminating antibiotics.

Tang et al. (2019) investigated a pilot-scale ozonation system post-treatment to reduce PhCs and toxicity in the effluent of a pilot-scale MBBR treating hospital wastewater. The reactor's HRT was 13.1 min with a flow rate of 1 L/min and a capacity of 13.1 L. The pretreated effluent was further refined by suspended biofilm carriers to eliminate potentially dangerous byproducts created during the ozonation process along with biodegradable organic compounds. In addition, as ozone dosage was raised, pharmaceutical concentrations declined, removing 90% of PhCs and DOC from the treated wastewater.

By combining two treatment methods (MBR and CW) with an advanced oxidation procedure in a semi-batch mode (O_3 and O_3/H_2O_2), Khan et al. (2021) investigated this relationship. Due to the complete removal of diclofenac and furosemide with both methods, this combination decreased the pharmaceuticals needed to address the inadequacies of common treatments. The lowest removal values were for ofloxacin and ibuprofen, albeit these values sharply increased with higher O_3 dosages without being affected by the hydrogen peroxide percentage. The MBR system in combination with the ozone process was found to be the most efficient among the four following functions for the pretreatment of HWW, based on the results.

Other research focuses on disinfecting HWW following biological wastewater treatment. Thus, vacuum UV (VUV) treatment was substantially more effective than UVC treatment in inactivating *Escherichia coli*, according to Moussavi et al. (2019). In the VUV and UVC photoreactors, which were run under similar circumstances with neutral solution pH, *E. coli* was inactivated by 6.4 and 3.7 logs, respectively, from an initial concentration of 1.09.1010 CFU/mL. Additionally, they investigated the procedure in a continuous-flow reactor while H_2O_2 was added to the procedure. In 10 minutes, this combined method (VUV/H_2O_2) achieved the best performance, total detergent and bacterial inactivation elimination, and a 94% TOC reduction. After being processed with the VUV/H_2O_2 method, the majority of trace compounds were broken down, and the biotoxicity significantly lowered. As a result, while MBRs often resulted in significant PhCs' elimination, most combination procedures resulted in higher trimethoprim degradation (93–100%). However, enhanced PhC degradation for those PhCs with high concentrations (analgesics like diclofenac, ibuprofen, and ketoprofen) was achieved when ozone was added to the system. Using a unique method that combined the biological treatment with a sono-photo-Fenton process, carbamazepine, diclofenac, and sulfamethoxazole—drugs known for their low biological degradation values—were completely eliminated. Additionally, biological wastewater treatment in conjugation with a membrane provided higher iohexol degradation values, taking the resistance of the contrast media into account. These combined treatments offer interesting alternatives for getting high PhCs' degradation at a cheaper cost because of the much lower oxidant dosages. Future research should, however, take into account a few alternative approaches, such as Fenton-like processes (Khan et al., 2020).

As already mentioned, people are at risk of infection due to the release of HWWs. Consequently, especially during the COVID-19 pandemic, reduced environmental and public health risks were essential. Up until now, most wastewater-based SARS-CoV-2 RNA surveillance focused on monitoring the disease in the general population by collecting samples from wastewater treatment plants (WWTPs) (D'Aoust et al., 2021).

Based on specific research, SARS-CoV-2 surveillance based on wastewater may be more effective at identifying severe cases in hospitals and providing more complete data. This suggests that wastewater sampling might be able to identify areas where COVID cases are increasing but are not being picked up by conventional techniques like individual tests. However, problems with the analysis and the intricate nature of the sample collecting still need to be addressed. Recent strategies prefer investigating single-facility assessments (Acosta et al., 2021; Gibas et al., 2021).

Subsequently being diluted in substantial amounts of wastewater, samples should be viewed as needing to be concentrated. As a result, efforts are now being made to develop appropriate methodologies. Examples include increasing the sensitivity of wastewater SARS-CoV-2 detection or improving sampling collection methods. For investigation, treatment techniques are required to guarantee virus-free treated water. The biological treatment removes more viruses than the main processes, which solely treat the wastewater by sedimentation as opposed to adsorbing it to the sludge for biological wastewater treatment. Although each procedure has a different mode of action, virus inactivation throughout most of the tertiary treatment can prevent infections. Given the widespread reports of SARS-CoV-2 RNA being shed from feces, drainage systems may be contaminated by HWW harboring virus particles. Existing wastewater treatment technologies can reduce the viral load even if more effective treatment systems still need to be created. To stop the spread of viruses, hospital wastewater could be treated at the point of production. There have only been a few investigations on SARS-CoV-2 inactivation so far. These trials demonstrated that the most effective treatments were UV radiation and chlorine disinfection. However, ozone disinfection's efficacy has not yet yielded positive results. Therefore, more research on ARBs, ARGs, and SARS-CoV-2 is still needed (Wang et al., 2020, Zhang et al., 2020b).

Hospitals are a significant source of viruses, ARGs/ARBs, and PhCs that are released into surface waterways. Additionally, if HWW is treated at the manufacturing site before entering drainage systems, viral transmission might be stopped, the levels of ARGs and ARBs could be reduced, and PhCs removed. In general, the issue cannot be resolved by any one method of controlling HWWs. HWW treatments require a lot of investigation because not all are effective at getting rid of all PhCs. In fact, a variety of tactics are frequently combined (Pariente et al., 2022).

At the pilot plant size, MBRs have produced intriguing results, with considerable removals ranging from 80% to 95% for analgesics and anti-inflammatories, cardiovascular PhCs, and several antibiotics. The treated wastewater still contains some resistant antibiotics, contrast agents, and psychiatric medications. Due to how they affect ARBs and ARGs, the latter are crucial. For the removal of PhCs from various families, including contrast media, which are resistant to biological and O_3 treatments, advanced oxidation techniques are particularly successful. However, because of HWW's extremely high flow rates, the price of the chemicals and the amount of energy consumed may be costly (Khan et al., 2020).

As a result, one could conclude that combination methods are recommended as the best way to remove PhCs, ARBs, ARGs, and even viruses like SARS-CoV-2 from hospital effluents. Prior to being released into the environment, the coupling approach may enhance the quality of the wastewater that has been treated and lessen the toxicity of the effluent. The most effective technique for the comprehensive pretreatment of hospital effluents appears to be the MBR in conjunction with ozonation treatment. Further strategies, including fungal bio-oxidation combined with Fenton-type treatments, need to be investigated at the pilot plant scale in order to validate the excellent results reached at the bench level (Pariente et al., 2022; Verlicchi, 2021).

The perfect technique recommended for eliminating PhCs, ARBs, ARGs, and even viruses like SARS-CoV-2 from hospital effluents is therefore the use of combination technologies. Before the effluent is discharged into the environment, the coupling technique may contribute to the enhancement of the wastewater that has already been treated and lessen its toxicity. The MBR in association with ozonation treatment appears to be the most efficient method for the thorough pretreatment of hospital effluents. To validate the great bench-level results, more approaches, such as fungal bio-oxidation paired with Fenton-type treatments, are required to be researched at the pilot plant scale (Pariente et al., 2022; Zhang et al., 2020a).

5.6 NUCLEAR WASTEWATER TREATMENT

As more nuclear technologies are used, radioactive wastewater is starting to become a serious problem. Such wastewater requires a much greater decontamination effort than typical industrial and municipal wastewater. Although there are many chemical, physical, and biological technologies currently in use, only a small number have been extensively tested. Decontamination of radioactive wastewater actually involves two challenges: 1) removing soluble radioactive nuclides; and 2) stabilizing/solidifying removed radioactive nuclides for disposal (Alkhadra et al., 2019).

Chiera et al. (2021) demonstrated the potential of a newly developed filtration system for the treatment of radioactive aqueous waste coming from the nuclear industry and other radiochemical processes. This material was developed on a blend of whey-protein fibrils and active charcoal. This filter membrane has previously been shown to eliminate a wide range of pollutants from wastewaters, under various chemical circumstances (Palika et al., 2021). Throughout this study, Chiera et al. (2021) concentrated on the treatment of wastewater containing 137Cs because the successful use of the above-mentioned filtering material in the cleaning of low- to medium-level radioactive aqueous waste containing other fission products can be viewed as a benchmark for the effective removal of this problematic fission product (e.g., 90Sr).

Gamma-spectrometric tests showed that the 137Cs content in 20 mL of radioactive water decreased by a factor of approximately 340 to approximately 700 Bq L1 after a single filtration cycle, with a decontamination effectiveness of 99.7%. The International Commission on Radiological Protection40 states that 10 Bq L1 is the recommended level limit for 137Cs in drinking water. Therefore, it only takes two filtering cycles in succession to make a starting sample of radioactively polluted water safe to drink (Bolisetty et al., 2020).

Alkhadra et al. (2019) used shock electrodialysis (SED), a cutting-edge electrokinetic deionization technique, to continually treat water polluted with radioactive ions. This work focused on the fundamental physics and design principles necessary to selectively remove cobalt (59Co) and cesium (133Cs) while recovering a significant part of the water injected into the process and minimizing its energy cost. The deionization shock wave phenomenon, which happens when a significant gradient in salt concentration spreads adjacent to an ion-selective surface like a metal electrodeposit or a cation exchange membrane, is the basis for this separation technique. This system also has a porous medium that is weakly charged to support an over-limiting current at which ion transport is quicker than by diffusion by itself as the conductivity of the solution declines approaching this surface (Baramanda et al., 2020). A concentrated and a deionized zone are created by the shock wave. The subsequent continuous separation of these zones is accomplished by forcing flow perpendicular to the applied electric field. With no physical obstructions in the flow direction, this technology can thus create electrically adjustable and "membraneless" separation within the porous material. Instead of common electrodialysis, where the over-limiting current is frequently maintained by chemical or hydrodynamic instabilities, surface conduction and electro-osmosis are electrokinetic phenomena that support the over-limiting current in SED at the scale of the pores (Alkhadra et al., 2019).

Along with radionuclides, low-level radioactive wastewater also contains a number of other inorganic and organic contaminants that are increasingly detrimental to both the environment and people. As a result, numerous technologies, such as chemical precipitation, evaporation and concentration, electrochemistry, and others, have been developed to purify radioactive wastewater. However, these approaches typically result in secondary pollution or insufficient purification. Membrane distillation, which has higher volume reduction efficiency than the aforementioned procedures, has become a more effective purifying technology to significantly increase the purifying effectiveness with little secondary contamination for radioactive wastewater. Membrane distillation's significant benefit encourages the development of nuclear power for a sustainable future and for the peaceful coexistence of man and nature. As a result, membrane distillation has rapidly improved, and it is now widely used to treat radioactive wastewater as well as the enormous amount of low-level radioactive wastewater discharged from nuclear power plants (Anvari et al., 2020; Jiang et al., 2022).

5.7 COMBINING MEMBRANE DISTILLATION WITH OTHER TECHNOLOGIES

Future research will concentrate on the skillful blending of MD technology with other approaches for greater and better efficiency. The following are some typical examples:

Membrane crystallization and distillation (MD-MC): The membrane distillation technique can be used to treat reverse osmosis (RO) brine that has high quantities of salt. Utilizing both inorganic salt crystallization and membrane distillation technology, it is feasible to effectively treat RO concentrated water by separating and extracting inorganic salts from concentrated inorganic salts.

Forward osmosis membrane: The feed and draw solutions are positioned on opposing sides of the separation membrane in the osmotically driven process of forward osmosis. The osmotic pressure on both sides of the membrane causes the water molecules to permeate from the feed solution to the draw solution. Extremely concentrated brine is typically used as a draw solution to ensure high osmotic pressure. A continuous positive osmosis desalination process requires water recovery from the draw solution. One of the membrane distillation process' most significant benefits is likely its effective handling of highly concentrated salt solutions, which enables the combined FO-MD process to guarantee the ongoing operation of forward osmosis. Findings on the hollow fiber membrane FO-MD combination process showed that the forward osmosis draw solution was effectively regenerated by membrane distillation when the oil-in-water emulsions (droplet diameter 1 p.m.) containing NaCl were treated by the combined FO-MD process with a water recovery rate of 90% (Guan et al., 2020).

Accelerated precipitation softening-membrane distillation (APS-MD) is the third method with the use of membrane distillation to treat concentrated brine. This strategy may recover RO concentrated brine and reduce scaling on membrane surfaces. According to the reported treatment of primary RO concentrated water via this method, the membrane flux in the combined APS-DCMD process decreases with time, but only by 20% within 300 h, in comparison to the single DCMD process. The resulting membrane distillation's treatment load and membrane contamination can be effectively reduced by APS (Jiang et al., 2022).

High-salt and low-level wastewater can be concentrated to a crystalline condition with a very high purification coefficient using membrane distillation. The direct separation of crystalline products from the solution using a membrane is now only possible with this method. Membrane distillation technology outperforms other documented methods for the treatment of low-level wastewater, particularly low-level wastewater with a high-salt concentration, thanks to these noteworthy advantages and distinctive qualities. The practical application of membrane distillation is currently hampered by a number of unresolved essential concerns, including the following: 1) The production cost of membrane distillation materials is relatively high without any cost benefit compared to other membrane separation technologies; 2) As a result of the prolonged membrane distillation process, the membrane surface becomes clogged and fouled, which eventually reduces the membrane flux and may even necessitate scraping the membrane; 3) The membrane distillation procedure uses excessive amounts of energy, which raises the cost of its operation. As a result, the following are some potential prospects for membrane distillation in the future (Jiang et al., 2022):

- Functional and effective combination of different membrane technologies and membrane distillation. For the treatment of concentrated liquids with a high-salt content, for instance, membrane distillation technology is introduced after the RO process, which can further reduce the volume of concentrated liquids and raise the concentration ratio. The front end of the membrane distillation equipment can be pretreated with microfiltration membrane technology to prevent membrane fouling and pore clogging, extending the useful life of the employed membrane.
- The optimization of membrane materials, membrane distillation process and system integration parameters, and the technique and effectiveness of waste heat utilization are all part of the development and enhancement of

membrane distillation technology. For instance, low-concentration nuclear power station wastewater has been treated using direct contact membrane distillation, which may make extensive use of the waste heat produced by these facilities, converting waste into treasure while conserving energy and lowering emissions.

- Significant reduction in energy usage for membrane distillation's industrial applications. In order to create a foundation for the logical selection of membrane fouling prevention measures, it is essential to get a thorough grasp of the mechanism of membrane fouling and common characteristics of membrane distillation. Looking ahead, the effective industrialization of membrane distillation would support environmentally friendly nuclear energy consumption and sustainable development for a peaceful coexistence between man and nature.

5.8 APPLICATION OF RISK ANALYSIS TO WASTEWATER TREATMENT TECHNOLOGIES

The WWTPs, which are a part of water management, are considered as essential parts of critical infrastructure. If their operations are stopped, effluent that has not been properly treated may be discharged into the environment. As a result, there may be health issues, soil, groundwater, and surface water contamination, etc. Establishing protection against potential risk activation, risk minimization, and improved risk-awareness are prerequisites for operational safety and the sustainability of WWTPs' operations (Łój-Pilch & Zakrzewska, 2019). According to ISO 31000, the risk assessment implementation process is divided into three parts (Vasilnakova, 2018). The identification of assets and the identification of risks were the two interconnected steps that comprise this step. Finding the assets was the first step. Reviewing the documentation served as the foundation for asset identification. The safety audit team members looked into the following internal documents: standards followed; audit results; commissioning; regular operation; shutdown occurrences; injury and accident reports; maintenance records (device inspections, testing, and inspections); characteristics of process materials (Falakh & Setiani, 2018; Tušer & Oulehlová, 2021). Identifying the internal and external hazards that influence or could influence the WWTPs' activities was the second phase. In the entire case study, this step was really important. Risks were found using a combination of document reviews, safety audits, checklists, and semi-structured interviews. Since not all the risks associated with the various subsystems at the WWTP site had been discovered during the first inquiry, the entire risk identification procedure was carried out iteratively. Working with an outside specialist on iteration was conducted. All risk identification techniques were used simultaneously throughout a total of three iterations. With the help of this strategy, the WWTPs' organization was able to implement thorough risk identification (Falakh & Setiani, 2018).

Then, hazards were identified and the effects of their operation explained. Successive iterations led to the discovery of circumstances and operational practices that could cause departures from the usual state, noncompliance with normative legal actions, injuries, and accidents. These versions made use of checklists and preplanned semi-structured interviews. A list of preplanned inquiries was sent to the WWTPs' staff by members of the safety audit. The scientists found 58 dangers, of which a sizable portion had to do with technology risk, health risk, nature risk, and human failure. Identified risk examples included: sewage and sludge discharge; decreased sludge pump performance; decreased cleaning effectiveness; electric shock; spread of contagious illnesses; fall; slip; malfunctioning software; heavy downpour (Łój-Pilch et al., 2019).

5.8.1 Risk Analysis

Understanding and defining the factors that activate risks was the aim of the risk analysis, as well as determining the degrees to which they are activated for each evaluated asset. Impacts were allocated to the levels after being described for each asset. Each risk's potential outcomes were assessed. These data were used to determine the probability of risk activation for every risk influencing the asset. In this step, semi-quantitative effect and probability description scales that were developed for the study were used. The particular degrees of risks and repercussions were assessed while accounting for the effectiveness of the measures currently in place (Tušer & Oulehlová, 2021).

The risk activation levels were established using archival data on the occurrence of risks over the preceding time period, each and every record in internal documents, and the researchers' experience. As a result, it was possible to calculate the subjective verbal probability of risk activation. In order to describe the impacts, the study considered the findings of the sensitivity, criticality, and vulnerability evaluations of specific assets. The implications were discussed verbally because the technology was complex, each of its components had a different price, and they were all necessary for effective wastewater treatment. A written explanation was also utilized because the financial evaluation of environmental damage was complicated. The length of therapy was included only in the assessment of the human health effect. Based on a mix of probability and impact, the risk assessment was done. The research team established four danger levels. A characteristic was given to each level in regard to the ensuing risk management or risk monitoring. Following the risk estimation, the results were added to the risk matrix. The R prefix denotes risk, and the number denotes the predicted risk's serial number (Łój-Pilch et al., 2019).

5.8.2 Risk Evaluation

The reference level, which was programmed to be nine in particular, was used as a comparison point between the risk analysis results and the predefined risk levels. The research team's estimate of the reference level's value based on the scope and purpose of the risk assessment were described in terms of scope, context, and criteria. Five hazards were found to be minimal and 17 risks to be acceptable as a consequence of the risk estimation. The operator was only urged to keep an eye on the dangers and continue the current countermeasures in line with the parameters of the established risk levels. Most assessed hazards exceeded the predetermined acceptable threshold.

The operator was advised to think about risk treatment alternatives for 25 unfavorable and 11 intolerable risks. The risk treatment phase must go through the risk treatment process (Tušer & Oulehlová, 2021).

5.9 ADVANCES IN INTEGRATED METHODS FOR WASTEWATER TREATMENT

The adoption of just one treatment method rarely results in adequate wastewater treatment (Kumar et al., 2022c). To achieve environmental sustainability, there is an urgent need to create a technology that is affordable, effective, and comprehensive, or to enhance the current approaches with certain methods for sufficient treatment and removal of contaminants from wastewater (Shah et al., 2022). Advanced integrated treatment methods, which combine two or more biological, physical, and chemical reactions, have recently attracted attention from all over the world for their ability to remove or clean up a variety of environmental contaminants from wastewater (Kumar et al., 2022c). The creation of a bio-based strategy for the remediation of white water using an indigenous bacterium isolate was the main topic of a study supplied by Verdel et al. (2021). For the breakdown of organic additives, the authors created a bacterial consortium consisting of *Aeromonas* sp. RES19-BTP, *Cellulosimicrobium* sp. AKD4-BF, *Sphingomonas* sp. BLA14-CF, and *Xanthomonadales* sp. CST37-CF. Even after 21 days, the consortia immobilized on carriers demonstrated a COD decrease of up to 88% in the white water in a pilot-scale 33-liter tubular reactor (Verdel et al., 2021).

Marathe et al. (2021) examined the efficacy of phytotreatment for the remediation of saline effluent produced by the pulp-paper, textile, and tannery industries. The Food and Agriculture Organization (FAO) states that wastewater with a high concentration of total dissolved solids (TDS; 2,000 mg/L) is severely restricted from being used for irrigation because it causes a complete loss in crop production via salt accumulation in crop roots. The authors employed lysimeters to treat saline wastewater with a high TDS (6,143.33 mg/L) that was helped by organic filter bedding raw material seeded with a promising plant, *Eucalyptus camaldulensis*. The plant species *E. camaldulensis* demonstrated excellent efficacy in treating wastewater while also accumulating significant amounts of salt, calcium, magnesium, and potassium. Plants generate more biomass and chlorophyll without having any adverse effects on growth as a result of wastewater. Coconut and rice husk used as bedding material in conjunction with forestry plant species showed the best TDS management efficiency when compared to other organic raw materials under investigation. The scientists concluded that using additives made from organic waste is an environmentally sound way to handle saline wastewater.

Toxic dyes used in the pharmaceutical, cosmetic, and textile sectors as well as other co-contaminants that represent a serious risk to people and the environment were reviewed by Mishra et al. (2022). Degradation, decolorization, and detoxification of contaminated wastewater discharges from the industry have thus been given top attention in response to public requests for decontaminated effluent to the receiving environment. The authors emphasize how diverse azo dyes are decolored and biodegraded by several bacterial enzymes. Additionally, the exact mechanism of bacterial enzymes in dye degradation can be explored thanks to the molecular docking of bacterial enzymes with dye molecules. The authors also talked about how isolates' molecular and physiological properties aided in bioremediation procedures. The integrated algae and bacterial methods for the degradation of tainted sewage water were summarized by Mathew et al. (2021). The standard procedures are inadequate as either an on-site or a centralized therapy because of the overwhelming blend of several refractory substances present, and specific therapies are urgently required. Because bacteria are adaptable and have dynamic metabolisms, they can be effectively used in integrated biodegradation processes. With the concurrent reduction of nutrients from wastewater, the direct application of algae also aids in the removal of hazardous heavy metals such as aluminum, nickel, copper, and others. Additionally, as they grow, their biomass may one day be converted into biofuels and other products. The wastewater's ability to be remedied was further improved by the addition of strong microorganisms. This occurred because bacteria used organic nutrients while the algal biomass assimilated the inorganic nutrients. Moreover, the intensification of algae's photosynthetic activity is aided by the mutualistic exchange of carbon dioxide and oxygen between bacterial and algal species, while bacterial oxidation-reduction processes remove nutrients from wastewater.

Generally, creating technological advancements aimed at the supply chain of sustainable processes is one of the main goals of research groups. It is imperative to create integrated treatment strategies for long-term development, applying alternative reuse techniques simultaneously to treat these residues and other harmful compounds that have developed resistance, in addition to the topic's importance in science, technology, and the environment. The most recent advances in integrated wastewater treatment systems offer a way to handle organic and emerging pollutants and boost process effectiveness, which increases the attraction of biotech and related industries because they use more renewable sources with added value. At the same time, it opens up new possibilities for biological treatment systems that can be used to treat these residues and other resistant aromatic chemicals. Improving the practicality of deployment in biorefineries involves enhancing adaptability to integrate novel processes and foster alignment with the principles of the circular bioeconomy (Kumar et al., 2022c). The new information presented by the publications will deepen our comprehension of the recent advancements in wastewater treatment using newly developed integrated methods, allowing us to build environmentally friendly treatment methods that produce safe wastewater.

5.10 CONCLUSION AND FUTURE PERSPECTIVES

Industrialization of societies and urban sprawl followed by increasing waste generation are unavoidable; and it appears they are always significant issues in the modern era. Hence, researchers and scientists contentiously develop various novel methods and upgrade traditional ones to increase waste treatment efficiency. However, more fundamental research, new policies, and innovations are required to modernize and overcome the considerable limitations of waste treatment approaches based on sustainability and the circular economy. Currently, many physicochemical and biological treatment methods are developed based on waste type and other significant parameters like economic effectiveness. It is noted that biological methods have grown at a rapid rate in comparison to physicochemical approaches, which indicates the significance of these cost-effective and innovative waste treatment methods. Moreover, during biological treatment of various wastes, the recovery of nutrients and other resources and conversion of them into value-added materials is possible. To sum up, finding innovative waste treatment methods, upgrading conventional processes, identifying new emerging wastes, assessing risks, and introducing operative waste management strategies help us and our societies to have cleaner and less polluted environments along with improved and sustainable productivity in different industries.

REFERENCES

Abuhasel, K., Kchaou, M., Alquraish, M., Munusamy, Y. & Jeng, Y. T. 2021. Oily wastewater treatment: Overview of conventional and modern methods, challenges, and future opportunities. *Water*, 13, 980.

Acosta, N., Bautista, M. A., Hollman, J., Mccalder, J., Beaudet, A. B., Man, L., Waddell, B. J., Chen, J., Li, C. & Kuzma, D. 2021. A multicenter study investigating SARS-CoV-2 in tertiary-care hospital wastewater. viral burden correlates with increasing hospitalized cases as well as hospital-associated transmissions and outbreaks. *Water Research,* 201, 117369.

Adetunji, A. I. & Olaniran, A. O. 2021. Treatment of industrial oily wastewater by advanced technologies: a review. *Applied Water Science,* 11, 1–19.

Agrawal, N., Kumar, V. & Shahi, S. K., 2021. Biodegradation and detoxification of phenanthrene in in-vitro and in-vivo conditions by a newly isolated ligninolytic fungus *Coriolopsis byrsina* strain APC5 and characterization of their metabolites for environmental safety. *Environmental Science and Pollution Research.* https://doi.org/10.1007/s11356-021-15271-w

Akpan, V. E. & Olukanni, D. O. 2020. Hazardous waste management: An African overview. *Recycling*, 5, 15.

Alkhadra, M. A., Conforti, K. M., Gao, T., Tian, H. & Bazant, M. Z. 2019. Continuous separation of radionuclides from contaminated water by shock electrodialysis. *Environmental Science & Technology,* 54, 527–536.

Anvari, A., Kekre, K. M. & Ronen, A. 2020. Scaling mitigation in radio-frequency induction heated membrane distillation. *Journal of Membrane Science,* 600, 117859.

Awasthi, M. K., Sarsaiya, S., Patel, A., Juneja, A., Singh, R. P., Yan, B., Awasthi, S. K., Jain, A., Liu, T. & Duan, Y. 2020. Refining biomass residues for sustainable energy and bioproducts: An assessment of technology, its importance, and strategic applications in circular bio-economy. *Renewable and Sustainable Energy Reviews,* 127, 109876.

Baramanda, T., Bakri, R. & Suseno, H. Cs-137 radionuclide bioaccumulation study in gold fish (*Cyprinus carpio*) through freshwater path with variations of potassium ion concentration (K+). *IOP Conference Series: Materials Science and Engineering*, 2020. IOP Publishing, 012056.

Baskar, C. 2020. *Handbook of Solid Waste Management: Sustainability Through Circular Economy*, Springer Singapore.

Bhat, S. A., Kumar, V., Kumar, S., Atabani, A. E., Štěpanec, L. & Juchelková, D., 2023. Supercapacitors Production from Waste: A new window for sustainable energy and waste management. *Fuel* 337, 127125. https://doi.org/10.1016/j.fuel.2022.127125

Blackman JR, W. C. 2016. *Basic Hazardous Waste Management*, CRC press.

Bolisetty, S., Coray, N. M., Palika, A., Prenosil, G. A. & Mezzenga, R. 2020. Amyloid hybrid membranes for removal of clinical and nuclear radioactive wastewater. *Environmental Science: Water Research & Technology,* 6, 3249–3254.

Cheremisinoff, N. P. 2003. *Handbook of Solid Waste Management and Waste Minimization Technologies*, Elsevier Science.

Chiera, N. M., Bolisetty, S., Eichler, R., Mezzenga, R. & Steinegger, P. 2021. Removal of radioactive cesium from contaminated water by whey protein amyloids–carbon hybrid filters. *RSC Advances*, 11, 32454–32458.

Christensen, T. 2011. *Solid Waste Technology and Management*, John Wiley & Sons.

D'Aoust, P. M., Mercier, E., Montpetit, D., Jia, J.-J., Alexandrov, I., Neault, N., Baig, A. T., Mayne, J., Zhang, X. & Alain, T. 2021. Quantitative analysis of SARS-CoV-2 RNA from wastewater solids in communities with low COVID-19 incidence and prevalence. *Water Research,* 188, 116560.

De Amorim, J. D. P., De Souza, K. C., Duarte, C. R., Da Silva Duarte, I., De Assis Sales Ribeiro, F., Silva, G. S., De Farias, P. M. A., Stingl, A., Costa, A. F. S. & Vinhas, G. M. 2020. Plant and bacterial nanocellulose: production, properties and applications in medicine, food, cosmetics, electronics and engineering. A review. *Environmental Chemistry Letters,* 18, 851–869.

Della-Flora, A., Wilde, M. L., Thue, P. S., Lima, D., Lima, E. C. & Sirtori, C. 2020. Combination of solar photo-Fenton and adsorption process for removal of the anticancer drug Flutamide and its transformation products from hospital wastewater. *Journal of Hazardous Materials,* 396, 122699.

Dolatabadi, M., Ahmadzadeh, S. & Ghaneian, M. T. 2020. Mineralization of mefenamic acid from hospital wastewater using electro-Fenton degradation: Optimization and identification of removal mechanism issues. *Environmental Progress & Sustainable Energy,* 39, e13380.

Elmobarak, W. F., Hameed, B. H., Almomani, F. & Abdullah, A. Z. 2021. A review on the treatment of petroleum refinery wastewater using advanced oxidation processes. *Catalysts,* 11, 782.

El Messaoudi, N., El Khomri, M., El Mouden, A., Bouich, A., Jada, A., Lacherai, A., Iqbal, H. M. N., Mulla, S. I., Kumar, V. & Pinê Américo-Pinheiro, J. H., 2022. Regeneration and reusability of non-conventional low-cost adsorbents to remove dyes from wastewaters in multiple consecutive adsorption-desorption cycles: A review. *Biomass Conversion and Biorefinery*. https://doi.org/10.1007/s13399-022-03604-9

Falakh, F. & Setiani, O. 2008. Hazard identification and risk assessment in water treatment plant considering environmental health and safety practice. *E3S web of conferences, 2018. EDP Sciences*, 06011.

Gadhave, R. V., Srivastava, S., Mahanwar, P. A. & Gadekar, P. T. 2019. Recycling and disposal methods for polyurethane wastes: A review. *Open Journal of Polymer Chemistry,* 9, 39–51.

Galdino Jr, C. J. S., Maia, A. D., Meira, H. M., Souza, T. C., Amorim, J. D., Almeida, F. C., Costa, A. F. & Sarubbo, L. A. 2020. Use of a bacterial cellulose filter for the removal of oil from wastewater. *Process Biochemistry,* 91, 288–296.

Gibas, C., Lambirth, K., Mittal, N., Juel, M. A. I., Barua, V. B., Brazell, L. R., Hinton, K., Lontai, J., Stark, N. & Young, I. 2021. Implementing building-level SARS-CoV-2 wastewater surveillance on a university campus. *Science of The Total Environment,* 782, 146749.

Gray, N. 2000. An introduction for environmental scientists and engineers. *Water Technology,* 93–231.

Guan, Q., Liu, Y., Ling, B., Zeng, G., Ji, H., Zhang, J. & Zhang, Q. 2020. Effect of magnetic field on sodium arsenate metastable zone width and crystal nucleation kinetics for crystallization. *International Journal of Chemical Kinetics,* 52, 463–471.

Gunarathne, V., Ashiq, A., Ramanayaka, S., Wijekoon, P. & Vithanage, M. 2019. Biochar from municipal solid waste for resource recovery and pollution remediation. *Environmental Chemistry Letters,* 17, 1225–1235.

Guo, S., Xu, D., Li, Y., Guo, Y., Wang, S. & Macdonald, D. D. 2021. Corrosion characteristics and mechanisms of typical Ni-based corrosion-resistant alloys in sub-and supercritical water. *Journal of Supercritical Fluids,* 170, 105138.

Hester, R. E. & Harrison, R. M. 2002. Environmental and health impact of solid waste management activities. *Royal Society of Chemistry*, 18, 2–36.

Hong, J., Han, X., Chen, Y., Wang, M., Ye, L., Qi, C. & Li, X. 2017. Life cycle environmental assessment of industrial hazardous waste incineration and landfilling in China. *International Journal of Life Cycle Assessment,* 22, 1054–1064.

Hou, Y., Duan, C., Zhu, G., Luo, H., Liang, S., Jin, Y., Zhao, N. & Xu, J. 2019. Functional bacterial cellulose membranes with 3D porous architectures: Conventional drying, tunable wettability and water/oil separation. *Journal of Membrane Science,* 591, 117312.

Hussain, Z., Sajjad, W., Khan, T. & Wahid, F. 2019. Production of bacterial cellulose from industrial wastes: A review. *Cellulose,* 26, 2895–2911.

Jiang, M., Fang, Z., Liu, Z., Huang, X., Wei, H. & Yu, C.-Y. 2022. Application of membrane distillation for purification of radioactive liquid. *Cleaner Engineering and Technology*, 12, 100589.

Khan, M. T., Shah, I. A., Ihsanullah, I., Naushad, M., Ali, S., Shah, S. H. A. & Mohammad, A. W. 2021. Hospital wastewater as a source of environmental contamination: An overview of management practices, environmental risks, and treatment processes. *Journal of Water Process Engineering,* 41, 101990.

Khan, N. A., Ahmed, S., Farooqi, I. H., Ali, I., Vambol, V., Changani, F., Yousefi, M., Vambol, S., Khan, S. U. & Khan, A. H. 2020. Occurrence, sources and conventional treatment techniques for various antibiotics present in hospital wastewaters: A critical review. *TrAC Trends in Analytical Chemistry,* 129, 115921.

Krause, M. J., Thoma, E., Sahle-Damesessie, E., Crone, B., Whitehill, A., Shields, E. & Gullett, B. 2022. Supercritical water oxidation as an innovative technology for PFAS destruction. *Journal of Environmental Engineering,* 148, 05021006.

Kumar, V., & Kumar, M. (Eds.). (2022). *Integrated Environmental Technologies for Wastewater Treatment and Sustainable Development*. Elsevier.

Kumar, V., Ameen, F., Islam, M. A., Agrawal, S., Motghare, A., Dey, A., Shah, M. P., Américo-Pinheiro, J. H. P., Singh, S. & Ramamurthy, P. C., 2022a. Evaluation of cytotoxicity and genotoxicity effects of refractory pollutants of untreated and biomethanated distillery effluent using Allium cepa. *Environmental Pollution*, 300, 118975. https://doi.org/10.1016/j.envpol.2022.118975 (IF 9.988)

Kumar, V., Agrawal, S., Bhat, S. A., Américo-Pinheiro, J. H. P., Shahi, S. K. & Kumar, S., 2022b. Environmental impact, health hazards, and plant-microbes synergism in remediation of emerging contaminants. *Cleaner Chemical Engineering*, 2, 100030. https://doi.org/10.1016/j.clce.2022.100030

Kumar, V., Bilal, M. & Ferreira, L. F. R., 2022c. Editorial: Recent trends in integrated wastewater treatment for sustainable development. *Frontiers in Microbiology*, 13, 846503. https://doi.org/10.3389/fmicb.2022.846503

Kumar, V., Shahi, S. K., Ferreira, L. F. R., Bilal, M., Biswas, J. K. & Bulgariu, L., 2021. Detection and characterization of refractory organic and inorganic pollutants discharged in biomethanated distillery effluent and their phytotoxicity, cytotoxicity, and genotoxicity assessment using *Phaseolus aureus* L. and *Allium cepa* L. *Environmental Research*, 201, 111551. https://doi.org/10.1016/j.envres.2021.111551

Lehtonen, J., Chen, X., Beaumont, M., Hassinen, J., Orelma, H., Dumée, L. F., Tardy, B. L. & Rojas, O. J. 2021. Impact of incubation conditions and post-treatment on the properties of bacterial cellulose membranes for pressure-driven filtration. *Carbohydrate Polymers,* 251, 117073.

Livingstone, D. 2001. Contaminant-stimulated reactive oxygen species production and oxidative damage in aquatic organisms. *Marine Pollution Bulletin,* 42, 656–666.

Łój-Pilch, M. & Zakrzewska, A. 2019. Analysis of risk assessment in a municipal wastewater treatment plant located in upper Silesia. *Water*, 12, 23.

Łój-Pilch, M., Zakrzewska, A. & Zielewicz, E. 2019. Risk assessment analysis in a municipal wastewater treatment plant. *Multidisciplinary Digital Publishing Institute Proceedings,* 16, 18.

Ma, D., Yi, H., Lai, C., Liu, X., Huo, X., An, Z., Li, L., Fu, Y., Li, B. & Zhang, M. 2021. Critical review of advanced oxidation processes in organic wastewater treatment. *Chemosphere,* 275, 130104.

Mackuľak, T., Grabic, R., Špalková, V., Belišová, N., Škulcová, A., Slavík, O., Horký, P., Gál, M., Filip, J. & Híveš, J. 2019. Hospital wastewaters treatment: Fenton reaction vs. BDDE vs. ferrate (VI). *Environmental Science and Pollution Research,* 26, 31812–31821.

Mahdavi, M., Mahvi, A. H., Salehi, M., Sadanid, M., Biglari, H. & Tashauoe, H. 2020. Wastewater reuse from hemodialysis section by combination of coagulation and Ultrafiltration processes: Case study in saveh-iran hospital. *Desalin Water Treat,* 193, 274–83.

Manikandan, S., Subbaiya, R., Saravanan, M., Ponraj, M., Selvam, M. & Pugazhendhi, A. 2022. A critical review of advanced nanotechnology and hybrid membrane based water recycling, reuse, and wastewater treatment processes. *Chemosphere,* 289, 132867.

Marathe, D., Raghunathan, K., Singh, A., Thawale, P. & Kumari, K. 2021. A Modified Lysimeter study for phyto-treatment of moderately saline wastewater using plant-derived filter bedding materials. *Frontiers in Microbiology*, 12.

Mathew, M. M., Khatana, K., Vats, V., Dhanker, R., Kumar, R., Dahms, H.-U. & Hwang, J.-S. 2021. Biological approaches integrating algae and bacteria for the degradation of wastewater contaminants—A review. *Frontiers in Microbiology*, 12.

Mead, L. 2021. "The Ocean Is Not a Dumping Ground" *Fifty Years of Regulating Ocean Dumping*. Accessed from www.iisd.org/articles/regulating-ocean-dumping

Medeiros, A. D. L. M. D., Silva Junior, C. J. G. D., Amorim, J. D. P. D., Durval, I. J. B., Costa, A. F. D. S. & Sarubbo, L. A. 2022. Oily wastewater treatment: Methods, challenges, and trends. *Processes,* 10, 743.

Mishra, A., Takkar, S., Joshi, N. C., Shukla, S., Shukla, K., Singh, A., Manikonda, A. & Varma, A. 2022. An integrative approach to study bacterial enzymatic degradation of toxic dyes. *Frontiers in Microbiology,* 12, 802544.

Moussavi, G., Fathi, E. & Moradi, M. 2019. Advanced disinfecting and post-treating the biologically treated hospital wastewater in the UVC/H_2O_2 and VUV/H_2O_2 processes: Performance comparison and detoxification efficiency. *Process Safety and Environmental Protection,* 126, 259–268.

Muddemann, T., Haupt, D., Sievers, M. & Kunz, U. 2019. Electrochemical reactors for wastewater treatment. *ChemBioEng Reviews,* 6, 142–156.

Nejad, Y. S., Jaafarzadeh, N., Ahmadi, M., Abtahi, M., Ghafari, S. & Jorfi, S. 2020. Remediation of oily sludge wastes using biosurfactant produced by bacterial isolate Pseudomonas balearica strain Z8. *Journal of Environmental Health Science and Engineering,* 18, 531–539.

Neksumi, M., Zishan, M., Sushmita, B. & Manzoor, U. 2022. The Global Menace of Hazardous Waste: Challenges and Management. In: Baskar, C., Ramakrishna, S., Baskar, S., Sharma, R., Chinnappan, A. & Sehrawat, R. (eds.) *Handbook of Solid Waste Management: Sustainability through Circular Economy*. Singapore: Springer Nature Singapore.

Obi, F., Ugwuishiwu, B. & Nwakaire, J. 2016. Agricultural waste concept, generation, utilization and management. *Nigerian Journal of Technology,* 35, 957–964-957–964.

Pahlevani, L., Mozdianfard, M. R. & Fallah, N. 2020. Electrochemical oxidation treatment of offshore produced water using modified Ti/Sb-SnO_2 anode by graphene oxide. *Journal of Water Process Engineering,* 35, 101204.

Palika, A., Armanious, A., Rahimi, A., Medaglia, C., Gasbarri, M., Handschin, S., Rossi, A., Pohl, M. O., Busnadiego, I. & Gübeli, C. 2021. An antiviral trap made of protein nanofibrils and iron oxyhydroxide nanoparticles. *Nature Nanotechnology,* 16, 918–925.

Pappu, A., Saxena, M. & Asolekar, S. R. 2007. Solid wastes generation in India and their recycling potential in building materials. *Building and Environment,* 42, 2311–2320.

Pariente, M., Segura, Y., Álvarez-Torrellas, S., Casas, J., De Pedro, Z., Diaz, E., García, J., López-Muñoz, M., Marugán, J. & Mohedano, A. 2022. Critical review of technologies for the on-site treatment of hospital wastewater: From conventional to combined advanced processes. *Journal of Environmental Management,* 320, 115769.

Pirsaheb, M., Mohamadisorkali, H., Hossaini, H., Hossini, H. & Makhdoumi, P. 2020. The hybrid system successfully to consisting of activated sludge and biofilter process from hospital wastewater: Ecotoxicological study. *Journal of Environmental Management,* 276, 111098.

Qasim, M. & Shareefdeen, Z. 2022. Advances in Land, Underground, and Ocean Disposal Techniques. In: Shareefdeen, Z. (ed.) *Hazardous Waste Management: Advances in Chemical and Industrial Waste Treatment and Technologies*. Cham: Springer International Publishing.

Rao, M. D., Singh, K. K., Morrison, C. A. & Love, J. B. 2020. Challenges and opportunities in the recovery of gold from electronic waste. *RSC Advances,* 10, 4300–4309.

Rao, M. N., Sultana, R., Kota, S. H., Shah, A. & Davergave, N. 2016. *Solid and Hazardous Waste Management: Science and Engineering*, Butterworth-Heinemann.

Samer, M. 2015. Biological and chemical wastewater treatment processes. *Wastewater Treatment Engineering,* 150, 212.

Sanghamitra, P., Mazumder, D. & Mukherjee, S. 2021. Treatment of wastewater containing oil and grease by biological method-a review. *Journal of Environmental Science and Health, Part A,* 56, 394–412.

Shah, A. V., Singh, A., Mohanty, S. S., Srivastava, V. K. & Varjani, S. 2022. Organic solid waste: Biorefinery approach as a sustainable strategy in circular bioeconomy. *Bioresource Technology*, 349, 126835.

Shah, A. V., Srivastava, V. K., Mohanty, S. S. & Varjani, S. 2021. Municipal solid waste as a sustainable resource for energy production: State-of-the-art review. *Journal of Environmental Chemical Engineering,* 9, 105717.

Sim, E. Y. S. & Wu, T. Y. 2010. The potential reuse of biodegradable municipal solid wastes (MSW) as feedstocks in vermicomposting. *Journal of the Science of Food and Agriculture,* 90, 2153–2162.

Singh, G. & Yadav, P. 2022. *Hazardous Waste Characteristics and Standard Management Approaches*. Hazardous Waste Management. Elsevier.

Soliman, N., Mohamed, H. S., Elsayed, R. H., Elmedny, N. M., Elghandour, A. H. & Ahmed, S. A. 2019. Removal of chromium and cadmium ions from aqueous solution using residue of Rumex dentatus L. plant waste. *Desalination and Water Treatment,* 149, 181–193.

Soliman, N. & Moustafa, A. 2020. Industrial solid waste for heavy metals adsorption features and challenges; A review. *Journal of Materials Research and Technology,* 9, 10235–10253.

Tang, K., Spiliotopoulou, A., Chhetri, R. K., Ooi, G. T., Kaarsholm, K. M., Sundmark, K., Florian, B., Kragelund, C., Bester, K. & Andersen, H. R. 2019. Removal of pharmaceuticals, toxicity and natural fluorescence through the ozonation of biologically-treated hospital wastewater, with further polishing via a suspended biofilm. *Chemical Engineering Journal,* 359, 321–330.

Tang, X., Zheng, Y., Liao, Z., Wang, Y., Yang, J. & Cai, J. 2021. A review of developments in process flow for supercritical water oxidation. *Chemical Engineering Communications,* 208, 1494–1510.

Tomić, T., Kremer, I. & Schneider, D. R. 2022. Economic efficiency of resource recovery—analysis of time-dependent changes on sustainability perception of waste management scenarios. *Clean Technologies and Environmental Policy,* 24, 543–562.

Tušer, I. & Oulehlová, A. 2021. Risk assessment and sustainability of wastewater treatment plant operation. *Sustainability,* 13, 5120.

Vallero, D. A. 2014. *Fundamentals of Air Pollution*, Academic press.

VanGuilder, C. 2018. Hazardous waste management: An introduction, Second edition. Mercury Learning and Information, Virginia, p 99–142.

Vasilnakova, A. 2018. Risk management in accredited testing laboratories. *Annals of DAAAM & Proceedings*, 29, 1071–1075.

Verdel, N., Rijavec, T., Rybkin, I., Erzin, A., Veliščček, Ž., Pintar, A. & Lapanje, A. 2021. Isolation, Identification, and Selection of Bacteria With Proof-of-Concept for Bioaugmentation of Whitewater From Wood-Free Paper Mills. *Frontiers in Microbiology*, 12, 1–15.

Vergara, S. E. & Tchobanoglous, G. 2012. Municipal solid waste and the environment: A global perspective. *Annual Review of Environment and Resources,* 37, 277–309.

Verlicchi, P. 2021. Trends, new insights and perspectives in the treatment of hospital effluents. *Current Opinion in Environmental Science & Health,* 19, 100217.

Vo, H. N. P., Koottatep, T., Chapagain, S. K., Panuvatvanich, A., Polprasert, C., Nguyen, T. M. H., Chaiwong, C. & Nguyen, N. L. 2019. Removal and monitoring acetaminophen-contaminated hospital wastewater by vertical flow constructed wetland and peroxidase enzymes. *Journal of Environmental Management,* 250, 109526.

Wang, J., Shen, J., Ye, D., Yan, X., Zhang, Y., Yang, W., Li, X., Wang, J., Zhang, L. & Pan, L. 2020. Disinfection technology of hospital wastes and wastewater: Suggestions for disinfection strategy during Coronavirus Disease 2019 (COVID-19) pandemic in China. *Environmental Pollution,* 262, 114665.

Wang, J., Shih, Y., Wang, P. Y., Yu, Y. H., Su, J. F. & Huang, C.-P. 2019a. Hazardous waste treatment technologies. *Water Environment Research,* 91, 1177–1198.

Wang, J., Tavakoli, J. & Tang, Y. 2019b. Bacterial cellulose production, properties and applications with different culture methods–A review. *Carbohydrate Polymers,* 219, 63–76.

Wang, S. & Wu, Y. 2021. Hyperthermophilic composting technology for organic solid waste treatment: Recent research advances and trends. *Processes,* 9, 675.

Wilson, D. C., Rodic, L., Scheinberg, A., Velis, C. A. & Alabaster, G. 2012. Comparative analysis of solid waste management in 20 cities. *Waste Management & Research,* 30, 237–254.

Xia, W., Jiang, Y., Chen, X. & Zhao, R. 2022. Application of machine learning algorithms in municipal solid waste management: A mini review. *Waste Management & Research,* 40, 609–624.

Yang, H. & Rose, N. L. 2003. Distribution of mercury in six lake sediment cores across the UK. *Science of the Total Environment,* 304, 391–404.

Zamri, M., Hasmady, S., Akhiar, A., Ideris, F., Shamsuddin, A., Mofijur, M., Fattah, I. R. & Mahlia, T. 2021. A comprehensive review on anaerobic digestion of organic fraction of municipal solid waste. *Renewable and Sustainable Energy Reviews,* 137, 110637.

Zhang, F., Zhao, Y., Wang, D., Yan, M., Zhang, J., Zhang, P., Ding, T., Chen, L. & Chen, C. 2021a. Current technologies for plastic waste treatment: A review. *Journal of Cleaner Production,* 282, 124523.

Zhang, J., Zhang, S. & Liu, B. 2020a. Degradation technologies and mechanisms of dioxins in municipal solid waste incineration fly ash: A review. *Journal of Cleaner Production,* 250, 119507.

Zhang, S., Huang, J., Zhao, Z., Cao, Y. & Li, B. 2020b. Hospital wastewater as a reservoir for antibiotic resistance genes: A meta-analysis. *Frontiers in Public Health,* 8, 574968.

Zhang, Y., Wang, L., Chen, L., Ma, B., Zhang, Y., Ni, W. & Tsang, D. C. 2021b. Treatment of municipal solid waste incineration fly ash: State-of-the-art technologies and future perspectives. *Journal of Hazardous Materials,* 411, 125132.

Zhao, C., Zhou, J., Yan, Y., Yang, L., Xing, G., Li, H., Wu, P., Wang, M. & Zheng, H. 2021. Application of coagulation/flocculation in oily wastewater treatment: A review. *Science of The Total Environment*, 765, 142795.

Zhao, H.-X., Zhou, F.-S., Lm, A. E., Liu, J.-L. & Zhou, Y. 2022. A review on the industrial solid waste application in pelletizing additives: Composition, mechanism and process characteristics. *Journal of Hazardous Materials,* 423, 127056.

Zhu, Y., Zhang, Y., Luo, D., Chong, Z., Li, E. & Kong, X. 2021. A review of municipal solid waste in China: Characteristics, compositions, influential factors and treatment technologies. *Environment, Development and Sustainability,* 23, 6603–6622.

Zueva, S., Corradini, V., Ruduka, E. & Veglio, F. 2020. Treatment of petroleum refinery wastewater by physicochemical methods. *E3S Web of Conferences*, EDP Sciences, 01042.

6 Contributions of Biodegradation and Biotransformation by a Microbial Nexus in Advanced Pharmaceutical Wastewater Treatment

Jyotsna Rani Sahu, Surendra Behera, Debapriya Sarkar, Kasturi Poddar, and Angana Sarkar

6.1 INTRODUCTION

Pharmaceuticals are bio-active compounds that are of high priority for the maintenance of public health as well as quality of life. From the beginning of the nineteenth century, natural compounds were the only source of therapeutics. These shrub and plant extracts are natural medicines that are well-used for healing purposes and pain relief. For the easy production of these herbal medicines, with time many pharmaceutical corporations focused their research on the development of counterfeit products for therapeutics (Tiwari et al., 2017). As a result, the number of pharmaceutical industries has increased globally in recent decades and has become one of the ascending and consistent pollutant-causing sectors causing massive damage to the ecosystem and are described as emerging contaminants (Narayanan et al., 2022). Due to heavy usage, the effluents of pharmaceutical industries are becoming the biggest problem for our natural habitat since many of these restorative compounds cannot be analogized and digested by the human body and hence, are released through urine and feces and are becoming part of municipal wastewater treatment plants (WWTPs) (Tiwari et al., 2017). Not only in WWTPs but pharmaceutical leftovers have been located in almost all eco-friendly models. This includes sludge, groundwater, and surface water (seawater, estuaries, rivers, lakes, and streams) and they are also found in the geosphere and biosphere as well as in polar regions (Patel et al., 2019). The major sources of pharmaceutical pollutants in the environment are hospitals, animal

 DOI: 10.1201/9781003441069-6

wastes, industries, utilization of therapeutic compounds in research activities, and the discharge of invalid medicines into the environment. Hormones, antibiotics, antipyretics, analgesics, antidepressants, and products that are used for chemotherapy purposes are released from the above working fields into the environment causing significant damage to human health as well as creating disturbances among aquatic flora and fauna. The waste generated by pharmaceutical plants during the processing of products in water streams pollutes water sources like groundwater, surface water, and water bodies. In many scientific studies it has been mentioned that the presence of these pharmaceutical pollutants in the environment in minimum concentrations does not trigger critical damage but constant exposure can cause acute or chronic toxicity. Endless exposure to these pharmaceutical contaminants in the aquatic environment not only changes species' traits but also changes the behavior of aquatic organisms.

About thirty million tons of prescribed drugs are produced worldwide including 50,000 varieties that have become a serious environmental issue. Hence, the removal of pharmaceuticals and personal care products (PPCPs) from the natural surroundings is an evolving concern (Xu et al., 2022). Numerous research studies and techniques featuring the degradation of pharmaceutical contaminants have been focused on. These techniques can be categorized into two specific groups; biotic and abiotic techniques. Abiotic techniques include physicochemical processes like adsorption, thermal degradation, oxidation, condensation, and many other electrochemical processes. These are normally very overpriced and detrimental to the environment as they release toxic sludge and gas (NOX) as byproducts. Few of them degrade the chemical structure of organic contaminants but ordinarily convert them into different forms, not dealing with the genuine problem. These processes are also unable to remediate the pollutants completely. On the other hand, biotic processes are economical and environmentally safe, and include processes like bioaccumulation, biosorption, biodegradation, and bioremediation (Costa et al., 2019).

Biodegradation is the breakdown or degradation process of complex organic matter through biological activity (Kumar et al., 2018a; Agrawal et al., 2021). This encompasses many microorganisms that make the process work through metabolic and enzymatic activities. Certain microorganisms like bacteria, fungi, and yeasts have a massive impact on the biodegradation process whereas the involvement reports of protozoa and algae in biodegradation are minimal (Joutey et al., 2013). The biodegradation of pharmaceutical contaminants with the help of many bacterial species such as *Rhizopus arrhizus*, *Bacillus subtilis*, *Pseudomonas putida*, and *Bacillus thuringiensis* has previously been studied (Al-Gheethi et al., 2019). Similarly, some fungal species namely *Penicillium oxalicum* (Olicón-Hernández et al., 2019), white-rot fungi (Jureczko et al., 2021), *Trichoderma harzanium*, and *Trichoderma asperellum* (Manasfi et al., 2020) have recently been studied for successful biodegradation of pharmaceuticals.

6.2 SOURCE

Pharmaceuticals have importance worldwide as they help maintain and restore human health through treatment and diagnosis. Hence, with the increasing population, the usage of PPCPs has also increased leading to the massive production of pharmaceuticals and their corresponding molecular intermediates throughout the world. Among PPCPs, hormones, antidepressants, antibiotics, nonsteroidal anti-inflammatory drugs, disinfectants, beta-blockers, sunscreen agents, and some preservatives are common groups that are released into the environment from pharmaceutical industries. Not only in pharmaceutical industries but the continuous discharge of pharmaceutical effluents into the environment also takes place from various sources (Fig. 6.1).

These sources are further divided into two groups, i.e., point sources and diffused sources (Nguyen et al., 2019). Point source indicates the appearance of pharmaceutical pollutants in the environment as byproducts of human and/or animal waste, e.g., excretions as well as domestic, industrial wastewater, and hospital wastes. Since pharmaceutical molecules are persistent organic pollutants, they remain in high concentrations in the environment in their parent form without any degradation (Kumar et al., 2022). For these reasons, an enormous quantity of these molecules is found in wastewater treatment plants (WWTPs) worldwide. A diffused source specifies the manifestation of these micropollutants in the natural habitat using agricultural, medical, industrial, and household waste, becoming an omnipresent reason for pharmaceutical pollutants in the environment. Wastewater discharge, landfill leaching, sewage treatment plants, and WWTPs are, without exception, the prominent sources of pharmaceutical pollutants in the environment (Li, 2014). In traditional treatment processes, the degradation of pharmaceuticals is significantly low; hence, they are found freely in the environment in moderate to high concentrations.

Municipal landfills are comprehensive sources of pharmaceutical pollutants of human, environmental, and wildlife health concern (Eggen et al., 2010). The deposited waste in landfills carries a different variety of contaminants that hypercritically affect

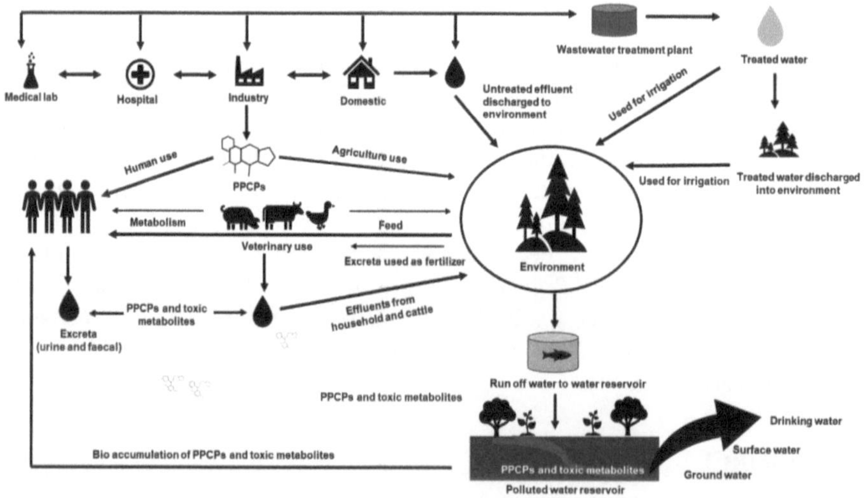

FIGURE 6.1 Different sources of pharmaceutical pollutants and their flow through the environment.

the health of the ecosystem. Municipal landfills also give rise to leachate, which is a solution of heavy metals and other organic pollutants. According to different studies, pharmaceuticals are present in high concentrations in wells close to landfills and it is tenacious in groundwater and subsequently in surface water due to the continuous flow of leachate (Li, 2014).

Similarly, wastewater released from different aspects like industries, hospitals, aquaculture, and domestic purposes has a significant role in the pollution of the aquatic environment. In particular, through the sewage system, most wastewater is transferred into wastewater treatment plants. During this transfer process, some of it invades directly into the environment by producing pharmaceuticals from PPCPs through different human activities like bathing, swimming, etc. Partial human metabolism and defecation into water streams also result in the production of pharmaceuticals in the aquatic environment. Domestic pathways including industrial and hospital effluents are also prime sources of pharmaceuticals released into the aquatic environment. Through irrigation and leaching, pharmaceuticals containing wastewater enter the groundwater; hence, affecting the quality of groundwater and surface water as well (Daughton & Ternes, 1999).

Another major source of pharmaceuticals in the environment is the septic tank which is a small-scale municipal wastewater treatment that collects domestic household waste. Further, the risk of leakage from this septic system accelerates soil and water pollution by releasing pharmaceutical pollutants into the environment (Godfrey et al., 2007). Similarly, sewage sludge is another important cause of pharmaceutical pollution in the natural environment (Lapworth et al., 2012). Sewage sludge is a residue of wastewater treatment that contains organic matter as well as nutrients and is called a biosolid. Biosolids are further used in soil due to their ability to improve soil quality, which leads to potential groundwater contamination (Li, 2014).

6.3 FATE

As soon as these micropollutants enter the environment they follow different avenues such as photolysis, dissolution, degradation, and binding to the substrate of solid particles, relying on their characteristics and physicochemical properties. These properties incorporate vapor pressure, chemical stability, water solubility, and water partitioning, which indicates that compounds with marginal water solubility are more stubborn, toxic, and bio-accumulative and hence, retain high concentrations in the environment. Alternatively, compounds that have excessive solubility can be easily dissolved and degraded; hence, are present in lower concentrations in the environment (Nguyen et al., 2019). Since pharmaceuticals cannot be completely removed by wastewater treatment plants, they eventually enter the waterbodies directly from the WWTPs. The fate of pharmaceuticals in the natural world is regulated by the adsorption potential ascribed to the biostatic and aquaphobic properties of the chemicals (Fent et al., 2006).

From all the major sources of PPCPs in the environment like industries (landfills and wastewater), domestic (disposals and excretions), and hospitals, pharmaceutical contaminants participate in the functional ecosystem and ultimately enter the food

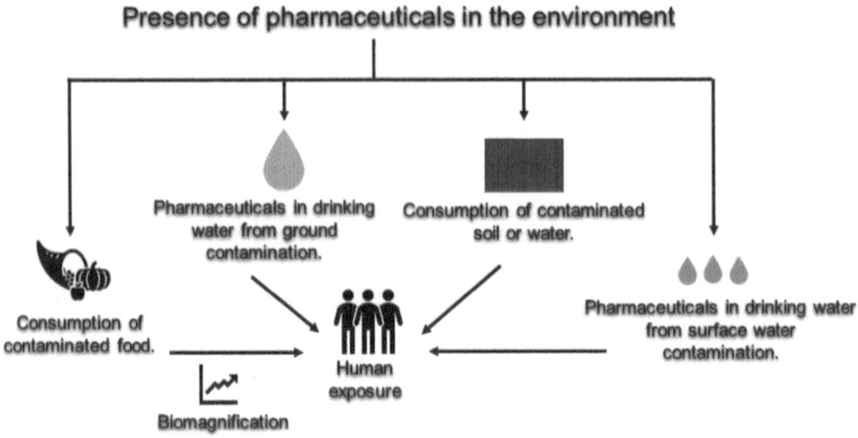

FIGURE 6.2 Fate and accumulation of pharmaceutical pollutants in the environment and paths of human exposure.

chain (Fig. 6.2). Similarly, disposals and excretions from humans and animals contain pharmaceuticals that cannot be degraded naturally and are found in WWTPs in high concentrations; entering into the food chain after being partially removed through basic filtering processes. From animal husbandry and aquaculture premises PPCPs are released directly into the natural habitat either from dead organisms or from the consumption and excretion of the animals thereafter entering into the natural food chain, eventually reaching higher order organisms through bioaccumulation and biomagnification (Piersma et al., 2009).

6.4 CATEGORIES OF PHARMACEUTICAL POLLUTANTS

Pharmaceutical pollutants can also be categorized into several groups based on their chemical configuration and their mode of action in addition to their various sources. Table 6.1 lists the different categories of pharmaceutical molecules along with their modes of action and the toxic effects of long-term exposure. Antibiotics are the most dangerous among them. Animal and human excretion, industrial effluents, drug disposals, and others are the principal sources of antibiotics in water. Tylosin, one antibiotic, is utilized as a growth promoter for cattle ranching in multiple nations (Macherius et al., 2014). Most prevalent antibiotics identified in various water bodies feature enoxacin, amoxicillin, cefalexin, ampicillin, ofloxacin, doxycycline, azithromycin, norfloxacin, penicillin, ciprofloxacin, tetracyclin, trimethoprim, and clavulanic acid (Jutkina et al., 2016; Grenni et al., 2018). These are generally functional and exceptionally durable chemicals. A relatively tiny amount of antibiotics is enough to disturb the microbial populations in the soil and underwater surroundings by removing susceptible inhabitants and producing a disparity in the microbial ecology. The most common issue regarding antibiotics as a pollutant is that they cause antibiotic resistance in bacterial strains. Antibiotic pollutants in the environment give a

TABLE 6.1
Categories of pharmaceutical pollutants

No	Categories	Activity	Compounds	Toxic effects	References
1	Antibiotics	An effect that is specifically bacteriostatic or antibacterial	Cefalexin Enoxacin Azithromycin Mupirocin Gentamycin Ampicillin Ofloxacin Amoxicillin Tetracyclin Doxycycline Ciprofloxacin Erythromycin Penicillin Norfloxacin Acid clavulanic Sulfamethoxazole Trimethoprim	1. Parallel transfer of resistance genes from a human pathogen to an ambient bacterial source 2. They affect the bacterial metabolic process involved in absorbing nutrients from alternate sources 3. Prevents human embryonic cells from proliferating 4. The bacterial population becomes resistant to antibiotics	Pomati et al. (2006); Kim and Aga (2007); Jutkina et al. (2016); Grenni et al. (2018)

(continued)

TABLE 6.1 (Continued)
Categories of pharmaceutical pollutants

No	Categories	Activity	Compounds	Toxic effects	References
2	Antidepressants	Treating anxiety, post-traumatic stress disorders, and similar obsessive conditions	Doxepin Nortriptyline Oxazepam Fluoxetine Venlafaxine Moclobemide Mianserin Sertraline Thioridazine Trimipramine Amitriptyline Agomelatine Desipramine Chlorpromazine Meprobamate Imipramine	1. Cardiopulmonary arrest or coronary artery disease (CAD) 2. Hyponatremia 3. An increase in seizure frequency 4. Gastrointestinal (GI) bleeding inclination	Fangmann et al. (2008); Coupland et al. (2011); Ramachandraih et al. (2011); Horowitz et al. (2015)
3	Nonsteroidal anti-inflammatories	Lowers body temperature, inflammation, and discomfort, prevents blood clotting	Etodolac Ketoprofen Phenazone Tolmetin Nabumetone Oxaprozin Sodium salicylate Paracetamol Diclofenac	1. Raising the risk of CAD 2. Causes membranous glomerulonephritis 3. Development of GI tract ulcers 4. Disrupts the body's normal temperature 5. Reduces the release of melatonin	Murphy et al. (1996); Crofford (2013); Day and Graham (2013)

			Indomethacin Nimesulide Celecoxib Naproxen Aspirin Ibuprofen		
4	Anticonvulsants	Reduces peripheral neuropathy and improves the state of mind	Clobazam Primidone Stiripentol Brivaracetam Diazepam Gabapentin Oxcarbazepine Acetazolamide Fosphenytoin Cenobamate	1. Motor nerves not functioning properly 2. Dizziness 3. Neurological coordination issues 4. Emotional problems accompanied by symptoms including nystagmus, tiredness, and hyperirritability 5. Morbilliform growth on the skin	Riss et al. (2008); Poddar and Sarkar (2020)
5	Colorants	Increases stability, makes dosage identification and formulation simple, and improves aesthetics	Indigo Carmine Tartrazine Amaranth Quinoline Yellow Sunset Yellow Allura Red	1. Metabolic disorders 2. Migraine, sleep disturbance, asthma, kidney disorders, cancers, and skin damage 3. Disturbance in the endocrine signaling system	Kale and Thorat (2014); Gilani et al. (2017); Poddar et al. (2022)

(continued)

TABLE 6.1 (Continued)
Categories of pharmaceutical pollutants

No	Categories	Activity	Compounds	Toxic effects	References
6	Preservatives	Keeps the active components from reacting with each other and growing in an undesirable way	Ethylparaben Methylparaben Sodium benzoate Phenoxyethanol Ethyl 4-hydroxybenzoate Butyl 4-hydroxybenzoate Isobutyl 4-hydroxybenzoate Propyl 4-hydroxybenzoate Isopropyl 4-hydroxybenzoate	1. Elicits contact dermatitis 2. Alters the hormone's course	Mowad (2000); Hafeez and Maibach (2013)
7	Steroids	Modify the course of the hormone to induce contact dermatitis	Estradiol Progesterone Norgestrel Oxytocin Androstenedione Insulin Thyroxin Estriol Mestranol Vasopressin Prostaglandins Testosterone	1. Liver injury 2. An early delay in bone growth 3. Cardiomyopathy 4. Emotional ups and downs and disturbed sleep 5. Sexual impotence 6. Cancer, acne, and baldness	Wang and Wang (2016); Wang et al. (2017); Kumar et al. (2018b)

selection advantage to strains of bacteria that have or have acquired resistive genes, and this circumstance further promotes the horizontal transmission of the resistive genes in a different group of species. As a result, harmful bacterial strains may develop resistance to certain antibiotics (Bottoni et al., 2010). In this manner, antibiotic pollutants pose a significant hazard to the ecosystem and human well-being.

Antidepressants are among the most prevalent pharmaceutical contaminants found in groundwater. This kind of contamination poses a serious hazard to all aquatic life, including vertebrates and invertebrates. Studies have revealed that antidepressants cause morphological imperfections, premature development, and behavioral disorders. According to Sehonova et al. (2018), certain antidepressants have an impact on both invertebrate and vertebrate species' capability for reproduction. Some of the most widespread antidepressants stated as pollutants include sertraline, oxazepam, maprotiline, desipramine, chlorpromazine, imipramine, trimipramine, diazepam, meprobamate, maprotiline, fluoxetine, nortriptyline, agomelatine, etc. (Fangmann et al., 2008; Ramachandraih et al., 2011; Horowitz et al., 2015).

Another category of chemical contaminants is anti-inflammatory drugs that consist of pharmaceutical compounds like paracetamol, aspirin, acetaminophen, celecoxib, phenazone, tolmetin, naproxen, nimesulide, etodolac, indomethacin, oxaprozin, salicylic acid, etc. (Crofford, 2013; Day & Graham, 2013). Almost all medicinal companies employ these compounds as painkillers. The majority catalyze the manufacture of various prostaglandins and are endocrine disruptors (Poddar & Sarkar, 2020). When present in surface water, pharmaceuticals have a negative impact, particularly on the aquatic population.

Anticonvulsants, also known as antiepileptic drugs consist of compounds like clobazam, primidone, stiripentol, cenobamate, gabapentin, etc. that help to reduce peripheral neuropathy and improve the state of mind. Anticonvulsants have a noxious effect on human health as they affect the function of motor nerves, cause neurological coordination issues, and give rise to emotional problems like nystagmus, tiredness, hyperirritability, etc. Apart from these, anticonvulsants also have toxic effects on human skin (Riss et al., 2008; Poddar & Sarkar, 2020).

Similarly, colorants are an additional group of pharmaceuticals that increase stability and make dosage identification simple. Colorants are mainly dyes, for example, indigo carmine, tartrazine, amaranth, quinoline yellow, sunset yellow, allura red, etc. (Kale & Thorat, 2014; Gilani et al., 2017).

Correspondingly, preservatives are an extra variety of pharmaceutical used mainly for storage purposes and to keep the active components from reacting with each other. Phosphate paraben, paraben ethyl, benzoate sodium, 4-hydroxybenzoate of ethanol, 4-hydroxypropyl benzoate, etc. are some preservatives that have toxic effects such as the alternation of hormone courses (Mowad, 2000; Hafeez & Maibach, 2013).

Steroids are an additional type of pharmaceutical that modify the course of hormones to induce contact dermatitis. Estradiol, progesterone, oxytocin, thyroxin, testosterone, and vasopressin are some steroid compounds that cause massive damage to human health including liver damage, cancer, acne, baldness, sexual impotence, bone growth delay, and emotional disturbance (Wang & Wang, 2016; Wang et al., 2017; Kumar et al., 2018b).

6.5 MICROBIAL BIODEGRADATION OF PHARMACEUTICAL POLLUTANTS

Environmental microorganisms play an important role in the degradation and removal of these emerging pollutants. These microorganisms are involved in the biodegradation process and degrade pollutants in metabolic and cometabolic pathways (Table 6.2). Biodegradation is considered the most sustainable process over

TABLE 6.2
Reports on microbial degradation of pharmaceutical pollutants

No	Compound's name	Microorganism name	Removal/ Degradation efficiency	References
1.	Oxacillin, Cloxacillin, Dicloxacillin	*Leptospaerulina* sp.	100% (6 days) 100% (8 days) 100% (8 days)	Copete-Pertuz et al. (2018)
2.	Norfloxacin and Ciprofloxacin	*Trametes versicolor*	90% (7 days)	Prieto et al. (2011)
3.	Sulfapyridine and Sulfathiazole	*Trametes versicolor*	80% (72 h)	Patel et al. (2019)
4.	Ofloxacin	*Trametes versicolor*	98.5% (8 days)	Gros et al. (2014)
5.	Salicylic acid	*Pseudomonas putida* (DSM 4478)	100% (8 h)	Combarros et al. (2014)
6.	Paracetamol and salicylic acid	*Chlorella sarokiniana*	67% and 73%, respectively (8 days)	Patel et al. (2019)
7.	Ibuprofen, metoprolol, acetaminophen, and diclofenac	*Chlorella sarokiniana*	60–100%	Patel et al. (2019)
8.	Naproxen and Ibuprofen	*Trametes versicolor*	100% (24 h)	Cruz-Morato et al. (2013)
9.	Carbamazepine	*Trichoderma hazardium*	72% (7 days)	Buchicchio et al. (2016)
		Plerotus osteratus	68% (7 days)	Buchicchio et al. (2016)
		Pseudomonas species	47%	Li et al. (2013)
		Phanerochaete chrysosporium	60–80% (100 days)	Zhang and Geiβen (2012)
10.	Plasma testosterone	*Carssius auratus*	50% (14 days)	Mimeault et al. (2005)
11.	Fluoxetine	*P. knackmussii* B-13	100% (72 h)	Khan and Murphy (2021)
12.	Tricyclic antidepressant	Anaerobic microbes and aerobic microbes	60–85% (16 days)	Choi et al. (2018)
13.	Clofibric acid	*Trametes versicolor*	97% (7 days)	Patel et al. (2019)

conventional methods for eliminating xenobiotics like pharmaceutical pollutants due to their nontoxic byproducts, higher degradation efficiency, and complete removal. The biodegradation process also requires very low maintenance, has operation costs with no complex infrastructures, and is eco-friendly. The degradation of different pharmaceutical molecules is discussed below.

Fungi and algae are the most capable microorganisms for the degradation of pharmaceutical pollutants in the wastewater treatment process. *Leptosperulina* species belonging to the white-rot fungi are the most promising agents for the removal of 100% oxacillin, cloxacillin, and dicloxacillin in less than eight days. The first significant decrease occurs during the initial two days, i.e., 80%, 53%, and 52% of oxacillin, cloxacillin, and dicloxacillin, respectively (Copete-Pertuz et al., 2018). However, norfloxacin and ciprofloxacin were transformed by about 90% in seven days by *Trametes versicolor*, belonging to the white-rot fungus at a concentration of 2 mg/L. This white-rot fungus produces a particular type of enzyme, i.e., laccase which catalyzes the oxidation of the substituted phenol group and aromatic thiols (Prieto et al., 2011). Apart from these, the fungus also successfully degraded sulfapyridine and sulfathiazole at a concentration of 10 mg/L (Patel et al., 2019). In addition, this strain effectively degraded 80% of the ofloxacin antibiotic (10 mg/L) by the Erlenmeyer flask method (Gros et al., 2014). Another group of white-rot fungi such as *Ganoderma lucidium*, *Trametes versicolor*, *Irpexlacteus* sp., and *Phanerochaete chrysoporium* completely degraded the ibuprofen compound in seven days from the incubation period (Cruz-Morató et al., 2013). *Ganoderma lucidium* degraded 47% of clofibric acid whereas *T. versicolor* degraded 58% of carbamazepine and 97% of clofibric acid (10 mg/L). In a hydroponic system, clofibric acid removals of 50% and 80% were achieved using *Typha* spp. in 48 h and 21 days, respectively (Patel et al., 2019).

Trichoderma harzianum actively metabolized clarithromycin and carbamazepine through the cometabolic oxidation pathway and removed 57% and 72% in seven days of treatment, while for the same days of treatment, *Plerotus osteratus* removed only 68%. Salicylic acid was degraded at 100% by *Pseudomonas putida* at 100 mg/L in only 8 h of treatment (Combarros et al., 2014), whereas carbamazepine was degraded at 47% by the same bacteria in optimal pH and temperature, i.e., 7 and 10 °C, respectively (Li et al., 2013). *Pseudomonas knackmussii* B-13 completely degraded fluoxetine within 72 h of treatment by using it as a sole carbon and energy source for their growth and produced fluoro metabolites (Khan & Murphy, 2021). Tricyclic antidepressants (TCAs) are endocrine-disrupting agents and recalcitrant molecules and about 60–85% were adsorbed immediately by the activated sludge (Choi et al., 2018). *Phenerochaete chrysosporium* was reported with a removal of 60–80% of carbamazepine (5 mg/L) after 100 days of operation (Zhang & Geiβen, 2012) along with the removal of 50–60% of diazepam and carbamazepine (Patel et al., 2019). The *P. chrysosporium* strain also degraded the naproxen compound in only 6 h of treatment at different concentrations. The removal rate increased to 80% by inducing the production of hydroxyl radicals. Iopromide molecules were degraded by *Trametes versicolor* through Erlenmeyer flask batches and a removal efficiency of 60% at 12 mg/L was reported. In the fluidized batch bioreactor, 87% iopromide molecules were removed with a sterile load and 65.4% with a nonsterile load of an initial 419.7 µg/L concentration (Patel et al., 2019). A high removal efficiency of NSAIDs (acetaminophen, ibuprofen, and diclofenac) through bioaccumulation by *Linum usitatissiumum* and *Armoracia rusticana* cell cultures and complete removal of acetaminophen along with 60–70% diclofenac removal by *Armoracia rusticana* after eight days were noted. Ibuprofen degradation reached 94% and 40% for both concentrations of 0.6mM and 1.2mM, respectively after eight days of treatment. *Linum usitatissiumum* removed 50% of acetaminophen in eight days, but complete removal occurred in ibuprofen and diclofenac in one and six days, respectively (Patel et al., 2019).

6.6 MICROBIAL NICHE NEXUS FOR SUSTAINABLE TREATMENT OF PHARMACEUTICAL WASTEWATER

In the nexus system, food, water, and the ecosystem can be interconnected and intertwined. The nexus concept is adopted for the treatment of wastewater to achieve the potential target of sustainable development of wastewater treatment. The microbial niche nexus performs a lead role in the efficient removal of known and new unknown pollutants. The selection of suitable functional microorganisms can achieve the removal of various types of emerging pollutants, recover energy, recycle resources, and also ensure the sustainable development of the nexus system.

In the context of the nexus system, paracetamol, which is a widely consumed analgesic drug, can be considered a source of recycling energy. As paracetamol contains an amide group in its structure and the bond can be broken by the enzyme amidase, the microbial genome contains amidase secreting gene, in response to providing acetate and 4-aminophenol. Further, this compound degrades into various intermediates that participate in various metabolic pathways. For instance, *Pseudomonas* species have a metabolic genome that is responsible for the secretion of amidase enzyme, and the metabolic product participates as an intermediate product in the TCA cycle (Rios-Miguel et al., 2022). In the view of the nexus system, these *Pseudomonas* species take this pollutant as a source of their nutrition from the ecosystem, the metabolic products are again involved in energy-making metabolic pathways and enter into the food chain through water or food.

Similarly, in the case of diclofenac, the degradation occurs by the bacterial strain *Labrys portucalensis* strain F11 and the degradation process includes several hydroxylations and oxygenation reactions to convert dihydroxy aromatic intermediates. These dihydroxy aromatic intermediates undergo ring fission to produce intermediates of Kreb's cycle (Cao et al., 2009). The production of DCF-11 and DCF-12 by *L. portucalensis* strain F11 shows the ability to perform the sulfation reaction, which is a well-documented metabolic pathway in mammals and essential for the detoxification and excretion of harmful compounds (Chapman et al., 2004). As a result, the link between food, water, and the ecosystem is maintained and these emerging pollutants can be removed from the wastewater without any toxic effects. The nexus system not only remediates wastewater with the removal of pharmaceutical micropollutants but is also a sustainable approach for maintaining the balance within the ecosystem.

6.7 CONCLUSION

With the advancement of medicine and pharmacology, the use of pharmaceuticals in different aspects of life has largely increased in the modern era. With the increase in application diversity, both the demand and production of different pharmaceuticals have increased significantly in recent decades. This elevated utility of pharmaceuticals has increased the environmental exposure of such compounds. In recent years, different studies on diverse pharmaceuticals in different models exhibited their destructive and destabilizing effects on the environment as micropollutants. To handle this challenge, different physicochemical approaches have been implemented exhibiting their incapability to treat trace amounts of such contaminants sustainably. Hence, the importance of biodegradation of such contaminants has drawn the attention of scientific society. Reports on the biodegradation of different pharmaceuticals have established the process as the most efficient, economical, and manageable technique. Moreover, the biodegradation approach is meant to stabilize the nexus of environment, energy, and food by removing stress from the environment through eliminating the pharmaceutical components and utilizing them as the energy source and flowing the energy through the food chain in the superior trophic levels. Hence, biodegradation has established itself as the most effective way to handle pharmaceutical pollutants in the environment.

REFERENCES

Agrawal, N., Kumar, V., and Shahi, S.K., 2021. Biodegradation and detoxification of phenanthrene in in-vitro and in-vivo conditions by a newly isolated ligninolytic fungus *Coriolopsis byrsina* strain APC5 and characterization of their metabolites for environmental safety. *Environmental Science and Pollution Research*. https://doi.org/10.1007/s11356-021-15271-w

Al-Gheethi, A., Noman, E., Mohamed, R.M.S.R., Ismail, N., and Kassim, A.H.M., 2019. Optimizing of pharmaceutical active compounds biodegradability in secondary effluents by β-lactamase from *Bacillus subtilis* using central composite design. *Journal of Hazardous Materials* 365:883–894. https://doi.org/10.1016/j.jhazmat.2018.11.068

Bottoni, P., Caroli, S., and Caracciolo, A.B., 2010. Pharmaceuticals as priority water contaminants. *Toxicological & Environmental Chemistry* 92:549–565. https://doi.org/10.1080/02772241003614320

Buchicchio, A., Bianco, G., Sofo, A., Masi, S. and Caniani, D., 2016. Biodegradation of carbamazepine and clarithromycin by *Trichoderma harzianum* and *Pleurotus ostreatus* investigated by liquid chromatography–high-resolution tandem mass spectrometry (FTICR MS-IRMPD). *Science of the Total Environment*, *557*, pp.733–739.

Cao, B., Nagarajan, K., and Loh, K.C., 2009. Biodegradation of aromatic compounds: current status and opportunities for biomolecular approaches. *Applied Microbiology and Biotechnology* 85:207–228. https://doi.org/10.1007/s00253-009-2192-4

Chapman, E., Best, M.D., Hanson, S.R., and Wong, C.H., 2004. Sulfotransferases: Structure, mechanism, biological activity, inhibition, and synthetic utility. *Angewandte Chemie International Edition* 43:3526–3548. https://doi.org/10.1002/anie.200300631

Choi, J.W., Zhao, Y., Bediako, J.K., Cho, C.W., and Yun, Y.S., 2018. Estimating environmental fate of tricyclic antidepressants in wastewater treatment plant. *Science of the Total Environment* 634:52–58. https://doi.org/10.1016/j.scitotenv.2018.03.278

Combarros, R.G., Rosas, I., Lavín, A.G., Rendueles, M. and Díaz, M., 2014. Influence of biofilm on activated carbon on the adsorption and biodegradation of salicylic acid in wastewater. *Water, Air, & Soil Pollution*, 225:1–12.

Copete-Pertuz, L.S., Plácido, J., Serna-Galvis, E.A., Torres-Palma, R.A., and Mora, A., 2018. Elimination of Isoxazolyl-Penicillins antibiotics in waters by the ligninolytic native Colombian strain Leptosphaerulina sp. considerations on biodegradation process and antimicrobial activity removal. *Science of the Total Environment* 630:1195–1204. https://doi.org/10.1016/j.scitotenv.2018.02.244

Costa, F., Lago, A., Rocha, V., Barros, O., Costa, L., Vipotnik, Z., Silva, B., and Tavares, T., 2019. A review on biological processes for pharmaceuticals wastes abatement—a growing threat to modern society. *Environmental Science & Technology* 53:7185–7202. https://doi.org/10.1021/acs.est.8b06977

Coupland, C., Dhiman, P., Morriss, R., Arthur, A., Barton, G., and Hippisley-Cox, J., 2011. Antidepressant use and risk of adverse outcomes in older people: Population based cohort study. *Bmj* 343. https://doi.org/10.1136/bmj.d4551

Crofford, L.J., 2013. Use of NSAIDs in treating patients with arthritis. *Arthritis Research & Therapy* 15:1–10. https://doi.org/10.1186/ar4174

Cruz-Morató, C., Ferrando-Climent, L., Rodriguez-Mozaz, S., Barceló, D., Marco-Urrea, E., Vicent, T., and Sarrà, M., 2013. Degradation of pharmaceuticals in non-sterile urban wastewater by *Trametes versicolor* in a fluidized bed bioreactor. *Water Research* 47:5200–5210. https://doi.org/10.1016/j.watres.2013.06.007

Daughton, C.G., and Ternes, T.A., 1999. Pharmaceuticals and personal care products in the environment: Agents of subtle change?. *Environmental Health Perspectives* 107:907–938. https://doi.org/10.1289/ehp.99107s6907

Day, R.O., and Graham, G.G., 2013. Non-steroidal anti-inflammatory drugs (NSAIDs). *Bmj* 346. https://doi.org/10.1136/bmj.f4310

Eggen, T., Moeder, M., and Arukwe, A., 2010. Municipal landfill leachates: A significant source for new and emerging pollutants. *Science of the Total Environment* 408:5147–5157. https://doi.org/10.1016/j.scitotenv.2010.07.049

Fangmann, P., Assion, H.J., Juckel, G., González, C.Á., and López-Muñoz, F., 2008. Half a century of antidepressant drugs: On the clinical introduction of monoamine oxidase inhibitors, tricyclics, and tetracyclics. Part II: tricyclics and tetracyclics. *Journal of Clinical Psychopharmacology* 28:1–4. https://doi.org/10.1097/jcp.0b013e3181bb617

Fent, K., Weston, A.A. and Caminada, D., 2006. Ecotoxicology of human pharmaceuticals. *Aquatic Toxicology* 76(2):122–159.

Gilani, R., Joshi, S., Shankari, V.S., Joshna, P.M., and Rao, V., 2017. Effect of brilliant blue FCF Food Colour on Plant and Human DNA. *International Journal of Molecular Genetics* 8:1–10.

Godfrey, E., Woessner, W.W., and Benotti, M.J., 2007. Pharmaceuticals in on-site sewage effluent and ground water, western Montana. *Groundwater* 45:263–271. https://doi.org/10.1111/j.1745-6584.2006.00288.x

Grenni, P., Ancona, V., and Caracciolo, A.B., 2018. Ecological effects of antibiotics on natural ecosystems: A review. *Microchemical Journal* 136:25–39. https://doi.org/10.1016/j.microc.2017.02.006

Gros, M., Cruz-Morato, C., Marco-Urrea, E., Longrée, P., Singer, H., Sarrà, M., Hollender, J., Vicent, T., Rodriguez-Mozaz, S., and Barceló, D., 2014. Biodegradation of the X-ray contrast agent iopromide and the fluoroquinolone antibiotic ofloxacin by the white rot fungus *Trametes versicolor* in hospital wastewaters and identification of degradation products. *Water Research* 60:228–241. https://doi.org/10.1016/j.watres.2014.04.042

Hafeez, F., and Maibach, H., 2013. An overview of parabens and allergic contact dermatitis. *Skin Therapy Letter* 18:5–7.

Horowitz, M.A., Wertz, J., Zhu, D., Cattaneo, A., Musaelyan, K., Nikkheslat, N., Thuret, S., Pariante, C.M., and Zunszain, P.A., 2015. Antidepressant compounds can be both pro-and anti-inflammatory in human hippocampal cells. *International Journal of Neuropsychopharmacology* 18. https://doi.org/10.1093/ijnp/pyu076

Joutey, N.T., Bahafid, W., Sayel, H., and El Ghachtouli, N., 2013. Biodegradation: involved microorganisms and genetically engineered microorganisms. In *Biodegradation-Life of Science*, ed. R. Chamy, and F. Rosenkranz, 289–320. Intech, Janeza Trdine, Croatia.

Jureczko, M., Przystaś, W., Krawczyk, T., Gonciarz, W., and Rudnicka, K., 2021. White-rot fungi-mediated biodegradation of cytostatic drugs-bleomycin and vincristine. *Journal of Hazardous Materials* 407:124632. https://doi.org/10.1016/j.jhazmat.2020.124632

Jutkina, J., Rutgersson, C., Flach, C.F., and Larsson, D.J., 2016. An assay for determining minimal concentrations of antibiotics that drive horizontal transfer of resistance. *Science of the Total Environment* 548:131–138. https://doi.org/10.1016/j.scitotenv.2016.01.044

Kale, R.V., and Thorat, P.R., 2014. Biodegradation of Allura Red AC (ARAC) by *Ochrobactrum anthropi* HAR08, isolated from textile dye contaminated soil. *International Journal of Advanced Research* 2:432–441.

Khan, M.F. and Murphy, C.D., 2021. Bacterial degradation of the anti-depressant drug fluoxetine produces trifluoroacetic acid and fluoride ion. *Applied Microbiology and Biotechnology*, 105(24):9359–9369.

Kim, S., and Aga, D.S., 2007. Potential ecological and human health impacts of antibiotics and antibiotic-resistant bacteria from wastewater treatment plants. *Journal of Toxicology and Environmental Health, Part B* 10:559–573. https://doi.org/10.1080/15287390600975137

Kumar, V., Shahi, S.K., and Singh, S., 2018a. Bioremediation: An eco-sustainable approach for restoration of contaminated sites. *Microbial Bioprospecting for Sustainable Development*, eds. J. Singh, D. Sharma, G. Kumar, and N. Sharma, Springer, Singapore. https://doi.org/10.1007/978-981-13-0053-0_6

Kumar, S.V., Rajan, C., Divya, P., and Sasikumar, S., 2018b. Adverse effects on consumer's health caused by hormones administered in cattle. *International Food Research Journal* 25:1–10.

Kumar, V., Agrawal, S., Bhat, S.A., Américo-Pinheiro, J.H.P., Shahi, S.K., and Kumar, S., 2022. Environmental impact, health hazards, and plant-microbes synergism in remediation of emerging contaminants. *Cleaner Chemical Engineering* 2, 100030. https://doi.org/10.1016/j.clce.2022.100030

Lapworth, D.J., Baran, N., Stuart, M.E., and Ward, R.S., 2012. Emerging organic contaminants in groundwater: A review of sources, fate and occurrence. *Environmental Pollution* 163:287–303. https://doi.org/10.1016/j.envpol.2011.12.034

Li, A., Cai, R., Cui, D., Qiu, T., Pang, C., Yang, J., Ma, F., and Ren, N., 2013. Characterization and biodegradation kinetics of a new cold-adapted carbamazepine-degrading bacterium, Pseudomonas sp. CBZ-4. *Journal of Environmental Sciences* 25:2281–2290. https://doi.org/10.1016/S1001-0742(12)60293-9

Li, W.C., 2014. Occurrence, sources, and fate of pharmaceuticals in aquatic environment and soil. *Environmental Pollution*, 187:193–201. https://doi.org/10.1016/j.envpol.2014.01.015

Macherius, A., Lapen, D.R., Reemtsma, T., Römbke, J., Topp, E., and Coors, A., 2014. Triclocarban, triclosan and its transformation product methyl triclosan in native earthworm species four years after a commercial-scale biosolids application. *Science of the Total Environment*, 472:235–238. https://doi.org/10.1016/j.scitotenv.2013.10.113

Manasfi, R., Chiron, S., Montemurro, N., Perez, S., and Brienza, M., 2020. Biodegradation of fluoroquinolone antibiotics and the climbazole fungicide by *Trichoderma* species. *Environmental Science and Pollution Research*, 27:23331–23341. https://doi.org/10.1007/s11356-020-08442-8

Mimeault, C., Woodhouse, A.J., Miao, X.S., Metcalfe, C.D., Moon, T.W., and Trudeau, V.L., 2005. The human lipid regulator, gemfibrozil bioconcentrates and reduces testosterone in the goldfish, *Carassius auratus. Aquatic Toxicology* 73:44–54. https://doi.org/10.1016/j.aquatox.2005.01.009

Mowad, C.M., 2000. Allergic contact dermatitis caused by parabens: 2 case reports and a review. *American Journal of Contact Dermatitis* 11:53–56. https://doi.org/10.1016/S1046-199X(00)90033-2

Murphy, P.J., Myers, B.L., and Badia, P., 1996. Nonsteroidal anti-inflammatory drugs alter body temperature and suppress melatonin in humans. *Physiology & Behavior* 59:133–139. https://doi.org/10.1016/0031-9384(95)02036-5

Narayanan, M., El-Sheekh, M., Ma, Y., Pugazhendhi, A., Natarajan, D., Kandasamy, G., Raja, R., Kumar, R.S., Kumarasamy, S., Sathiyan, G., and Geetha, R., 2022. Current status of microbes involved in the degradation of pharmaceutical and personal care products (PPCPs) pollutants in the aquatic ecosystem. *Environmental Pollution* 300:118922. https://doi.org/10.1016/j.envpol.2022.118922

Nguyen, P.M., Afzal, M., Ullah, I., Shahid, N., Baqar, M., and Arslan, M., 2019. Removal of pharmaceuticals and personal care products using constructed wetlands: effective plant-bacteria synergism may enhance degradation efficiency. *Environmental Science and Pollution Research* 26:21109–21126. https://doi.org/10.1007/s11356-019-05320-w

Olicón-Hernández, D.R., Camacho-Morales, R.L., Pozo, C., González-López, J., and Aranda, E., 2019. Evaluation of diclofenac biodegradation by the ascomycete fungus Penicillium oxalicum at flask and bench bioreactor scales. *Science of The Total Environment* 662:607–614. https://doi.org/10.1016/j.scitotenv.2019.01.248

Patel, M., Kumar, R., Kishor, K., Mlsna, T., Pittman Jr, C.U., and Mohan, D., 2019. Pharmaceuticals of emerging concern in aquatic systems: Chemistry, occurrence, effects, and removal methods. *Chemical Reviews* 119:3510–3673. https://doi.org/10.1021/acs.chemrev.8b00299

Piersma, A.H., Luijten, M., and Popov, V., et al 2009. Pharmaceuticals. In *Endocrine-Disrupting Chemicals in Food*, ed. I. Shaw, 459–518. Woodhead Publishing.

Poddar, K., and Sarkar, A., 2020. Emerging treatment strategies of pharmaceutical pollutants: Reactive physiochemical and innocuous biological approaches. In *Removal of Toxic Pollutants Through Microbiological and Tertiary Treatment*, ed. M. P. Shah, 431–451. Elsevier.

Poddar, K., Sarkar, D., and Sarkar, A., 2023. Construction of bacterial consortium for efficient degradation of mixed pharmaceutical dyes. *Environmental Science and Pollution Research*, 30:25226–25238. https://doi.org/10.1007/s11356-021-18217-4

Pomati, F., Castiglioni, S., Zuccato, E., Fanelli, R., Vigetti, D., Rossetti, C., and Calamari, D., 2006. Effects of a complex mixture of therapeutic drugs at environmental levels on human embryonic cells. *Environmental Science & Technology* 40:2442–2447. https://doi.org/10.1021/es051715a

Prieto, A., Möder, M., Rodil, R., Adrian, L., and Marco-Urrea, E., 2011. Degradation of the antibiotics norfloxacin and ciprofloxacin by a white-rot fungus and identification of degradation products. *Bioresource Technology* 102:10987–10995. https://doi.org/10.1016/j.biortech.2011.08.055

Ramachandraih, C.T., Subramanyam, N., Bar, K.J., Baker, G., and Yeragani, V.K., 2011. Antidepressants: From MAOIs to SSRIs and more. *Indian Journal of Psychiatry* 53:180–182. doi: 10.4103/0019-5545.82567

Rios-Miguel, A.B., Smith, G.J., Cremers, G., van Alen, T., Jetten, M.S., den Camp, H.J.O., and Welte, C.U., 2022. Microbial paracetamol degradation involves a high diversity of novel amidase enzyme candidates. *Water Research X* 16:100152. https://doi.org/10.1016/j.wroa.2022.100152

Riss, J., Cloyd, J., Gates, J., and Collins, S., 2008. Benzodiazepines in epilepsy: Pharmacology and pharmacokinetics. *Acta neurologica scandinavica* 118 :69–86. https://doi.org/10.1111/j.1600-0404.2008.01004.x

Sehonova, P., Svobodova, Z., Dolezelova, P., Vosmerova, P., and Faggio, C., 2018. Effects of waterborne antidepressants on non-target animals living in the aquatic environment: A review. *Science of the Total Environment* 631:789–794. https://doi.org/10.1016/j.scitotenv.2018.03.076

Tiwari, B., Sellamuthu, B., Ouarda, Y., Drogui, P., Tyagi, R.D., and Buelna, G., 2017. Review on fate and mechanism of removal of pharmaceutical pollutants from wastewater using biological approach. *Bioresource Technology* 224:1–12. https://doi.org/10.1016/j.biortech.2016.11.042

Wang, J., and Wang, S., 2016. Removal of pharmaceuticals and personal care products (PPCPs) from wastewater: A review. *Journal of Environmental Management* 182:620–640. https://doi.org/10.1016/j.jenvman.2016.07.049

Wang, Y., Liu, J., Kang, D., Wu, C., and Wu, Y., 2017. Removal of pharmaceuticals and personal care products from wastewater using algae-based technologies: A review. *Reviews in Environmental Science and Bio/Technology* 16:717–735. https://doi.org/10.1007/s11157-017-9446-x

Xu, W., Zou, R., Jin, B., Zhang, G., Su, Y., and Zhang, Y., 2022. The ins and outs of pharmaceutical wastewater treatment by microbial electrochemical technologies. *Sustainable Horizons* 1:100003. https://doi.org/10.1016/j.horiz.2021.100003

Zhang, Y., and Geißen, S.U., 2012. Elimination of carbamazepine in a non-sterile fungal bioreactor. *Bioresource Technology* 112:221–227. https://doi.org/10.1016/j.biortech.2012.02.073

7 Recovery of Value-added Materials from Wastewater

Developments, Challenges, and Opportunities in Developing Countries

Sisanda Dlova, Olusola Olaitan Ayeleru, Lukhanyo Mekuto, and Peter Apata Olubambi

7.1 INTRODUCTION

The fast-rising human population is predicted to reach ~9.8 billion in 2050 from ~7.6 billion people in 2017. Urbanization has increased due to population growth over the last few decades (Singh, 2021). This has, in turn, increased fresh water usage at twice the rate of the population over the last two decades (Devda et al., 2021), while the rate of municipal wastewater production (MWW) is also rising (Singh, 2021). Additionally, plastic waste pollution, which is detrimental to the environment and human race, is also increasing in the environment (Iroegbu et al., 2020). There is an urgent need to properly manage waste generated not only from municipal wastewater as freshwater resources are being overburdened (Singh, 2021) but also from municipal solid waste streams since the bulk of these wastes (e.g., plastic wastes) are nonbiodegradable. In addition to the stressed freshwater resources, energy is also a problem, hence many countries including, the Republic of South Africa are experiencing load shedding for over eight hours daily. Moreover, climate change has not stopped its continuous impact on Earth, and as a result, it has been projected to lead to a ~10% reduction in freshwater supply to ~685 million people living in over 570 cities by 2050 (Devda et al., 2021). Several studies have been carried out globally to investigate ways by which water wastage can be combated, to improve water availability, and to reduce municipal solid waste generation while recovering resources from it. One of the activities that is aiding the current freshwater strain is agricultural activity or irrigation, which makes use of MWW. This activity results in energy strain from fertilizer production and contamination of groundwater via the usage of conventional fertilizers. There have been many concerns over human health as untreated MWW and sewage sludge containing pathogens, diseases, heavy metals, and other

 DOI: 10.1201/9781003441069-7

contaminants are continuously generated and released into the environment. Some of the ideas and studies that are being explored around the world include the possibility of using treated wastewater and sludge for agricultural purposes while biosolids are used for energy generation and biomaterial production, and further treated wastewater for reuse as potable water. These ideas would be beneficial as environmental pollution could be reduced, agricultural income increased, and wastewater release into the environment minimized while augmenting freshwater reserves (Singh, 2021).

7.2 RESOURCES FROM WASTEWATER

7.2.1 Reusable Water for Irrigation

Reusable or reclaimed water refers to the process of treating municipal wastewater of solids and other impurities to be further utilized for various purposes such as recharging the ground basin, agriculture and landscape irrigation, industrial processes, and toilet flushing (Chaudhary et al., 2019). Reclaimed water is also called recycled water, meaning, it is water recovered from domestic, municipal, and industrial wastewater treatment plants (Kehrein et al., 2020). This kind of water can be used as an alternative source of water supply for various purposes such as agricultural activities (irrigation purposes) in regions with water scarcity (Tzanakakis et al., 2020). The use of reclaimed water for irrigation is commended to ease the stress on freshwater resources and conventional fertilizer (Lahlou et al., 2021). With the continuous population growth and a sixfold increase in demand over the past fifty years, the existing global fertilizer production capacity cannot keep up with current demands (Chojnacka et al., 2020), hence, many countries with water scarcity problems have adopted the use of reclaimed water for agricultural purposes while some are still adjusting to the idea of conventional fertigation, which involves the use of partially treated or untreated wastewater for irrigation (Cipolletta et al., 2021). As a result, changing to the reuse of wastewater fertigation for agricultural purposes is a good idea because it gives governments all over the world a chance to lessen the strain on freshwater supplies and the demand for conventional fertilizer (Lahlou et al., 2020). However, despite the fact that treated wastewater has a higher concentration of nutrients than conventional water sources, reusing treated wastewater could pose risks to public health and the environment (Chaudhary et al., 2019; Lahlou et al., 2021). The advantages and disadvantages of this application were discussed by Chojnacka et al. (2020) and guidelines for recycled water quality for irrigation purposes were given by Chaudhary et al. (2019) and Chojnacka et al. (2020).

7.2.1.1 Benefits and Drawbacks of Using Reusable Water for Irrigation

Different nations have different standards for acceptable usable wastewater quality and tolerance. The tolerance of risks to public health, the environment, and the economy also varies from country to country. All these factors are influenced by how each nation is affected by water scarcity and the policies that govern the use of wastewater for agricultural purposes. In addition, farmers' perceptions of the idea of using wastewater and their level of knowledge play a significant role in determining how much wastewater is used. According to Brar and Rawat (2022), despite concerns about risks to public health and its negative effects on the environment, the practice is gaining popularity worldwide due to its numerous benefits. Some of the benefits include but are not limited to the accessibility to a great quantity of water throughout the year that is unaffected by environmental state and a nutrient-dense medium that can cutdown the utilization of chemical composts, increase yield on infertile lands, and lessen the loss of freshwater in the environment caused by eutrophication and algal blooms (Ungureanu & Vladut, 2018). The soil also gains important macro- and micronutrients from wastewater irrigation. Due to its composition, wastewater is an excellent source of nitrogen (N), potassium (K), phosphorus (P), zinc (Zn), iron (Fe), manganese (Mn), and copper (Cu) (Ofori et al., 2021). The use of treated wastewater for irrigation farming can increase the amount of nitrogen in the soil and act as a fertilizer because it is a good source of inorganic nitrogen (ammonium and nitrates) and therefore, wastewater makes soil nitrogen more accessible. In addition, the utilization of wastewater for irrigation purposes speeds up mineralization and makes it easier for crops to absorb nitrogen (Quemada et al., 2016). The formation of important genetic information carriers (DNA and RNA) and biological molecules requires the availability of nitrogen which is obtained from wastewater treated for irrigation (Aczel, 2019). As there are many benefits of using wastewater, so also are there drawbacks which are not limited to several ailments in farmers and buyers of food from wastewater-irrigated crops, build up of heavy metals, salts, antibiotics, growth hormones, and other unsafe substances in the soil, low hydraulic conductivity due to congestion of the vents with suspended solids from wastewater, and weakening of the quality of crops due to tainted soil (Maaß & Grundmann, 2018). Wastewater contains pathogenic microorganisms that have the potential to spread disease, including bacteria, viruses, and parasites. Human parasites like protozoa and helminth eggs, which are very complicated to treat, are associated with a type of transmittable gastrointestinal illness in both rich and impoverished nations (Brar & Rawat, 2022). Therefore, the is a need to regulate the use of wastewater (treated or untreated) to lessen public health issues and environmental concerns. Such legislation can include the use of effective disinfection practices to remove pathogenic microorganisms such that public health concerns can be ameliorated, among others.

7.2.1.2 Safe Drinking/Potable Water

Researchers across the globe have shared similar views concerning the treatment and conversion of municipal effluent wastewater to drinking water, with the main concerns being public health safety, public perceptions, dangers around the treated drinking water, ways to augment drinking water supply, and causes of water scarcity (Tortajada & van Rensburg, 2020). Other scholars have highlighted population growth and climate change as the main causes for the scarcity of drinkable water, thereby forcing the global community to search of alternative sources of water. The drinking of ultrapurified treated municipal wastewater, known as reused water, is growing in several countries around the world (Tortajada & van Rensburg, 2020). Marron et al. (2019) referred to reused water as potable water, which is a product of highly designed water treatment systems with the capacity to purify municipal wastewater effluent to drinking water. This process is dependent upon reverse osmosis (RO) treatment with ultraviolet (UV) light as the secondary stage to RO and the addition of hydrogen peroxide (H_2O_2) (Marron et al., 2019). All the efforts to get a sustainable means to expand the drinking water supply are let down by negativity due to less education and minimal research on the subject, anxieties associated with public health risks while often aggravated by dramatic media coverage. Some concerns include that reused water is contaminated by pathogens and chemicals and this has stalled many projects involving wastewater treatment to potable water reuse (Luthy et al., 2020; Tortajada & van Rensburg, 2020). In addition, many countries around the world have become skeptical of investing in such projects, however, there are growing success stories including the Goreangab Reclamation Plant in Windhoek, in Namibia, Orange County Water District in California, Western United States of America, and the NEWater project in Singapore (Luthy et al., 2020; Tortajada & Nambiar, 2019). On the contrary, in Queensland Australia, residents successfully resisted the Toowoomba reuse potable water project in 2006 and the Western Corridor Recycled Water Scheme (Caball & Malekpour, 2019). Apart from the successes and few oppositions, other countries such as Australia, Belgium, France, the United Kingdom, and South Africa followed suit by making potable water reuse a trend (Tortajada & Nambiar, 2019). The water quality in successful projects is measured by applying processes like the multiple barrier principle (MBP) that relies largely on RO as a primary stage accompanied by UV treatment at the last stage with chemical treatment via H_2O_2 (Menge, 1997). Capodaglio (2020) defined the principle as multi-barrier treatment (MBT). The author first described multi-barrier as a term that was invented to cover the broader spectrum of topics related to water sources, watershed safety, water treatment, and sterilization to provide a safe source for dissemination. MBT denotes a narrow stream that describes a series of processes for the treatment and purification of drinking water (Capodaglio, 2020). Although, the development of drinking water treatment was first intended to address the dangers of waterborne infections, it has since been expanded to encompass the elimination of chemical compounds that have an aesthetic influence and may be hazardous to human health (Capodaglio, 2020; Marron et al., 2019). The same idea could be implemented for water intended for farming purposes, industrial, nonpotable, or any other general applications that may have an indirect impact on human health (Chaudhary et al., 2019). As there is no once-off treatment that may well offer an utter barricade to all chemicals, the fundamental premise of MBT is that subsequent treatments will eliminate pollutants that have eluded previous treatment depending on their physico-chemical characteristics. In the end, only trace amounts of contaminants will be able to bypass the treatment process due to the use of increasingly selective procedures. After a few decades of MBT's use in the reuse of potable water, this assumption can be deemed safe. Therefore, MBT or MBP principles' development was based on known incidents and may be futile against yet to be experienced phenomena. In general, MBP or MBT comprises the following steps:

- Nontreatment hurdles, which involve change of industrial effluents to several drainage areas and regulating of industrial discharges. Then, there must be rigorous, ongoing quality observation of the raw and preserved water (both

TABLE 7.1
Process selection for specific treatment targets in drinking water reclamation (various source supply)

Treatment purpose	Processes	Target contaminants
Aestetic	Chemical dissolve (CD) + dissolved air flotation (DAF) + sand filtration (SF) + AC	Suspended and dissolved (partial) solids, color, NOM, taste
Solids	UF or other media filtration	Particles to the desired size
Pathogens	UF + O_3 (or other AOPs/AORPs)	Bacteria, viruses, gardia, cyptosporidium
Organics	Ozonation + biological AC + GAC (PAC)	Organic matter and micropollutants
DBPs	Coagulation, chlorination improvement, alternate processes (AOPs/AORPs)	Distinction byproducts
Other residuals	Sodium hydroxide/permanganate + SF	Iron, manganese
Micropollutants	AOPs, ARPs, AORPs	CECs, PPCPs, EDCs

Source: Capodaglio (2020); Menge (1997).

online and in the lab) so that corrective action can be taken to protect the consumer. Finally, reclaimed water must be blended with conventional sources to a maximum of 35% and ultimately, catchment management policies put in place.

- Treatment barriers are always guarded against specific contaminants. WARNING: They are not completely "dead-stop"! They are categorized as either partial or complete, and the responses of various contaminants to various treatment techniques vary. Each contaminant has a unique effect when something goes wrong (the generic summary in Table 7.1 explains drinking water recovery and Figure 7.1 shows industrial reutilization).
- Operational hurdles (e.g., quantity of powered activated carbon (PAC)) are not part of the process training on a regular basis but can be used as a backup or to add capacity (Marron et al., 2019; Menge, 1997; Capodaglio, 2020).

7.2.2 Polyhydroxyalkanoates (PHAs)

Prokaryotic organisms use renewable resources that make biodegradable polymers known as polyhydroxyalkanoates (PHAs), also referred to as biological polyester (Castilho et al., 2009). PHAs are a class of intracellular biopolyesters, according to Chen et al., (2020), while Crutchik et al. (2020) define PHAs as biodegradable polyester produced by many bacteria that build up as intracellular carbon reserves when nutrients are limited (Crutchik et al., 2020). PHAs are also known as polyesters with a distinctive link of esters that are grouped together as carbon and energy reserves with a controlled source of nitrogen generation that helps in energy delivery (Phas et al.,

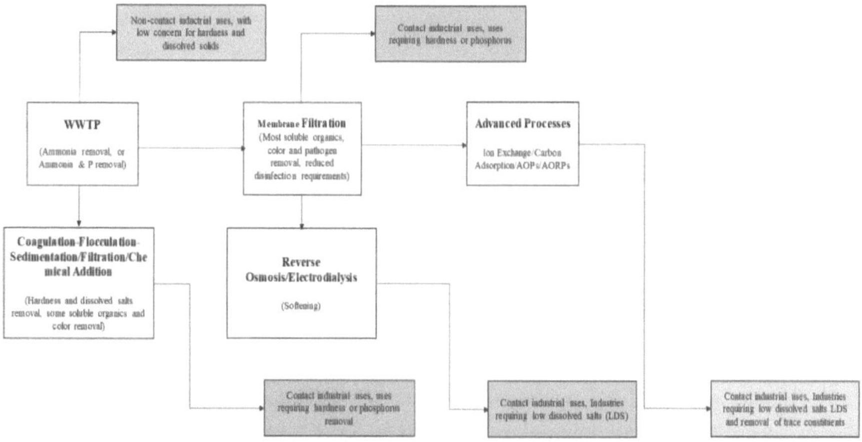

FIGURE 7.1 Potential cycle of treatment technologies to go through industrial water uses (Reprinted from Capodaglio (2020), Copyright (2020), with permission from Taylor & Francis).

2021). Generally, PHAs are produced through microbial fermentation (Castilho et al., 2009). PHAs can be formed by feeding blended microbial cultures with feedstocks of the second generation, like solid wastes and wastewater. This prevents PHAs from competing with crops and encourages waste valorization (Almeida et al., 2021). PHAs can also be produced from municipal wastewater treatment plants (MWWTPs) sludge and any food related industry waste, e.g., hotel waste and alcohol fermentation industries (Crutchik et al., 2020; Gecim et al., 2021). Mannina et al. (2019), in their study on PHAs, confirmed that PHAs are a type of biobased biodegradable polymer that are particularly intriguing because they can be made via bacterial fermentation from a variety of complex organic substrates, comprising those found in waste streams like agro-industrial wastewater and sewage sludge. Consequently, resource recovery from wastewater treatment operations can contribute to the circular economy of plastic in this context (Mannina et al., 2019). Several methods that are commonly used to produce PHAs at the industrial level include pure culture or engineered microbial strains and mixed microbial culture (MMC). The main disadvantage of the use of pure culture for the production of PHAs is the prerequisite to have a sterilized environment which results in it being steep and complex (Mannina et al., 2020). Once the carbon substrates go through the bacterial cells, via diffusion across the cytoplasmic membrane, PHA synthesis takes place within the cell. There are generally three elementary metabolic pathways conducted to yield short-chain length PHAs (scl-PHA; 3–5 carbon atoms) or medium-chain length PHAs (ml-PHA) which consist of 6–14 carbon atoms (Chen et al., 2015). The three pathways are normally linked to genetic engineering which is utilized to modify the PHA biosynthesis mechanisms and genes of the PHA synthetic enzymes for diversifying PHAs' biopolymers (Chen et al., 2015; Nguyenhuynh et al., 2021). The glycolytic process, also known as pathway 1 (acetyl-CoA to 3-hydroxybutyryl-CoA), is used by the PHA-producing strain *Ralstonia eutropha* (also known as *Cupriavidus necator* or *Alcaligenes eutrophus*). Sugars, fatty acids, or amino acids are carbon substrates for pathway 1. Enter-Doudoroff for mannose, galactose, and glucose, and Pento Phosphasphate Shunt for xylose and arabinose are the first steps in the conversion of the sugar-based substrates to pyruvate. The pyruvate component is subsequently transformed into acetyl-CoA, which is used as the initial substrate to produce scl-PHA. The acetoacetyl-CoA synthetase directly converts the acetoacetate substrate to acetyl-CoA (Mannina et al., 2020; Nguyenhuynh et al., 2021; Phas et al., 2021). Acetyl-CoA is then converted to acetoacetyl-CoA, (R)-3-hydroxybuanoyl-CoA, or 3HB monomer, and then polymerized into poly(3-hydroxybutyrate) or PHB (scl-PHA) with the help of PHA storage enzymes (pha A, pha B, and pha C) as shown in Figure 7.2.

The other two pathways—oxidation and in situ fatty acid synthesis—produce mcl-PHA. By consuming fatty acids, the bacteria *Pseudomonas putida*, *P. oleovorans*, and *P. aeruginosa* create enoyl-CoA, which is then converted by the enzyme R-3-hydroxyacyl-CoA hydratase into R3-hydroxyacyl-CoA (precursor). mcl-PHA synthase catalyzes the polymerization of mcl-PHA using this precursor (Novelli et al., 2021), this is shown in Figure 7.3 and Figure 7.4.

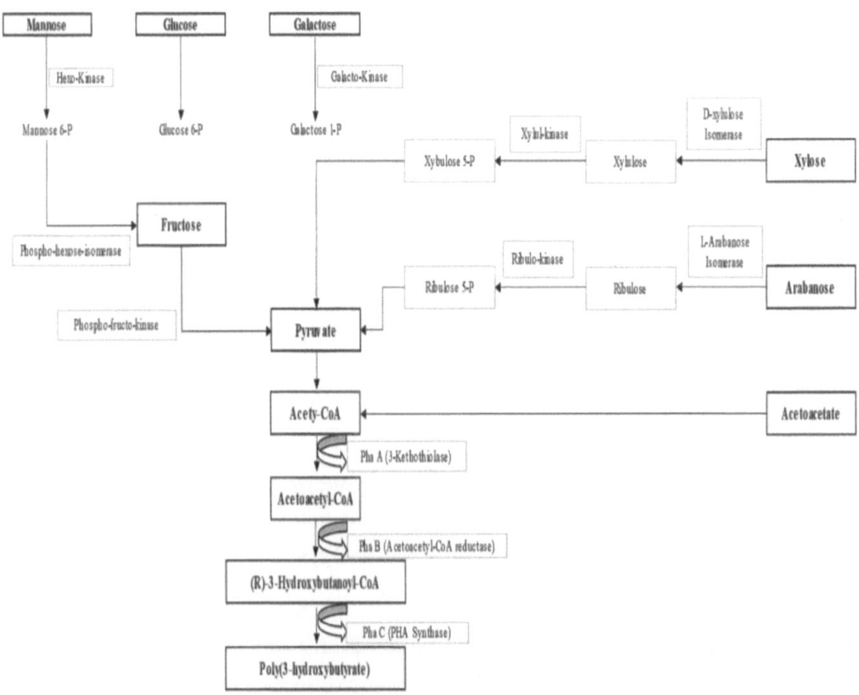

FIGURE 7.2 Conversion in pathway 1: Acetyl-CoA to 3-Hydroxybutyryl-CoA to produce PHB (scl-PHA) (Reprinted from Nguyenhuynh et al. (2021), Copyright (2021), with permission from Elsevier Inc.).

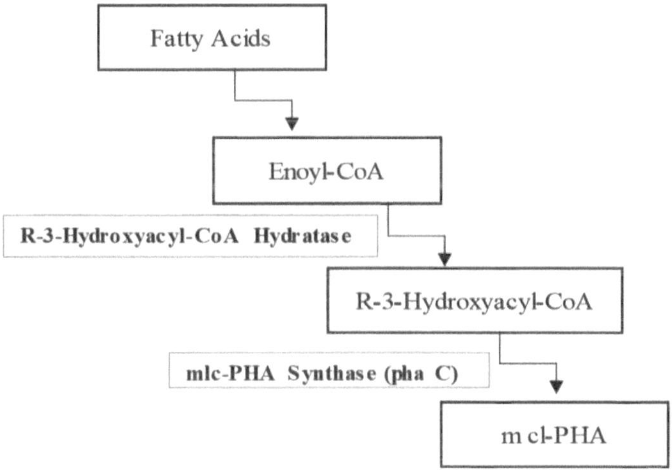

FIGURE 7.3 Pathway 2 where mcl-PHA is formed through β-oxidation (Reprinted from Nguyenhuynh et al. (2021), Copyright (2021), with permission from Elsevier Inc.).

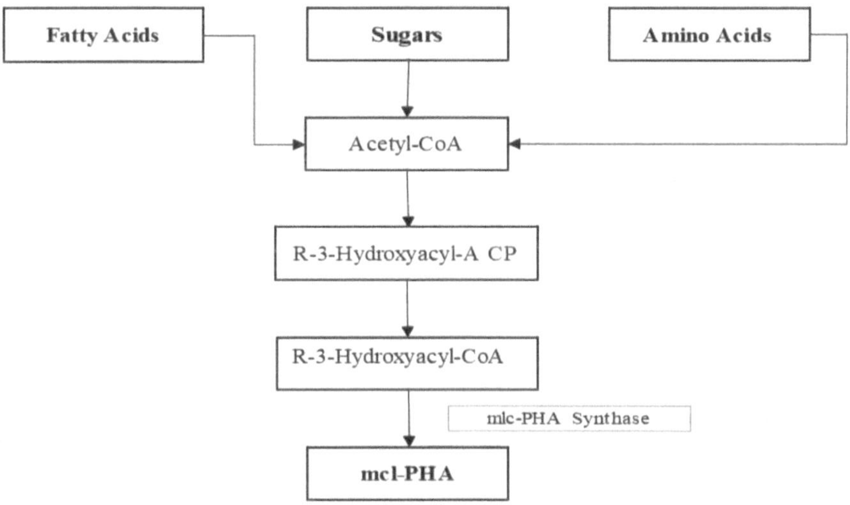

FIGURE 7.4 Pathway 3 where mcl-PHA is formed via in situ fatty acids (Reprinted from Nguyenhuynh et al. (2021), Copyright (2021), with permission from Elsevier Inc.).

PHA production by MMCs is rooted in natural principles and there are requirements to ensure efficiency in selection of PHA producers. The following are prerequisites for PHA production:

- A high rate of cell development and polymer production
- High conversion efficiency to PHA substrate

- Full capacity for polymer growth
- Ability to use waste streams as feedstock
- The potential for PHA generation in environments that enable high productivity with little energy consumption, such as continuous bioprocesses in a nonsterile, open environment (Mannina et al., 2020).

Subsequent steps for production of PHAs from mixed cultures include: (i) Acidogenic fermentation to create volatile fatty acids (VFAs) from biodegradable organics; (ii) Choosing PHAs-loading biomass in a cycling or sequencing batch reactor; and (iii) Batch phase to expand PHA growth in the bacterial cells. The enhancement and long-term farming of PHA-creating bacteria and communities have been found to benefit from the carbon control strategy under feast and starvation conditions, while the nitrogen curb is a fruitful strategy that can be used to achieve an elevated PHA medium during the PHA production step (Bagatella et al., 2022)

7.2.2.1 Comparison of Culture and MMC

MMCs possess several merits and demerits and these are outlined in Table 7.2 (Gecim et al., 2021; Mannina et al., 2020; Pakalapati et al., 2018).

7.2.2.2 PHA Properties

PHAs' chemical and physical properties differ significantly from the monomer structure. PHAs do not dissolve in water (Phas et al., 2021). Ordinarily, polymers are categorized by utilizing parameters such as hydrophobicity, melting point, glass transition temperature, and degree of crystallinity. PHAs possess various mechanical properties ranging from hard crystalline to elastic (Cavaliere et al., 2021). One extremely unique property of PHAs is biodegradability (Crutchik et al., 2020). PHA bioplastics have a molecular weight between 500,000 and over 1,000,000 Da, which is higher than that of traditional polymer plastics because

TABLE 7.2
Disadvantages and advantages of pure and mixed microbial culture

Condition	Pure culture	Mixed microbial culture
Advantages	It produces more polymer content	Ability to use bigger range of substrate
		Reactor sterilization not required
		Full utilization of substrate
		Less process control requirements
		More economical compared to pure culture
Disadvantages	Reactor sterilization required	Low polymer yield
	Complex and steep	Limited PHA storage compared to pure culture
	High risk of contamination	Unable to store PHAs from carbohydrate substrates

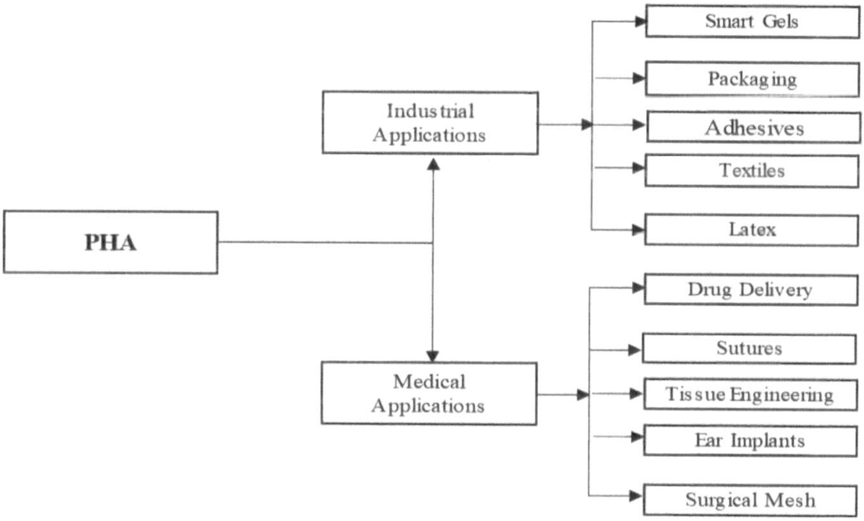

FIGURE 7.5 Demonstration of various PHA applications (Reprinted from Pakalapati et al. (2018), Copyright (2018), with permission from Elsevier Inc.).

they have a high degree of polymerization from 105 to 107 C/m^2 (Nguyenhuynh et al., 2021). PHA bioplastics can therefore be processed further by mixing with other petroleum-based polymers to overcome their shortcomings, such as brittleness and low thermal stability (Akaraonye et al., 2010). The treating temperature and delicacy of PHA-based plastics are reduced by the processability of PHB and polyethylene glycol blends (Bugnicourt et al., 2014). Figure 7.5 shows the various applications of PHAs.

7.2.3 Biosolids

Paz-Ferreiro et al. (2018) describe biosolids as the treated sewage sludge and sewage sludge as the remaining filter cake after filtering wastewater. This treatment is conducted to comply with local regulations for heavy metals and pathogens based on the possibility of landfill disposal (Paz-Ferreiro et al., 2018). The product of properly treated wastewater sludge after dewatering is called biosolids and it results from the wastewater treatment process (Mohajerani et al., 2019). Elsewhere, biosolids are referred to as treated sewage sludge for appropriate utilization in land applications (Bhatt et al., 2018). According to Al-Gheethi et al. (2018), biosolids are treated sewage sludge that has undergone sophisticated processes such as aerobic and anaerobic, heat or lime treatment and has achieved standards necessary for beneficial usage. Depending on their origin (human, vegetable, or animal) and the type of treatment they underwent, biosolids' specific features change, including physical, chemical, or biological, anaerobic, or aerobic treatment, alkaline treatment by lime, etc. Biosolids' organic and inorganic components are crucial for plant growth and soil enrichment (Al-Gheethi et al., 2018; Mohajerani et al., 2017). Due to constraints on ocean disposal, expensive landfill disposal, and social conflicts concerning incineration, terrestrial application of biosolids is a desirable choice for disposal due to their organic matter and nutrient value. However, before it can be used on land and referred to as biosolids, wastewater sludge must be adequately treated and of high quality (Fisher et al., 2019). Biosolids are rich in nutrients (Brown et al., 2020). Wastewater solids are stabilized to minimize pathogens in order to be within the acceptable specifications. Biosolid stabilization processes include anaerobic or aerobic digestion, composting, thermal drying, and alkaline stabilization. Composting combines digestion with disinfection, requiring a decrease in organic matter and an increase in temperature brought about by the aerobic decomposition of waste. To reduce the pathogenic load by disinfection or create an environment that is unfavorable for microbial development in the sludge, thermal treatment and alkaline stabilization are used. To lower the pathogenic load, the former uses high temperatures and the removal of water, whereas the latter uses an increase in pH and occasionally an increase in temperature (Fisher et al., 2019). Biosolids' management or disposal processes include incineration and landfill which are not environmentally friendly, and harmful to the ecosystem. Some of the issues hindering the use of biosolids on a bigger scale are concerns about contamination (Bhatt et al., 2018). Additionally, biosolid management processes are based on disposal or the use of the three types, including:

i. Clay land application technologies: Land reclamation, forestry enhancement, direct agricultural use, and fired clay bricks (Al-Rumaihi et al., 2020)
ii. Energy source production/recovery: Direct/hydrothermal liquefaction, gasification, pyrolysis, anaerobic digestion, synthesis of bioethanol, and hydrogen production (Elkhalifa et al., 2022)
iii. Material application: Water treatment absorbents, adsorbents, filtration media, composites, cement, construction, asphalt manufacturing, and building blocks (Elkhalifa et al., 2022).

7.2.4 Biogas

Biogas is a form of renewable energy gas that is produced by breaking down organic waste as feedstock in the absence of oxygen. It is made up of methane (CH_4), carbon dioxide (CO_2), and nitrogen (N_2) (Obaideen et al., 2022; Ullah Khan et al., 2017). Patinvoh and Taherzadeh (2019) describe biogas as a sustainable environmentally friendly and renewable energy that can address waste management and energy shortage problems (Patinvoh & Taherzadeh, 2019). It has the potential to replace natural gas and fossil fuels due to its cost-effectiveness (Ullah Khan et al., 2017). Biogas production is also known as "anaerobic digestion," where various microbes use a variety of metabolic pathways to break down organic materials, producing biogas as its byproduct. Biogas is produced through the anaerobic digestion (AD) process from different organic feedstocks in oxygen-free environments. It is common knowledge that AD recycles carbon in a variety of environments, including wetlands, rice fields, the intestines of animals, aquatic sediments, and manure (Kougias & Angelidaki, 2018). It is a true reflection of "reduce, reuse, and recycle" as it plays a vital role in waste reduction by generating energy while reducing the quantity of sludge.

However, when wastewater is the only feedstock used to produce methane through AD, it limits the benefits due to its low C/N ratio and low digestion resulting in low methane yield (Cardona et al., 2019). This setback could be improved by blending wastewater sludge with other feedstocks that are richer in carbon. This approach of blending various feedstocks has multiple benefits, which include:

- Improved digester performance that treats sludge
- Simultaneously treating different types of waste
- Reducing inhibition risk that occurs during mono-digestion (Cardona et al., 2019; Mata-Alvarez et al., 2014)

Biogas is a renewable gas that could be utilized to produce electricity and heat (Scarlat et al., 2018) and as biomethane when upgraded before blending to the gas grid or used as a vehicle fuel (Mirmohamadsadeghi et al., 2019). In some regions of the world, biogas is used to support isolated populations that are off the grid, and it is perfect for grid decentralization (Obaideen et al., 2022). Table 7.3 presents the various feedstocks used in the production of biogas and Table 7.4 shows the composition of biogas and its properties.

TABLE 7.3
Various feedstocks

Agriculture	Industrial	Municipal
Animal waste, e.g., manure	Wastewater	Sewage sludge
Dedicated energy crop	Sludge	Municipal solid waste
Crop waste	Byproduct	

Source: Balat & Balat (2009).

TABLE 7.4
Composition of biogas and its properties

Component	%Volume	Properties
Methane (CH_4)	55–65	Energy carrier
Carbon dioxide (CO_2)	35–45	Reduces heat value, causes corrosion in the presence of moisture
Hydrogen sulfide (H_2S)	0.005–2	Cause corrosion, SO_2 during combustion
Nitrogen (N_2)	0–2	Reduces heat value
Oxygen (O_2)	0–2	Facilitates combustion
Hydrogen (H_2)	0–1	Energy carrier
Ammonia (NH_3)	0–1	NOX emissions during combustion
Water (H_2O)	~10	Increase corrosion in the presence of CO_2 and SO_2

Source: Katariya & Patolia (2021).

7.2.4.1 Operating Parameters

This section presents the various operating parameters needed to produce biogas. These include:

i. pH
pH is the most important variable that needs to be tracked and managed. The appropriate pH range for digestion is, in theory, between 5.5–8.5. However, most methanogens can only function in the pH range of 6.7–7.4. A declining pH may indicate acid build up and digester variability. The only variable that demonstrates digester variability more quickly than pH is gas output. The quantity of VFAs and acetic acid must be below 2,000 mg/l for an anaerobic fermentation to proceed normally (Balat & Balat, 2009; Nabaterega et al., 2021).

ii. Temperature
Bacteria are active within a controlled range of temperatures. The bacteria productivity range is 25–40 °C under mesophilic conditions with a digestor temperature maintained between 30–35 °C, while under thermophilic conditions at 50–65 °C, the digester temperature must be kept at 50 °C (Mirmohamadsadeghi et al., 2019; Balat & Balat, 2009).

iii. Carbon/nitrogen ratio
To ensure that the C/N ratio in the input stays within the desired range, it is essential to retain the right structure of the feedstock for optimal plant operation. The widespread consensus is that during AD, carbon is utilized by microorganisms, at least 25–30 times more quickly than nitrogen. Microbes therefore require a 20–30:1 ratio of carbon to nitrogen, with most of the carbon being easily degradable, to meet this requirement. Low gas yield is caused by methanogens' prompt consumption of nitrogen, which is shown by a high C/N ratio. A declining C/N ratio, on the other hand, results in a build up of ammonia and pH levels higher than 8.5, which are risky to methanogenic bacteria. By combining the ingredients, the digester materials can have the best C/N ratios (Balat & Balat, 2009).

iv. Retention time
The batch and the continuous process are the two types of reactors utilized in AD technology. In the batch method, the substrate is added to the reactor at the start of the decomposition phase and kept shut during digestion. The processing of biogas follows a bell curve because all the reaction phases take place roughly in that order. Only about one-third of the reservoir volume is consumed for active digestion during retention periods of 30–60 days. The retention period must be less than the industry norm of 20 days if AD is to be competitive with alternative MSW disposal methods (Balat & Balat, 2009; Sarker et al., 2019).

7.2.5 Biopolymers

Biopolymers are some of the developing materials with the potential for wastewater treatment. They are better described as a group of naturally occurring polymers

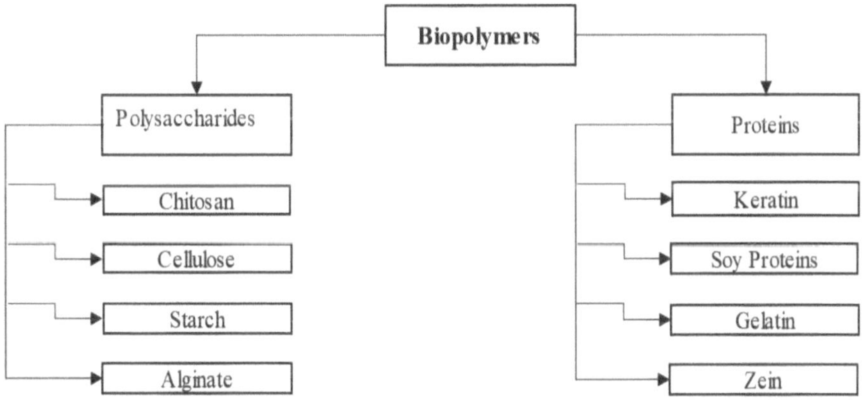

FIGURE 7.6 Biopolymers that are commonly used in wastewater treatment (Reprinted from Zubair & Ullah (2021), Copyright (2021), with permission from Elsevier Inc.).

such as polysaccharides, lipids, proteins, etc. (Zubair & Ullah, 2021). Algade et al. (2021) defined biopolymers as bioplastics that are produced from renewable biomass as opposed to petroleum counterparts. Biopolymers are obtained from the first generation of feedstock such as sugar beet, corn, or second feedstock generation like lignocellulose materials (Algade et al., 2021). It has been highlighted in some papers that currently there is no standardized definition for biopolymers. Nandakumar et al. (2021) refer to bioplastics as polymers that are biobased and which are biodegradable. However, they indicated that not all bioplastics are biodegradable, e.g., poly (butylene adipate-co-terephthalate), PBAT. The molecular weight of a biopolymer should not exceed 5000 Da. Some biopolymers are named so because they are derived from biobased raw material, which are however, nondegradable (Nandakumar et al., 2021). Alternatively, biopolymers are defined as materials that have been made from living things that have linear or branched structured molecules and are made up of monomeric units. The molecules that come from protein sources and contain nucleic acids, saccharides, or amino acids are referred to as monomeric units (Phas et al., 2021). They can be categorized as natural, chemically synthesized, and microbial biopolymers (Prabu et al., 2021). Biopolymers are favored over traditional polymers due to their renewability and environmentally friendly properties. In addition, they are cost-effective because they are directly produced from the environment (George et al., 2020). Figure 7.6 shows the various types of biopolymers that are employed in wastewater treatment.

In the family of biopolymers, polysaccharide and protein developed materials are the most commonly used for water and wastewater treatment (Zubair & Ullah, 2021).

7.2.5.1 Classes of Biopolymers

Biopolymers are broadly divided into two major groups namely, biodegradable and nonbiodegradable. The group is the driving force behind meticulous biopolymer

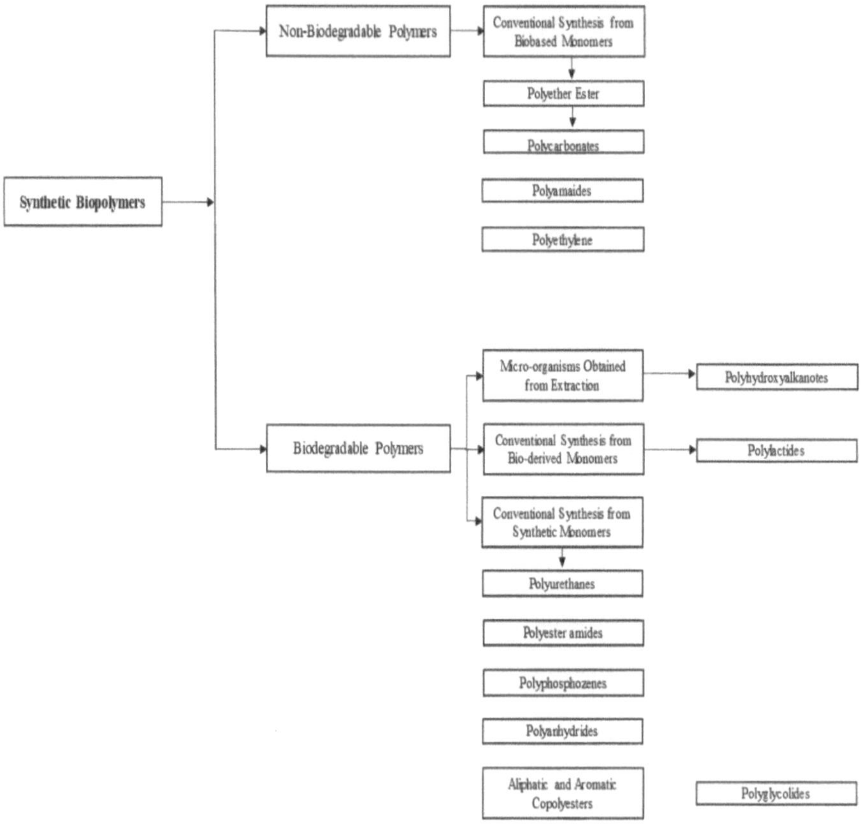

FIGURE 7.7 Flow that shows biopolymers based on biodegradability (Reprinted from George et al. (2020), Copyright (2020) with permission from Elsevier Inc.).

research because it is crucial to create novel products that do not harm the environment (Baranwal et al., 2022; Buckner et al., 2016; Ibrahim et al., 2021). Biopolymers fall into two categories based on where they came from; they are either from natural sources (like plants, animals, and microorganisms) or from fossil fuels and both categories are equally important. In terms of their usability in several functional areas, nondegradable biobased biopolymers outperformed biobased degradable ones when compared numerically (George et al., 2020). Additionally, they are classified based on their ability to withstand thermal conditions, and such biopolymers are called elastomers, thermosets, and thermoplastics. Biopolymers are clustered into three classes based on their composition in the form of blends, laminates, or composites. Figure 7.7 shows biopolymers based on their biodegradability.

7.2.5.2 Production of Biopolymers

In general, there are three different forms of bioplastics: (i) polymers made straight from biomass; (ii) polymers made through bioderived intermediates; and

(iii) polymers made by microbes (Di Bartolo et al., 2021; Nanda et al., 2022; Rosenboom et al., 2022). Bioplastics made from biomass include polymers like those made from lignin, cellulose, starch, chitin, and even milk. It is a common practice to create intermediates from polymers based on lignin, cellulose, and starch to blend them with other plastics (Abe et al., 2021; Kumar & Thakur, 2017) The byproducts of the timber industry and agricultural areas appear to be the major sources of biomass in the world and 90% of this plant biomass is made up of lignocellulose, and cellulose makes up about half of the lignocellulosic material's carbon content (Nandakumar et al., 2021). Polymers made of cellulose, such as cellulose acetate and cellulose, are produced through procedures like acylation, acetylation, and esterification with additional ingredients like carboxylic acids and their derivatives. Cellulose-based polymers including cellulose acetate, cellulose nitrate, and methyl cellulose can be produced (Nandakumar et al., 2021; Razali et al., 2022). The extraction of cellulose is the key consideration during processing since it is essential to preserve the cellulose while removing other polymers including lignin, pectin, and hemicelluloses. In terms of both yield and cost, this step serves as the rate-limiting step in the conversion of biomass to cellulose (Kumar & Chauhan, 2022; Nandakumar et al., 2021). The Kraft method, processing with caustic soda, and the sulfite method are the three most popular techniques for cellulose extraction. Each of the three procedures calls for combining wood chips with chemicals such as sodium hydroxide and sodium sulfite (Østby et al., 2020).

7.2.5.3 Properties of Biopolymers or Bioplastics

It is distinct for biopolymers to have properties that vary into three primary groups, namely: (i) relative properties, which refer to the characteristics of the polymer as a whole; (ii) synthesizing properties, which describe the behaviors during the formation stage; and (ii) component properties, which describe the polymer's functional capabilities.

These properties for all three categories are linked together, and the correlation is demonstrated in Figure 7.8 (George et al., 2020).

7.2.5.4 Applications of Biopolymers

Bioplastics or biopolymers such as PHAs can replace fossil fuels as they are produced from biomass which signifies high chances of carbon neutrality, hence the rise in the utilization of these materials (Atiwesh et al., 2021; Kumar et al., 2021; Moshood et al., 2022). They are usually used as blends or composites of synthetic polymers that are degradable or nondegradable. This concept is not new as companies like Ford derived machine parts from corn and soybean for the Ford Model T (Mangaraj et al., 2019; Vieyra et al., 2022). Currently, biopolymers are mostly utilized for food packaging, utensils, and film wrap as a single-use material. They have also found applications in biomedical, agricultural, and automotive industries (George et al., 2020) and in wastewater treatment with other reinforcement materials to improve water quality (Perspectives, 2021).

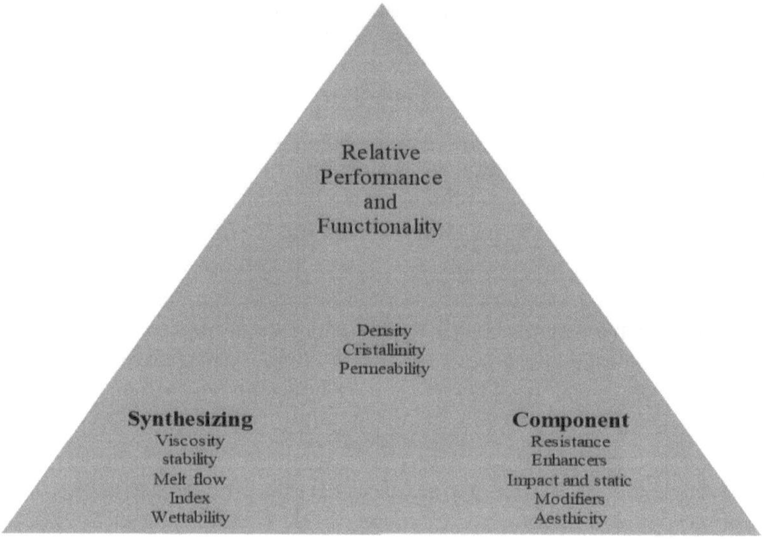

FIGURE 7.8 Triangle showing biopolymer properties (Reprinted from George et al. (2020), Copyright (2020) with permission from Elsevier Inc.).

7.3 CHALLENGES OF WASTEWATER AND OPPORTUNITIES FROM EFFECTIVE WASTEWATER MANAGEMENT

Rapid population growth and industrialization have contributed immensely to the rapid proliferation of wastewater (Kakavandi & Ahmadi, 2019). Wastewater comprises complex and heterogeneous contaminants which makes it a bit difficult to manage, thereby requiring substantially effective equipment to provide water of high quality for the masses (Obaideen et al., 2022). Several traditional methods of treatment have been employed but many of the contaminants were difficult to trap (Huang et al., 2019). Hence, recent research has focused on the use of advanced oxidation processes (AOPs) which include ozonation, photocatalysis, oxidation, etc. for the removal of pollutants and effective treatment of wastewater (Asghar et al., 2015; Zhu et al., 2017). Proper handling of wastewater contributes to economic growth and protection of the ecological environment, and assists in lessening threats associated with agricultural, industrial, and municipal products. It has been estimated that sanitation infrastructure has the potential to reward capital investment up to ~5 times while the dearth of infrastructure can drain ~8% of the gross domestic product yearly (Hung, 2015; United Nations University, 2013).

7.4 CONCLUSION

The scarcity of freshwater resources has encouraged many activities around wastewater. This has resulted in the development of valuable materials from wastewater and new economic resources. Obviously, there are challenges as much as there are opportunities. Apart from potable water, other valuable resources from wastewater include biosolids, bioplastics, biopolymers, biogas, and PHAs. Bioplastics, and particularly PHAs, will undoubtedly play a significant role in the plastics market in the future, though their production costs are still too high compared to traditional, petroleum-based polymers today. Biosolids have high potential in agricultural application and land reclamation, provided enough treatment has been done to reduce the pathogenic load and bacteria, thereby avoiding long-term effects on humans. Biogas operation can be improved by utilizing co-digestion and a stage digester is recommended for a stable process. One important factor is to guard over the type of feedstock as they differ with carbon loading. Operational parameters must also be checked for a greater methane yield and a much more stable process.

With the aggravated global warming and decreasing supplies of fresh water and (river) stream water, wastewater can partly meet the requirements of water crops. However, it is important to keep in mind that unprocessed wastewater can have detrimental consequences on public health. Employing only waste material of the proper quality for agricultural irrigation is the safest option given the growing practice of wastewater recovery and the longing to reduce risks to public health and the ecosystem. It is therefore crucial to educate both farmers and the public (consumers) on the benefits and drawbacks of using wastewater to significantly change farmers' attitudes since they are the primary stakeholders. In addition, empowering farmers with knowledge is significant in influencing wastewater acceptance. The treatment conditions, regulations, and quality standards of wastewater for irrigation are very important for the safety of public health and the environment. Finally, it is recommended that the implementation of a multiple barrier approach and the use of RO as the primary and last stage of quality assurance should be regulated.

ACKNOWLEDGMENT

The first author would like to thank the University of Johannesburg, Global Excellence Stature for the funding provided for her study.

REFERENCES

Abe, M. M., Martins, J. R., Sanvezzo, P. B., Macedo, J. V., Branciforti, M. C., Halley, P., Botaro, V. R., and Brienzo, M. (2021). Advantages and disadvantages of bioplastics production from starch and lignocellulosic components. *Polymers*, *13*(15). https://doi.org/10.3390/polym13152484

Aczel, M. R. (2019). What is the nitrogen cycle and why is it key to life? *Frontiers for Young Minds*, *7*(March), 1–9. https://doi.org/10.3389/frym.2019.00041

Akaraonye, E., Keshavarz, T., and Roy, I. (2010). Production of polyhydroxyalkanoates: The future green materials of choice. *Journal of Chemical Technology and Biotechnology*, *85*(6), 732–743. https://doi.org/10.1002/jctb.2392

Al-Gheethi, A. A., Efaq, A. N., Bala, J. D., Norli, I., Abdel-Monem, M. O., and Ab. Kadir, M. O. (2018). Removal of pathogenic bacteria from sewage-treated effluent and biosolids for agricultural purposes. *Applied Water Science*, *8*(2), 1–25. https://doi.org/10.1007/s13201-018-0698-6

Al-Rumaihi, A., McKay, G., Mackey, H. R., and Al-Ansari, T. (2020). Environmental impact assessment of food waste management using two composting techniques. *Sustainability (Switzerland)*, *12*(4). https://doi.org/10.3390/su12041595

Algade, A., Qiu, S., Ge, S., Naa, G., Addico, D., Komla, G., Yu, Z., Xia, W., Abbew, A., Shao, D., Champagne, P., and Wang, S. (2021). Science of the Total Environment A review of biopolymer (Poly- β -hydroxybutyrate) synthesis in microbes cultivated on wastewater. *Science of the Total Environment*, *756*, 143729. https://doi.org/10.1016/j.scitotenv.2020.143729

Almeida, J. R., Serrano, E., Fernandez, M., Fradinho, J. C., Oehmen, A., and Reis, M. A. M. (2021). Polyhydroxyalkanoates production from fermented domestic wastewater using phototrophic mixed cultures. *Water Research*, *197*, 117101. https://doi.org/10.1016/j.watres.2021.117101

Asghar, A., Raman, A. A. A., and Daud, W. M. A. W. (2015). Advanced oxidation processes for in-situ production of hydrogen peroxide/hydroxyl radical for textile wastewater treatment: A review. *Journal of Cleaner Production*, 87, 826–838.

Atiwesh, G., Mikhael, A., Parrish, C. C., Banoub, J., and Le, T. A. T. (2021). Environmental impact of bioplastic use: A review. *Heliyon*, *7*(9), e07918. https://doi.org/10.1016/j.heliyon.2021.e07918

Bagatella, S., Ciapponi, R., Ficara, E., Frison, N., and Turri, S. (2022). *Production and Characterization of Polyhydroxyalkanoates from Wastewater via Mixed Microbial Cultures and Microalgae*. 1–19. C:\Users\Sisanda\Documents\Research\Resource from wastewater 240922\Polyhydroxyalkanoates\Production and Characterization of PHA from WW.pdf

Balat, M., and Balat, H. (2009). Biogas as a renewable energy source a review. *Energy Sources, Part A: Recovery, Utilization and Environmental Effects*, *31*(14), 1280–1293. https://doi.org/10.1080/15567030802089565

Baranwal, J., Barse, B., Fais, A., Delogu, G. L., and Kumar, A. (2022). Biopolymer: A sustainable material for food and medical applications. *Polymers*, *14*(5), 1–22. https://doi.org/10.3390/polym14050983

Bhatt, D., Shrestha, A., Dahal, R. K., Acharya, B., Basu, P., and MacEwen, R. (2018). Hydrothermal carbonization of biosolids from Waste water treatment plant. *Energies*, *11*(9), 1–10. https://doi.org/10.3390/en11092286

Brar, B., and Rawat, J. (2022). Wastewater reuse for irrigation of a vegetable crops and its impacts. The Pharma Innovation, *11*(7), 111–117. C:\Users\Susanda\Documents\Research\Resource from wastewater 240922\Irrigation\Wastewater Reuse for Irrigation of Vegetable Crops and Impacts.pdf

Brown, S., Ippolito, J. A., Hundal, L. S., and Basta, N. T. (2020). Municipal biosolids — A resource for sustainable communities. *Current Opinion in Environmental Science and Health*, *14*, 56–62. https://doi.org/10.1016/j.coesh.2020.02.007

Buckner, C. A., Lafrenie, R. M., Dénommée, J. A., Caswell, J. M., Want, D. A., Gan, G. G., Leong, Y. C., Bee, P. C., Chin, E., Teh, A. K. H., Picco, S., Villegas, L., Tonelli, F., Merlo, M., Rigau, J., Diaz, D., Masuelli, M., Korrapati, S., Kurra, P., … Mathijssen, R. H. J. (2016). We are IntechOpen, the world ' s leading publisher of Open Access books Built by scientists, for scientists TOP 1 %. *Intech*, *11*, 13. www.intechopen.com/books/advanced-biometric-technologies/liveness-detection-in-biometrics

Bugnicourt, E., Cinelli, P., Lazzeri, A., and Alvarez, V. (2014). Polyhydroxyalkanoate (PHA): Review of synthesis, characteristics, processing and potential applications in packaging. *Express Polymer Letters*, *8*(11), 791–808. https://doi.org/10.3144/expresspolymlett.2014.82

Caball, R., and Malekpour, S. (2019). Decision making under crisis: Lessons from the Millennium Drought in Australia. *International Journal of Disaster Risk Reduction*, *34*(July 2018), 387–396. https://doi.org/10.1016/j.ijdrr.2018.12.008

Capodaglio, A. G. (2020). Fit-for-purpose urban wastewater reuse: Analysis of issues and available technologies for sustainable multiple barrier approaches. *Critical Reviews in*

Environmental Science and Technology, *51*(15), 1–48. https://doi.org/10.1080/10643389.2020.1763231

Cardona, L., Levrard, C., Guenne, A., Chapleur, O., and Mazéas, L. (2019). Co-digestion of wastewater sludge: Choosing the optimal blend. *Waste Management*, *87*, 772–781. https://doi.org/10.1016/j.wasman.2019.03.016

Castilho, L. R., Mitchell, D. A., and Freire, D. M. G. (2009). Production of polyhydroxyalkanoates (PHAs) from waste materials and by-products by submerged and solid-state fermentation. *Bioresource Technology*, *100*(23), 5996–6009. https://doi.org/10.1016/j.biortech.2009.03.088

Cavaliere, C., Capriotti, A. L., Cerrato, A., Lorini, L., Montone, C. M., Valentino, F., Laganà, A., and Majone, M. (2021). Identification and quantification of polycyclic aromatic hydrocarbons in polyhydroxyalkanoates produced from mixed microbial cultures and municipal organic wastes at pilot scale. *Molecules*, *26*(3). https://doi.org/10.3390/molecules26030539

Chaudhary, J., Chadik, P. A., and Osborne, T. (2019). *Application of Reclaimed Water for Irrigation: A Review*. November. C:\Users\Sisanda\Documents\Research\Resource from wastewater 240922\Irrigation\Chaudhary_Juhi_Immediate_Release.pdf

Chen, G., Chen, X., Wu, F., and Chen, J. (2020). Advanced Industrial and Engineering Polymer Research Polyhydroxyalkanoates (PHA) toward cost competitiveness and functionality. *Advanced Industrial and Engineering Polymer Research*, *3*(1), 1–7. https://doi.org/10.1016/j.aiepr.2019.11.001

Chen, G. Q., Hajnal, I., Wu, H., Lv, L., and Ye, J. (2015). Engineering biosynthesis mechanisms for diversifying Polyhydroxyalkanoates. *Trends in Biotechnology*, *33*(10), 565–574. https://doi.org/10.1016/j.tibtech.2015.07.007

Chojnacka, K., Witek-Krowiak, A., Moustakas, K., Skrzypczak, D., Mikula, K., and Loizidou, M. (2020). A transition from conventional irrigation to fertigation with reclaimed wastewater: Prospects and challenges. *Renewable and Sustainable Energy Reviews*, *130*, 109959. https://doi.org/10.1016/j.rser.2020.109959

Cipolletta, G., Ozbayram, E. G., Eusebi, A. L., Akyol, Ç., Malamis, S., Mino, E., and Fatone, F. (2021). Policy and legislative barriers to close water-related loops in innovative small water and wastewater systems in Europe: A critical analysis. *Journal of Cleaner Production*, *288*. https://doi.org/10.1016/j.jclepro.2020.125604

Crutchik, D., Franchi, O., Caminos, L., Jeison, D., and Belmonte, M. (n.d.). *Polyhydroxyalkanoates (PHAs) Production: A Feasible Economic Option for the Treatment of Sewage Sludge in Municipal Wastewater*. 1–12.

Di Bartolo, A., Infurna, G., and Dintcheva, N. T. (2021). A review of bioplastics and their adoption in the circular economy. *Polymers*, *13*(8). https://doi.org/10.3390/polym13081229

Devda, V., Chaudhary, K., Varjani, S., Pathak, B., Patel, A. K., Singhania, R. R., Taherzadeh, M. J., Ngo, H. H., Wong, J. W. C., Guo, W., and Chaturvedi, P. (2021). Recovery of resources from industrial wastewater employing electrochemical technologies: Status, advancements and perspectives. *Bioengineered*, *12*(1), 4697–4718. https://doi.org/10.1080/21655979.2021.1946631

Elkhalifa, S., Mackey, H. R., Al-Ansari, T., and McKay, G. (2022). Pyrolysis of biosolids to produce biochars: A review. *Sustainability (Switzerland)*, *14*(15). https://doi.org/10.3390/su14159626

Fisher, R. M., Alvarez-Gaitan, J. P., and Stuetz, R. M. (2019). Review of the effects of wastewater biosolids stabilization processes on odor emissions. *Critical Reviews in Environmental Science and Technology*, *49*(17), 1515–1586. https://doi.org/10.1080/10643389.2019.1579620

Gecim, G., Aydin, G., Tavsanoglu, T., Erkoc, E., and Kalemtas, A. (2021). Journal of Water Process Engineering Review on extraction of polyhydroxyalkanoates and astaxanthin from food and beverage processing wastewater United States of America. *Journal of Water Process Engineering*, *40*, 101775. https://doi.org/10.1016/j.jwpe.2020.101775

George, A., Sanjay, M. R., Srisuk, R., Parameswaranpillai, J., and Siengchin, S. (2020). A comprehensive review on chemical properties and applications of biopolymers and their composites. *International Journal of Biological Macromolecules*, *154*, 329–338. https://doi.org/10.1016/j.ijbiomac.2020.03.120

Huang, Y., Luo, M., Xu, Z., Zhang, D., and Li, L. (2019). Catalytic ozonation of organic contaminants in petrochemical wastewater with iron-nickel foam as catalyst. *Separation and Purification Technology*, *211*, 269–278.

Hung, H. P. (2015, 16 February 2015). Utilizing wastewater presents environmental and economic opportunities. Retrieved from https://breakingenergy.com/2015/02/16/utilizing-wastewater-presents-environmental-and-economic-opportunities/

Ibrahim, N. I., Shahar, F. S., Hameed Sultan, M. T., Md Shah, A. U., Azrie Safri, S. N., and Mat Yazik, M. H. (2021). Overview of bioplastic introduction and its applications in product packaging. *Coatings*, *11*(11). https://doi.org/10.3390/COATINGS11111423

Iroegbu, A. O. C., Sadiku, R. E., Ray, S. S., and Hamam, Y. (2020). Plastics in municipal drinking water and wastewater treatment plant effluents: challenges and opportunities for South Africa—A review. *Environmental Science and Pollution Research*, *27*(12), 12953–12966. https://doi.org/10.1007/s11356-020-08194-5

Kakavandi, B., and Ahmadi, M. (2019). Efficient treatment of saline recalcitrant petrochemical wastewater using heterogeneous UV-assisted sono-Fenton process. *Ultrasonics Sonochemistry*, *56*, 25–36.

Katariya, H. G., and Patolia, H. P. (2021). Advances in biogas cleaning, enrichment, and utilization technologies: A way forward. *Biomass Conversion and Biorefinery*, *0123456789*. https://doi.org/10.1007/s13399-021-01750-0

Kehrein, P., Van Loosdrecht, M., Osseweijer, P., Garfí, M., Dewulf, J., and Posada, J. (2020). A critical review of resource recovery from municipal wastewater treatment plants-market supply potentials, technologies and bottlenecks. *Environmental Science: Water Research and Technology*, *6*(4), 877–910. https://doi.org/10.1039/c9ew00905a

Kougias, P. G., and Angelidaki, I. (2018). Biogas and its opportunities — A review Keywords. *Frontiers in Environmental Science*, *12*, 1–22.

Kumar, R., and Chauhan, S. (2022). Cellulose nanocrystals based delivery vehicles for anticancer agent curcumin. *International Journal of Biological Macromolecules*, *221*(September), 842–864. https://doi.org/10.1016/j.ijbiomac.2022.09.077

Kumar, S., and Thakur, K. (2017). Bioplastics – classification, production and their potential food applications. *Journal of Hill Agriculture*, *8*(2), 118. https://doi.org/10.5958/2230-7338.2017.00024.6

Kumar, V., Srivastava, S., and Thakur, I.S., 2021. Enhanced recovery of polyhydroxyalkanoates from secondary wastewater sludge of sewage treatment plant: Analysis and process parameters optimization. *Bioresource Technology Reports*, 15, 100783.

Lahlou, F. Z., Mackey, H. R., and Al-Ansari, T. (2021). Wastewater reuse for livestock feed irrigation as a sustainable practice: A socio-environmental-economic review. *Journal of Cleaner Production*, *294*, 126331. https://doi.org/10.1016/j.jclepro.2021.126331

Lahlou, F. Z, Namany, S., Mackey, H. R., and Al-Ansari, T. (2020). Treated Industrial wastewater as a water and nutrients source for tomatoes cultivation: An optimisation approach. *Computer Aided Chemical Engineering*, *48*, 1819–1824. https://doi.org/10.1016/B978-0-12-823377-1.50304-9

Luthy, R. G., Wolfand, J. M., and Bradshaw, J. L. (2020). Urban water revolution: Sustainable water futures for California cities. *Journal of Environmental Engineering*, *146*(7), 1–15. https://doi.org/10.1061/(asce)ee.1943-7870.0001715

Maaß, O., and Grundmann, P. (2018). Governing transactions and interdependences between linked value chains in a circular economy: The case of wastewater reuse in Braunschweig (Germany). *Sustainability*, *10*(4), 1–29. https://doi.org/10.3390/su10041125

Mangaraj, S., Yadav, A., Bal, L. M., Dash, S. K., and Mahanti, N. K. (2019). Application of biodegradable polymers in food packaging industry: A comprehensive review. *Journal of Packaging Technology and Research*, *3*(1), 77–96. https://doi.org/10.1007/s41783-018-0049-y

Mannina, G., Presti, D., Montiel-jarillo, G., Carrera, J., and Suárez-ojeda, M. E. (2020). Bioresource Technology Recovery of polyhydroxyalkanoates (PHAs) from wastewater: A review. *Bioresource Technology*, *297*, 122478. https://doi.org/10.1016/j.biortech.2019.122478

Mannina, G., Presti, D., Montiel-jarillo, G., and Suárez-ojeda, M. E. (2019). Bioresource Technology Bioplastic recovery from wastewater: A new protocol for polyhydroxyalkanoates (PHA) extraction from mixed microbial cultures. *Bioresource Technology*, *282*, 361–369. https://doi.org/10.1016/j.biortech.2019.03.037

Marron, E. L., Mitch, W. A., Gunten, U. Von, and Sedlak, D. L. (2019). A tale of two treatments: The multiple barrier approach to removing chemical contaminants during potable water reuse. *Accounts of Chemical Research*, *52*(3), 615–622. https://doi.org/10.1021/acs.accounts.8b00612

Mata-Alvarez, J., Dosta, J., Romero-Güiza, M. S., Fonoll, X., Peces, M., and Astals, S. (2014). A critical review on anaerobic co-digestion achievements between 2010 and 2013. *Renewable and Sustainable Energy Reviews*, *36*, 412–427. https://doi.org/10.1016/j.rser.2014.04.039

Menge, J. (1997). *Treatment of ww for Re-use in the Drinking Water System of Windhoek*, NAMIBIA.

Mirmohamadsadeghi, S., Karimi, K., Tabatabaei, M., and Aghbashlo, M. (2019). Biogas production from food wastes: A review on recent developments and future perspectives. *Bioresource Technology Reports*, *7*, 100202. https://doi.org/10.1016/j.biteb.2019.100202

Mohajerani, A., Lound, S., Liassos, G., Kurmus, H., Ukwatta, A., and Nazari, M. (2017). Physical, mechanical and chemical properties of biosolids and raw brown coal fly ash, and their combination for road structural fill applications. *Journal of Cleaner Production*, *166*, 1–11. https://doi.org/10.1016/j.jclepro.2017.07.250

Mohajerani, A., Ukwatta, A., Jeffrey-Bailey, T., Swaney, M., Ahmed, M., Rodwell, G., Bartolo, S., Eshtiaghi, N., and Setunge, S. (2019). A proposal for recycling theworld's unused stockpiles of treated wastewater sludge (biosolids) in fired-clay bricks. *Buildings*, *9*(1). https://doi.org/10.3390/buildings9010014

Moshood, T. D., Nawanir, G., Mahmud, F., Mohamad, F., Ahmad, M. H., and AbdulGhani, A. (2022). Sustainability of biodegradable plastics: New problem or solution to solve the global plastic pollution? *Current Research in Green and Sustainable Chemistry*, *5*(November 2021). https://doi.org/10.1016/j.crgsc.2022.100273

Nabaterega, R., Kumar, V., Khoei, S., and Eskicioglu, C. (2021). A review on two-stage anaerobic digestion options for optimizing municipal wastewater sludge treatment process. *Journal of Environmental Chemical Engineering*, *9*(4), 105502. https://doi.org/10.1016/j.jece.2021.105502

Nanda, S., Patra, B. R., Patel, R., Bakos, J., and Dalai, A. K. (2022). Innovations in applications and prospects of bioplastics and biopolymers: A review. *Environmental Chemistry Letters*, *20*(1), 379–395. https://doi.org/10.1007/s10311-021-01334-4

Nandakumar, A., Chuah, J. A., and Sudesh, K. (2021). Bioplastics: A boon or bane? *Renewable and Sustainable Energy Reviews*, *147*, 111237. https://doi.org/10.1016/j.rser.2021.111237

Nguyenhuynh, T., Yoon, L. W., Chow, Y. H., and Chua, A. S. M. (2021). An insight into enrichment strategies for mixed culture in polyhydroxyalkanoate production: feedstocks, operating conditions and inherent challenges. *Chemical Engineering Journal*, *420*(P3), 130488. https://doi.org/10.1016/j.cej.2021.130488

Novelli, L. D. D., Sayavedra, S. M., and Rene, E. R. (2021). Bioresource Technology Polyhydroxyalkanoate (PHA) production via resource recovery from industrial waste streams: A review of techniques and perspectives. *Bioresource Technology*, *331*, 124985. https://doi.org/10.1016/j.biortech.2021.124985

Obaideen, K., Abdelkareem, M. A., Wilberforce, T., Elsaid, K., Sayed, E. T., Maghrabie, H. M., and Olabi, A. G. (2022). Biogas role in achievement of the sustainable development goals: Evaluation, Challenges, and Guidelines. *Journal of the Taiwan Institute of Chemical Engineers*, *131*, 104207. https://doi.org/10.1016/j.jtice.2022.104207.

Ofori, S., Puškáčová, A., Růžičková, I., and Wanner, J. (2021). Treated wastewater reuse for irrigation: Pros and cons. *Science of the Total Environment*, *760*. https://doi.org/10.1016/j.scitotenv.2020.144026

Østby, H., Hansen, L. D., Horn, S. J., Eijsink, V. G. H., and Várnai, A. (2020). Enzymatic processing of lignocellulosic biomass: Principles, recent advances and perspectives. In *Journal of Industrial Microbiology and Biotechnology* 47(9–10). Springer International Publishing. https://doi.org/10.1007/s10295-020-02301-8

Pakalapati, H., Chang, C. K., Show, P. L., Arumugasamy, S. K., and Lan, J. C. W. (2018). Development of polyhydroxyalkanoates production from waste feedstocks and applications. *Journal of Bioscience and Bioengineering*, *126*(3), 282–292. https://doi.org/10.1016/j.jbiosc.2018.03.016

Patinvoh, R. J., and Taherzadeh, M. J. (2019). Challenges of biogas implementation in developing countries. *Current Opinion in Environmental Science and Health*, *12*, 30–37. https://doi.org/10.1016/j.coesh.2019.09.006

Paz-Ferreiro, J., Nieto, A., Méndez, A., Askeland, M. P. J., and Gascó, G. (2018). Biochar from biosolids pyrolysis: A review. *International Journal of Environmental Research and Public Health*, *15*(5). https://doi.org/10.3390/ijerph15050956

Perspectives, F. (2021). *Sustainable Removal of Contaminants by Biopolymers: and Future Perspectives*. C:\Users\Sisanda\Documents\Research\Resource from wastewater 240922\Biopolymers\Sustainable Removal.pdf

Phas, P., Nanoparticles, P., Samrot, A. V, Samanvitha, S. K., Shobana, N., Renitta, E. R., Senthilkumar, P., Kumar, S. S., Abirami, S., Dhiva, S., Bavanilatha, M., Prakash, P., Saigeetha, S., and Shree, K. S. (2021). *The Synthesis, Characterization and Applications of*. 1–29. C:\Users\Sisanda\Documents\Research\Resource from wastewater 240922\Polyhydroxyalkanoates\polymers-13-03302-v2.pdf

Prabu, G., Muthusamy, S., Selvaganesh, B., Sivarajasekar, N., Rambabu, K., Sivamani, S., Sivakumar, N., Maran, J. P., and Hosseini-bandegharaei, A. (2021). Ecofriendly biopolymers and composites: Preparation and their applications in water-treatment. *Biotechnology Advances*, *52*, 107815. https://doi.org/10.1016/j.biotechadv.2021.107815

Quemada, M., Delgado, A., Mateos, L., and Villalobos, F. J. (2016). Principles of agronomy for sustainable agriculture. *Principles of Agronomy for Sustainable Agriculture*. https://doi.org/10.1007/978-3-319-46116-8

Razali, N. A. M., Sohaimi, R. M., Othman, R. N. I. R., Abdullah, N., Demon, S. Z. N., Jasmani, L., Yunus, W. M. Z. W., Ya'acob, W. M. H. W., Salleh, E. M., Norizan, M. N., and Halim, N. A. (2022). Comparative study on extraction of cellulose fiber from rice straw waste from chemo-mechanical and pulping method. *Polymers*, *14*(3). https://doi.org/10.3390/polym14030387

Rosenboom, J. G., Langer, R., and Traverso, G. (2022). Bioplastics for a circular economy. *Nature Reviews Materials*, *7*(2), 117–137. https://doi.org/10.1038/s41578-021-00407-8

Sarker, S., Lamb, J. J., Hjelme, D. R., and Lien, K. M. (2019). A review of the role of critical parameters in the design and operation of biogas production plants. *Applied Sciences*, *9*(9). https://doi.org/10.3390/app9091915

Scarlat, N., Dallemand, J. F., and Fahl, F. (2018). Biogas: Developments and perspectives in Europe. *Renewable Energy*, *129*, 457–472. https://doi.org/10.1016/j.renene.2018.03.006

Singh, A. (2021). A review of wastewater irrigation: Environmental implications. *Resources, Conservation and Recycling*, *168*, 105454. https://doi.org/10.1016/j.resconrec.2021.105454

Tortajada, C., and Nambiar, S. (2019). Communications on technological innovations: Potable water reuse. *Water*, *11*(2), 1–29. https://doi.org/10.3390/w11020251

Tortajada, C., and van Rensburg, P. (2020). Drink more recycled wastewater. *Nature*, *577*(7788), 26–28. https://doi.org/10.1038/d41586-019-03913-6

Tzanakakis, V. A., Paranychianakis, N. V., and Angelakis, A. N. (2020). Water supply and water scarcity. *Water (Switzerland)*, *12*(9), 1–16. https://doi.org/10.3390/w12092347

Ullah Khan, I., Hafiz Dzarfan Othman, M., Hashim, H., Matsuura, T., Ismail, A. F., Rezaei-DashtArzhandi, M., and Wan Azelee, I. (2017). Biogas as a renewable energy fuel – A review of biogas upgrading, utilisation and storage. *Energy Conversion and Management*, *150*, 277–294. https://doi.org/10.1016/j.enconman.2017.08.035

Ungureanu, N., and Vladut, V. (2018). Current state of wastewater use in irrigated agriculture. *Analele Universității Din Craiova*, *XLVIII*, 417–424. http://anale.agro-craiova.ro/index.php/aamc/article/view/853/808

United Nations University. (2013, 9 September 2013). UN: Rising reuse of wastewater in forecast but world lacks data on "Massive Potential Resource". Retrieved from https://unu.edu/media-relations/releases/rising-reuse-of-wastewater-in-forecast-but-world-lacks-data.html

Vieyra, H., Molina-Romero, J. M., Calderón-Nájera, J. de D., and Santana-Díaz, A. (2022). Engineering, recyclable, and biodegradable plastics in the automotive industry: A review. *Polymers*, *14*(16). https://doi.org/10.3390/polym14163412

Zhu, Q., Chen, F., Guo, S., Chen, X., and Chen, J. (2017). Variation of catalyst structure and catalytic activity during catalyst preparation for catalytic ozonation of heavy oil produced water. *Water, Air, and Soil Pollution*, *228*, 1–11.

Zubair, M., and Ullah, A. (2021). Chapter 14 – Biopolymers in environmental applications: Industrial wastewater treatment. In *Biopolymers and Their Industrial Applications*. Elsevier Inc. https://doi.org/10.1016/B978-0-12-819240-5.00014-6

8 The Intervention of Microbial Enzymes in Wastewater Management

An Update

Kanchan Yadav, Shruti Dwivedi, Supriya Gupta, Aiman Tanveer, and Dinesh Yadav

8.1 INTRODUCTION

The natural resources on Earth and their decreasing quality are a major global environmental catastrophe. With 2.5% of the total supply, water is the most important natural resource on Earth. From the available quotient, only one third is readily accessible to humans. For each living thing to survive, clean water is necessary. The declining quality of water is directly proportional to the increase in anthropogenic activities. Industrialization and rapidly increasing population growth are vigorously hampering water health. The agricultural, industrial, and domestic sectors have seen a sharp increase in water demand, consuming 70%, 20%, and 8%, respectively, of the available fresh water. This further results in the production of significant amounts of wastewater containing various pollutants like dyes, chemical compounds, etc. (Aitken et al., 1994; Geyer et al., 2017; Maloney et al., 2015; Santhosh et al., 2016). Certain recalcitrant compounds and other hard to remove pollutants include food wastes, oil, fat, grease, heavy metals, plastics, industrial chemicals, and pesticides (Ahmad et al., 2020; Feng et al., 2021; Meng et al., 2015).

Filamentous microbes may develop as a result of floating oil and grease, causing microbial bloom. This further prohibits sludge from floating, resulting in poor sedimentation, and a decrease in the biomass of activated sludge (Wee et al., 2016; Al-Maqdi et al., 2021; Cammarota & Freire, 2006; Raouf et al., 2019; Esposito & Durán, 2000). These methods have been highly industrially efficient for years but rewarded the environment with heavy pollutants and low-quality water (Mohammadi et al., 2015; Villegas et al., 2016). Biological methods are replacements for these conventional approaches. They are equally efficient and do not harm the environment (Al-Maqdi et al., 2021; Alshabib & Onaizi, 2019; Müller et al., 2023).

 DOI: 10.1201/9781003441069-8

These technologies can either degrade pollutants or use organisms to assimilate them. While some harmful chemicals in wastewater can be poisonous to microorganisms and plants, enzymes can function swiftly and selectively (Ebele et al., 2017; Gianfreda et al., 2016; Kesari et al., 2021; Feng et al., 2021). Particularly in mild settings, enzymes function as biocatalysts to biodegrade compounds. To reduce the activation energy during enzymatic reactions, enzymes include active sites that can bind to certain substrates. As a result, the specificity and kinetics of these processes are very high. Enzymes may also hasten the transport of substrates into cells, increasing the efficiency of those cells. The two enzyme classes that are utilized in wastewater treatment most frequently are hydrolases and oxidoreductases. The majority of contaminants in wastewater can be biocatalyzed by them because of the vast range of compounds they can bind (Pandey et al., 2017; Mishra et al., 2020; Feng et al., 2021).

Commercial applications exist today for enzymes like lipase, laccase, and peroxidase. Laccase and peroxidase are frequently employed to remove some organic micropollutants because they have broad substrate specificities or are promiscuous. The "green chemistry" trend is met by enzyme technology (García-Molina et al., 2022; Ji et al., 2016; Zerva et al., 2017; Varga et al., 2019). Since enzymes are not consumed by reactions and can be regenerated later, enzymatic processes are more efficient than typical chemical ones in terms of water and energy use. Enzymes are also not in competition with other microbes, unlike bacteria or other biological processes. The removal of pollutants including oil, fat, and grease as well as various other organic micropollutants from wastewater is therefore possible using enzymatic methods (Summerscales, 2021; de Cazes et al., 2014; Manmohit Kalia, 2015; Feng et al., 2021). This chapter provides a brief overview of the current status of the application of microbial enzymes in wastewater management.

8.2 WASTEWATER FROM DIVERSE SOURCES

8.2.1 Agro-based Wastewater

Pesticides, such as herbicides, insecticides, and fungicides, are currently used extensively for crop protection around the world, and their use is anticipated to rise in the coming years. Due to their high residue levels, low efficacy, and significant wastage, pesticides constitute a substantial hazard to aquatic ecosystems and human health (Rao et al., 2010; Maier et al., 2015; Marican & Durán-Lara, 2018; Feng et al., 2021). As a result of runoff or percolation, residues from highly toxic pesticides enter water bodies, destroying water quality, altering the metabolism, regulation, and biochemical processes of aquatic lifeforms, and upsetting the equilibrium of aquatic ecosystems (Al-Maqdi et al., 2021; Villarrocl ct al., 2009; Pal ct al., 2014; Moraes et al., 2007). Pesticides with an organophosphorus or carbamate chemical structure are currently employed regularly (Jiang et al., 2019; Nguyen et al., 2014). These pesticides threaten not just animal health but also the ecosystem since they inhibit cholinesterase. Other life-threatening consequences have also been documented, in addition to cholinesterase toxicities (Sultatos, 1994). Triazophos, methamidophos, and carbofuran are some examples of highly hazardous pesticides that are still in high

FIGURE 8.1 The diverse sources of wastewater production.

demand, particularly in poor nations (Zhang et al., 2020; Feng et al., 2021). Multiple sources of wastewater creation are shown in Figure 8.1.

8.2.2 Cosmetic-chemical Based Wastewater

In the modern world, personal care products (PCPs) including shampoos, toothpastes, facial creams, shaving creams, sunscreen lotions, detergents, UV filters, deodorants, perfumes, lipsticks, lip balms, and other items are used frequently by millions of people (Balmer et al., 2005). Numerous PCPs are unintentionally and intentionally introduced into the aquatic environment (Kumar et al., 2022). Inadequate removal of these contaminants by wastewater treatment facilities and human activities such as bathing, washing, cleaning, and laundering may result in an indirect source of PCPs. Many dangerous chemical substances are frequently found in PCPs, such as triclosan (TCS), octocrylene, benzophenone-3, and oxybenzone (Lee et al., 2020; Sheng et al., 2021; Feng et al., 2021; Sendra et al., 2017).

8.2.3 Pharmaceutical Industry-based Wastewater

Pharmaceuticals have been identified as a frequent organic micropollutant in aquatic settings in recent years. According to reports (Wang et al., 2016; Rao et al., 2010), the widespread use of pharmaceuticals has led to the detection of over 200 different pharmacologically active chemicals in water bodies. Pharmaceuticals commonly detected in wastewater include antibiotics, cardiovascular and nervous system drugs (such as analgesics), metabolic and hormonal drugs (such as estrogens), and psychotherapeutic drugs (such as anti-depressants) (Varga et al., 2019; Al-Maqdi et al., 2021). By using traditional activated sludge systems, pharmaceuticals cannot be successfully removed. Only 4 of 35 pharmaceuticals, or less than 50% of them, could be eliminated using activated sewage sludge from municipal wastewater treatment plants, according to a report (Joss et al., 2006). Additionally, one study (Watkinson et al., 2007) revealed that wastewater treatment facilities could only passively remove 20–90% of antibiotic residues. Antibacterial medications like cephalosporins and tetracyclines have difficulty degrading naturally (Guo & Chen, 2015). Medications are widely used, resulting in the ongoing release of leftover medications into aquatic environments. They have been detected in surface water, groundwater, and even drinking water. Pharmaceuticals like diclofenac (DCF) and carbamazepine (CBZ) are ecotoxic (Lonappan et al., 2016; Stadlmair et al., 2017). Pharmaceutics' bioaccumulation and ecotoxicological characteristics make them a significant threat to aquatic ecosystems.

8.2.4 Textile Industry-based Wastewater

Textile industries release dyes in their effluents. The discharge of dyes into the environment only makes up a minor part of total water pollution when compared to other chemical pollutants. In contrast to the estimated yearly output of these dyes, 15–20% of dye production is lost in water bodies. These dyes like azo, nitro, or sulpho-groups, are resistant to microbial destruction. They cause a rise in chemical oxygen demand (COD) and biochemical oxygen exchange (BOD), which slows down photosynthesis and plant growth. In the long run, dye-contaminated water causes recalcitrance and bioaccumulation. Additionally, it develops toxicity, and carcinogenicity and causes mutations in living systems upon utilization. Aquatic life, flora, and fauna are adversely affected, and humans are also harmed when they consume seafood that has been raised in such contaminated water. Environmental regulatory bodies are concerned about the entire cycle of health issues. According to various scientific publications that have been published on the topic, the removal of color from effluents has attracted scientific attention on a global scale. Only a small number of physio-chemical methods for the decolorization of textile wastewater have been described in recent years, even though many of them are technically feasible for use by the textile industry (Dahiya & Nigam, 2020).

Conventional methods have been extensively used in removing dyes from wastewater. However, the lesser efficiency, high pricing, and inapplicability of methods to treat the wide range of synthetic dyes present makes conventional methods unprofitable. For the remediation of textile wastewater, some frequently utilized techniques include oxidation, precipitation, coagulation, adsorption, filtration, and ion-exchange affinity-adsorption. Above the other treatments, the adsorption method seems to have the best chances. These systems are typically expensive, and although the concentration of dyes is either lowered or removed from the wastewater, the resulting sludge raises another issue about their safe and practical disposal following the treatment process. Several azo and anthraquinone dye types have been observed to degrade due to microbes in the past. By using microbial systems, bacteria could only function in anaerobic environments, reducing azo dyes to their corresponding amines, which are difficult to entirely degrade aerobically (Saeed et al., 2022).

8.2.5 Wastewater from the Food Industry

The food industry consumes water and produces wastewater equal to the textile industry. The washing, heating, and chilling processes required during the utilization of raw materials all use water (Dhanker et al., 2022). The effluents from the food processing industry contain complicated chemicals that are challenging to remove, necessitating effective treatment technology. A heavy amount of wastewater effluents with a high organic content is produced by each food sector. Salts, fats, dyes, greases, oils, total dissolved solids (TDS), and total suspended solids (TSS) are a few examples of COD. Depending on the raw materials and processing techniques, there can potentially be other contaminants.

Soybean processing industries release amino acids, saponins, lipids, and oligosaccharides which can increase COD levels up to 20 g/L. Dairy industries release higher concentrations of dissolved solids and wastes, increasing oxygen demand up to 10 g/L. Oil mills release lesser nitrogenous wastes but significantly higher amounts of phenolics. Extreme peaks of carbon, phosphorous, and nitrogenous wastewater are released from starch processing industries. This heavy nitrogenous wastewater can increase COD up to 50 g/L, which is alarming. The number of mineral compounds is greater in wastewater effluent gathered from other businesses like breweries, wineries, and meat production. Additionally, the direct release of food industries' leftovers into water bodies has a hyper negative response. Therefore, it is evident that wastewater produced by all areas of the food business must be dealt with utilizing extremely effective treatment procedures (Pervez et al., 2021).

8.2.6 Wastewater from Paper and Pulp Industries

The highest effluents are released from paper and pulp industries. An enormous number of pollutants, including suspended solids (SS), carbon dioxide (COD), toxins, and BOD, are produced throughout the pulp and paper manufacturing process (Kumar et al., 2020). Dissolved organic matter (residual lignin, hemicelluloses, and extractives) is released from the pulping and bleaching processes. Lignin and hemicelluloses are removed from the wood chips by alkaline NaOH or Na_2S treatment during wood chip preparation and washing, and the pulping process (kraft process) adds in effluents. Black liquor, a lignin-rich effluent, is produced at this point. The pulp is treated with harmful chemicals throughout the bleaching process. This process releases chlorine ozone, calcium oxides hydrogen peroxide, etc., increasing the toxicity of the collected effluents (Virkutyte, 2017).

The type of raw materials and chemicals used, the kind of paper products, and the level of water recovery all affect the amount and pollutant load of the effluent that is generated during production. Ecosystems on land and in water are impacted by the toxins released by the paper industry. Prior to the 1970s, wastewater from pulp and paper mills was typically dumped directly into lakes or rivers without any kind of treatment, not even a basic primary treatment. Numerous alterations in fish physiology and reproduction were caused by the high organic loads and solid content in the effluents, in addition to localized damage to the benthic fauna and oxygen deprivation in wide areas. The development of chlorinated chemicals in bleaching plants is the primary effluent problem (Cabrera, 2017; Mehmood et al., 2019).

8.3 CONVENTIONAL METHODS FOR WASTEWATER MANAGEMENT

Wastewater can be treated in a variety of ways to remove toxic contaminants. Wastewater treatment methods are chosen based on their ability to bring wastewater quality to a permissible level, process cost, control flexibility, and environmental compatibility. Table 8.1 lists the major methods of conventional wastewater management.

8.3.1 Biological Wastewater Management

Conventional methods are quick and require much more energy but are also harmful to animals, plants, and humans; thus, biological methods are considered an efficient and sustainable way to degrade dissolved, suspended, and toxic organic chemicals and simultaneously produce energy.

There are two approaches to the management of wastewater on the basis of their microbes and their products (Sauvé & Desrosiers, 2014; Saravanan et al., 2021; Wu et al., 2022). Figure 8.2 shows the intervention of microbes and microbial enzymes on wastewater.

8.3.2 Microbial Remediated Wastewater Management

Nearly 80% of biodegradable material in polluted water is removed by the activated sludge process. Nature has historically endowed these bodies of water with bacteria that can degrade such organic pollutants to some extent. However, increased organic wastes in water require more oxygen for these bacteria, which lowers the amount of dissolved oxygen in the water, hurting other aquatic creatures and generally, degrading the water quality. As an organic contaminant breaks down in wastewater treatment facilities, microbial remediation makes use of microorganisms such as bacteria, fungi, algae, protozoa, and rotifers (Kumar et al., 2018). Organic compounds in wastewater are broken down by bacteria using oxidation and biosynthesis. The decrease of lipids, suspended particles, oils, greases, BOD, and COD, as well as wastewater that can be reused for a variety of purposes, can be facilitated by microorganisms (Hayat et al., 2017). Bacteria, fungi, protozoa, and other microorganisms are employed in wastewater control. In the process of bioremediation, biological organisms are used to remove or neutralize environmental pollution through a metabolic reaction (Paliwal et al., 2012; Przystaś et al., 2018). Microorganisms like fungi, algae, and bacteria,

TABLE 8.1
Physical and chemical methods of wastewater treatment

Method	Type of methodology	Definition	Remarks	References
PHYSICAL METHODS	Sedimentation	Gravitation forces separate the contaminant particles from fluids	Organic chemical of larger density particles than water No energy consumption and a selective process	Nystrom et al. (2020)
	Degasification	Removal of dissolved gases from liquid through pH, high temperature, pressure, ultrasonic energy, and frequency	Inadequate capacity for the pollutant like dissolved CO_2 removal	Saravanan et al. (2021)
	Filtration	Separation of solid pollutant on the basis of size, shape, quantity texture, and density	It also removes suspended particles, microbes such as bacteria, fungi, viruses etc. and other chemical pollutants	Medeiros et al. (2020)
CHEMICAL METHODS	Flocculation and coagulation	Flocculation used to destabilize particles through mechanical agitation In coagulation, charged suspended collides are destabilized by neutralization	It is generally used for separation of natural and synthetic materials, e.g., starch, chitonas, ferrous sulfate etc.	Nystrom et al. (2020)
	Chemical precipitation	Removal of heavy metal from water using hydroxide precipitation, carbonate precipitation, sulfide precipitation methods	Many heavy metals are carcinogenic, so difficulty with sludge disposal	Son et al. (2020)
	Ozonation	It is the only method by which removal of odor, color, inorganic compounds is done, taste by disinfection and a change their oxidation state observed	Its excessive use may produce hazardous products	Guo et al. (2019)
	Adsorption	Removes organic and natural pollutants by interaction between two phases, i.e., adsorbents and adsorbate	Relatively high efficiency, nontoxic process with pollutant disposal problems	Zhou et al. (2019)
	Ion exchange	Exchange of ions	Highly contaminated charged molecules exchange with safe or low toxic charged molecules	Liu et al. (2020)

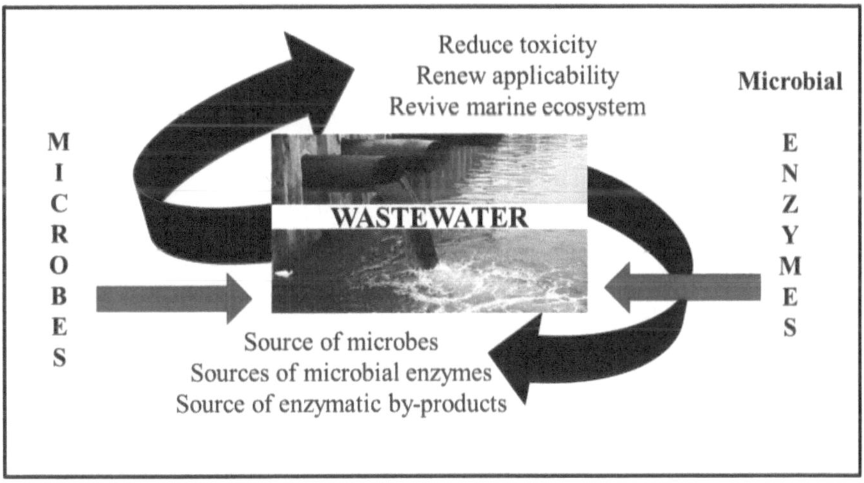

FIGURE 8.2 Wastewater management using microbes and microbial enzymes.

as well as remediation—handling the situation—are included in the biological organisms. Microorganisms thrive in a variety of settings throughout the Earth's biosphere. They may grow in a variety of environments, including ice, deep water, plants, and animals. Both autotrophs and heterotrophs can bioremediate wastewater using the following techniques: phytoremediation, rhizofiltration, bioaugmentation, and biostimulation, which includes the use of bacteria, fungi, and the enzymes that are produced by them, along with several types of traditional aerobic and anaerobic treatment (Danso et al., 2019; Dhal et al., 2013; Feng et al., 2021).

8.3.3 Microbial Enzymes over Microbes for Wastewater Management

Natural catalysts known as enzymes are present in every living thing. They can be employed to either selectively break down a mixture of larger molecules or to construct more complex compounds from simpler ones. Over 1,000 distinct enzymes can be found in only one microbe (Cheng et al., 2020; Jun et al., 2019). Enzyme technology offers fresh options for cleanup techniques at locations where other approaches fall short because of the toxicity or recalcitrant nature of the pollution. Catabolic enzymes are frequently produced by microorganisms that live in harsh settings and make use of nearby chemicals. Higher enzyme synthesis and biomass activities are provided by mixed cultures (Lee et al., 2019; Malomo et al., 2018; Naha et al., 2023).

One crucial aspect that all biotechnology applications have in common is the requirement that the application design must successfully account for the rate-limiting impacts of numerous parameters, some of which may be microbial or environmental in origin. The lower availability of microbes in specific areas or their ability to degrade the pollutant is the biggest limitation of microbial intervention. These microbes survive in such conditions but the temperature, moisture, and oxygen level required for new cultures to thrive is absent or less. The soil biochemistry affects microbial growth. As a result of the difficulty of retaining active cells during transport to the polluted region, whole-cell detoxification technologies are further restricted. The amount of pollutant will determine the quantity of microbe for remediations. It is possible for microbial cells to be irreversibly damaged or metabolically inactivated by harsh environmental conditions, including chemical shock, extreme pH and temperature, toxins, predators, and high concentrations of pollutants. And lastly, the disturbance in soil population ecology affects the degree of microbial bioremediation. Therefore, there are several restrictions on the use of microbes for bioremediation (Rabbani et al., 2021).

Using enzymes for environmental applications has the following advantages: enzymes themselves are biodegradable, so they are easily absorbed into the environment. They are highly specific, resulting in fewer unwanted side effects and byproducts in the production process A further benefit of utilizing enzymes over microbes is in situations where disposing of the sludge produced by biomass in bioremediation methods using living cells, primarily aerobic treatments, is a concern. In underdeveloped nations where the need for water purification is great, its management can be challenging (Bugg & Rahmanpour, 2015; Falade et al., 2017). The way will be opened for novel enzyme uses in a variety of industries as environmental preservation and resource conservation take on greater importance. Some locations, such as those contaminated by deeply buried and widely spread subterranean halogenated hydrocarbons, may only be remedied using enzymes, according to some theories (Pandey et al., 2017).

Although it can be challenging to determine the costs of enzyme-enhanced bioremediation methods, some businesses have asserted that these methods may be more affordable than ex situ techniques. The two primary kinds of microbial enzymes that can break down organic contaminants found in wastewater are hydrolases and oxidoreductases:

a. Cellulase, hemicellulase, and glycosidase are examples of hydrolase enzymes that act on the chemical links in complex harmful pollutants and break them down into simpler forms:
    i. Lipase: Lipase enzymes hydrolyze triacylglycerols to glycerol and fatty acids, breaking down the lipids found in plant, animal, and microbial wastes. Diverse lipases capable of esterification, alcoholysis, and aminolysis are produced by bacteria and actinomycetes, respectively.
    ii. Cellulase: Cellulases are produced by microorganisms and can degrade cellulose into less complex reduced sugars. They can be cell-bound, extracellular, or cell envelope-associated. Cellulases are a group of enzymes that work together to hydrolyze cellulose, including endoglucanase, exoglucanase, and ß-glucosidase.
    iii. Protease: Waste from sectors like food, medicine, leather, detergent, fishing, and poultry contains proteins that microbial proteases can catalyze. All waste proteins are hydrolyzed by bacterial proteases, and the bacteria then eat the broken-down molecules.

b. To break down xenobiotics, phenolic chemicals, and azo dyes, oxidoreductase enzymes are used. An oxidoreductase-producing fungus is used to release ligninolytic enzymes including laccases, lignin peroxidase, and other enzymes because it has a larger surface area to interact with contaminants:
   i. Laccases: Laccases are oxidizers of reduced phenolic and aromatic substrates produced by bacteria, fungi, insects, and plants. They can oxidize lignin, aryl diamines, aminophenols, polyphenols, polyamines, ortho- and para-diphenols, as well as phenolic and methoxyphenolic acids. Additionally, these enzymes can decarboxylate and demethylate.
   ii. Peroxidases: The hemoprotein peroxidase is responsible for catalyzing processes involving hydrogen peroxide and is largely produced by microorganisms and plants. This enzyme works well for treating wastewater because it can operate in a wide pH and temperature range, potentially removing some chemicals that are not directly catalyzed by the peroxidase enzyme through the induction of mixed polymer synthesis. For wastewater that frequently contains numerous contaminants, this has significant ramifications.

Numerous wastewater treatment applications involve enzymes produced by bacteria, fungi, and plants. Table 8.2 lists the different enzymes and their applications in wastewater management.

8.4 CONCLUSION

Microorganisms provide access to a wide range of biocatalysts for wastewater management. Microbial enzymes are more effective solutions for the cleanup of polluted water because of their high specificity and rapid activity. The recalcitrant pollutants using enzymatic action can be converted into less hazardous products. Enzymes can also alter the features of waste, making it more amenable to treatment or assisting in the bioconversion of waste into products with added value. The number of processes that can be catalyzed by microbes is yet to be thoroughly screened or investigated, and the number of reactions that can be catalyzed by enzymes will undoubtedly increase. Wastewater can also serve as a source of uncultivated microorganisms. Finding new sources of existing enzymes, as well as new enzymes for certain reactions, can be performed using novel technologies such as metagenomics and wastewater genomics.

Recent developments in expression systems, proteomics, and recombinant technology have made it easier to improve and find new microbial enzymes. The creation of more efficient systems that use less energy, water, and chemicals to work at their best may be a trend in the future. The application of contemporary biotechnology methods will result in enzyme systems with enhanced effects at various physiological temperatures and pH conditions, resulting in increased efficiency, reduced energy consumption, and lower prices. Efficient enzyme immobilization can alleviate some of these difficulties by improving the stability of many enzymes and expanding the range of pH and temperature. To address some of the issues with support that are now on the market, more research is still required to create new and/or hybrid materials.

TABLE 8.2
Microbial enzymatic based management of wastewater from different sources

S.no	Source of microbe	Microbe	Enzyme screening	Production of enzyme	Wastewater source	Application	References
1.	Domestic wastewater sample	Consortium of bacterial isolates MCSt-1	Cellulases	Bioreactor	--	Removal of COD, nitrate, ammonia, and phosphorus, exhibited better biofilm formation, hence results in significant organic load and nutrient removal from sewage water	Jha et al. (2021)
2.	Wastewater	*Talaromyces cellulolyticus*	Cellulase	Yeast broth culture medium	Wastewater treatment plant	Removal of COD and chromaticity of textile wastewater	Zhang et al. (2021)
3.	Soil	*Arthrobacter woluwensis* TDS9	Cellulase	Cell culture	Paper mill	Bioconversion of sludge	Das et al. (2022)
4.	Ramie biodegumming wastewater	*Diaphorobacter nitroreducen*	Cellulase Laccase	Fermentation based	Ramie fiber mill	Bioflocculation and removal of degumming wastewater	Zhong et al. (2020)
5.	Lab procured	*Hypocrea* sp.	Cellulase Xylanase	Liquid culture	Sewage wastewater Treatment	Waste sludge from water as substrate to enhance enzyme production	Liang et al. (2020)
6.	Rumen fluid	Unidentified microbial diversity	Endoglucanase activity	Culture based	Slaughterhouse	Use of rumen fluid as microbial source will reduce waste load	Takizawa et al. (2020)
7.	-	*Trametes pubescence*	Laccase	Culture based	In vivo aqueous solution of heavy metals	Removal of heavy metals from wastewater	Enayatizamir et al. (2020)

8.	Mushroom cultivation manor	*Ganoderma lucidum*	Laccase	Culture based	Activated sludge tank wastewater of petroleum cracking tank	Removal of azo dyes found in microbial fuel cells and sludge cleaning	Liu et al. (2020)
9.	-	*T. versicolor*	Laccase	Immobilized		Degradation of toxic chemicals in wastewater	Maryšková et al. (2020)
10.		*Trametes versicolor*	Laccase	Submerged fermentation	Blue wastewater from aircraft in an airlift reactor	Reduced COD and toxicity of blue wastewater	Atilano-Camino et al. (2020)
11.	--	*Pycnoporus* sp.	Laccase	Bioreactor	Textile wastewater	Decolorization efficiency of 11 disperse dyes	Wang et al. (2022)
12.	--	*Phanerochaete chryosporium*	Laccase	Solid state fermentation	-	Dye decolorization for dye removal from wastewater	Singh et al. (2022)
13.	--	*Alcaligenes faecalis*	Laccase	Immobilized		Decolorization of effluent	Mehandia et al. (2022)
14.	Waster from poultry processing factories	*Acinetobacter* baumannii	Lipase	-	Poultry processing factory	Oil hydrolysis of oily wastewater	Bunmadee et al. (2022)
15.	Olive oil mill soil	*Bacillus stearothermophilus*	Lipase	Immobilized	Water from Wadi Hanifah	Hydrolysis of oily wastewater	Ben Bacha et al. (2022)
16.	Soil	*Bacillus velezensis*	Lignin peroxidase	Culture based	Industrial wastewater	Reduction of methylene blue absorption	Pham et al. (2022)
17.	-	*Schizophyllum commune*	Lignin peroxidase	Solid state fermentation	Industrial wastewater	Dye decolorization	Parveen et al. (2021)

Additionally, there is still a significant gap between research conducted in laboratories, field studies, and scaling-up and bioreactor applications of these enzymes. The practical use of enzymes in wastewater treatment plants should be the primary focus of future research.

REFERENCES

Ahmad, T., T. Belwal, L. Li, S. Ramola, R.M. Aadil, Abdullah, Y. Xu, and L. Zisheng. 2020. Utilization of Wastewater from Edible Oil Industry, Turning Waste into Valuable Products: A Review. *Trends in Food Science and Technology* 99, no. January: 21–33.

Aitken, M.D., I.J. Massey, T. Chen, and P.E. Heck. 1994. Characterization of Reaction Products from the Enzyme Catalyzed Oxidation of Phenolic Pollutants. *Water Research* 28, no. 9: 1879–1889.

Al-Maqdi, K.A., N. Elmerhi, K. Athamneh, M. Bilal, A. Alzamly, S.S. Ashraf, and I. Shah. 2021. Challenges and Recent Advances in Enzyme-Mediated Wastewater Remediation—A Review. *Nanomaterials* 11, no. 11: 3124.

Alshabib, M., and S.A. Onaizi. 2019. A Review on Phenolic Wastewater Remediation Using Homogeneous and Heterogeneous Enzymatic Processes: Current Status and Potential Challenges. *Separation and Purification Technology* 219, no. March: 186–207. https://doi.org/10.1016/j.seppur.2019.03.028.

Atilano-Camino, M. M., Álvarez-Valencia, L. H., García-González, A., and García-Reyes, R. B. 2020. Improving Laccase Production from Trametes Versicolor Using Lignocellulosic Residues as Cosubstrates and Evaluation of Enzymes for Blue Wastewater Biodegradation. *Journal of Environmental Management* 275: 111231. https://doi.org/10.1016/j.jenvman.2020.111231

Balmer, M.E., H.R. Buser, M.D. Müller, and T. Poiger. 2005. Occurrence of Some Organic UV Filters in Wastewater, in Surface Waters, and in Fish from Swiss Lakes. *Environmental Science and Technology* 39, no. 4: 953–962.

Ben Bacha, A., Alonazi, M., Alanazi, H., Alharbi, M. G., Jallouli, R., and Karray, A. 2022. Biochemical Study of *Bacillus stearothermophilus* Immobilized Lipase for Oily Wastewater Treatment. *Processes* 10, no. 11: 2220. https://doi.org/10.3390/pr10112220

Bugg, T.D.H., and R. Rahmanpour. 2015. Enzymatic Conversion of Lignin into Renewable Chemicals. *Current Opinion in Chemical Biology* 29: 10–17. http://dx.doi.org/10.1016/j.cbpa.2015.06.009.

Bunmadee, S., Teeka, J., Lomthong, T., Kaewpa, D., Areesirisuk, P., and Areesirisuk, A. 2022. Isolation and Identification of a Newly Isolated Lipase-Producing Bacteria (*Acinetobacter baumannii* RMUTT3S8-2) from Oily Wastewater Treatment Pond in a Poultry Processing Factory and Its Optimum Lipase Production. *Bioresource Technology Reports* 20: 101267. https://doi.org/10.1016/j.biteb.2022.101267

Cabrera, M. N. 2017 Pulp Mill Wastewater: Characteristics and Treatment. *Biological Wastewater Treatment and Resource Recovery* 2, 119–139. doi: 10.5772/67537

Cammarota, M.C., and D.M.G. Freire. 2006. A Review on Hydrolytic Enzymes in the Treatment of Wastewater with High Oil and Grease Content. *Bioresource Technology* 97, no. 17: 2195–2210.

de Cazes, M., R. Abejón, M.P. Belleville, and J. Sanchez-Marcano. 2014. Membrane Bioprocesses for Pharmaceutical Micropollutant Removal from Waters. *Membranes* 4, no. 4: 692–729.

Cheng, D., Y. Liu, H.H. Ngo, W. Guo, S.W. Chang, D.D. Nguyen, S. Zhang, G. Luo, and Y. Liu. 2020. A Review on Application of Enzymatic Bioprocesses in Animal Wastewater and Manure Treatment. *Bioresource Technology* 313: 123683. https://doi.org/10.1016/j.biortech.2020.123683.

Dahiya D and Nigam PS. 2020. Waste Management by Biological Approach Employing Natural Substrates and Microbial Agents for the Remediation of Dyes' wastewater. *Applied Sciences*. 24, 10 no. 8: 2958. https://doi.org/10.3390/app10082958.

Danso, D., J. Chow, and W.R. Streita. 2019. Plastics: Environmental and Biotechnological Perspectives on Microbial Degradation. *Applied and Environmental Microbiology* 85, no. 19.

Das, T., Ali, F., and Rahman, M. (2022). Cellulase Activity of a Novel Bacterial Strain Arthrobacter Woluwensis TDS9: Its Application on Bioconversion of Paper Mill Sludge. *Journal of Genetic Engineering and Biotechnology* 20, no. 1: 1–16. https://doi.org/10.1186/s43141-022-00373-w

Dhal, B., Thatoi, H. N., Das, N. N., and Pandey, B. D. 2013. Chemical and Microbial Remediation of Hexavalent Chromium from Contaminated Soil and Mining/Metallurgical Solid Waste: A Review. *Journal of Hazardous Materials* 250: 272–291. https://doi.org/10.1016/j.jhazmat.2013.01.048

Dhanker, R., Rawat, S., Chandna. V., Deepa, Das, S., Sharma, A., Kumar, V., Kumar, R., 2022. Recovery of silver nanoparticles and management of food wastes: obstacles and opportunities. *Environmental Advances* 9, 100303. https://doi.org/10.1016/j.envadv.2022.100303

Ebele, A.J., M. Abou-Elwafa Abdallah, and S. Harrad. 2017. Pharmaceuticals and Personal Care Products (PPCPs) in the Freshwater Aquatic Environment. Emerging Contaminants 3, no. 1: 1–16. http://dx.doi.org/10.1016/j.emcon.2016.12.004.

Enayatizamir, N., Liu, J., Wang, L., Lin, X., and Fu, P. 2020. Coupling Laccase Production from Trametes Pubescence with Heavy Metal Removal for Economic Waste Water Treatment. *Journal of Water Process Engineering* 37: 101357. https://doi.org/10.1016/j.jwpe.2020.101357

Esposito, E., and N. Durán. 2000. Potential Applications of Oxidative Enzymes and Phenoloxidase-like Compounds in Wastewater and Soil Treatment: A Review. *Applied Catalysis B: Environmental* 28, no. 2: 83–99.

Falade, A.O., U.U. Nwodo, B.C. Iweriebor, E. Green, L. V. Mabinya, and A.I. Okoh. 2017. Lignin Peroxidase Functionalities and Prospective Applications. *Microbiology Open* 6, no. 1: 1–14.

Feng, S., H. Hao Ngo, W. Guo, S. Woong Chang, D. Duc Nguyen, D. Cheng, S. Varjani, Z. Lei, and Y. Liu. 2021. Roles and Applications of Enzymes for Resistant Pollutants Removal in Wastewater Treatment. *Bioresource Technology* 335: 125278. https://doi.org/10.1016/j.biortech.2021.125278.

García-Molina, P., F. García-Molina, J.A. Teruel-Puche, J.N. Rodríguez-López, F. García-Cánovas, and J.L. Muñoz-Muñoz. 2022. Considerations about the Kinetic Mechanism of Tyrosinase in Its Action on Monophenols: A Review. *Molecular Catalysis* 518: 112072. https://www.sciencedirect.com/science/article/pii/S246882312100688X.

Geyer, R., J.R. Jambeck, and K.L. Law. 2017. Production, Use, and Fate of All Plastics Ever Made CA 93106; Bren School of Environmental Science and Management. University of California, Santa Barbara, Santa Barbara, USA. no. July: 25–29.

Gianfreda, L., M.A. Rao, R. Scelza, and M. de la Luz Mora. 2016. Chapter 6 – Role of Enzymes in Environment Cleanup/Remediation. In, ed. G.S. Dhillon and S.B.T.-A.-I.W. as F. for E.P. Kaur, 133–155. San Diego: Academic Press. https://www.sciencedirect.com/science/article/pii/B978012802392100006X.

Guo, R., and J. Chen. 2015. Application of Alga-Activated Sludge Combined System (AASCS) as a Novel Treatment to Remove Cephalosporins. *Chemical Engineering Journal* 260: 550–556. http://dx.doi.org/10.1016/j.cej.2014.09.053.

Guo, Y., Zhu, S., Wang, B., Huang, J., Deng, S., Yu, G., and Wang, Y. 2019. Modelling of Emerging Contaminant Removal During Heterogeneous Catalytic Ozonation Using Chemical Kinetic Approaches. *Journal of Hazard Materials* 380: 120888. https://doi.org/ 10.1016/j.jhazmat.2019.120888.

Hayat, K., Menhas, S., Bundschuh, J., and Chaudhary, H. J. 2017. Microbial Biotechnology as An Emerging Industrial Wastewater Treatment Process for Arsenic Mitigation: A Critical Review. *Journal of Cleaner Production* 151: 427–438. https://doi.org/10.1016/j.jclepro.2017.03.084

Ji, C., J. Hou, K. Wang, Y. Zhang, and V. Chen. 2016. Biocatalytic Degradation of Carbamazepine with Immobilized Laccase-Mediator Membrane Hybrid Reactor. *Journal of Membrane Science* 502: 11–20. http://dx.doi.org/10.1016/j.memsci.2015.12.043.

Jiang, B., N. Zhang, Y. Xing, L. Lian, Y. Chen, D. Zhang, G. Li, G. Sun, and Y. Song. 2019. Microbial Degradation of Organophosphorus Pesticides: Novel Degraders, Kinetics, Functional Genes, and Genotoxicity Assessment. *Environmental Science and Pollution Research* 26, no. 21: 21668–21681.

Jha, V., Dafale, N. A., Hathi, Z., and Purohit, H. 2021. Genomic and Functional Potential of the Immobilized Microbial Consortium MCSt-1 for Wastewater Treatment. *Science of the Total Environment 777*: 146110. doi:10.1016/j.scitotenv.2021.146110

Joss, A., S. Zabczynski, A. Göbel, B. Hoffmann, D. Löffler, C.S. McArdell, T.A. Ternes, A. Thomsen, and H. Siegrist. 2006. Biological Degradation of Pharmaceuticals in Municipal Wastewater Treatment: Proposing a Classification Scheme. *Water Research* 40, no. 8: 1686–1696.

Jun, L.Y., L.S. Yon, N.M. Mubarak, C.H. Bing, S. Pan, M.K. Danquah, E.C. Abdullah, and M. Khalid. 2019. An Overview of Immobilized Enzyme Technologies for Dye and Phenolic Removal from Wastewater. *Journal of Environmental Chemical Engineering* 7, no. 2: 102961. https://doi.org/10.1016/j.jece.2019.102961.

Kesari, K.K., R. Soni, Q.M.S. Jamal, P. Tripathi, J.A. Lal, N.K. Jha, M.H. Siddiqui, P. Kumar, V. Tripathi, and J. Ruokolainen. 2021. Wastewater Treatment and Reuse: A Review of Its Applications and Health Implications. *Water, Air, and Soil Pollution* 232, no. 5.

Kumar, V., Agrawal, S., Bhat, S.A., Américo-Pinheiro, J.H.P., Shahi, S.K., and Kumar, S. 2022. Environmental Impact, Health Hazards, and Plant-Microbes Synergism in Remediation of Emerging Contaminants. *Cleaner Chemical Engineering* 2, 100030. https://doi.org/10.1016/j.clce.2022.100030

Kumar, V., Shahi, S.K., and Singh, S. 2018. Bioremediation: An Eco-Sustainable Approach for Restoration of Contaminated Sites. In: Singh J., Sharma D., Kumar G., Sharma N. (eds). *Microbial Bioprospecting For Sustainable Development*. Springer, Singapore. https://doi.org/10.1007/978-981-13-0053-0_6

Kumar, V., Thakur, I.S., and Shah, M.P. 2020. Bioremediation Approaches for Pulp and Paper Industry Wastewater Treatment: Recent Advances and Challenges. In: Shah, M.P. (Ed.) *Microbial Bioremediation & Biodegradation*. Springer, Singapore. https://doi.org/10.1007/978-981-15-1812-6_1

Lee, S.H., J.Q. Xiong, S. Ru, S.M. Patil, M.B. Kurade, S.P. Govindwar, S.E. Oh, and B.H. Jeon. 2020. Toxicity of Benzophenone-3 and Its Biodegradation in a Freshwater Microalga Scenedesmus Obliquus. *Journal of Hazardous Materials* 389, no. November 2019: 122149. https://doi.org/10.1016/j.jhazmat.2020.122149.

Lee, Z.S., S.Y. Chin, J.W. Lim, T. Witoon, and C.K. Cheng. 2019. Treatment Technologies of Palm Oil Mill Effluent (POME)and Olive Mill Wastewater (OMW): A Brief Review.

Environmental Technology and Innovation 15: 100377. https://doi.org/10.1016/j.eti.2019.100377.

Lonappan, L., S.K. Brar, R.K. Das, M. Verma, and R.Y. Surampalli. 2016. Diclofenac and Its Transformation Products: Environmental Occurrence and Toxicity – A Review. *Environment International* 96: 127–138. http://dx.doi.org/10.1016/j.envint.2016.09.014.

Liang, C., Xu, Z., Wang, Q., Wang, W., Xu, H., Guo, Y., and Wang, Z. 2020. Improving β-Glucosidase and Xylanase Production in a Combination of Waste Substrate from Domestic Wastewater Treatment System and Agriculture Residues. *Bioresource Technology* 318: 124019. https://doi.org/10.1016/j.biortech.2020.124019.

Liu, Z., Lompe, K.M., Mohseni, M., Berube, P.R., Sauve, S., and Barbeau, B. 2020. Biological Ion Exchange as an Alternative to Biological Activated Carbon for Drinking Water Treatment. *Water Research* 168, 115148. https://doi.org/10.1016/j.watres.2019.115148.

Liu, S. H., Tsai, S. L., Guo, P. Y., and Lin, C. W. (2020). Inducing Laccase Activity in White Rot Fungi Using Copper Ions and Improving the Efficiency of Azo Dye Treatment with Electricity Generation Using Microbial Fuel Cells. *Chemosphere* 243: 125304. https://doi.org/10.1016/j.chemosphere.2019.125304

Maier, D., L. Blaha, J.P. Giesy, A. Henneberg, H.R. Köhler, B. Kuch, R. Osterauer, et al. 2015. Biological Plausibility as a Tool to Associate Analytical Data for Micropollutants and Effect Potentials in Wastewater, Surface Water, and Sediments with Effects in Fishes. *Water Research* 72: 127–144.

Malomo, G.A., A.S. Madugu, and S.A. Bolu. 2018. *Sustainable Animal Manure Management Strategies and Practices*. Agricultural Waste and Residues.

Maloney, A.J., C. Dong, A.S. Campbell, and C.Z. Dinu. 2015. Emerging Enzyme-Based Technologies for Wastewater Treatment. *ACS Symposium Series* 1192: 69–85.

Manmohit Kalia, P.K. 2015. Pectin Methylesterases: A Review. *Journal of Bioprocessing & Biotechniques* 05, no. 05.

Marican, A., and E.F. Durán-Lara. 2018. A Review on Pesticide Removal through Different Processes. *Environmental Science and Pollution Research* 25, no. 3: 2051–2064.

Maryšková, M., Schaabová, M., Tomankova, H., Novotný, V., and Rysová, M. 2020. Wastewater Treatment by Novel Polyamide/Polyethylenimine Nanofibers with Immobilized Laccase. *Water* 12: 588. https://doi.org/10.3390/w12020588

Mehmood, K., Rehman, S. K. U., Wang, J., Farooq, F., Mahmood, Q., Jadoon, A. M., .and Ahmad, I. 2019. Treatment of Pulp and Paper Industrial Effluent Using Physicochemical Process for Recycling. *Water* 11, no. 11: 2393. https://doi.org/10.3390/w11112393

Mehandia, S., Ahmad, S., Sharma, S. C., and Arya, S. K. 2022. Decolorization and Detoxification of Textile Effluent by Immobilized Laccase-ACS into Chitosan-Clay Composite Beads Using a Packed Bed Reactor System: An Ecofriendly Approach. *Journal of Water Process Engineering* 47: 102662. https://doi.org/10.1016/j.jwpe.2022.102662

Medeiros, R.C., de M N Fava, N., Freitas, B.L.S., Sabogal-Paz, L.P., Hoffmann, M.T., Davis, J., Fernandez-Ibanez, P., and Byrne, J.A. 2020. Drinking Water Treatment by Multistage Filtration on a Household Scale: Efficiency and Challenges. *Water Research* 178: 115816. https://doi.org/10.1016/j.watres.2020.115816.

Meng, Y., S. Li, H. Yuan, D. Zou, Y. Liu, B. Zhu, A. Chufo, M. Jaffar, and X. Li. 2015. Evaluating Biomethane Production from Anaerobic Mono- and Co-Digestion of Food Waste and Floatable Oil (FO) Skimmed from Food Waste. *Bioresource Technology* 185: 7–13. http://dx.doi.org/10.1016/j.biortech.2015.02.036.

Mishra, B., S. Varjani, D.C. Agrawal, S.K. Mandal, H.H. Ngo, M.J. Taherzadeh, J.S. Chang, S. You, and W. Guo. 2020. Engineering Biocatalytic Material for the Remediation of Pollutants: A Comprehensive Review. *Environmental Technology and Innovation* 20: 101063. https://doi.org/10.1016/j.eti.2020.101063.

Mohammadi, S., A. Kargari, H. Sanaeepur, K. Abbassian, A. Najafi, and E. Mofarrah. 2015. Phenol Removal from Industrial Wastewaters: A Short Review. *Desalination and Water Treatment* 53, no. 8: 2215–2234.

Moraes, B.S., V.L. Loro, L. Glusczak, A. Pretto, C. Menezes, E. Marchezan, and S. de Oliveira Machado. 2007. Effects of Four Rice Herbicides on Some Metabolic and Toxicology Parameters of Teleost Fish (*Leporinus obtusidens*). *Chemosphere* 68, no. 8: 1597–1601.

Müller, F.M., D. de Oliveira, and C. Michels. 2023. Current Status, Gaps and Challenges of Rendering Industries Wastewater. *Journal of Water Process Engineering* 52, no. January.

Naha, A., Antony, S., Nath, S., Sharma, D., Mishra, A., Biju, D. T., and Sindhu, R. 2023. A Hypothetical Model of Multi-Layered Cost-Effective Wastewater Treatment Plant Integrating Microbial Fuel Cell and Nanofiltration Technology: A Comprehensive Review on Wastewater Treatment and Sustainable Remediation. *Environmental Pollution*, 121274. https://doi.org/10.1016/j.envpol.2023.121274

Nguyen, T.P.O., D.E. Helbling, K. Bers, T.T. Fida, R. Wattiez, H.P.E. Kohler, D. Springael, and R. De Mot. 2014. Genetic and Metabolic Analysis of the Carbofuran Catabolic Pathway in *Novosphingobium* Sp. KN65.2. *Applied Microbiology and Biotechnology* 98, no. 19: 8235–8252.

Nystrom, F., Nordqvist, K., Herrmann, I., Hedstrom, A.,and Viklander, M. 2020. Removal of Metals and Hydrocarbons from Stormwater Using Coagulation and Flocculation. *Water Research* Inpress Accepted Proof. https://doi.org/10.1016/ j.watres.2020.115919.

Pal, A., Y. He, M. Jekel, M. Reinhard, and K.Y.H. Gin. 2014. Emerging Contaminants of Public Health Significance as Water Quality Indicator Compounds in the Urban Water Cycle. *Environment International* 71: 46–62. http://dx.doi.org/10.1016/j.envint.2014.05.025.

Paliwal, R., A.P. Rawat, M. Rawat, and J.P.N. Rai. 2012. Bioligninolysis: Recent Updates for Biotechnological Solution. *Applied Biochemistry and Biotechnology* 167, no. 7: 1865–1889.

Pandey, K., B. Singh, A.K. Pandey, I.J. Badruddin, S. Pandey, V.K. Mishra, and P.A. Jain. 2017. Application of Microbial Enzymes in Industrial Waste Water Treatment. *International Journal of Current Microbiology and Applied Sciences* 6, no. 8: 1243–1254.

Parveen, S., Asgher, M., and Bilal, M. 2021. Lignin Peroxidase-Based Cross-Linked Enzyme Aggregates (LiP-CLEAs) as Robust Biocatalytic Materials for Mitigation of Textile Dyes-Contaminated Aqueous Solution. *Environmental Technology & Innovation* 21: 101226. https://doi.org/10.1016/j.eti.2020.101226.

Pervez M, Mishu MR, Stylios GK, Hasan SW, Zhao Y, Cai Y, Zarra T, Belgiorno V, and Naddeo V. 2021. Sustainable Treatment of Food Industry Wastewater Using Membrane Technology: A Short Review. *Water* 13, no. 23: 3450. https://doi.org/10.3390/w13233450.

Pham, V. H. T., Kim, J., Chang, S., and Chung, W. 2022. Biodegradation of Methylene Blue Using a Novel Lignin Peroxidase Enzyme Producing Bacteria, Named *Bacillus* sp. react3, as a Promising Candidate for Dye-Contaminated Wastewater Treatment. *Fermentation* 8, no. 5: 190. https://doi.org/10.3390/fermentation8050190

Przystaś, W., E. Zabłocka-Godlewska, and E. Grabińska-Sota. 2018. Efficiency of Decolorization of Different Dyes Using Fungal Biomass Immobilized on Different Solid Supports. *Brazilian Journal of Microbiology* 49, no. 2: 285–295.

Rabbani, A., Zainith, S., Deb, V. K., Das, P., Bharti, P., Rawat, D. S., and Saxena, G. 2021. Microbial Technologies for Environmental Remediation: Potential Issues, Challenges, and Future Prospects. *Microbe Mediated Remediation of Environmental Contaminants* 271–286. https://doi.org/10.1016/B978-0-12-821199-1.00022-5

Rao, M.A., R. Scelza, R. Scotti, and L. Gianfreda. 2010. Role of Enzymes in the Remediation of Polluted Environments. *Journal of Soil Science and Plant Nutrition* 10, no. 3: 333–353.

Raouf, M.E.A., N.E., Maysour, R.K., Farag, and A.M. Abdul-Raheim, 2019. Wastewater Treatment Methodologies, Review Article. *International Journal of Environmental & Agricultural Science 3*: 018.

Saeed, M. U., Hussain, N., Sumrin, A., Shahbaz, A., Noor, S., Bilal, M., ... and Iqbal, H. M. 2022. Microbial Bioremediation Strategies with Wastewater Treatment Potentialities–A Review. *Science of the Total Environment*, 818, 151754. https://doi.org/10.1016/j.scitotenv.2021.151754

Santhosh, C., V. Velmurugan, G. Jacob, S.K. Jeong, A.N. Grace, and A. Bhatnagar. 2016. Role of Nanomaterials in Water Treatment Applications: A Review. *Chemical Engineering Journal* 306: 1116–1137. http://dx.doi.org/10.1016/j.cej.2016.08.053.

Saravanan, A., P.S., Kumar, S., Varjani, S., Jeevanantham, P.R., Yaashikaa, P., Thamarai, B. Abirami, and C.S. George, 2021. A review on algal-bacterial symbiotic system for effective treatment of wastewater. *Chemosphere* 271: 129540.

Sauvé, S., and M. Desrosiers. 2014. A Review of What Is an Emerging Contaminant. *Chemistry Central Journal* 8, no. 1: 1–7.

Sendra, M., M.G. Pintado-Herrera, G. V. Aguirre-Martínez, I. Moreno-Garrido, L.M. Martin-Díaz, P.A. Lara-Martín, and J. Blasco. 2017. Are the TiO2 NPs a "Trojan Horse" for Personal Care Products (PCPs) in the Clam Ruditapes Philippinarum? *Chemosphere* 185: 192–204. http://dx.doi.org/10.1016/j.chemosphere.2017.07.009.

Sheng, C., S. Zhang, and Y. Zhang. 2021. The Influence of Different Polymer Types of Microplastics on Adsorption, Accumulation, and Toxicity of Triclosan in Zebrafish. *Journal of Hazardous Materials* 402: 123733. https://doi.org/10.1016/j.jhazmat.2020.123733.

Singh, S., Adeyemi, K. T., and Vernwal, S. 2022. Adsorption-Microbial Fermentation Based Multi-Step Approach to Dye Remediation for Safe and Environment Compatible Waste Water Treatment. *Environmental Challenges* 7: 100459. https://doi.org/10.1016/j.envc.2022.100459

Son, D.J., Kim,W.Y., Jung, B.R., Chang, D., and Hong, K.H. 2020. Pilot-Scale Anoxic/Aerobic Biofilter System Combined with Chemical Precipitation for Tertiary Treatment of Wastewater. *Journal of Water Processing Engineering* 35: 101224. https://doi.org/10.1016/ j.jwpe.2020.101224.

Stadlmair, L.F., T. Letzel, J.E. Drewes, and J. Graßmann. 2017. Mass Spectrometry Based in Vitro Assay Investigations on the Transformation of Pharmaceutical Compounds by Oxidative Enzymes. *Chemosphere* 174: 466–477. http://dx.doi.org/10.1016/j.chemosphere.2017.01.140.

Sultatos, L.G. 1994. Mammalian Toxicology of Organophosphorus Pesticides. *Journal of Toxicology and Environmental Health* 43, no. 3: 271–289.

Summerscales, J. 2021. A Review of Bast Fibres and Their Composites: Part 4 ~ Organisms and Enzyme Processes. *Composites Part A: Applied Science and Manufacturing* 140, no. October 2020: 106149. https://doi.org/10.1016/j.compositesa.2020.106149.

Takizawa, S., Abe, K., Fukuda, Y., Feng, M., Baba, Y., Tada, C., and Nakai, Y. 2020. Recovery of the Fibrolytic Microorganisms from Rumen Fluid by Flocculation for Simultaneous Treatment of Lignocellulosic Biomass and Volatile Fatty Acid Production. *Journal of Cleaner Production* 257: 120626. https://doi.org/10.1016/j.jclepro.2020.120626

Varga, B., V. Somogyi, M. Meiczinger, N. Kováts, and E. Domokos. 2019. Enzymatic Treatment and Subsequent Toxicity of Organic Micropollutants Using Oxidoreductases – A Review. *Journal of Cleaner Production* 221: 306–322.

Villarroel, M.J., E. Sancho, E. Andreu-Moliner, and M.D. Ferrando. 2009. Biochemical Stress Response in Tetradifon Exposed Daphnia Magna and Its Relationship to Individual

Growth and Reproduction. *Science of the Total Environment* 407, no. 21: 5537–5542. http://dx.doi.org/10.1016/j.scitotenv.2009.06.032.

Villegas, L.G.C., N. Mashhadi, M. Chen, D. Mukherjee, K.E. Taylor, and N. Biswas. 2016. A Short Review of Techniques for Phenol Removal from Wastewater. *Current Pollution Reports* 2, no. 3: 157–167. http://dx.doi.org/10.1007/s40726-016-0035-3.

Virkutyte, J., 2017. Aerobic treatment of effluents from pulp and paper industries. In *Current developments in biotechnology and bioengineering* (pp. 103–130). Elsevier.

Wang, B., Chen, Y., Guan, J., Ding, Y., He, Y., Zhang, X., and Zhu, M. 2022. Biodecolorization and Ecotoxicity Abatement of Disperse Dye-Production Wastewater Treatment with Pycnoporus Laccase. *International Journal of Environmental Research and Public Health* 19, no. 13: 7983. https://doi.org/10.3390/ijerph19137983

Wang, Y., S.H. Ho, C.L. Cheng, W.Q. Guo, D. Nagarajan, N.Q. Ren, D.J. Lee, and J.S. Chang. 2016. Perspectives on the Feasibility of Using Microalgae for Industrial Wastewater Treatment. *Bioresource Technology* 222: 485–497. http://dx.doi.org/10.1016/j.biortech.2016.09.106.

Watkinson, A.J., E.J. Murby, and S.D. Costanzo. 2007. Removal of Antibiotics in Conventional and Advanced Wastewater Treatment: Implications for Environmental Discharge and Wastewater Recycling. *Water Research* 41, no. 18: 4164–4176.

Wee, T., C. Kim, and M. Hanif. 2016. Review on Wastewater Treatment Technologies. *International Journal of Applied Environmental Sciences* 11, no. 1: 111–126.

Wu, X., X. Zhao, R. Chen, P. Liu, W. Liang, J. Wang, M. Teng, X. Wang, and S. Gao. 2022. Wastewater Treatment Plants Act as Essential Sources of Microplastic Formation in Aquatic Environments: A Critical Review. *Water Research* 221, no. July: 118825. https://doi.org/10.1016/j.watres.2022.118825.

Zerva, A., G.I. Zervakis, P. Christakopoulos, and E. Topakas. 2017. Degradation of Olive Mill Wastewater by the Induced Extracellular Ligninolytic Enzymes of Two Wood-Rot Fungi. *Journal of Environmental Management* 203: 791–798. http://dx.doi.org/10.1016/j.jenvman.2016.02.042.

Zhang, Z., Zhao, Y., Yang, J., Guo, J., and Li, J. 2021. Talaromyces Cellulolyticus as a Promising Candidate for Biofilm Construction and Treatment of Textile Wastewater. *Bioresource Technology*, 340, 125718. https://doi.org/10.1016/j.biortech.2021.125718

Zhang, Y., Z. Xu, Z. Chen, and G. Wang. 2020. Simultaneous Degradation of Triazophos, Methamidophos and Carbofuran Pesticides in Wastewater Using an Enterobacter Bacterial Bioreactor and Analysis of Toxicity and Biosafety. *Chemosphere* 261: 128054. https://doi.org/10.1016/j.chemosphere.2020.128054.

Zhong, C., Sun, S., Zhang, D., Liu, L., Zhou, S., and Zhou, J. 2020. Production of a Bioflocculant from Ramie Biodegumming Wastewater Using a Biomass-Degrading Strain and its Application in the Treatment of Pulping Wastewater. *Chemosphere* 253: 126727. https://doi.org/10.1016/j.chemosphere.2020.126727

Zhou, Y., J., Lu, Y. Zhou, and Y. Liu 2019. Recent advances for dyes removal using novel adsorbents: a review. *Environmental Pollution*, 252: 352–365.

9 The Microbial Nexus in Effluent Treatment Plants

Trends and Perspectives

Heba Saad Taher, Sara Fareed Mohamed Wahdan, Rania Sayed, Amira Ouda, and Hesham Abdulla

9.1 THE MICROBIAL NEXUS IN EFFLUENT TREATMENT PLANTS

Most treated wastewater still contains some residues, which consist of biochemical oxygen demand (BOD), chemical oxygen demand (COD), and a certain salt load, in addition to bacteria, and modern technologies may eliminate these residues. The discharge of effluents containing these residues may damage the receiving ecosystems, affecting the self-purification capacity of the receiving water, the reproduction of wild fauna and flora, while also negatively affecting human health (Sharma et al., 2023; Kumar et al., 2022). To improve wastewater treatment efficiencies and purification techniques, detailed biochemical knowledge and metabolic pathways are required (Lu et al., 2023; Gallert & Winter, 2005). Organic and inorganic compounds, both suspended and dissolved, are present in effluent water in small amounts. Most organic substances are suspended and are comprised of carbohydrates, proteins, fats, and their breakdown products, while the inorganics are soluble. Several elements such as N, P, S, K, Ca, Mg, B, Si, Na, Fe, Mn, Co, Ni, and Sn are present in effluent, while their concentrations depend on the treatment process (Lu et al., 2023).

The functional diversity of the microbial content influences the effluent quality. The main pathway for total and partial decomposition of organic compounds in wastewater is the catabolic processes of microorganisms including algae, yeasts, and fungi. Biodegradation of organic compounds occurs in the presence of O_2 by respiration or anaerobically by denitrification, methanogenesis, or sulfidogenesis (Kayser, 2005). Microorganisms only take up and metabolize soluble, low molecular-weight compounds (Chandra & Kumar, 2015). The presence of biopolymers induces microorganisms to excrete exoenzymes that adsorb to and hydrolyze these polymers into monomers to make them accessible for microbial metabolism. Glycolysis, hydrolysis, oxidation, and/or deamination start in cells once the soluble monomers are absorbed (Gallert & Winter, 2005).

DOI: 10.1201/9781003441069-9

Carbohydrates are soluble forms or particulate of homo- or heteropolymers. Hydrolysis of carbohydrates favors a slightly acidic pH, and their fermentation lowers the pH (He et al., 2006). Proteins are large soluble or solid biomolecules, the soluble molecules of which are precipitated by enzymes or in acidic conditions. Proteases hydrolyze proteins optimally at neutral or slightly alkaline pH and amino acid fermentation does not change pH due to ammonia, bicarbonate, and carbonate buffering (McInerney, 1988).

Another group of biopolymers that raise the COD significantly is fats and lipids (Winter et al., 1992; Broughton et al., 1998). Fats and lipids must be emulsified to be lipolyzed by lipases or phospholipases. Phospholipid lipolysis generates fatty acids, glycerol, alcohols, and phosphate. In low-loaded systems, glycerol and sugar fractions can be degraded to methane and CO_2 by methanogenic and fermentative bacteria, while by acetogenic, fermentative, and methanogenic bacteria in systems with high load (Lu et al., 2023).

9.1.1 Carbon Metabolism in Effluent

The organic carbon in wastewater includes solids as well as many solubles. Organic matter concentration is assessed as BOD, COD, or total organic carbon (TOC). BOD includes only biodegradable organics, while COD includes both biodegradable and nondegradable. Microorganisms require several elements to degrade organic carbon, including N, P, Ca, Na, Mg, Fe, and other essential trace elements (Lu et al., 2023; Gallert & Winter, 2005).

Hydrolysis of organic compounds, absorption, and primary metabolism are similar in aerobic and anaerobic bacteria. Respiration of soluble organic compounds results in CO_2, water, and a large amount of residual sludge. Degradation of amino acids containing sulfur or heterocyclic compounds forms ammonia and hydrogen sulfide. Anaerobically, soluble carbon compounds are gradually decomposed to CH_4, CO_2, H_2S, and NH_3 syntrophically by fermentative, acetogenic bacteria, and methanogens or sulfate reducers (Gallert & Winter, 2005). Anaerobes, such as *Clostridium* sp. liberate H_2 or convert pyruvate or acetate to highly reduced products that undergo anaerobic oxidation by acetogens. Methanogens or sulfate reducers scavenge the hydrogen and then the acetogens are allowed to grow (Bryant, 1979).

9.1.2 Nitrogen Metabolism in Effluent

Nitrogen compounds such as proteins, nucleic acids, amino acids, urea, amines, and ammonia are the major nitrogen compounds in wastewater. Ammonia is principally derived from urine through the breakdown of urea by ureases, while proteolysis and deamination provide very low ammonia due to the short residence time. The presence of oxidized forms of nitrogen such as nitrite and nitrate is not significant (Gallert & Winter, 2005).

Autotrophic nitrifiers are slow-growing aerobic bacteria that oxidize ammonia into nitrite and nitrate; they cannot compete with the fast growers. The genera that oxidize ammonia to nitrite are *Nitrosomonas*, *Nitrosococcus*, *Nitrosospira*, *Nitrosolobus*, and *Nitrosovibrio*, while those that transform nitrite to nitrate are *Nitrobacter*,

Nitrococcus, and *Nitrospira* (Prosser et al., 2014). Heterotrophic bacteria such as *Arthrobacter*, *Thiosphaera*, and *Flavobacterium* catalyze nitrification of organic compounds containing nitrogen (Schlegel & Zaborosch, 1993).

Nitrate is removed through denitrification, which is carried out by the switched oxidative metabolism into nitrate respiration of heterotrophic aerobic bacteria. Under unfavorable conditions, denitrifying bacteria release intermediates (NO, NO_2^-, or N_2O) during the denitrification of nitrate or use them as final electron acceptors (Conrad, 1996).

9.1.3 Metal Ions' Removal

Heavy metal ions can be removed by microorganisms, and after removal, the biomass must be separated and then the metals can be remobilized. Many bacteria need metal ions as electron acceptors such as Fe^{3+}, Mn^{4+}, Cr^{6+}, As^{5+}, Se^{6+}, Hg^{2+}, Pb^{2+}, or U^{6+} for anaerobic respiration (Bueno et al., 2012). Microorganisms can actively (energy-dependent) accumulate metal ions intracellularly. Bioaccumulation of metal ions requires living cells that are metabolically active, whereas inactive cell material passively adsorbs metal ions (Unz & Shuttleworth, 1996; Kumar, 2018). Sulfate-reducing bacteria generate sulfide which reduces metal ions. Additionally, microorganisms alter the redox state of ions leading to precipitation (Lovley & Coates, 1997). Fungi are capable of leaching insoluble metal salts by organic acids and forming complexes (Brynhildsen & Rosswall, 1997). Living organisms require Fe and aerobic bacteria excrete siderophores that chelate Fe^{3+} by phenolate or hydroxamate moiety (Straub et al., 1996). Members of *Geobacteriaceae* gain their required energy by reducing Fe(III) or Mn(IV) (Lloyd, 2003).

9.1.4 Degradation of Xenobiotics

Many xenobiotic substances can be degraded aerobically or anaerobically. Xenobiotic degradation is slow, and the short time of residence in wastewater treatment plants prevents the selection of bacteria or adaptation for degradation. The most suitable microorganisms selected naturally by adaptation processes are usually used, while effective use of transgenic bacteria or transfer of degradative enzyme genes is rare (Van Limbergen et al.,1998). The limitations of using transgenic bacteria to degrade xenobiotics in wastewater treatment plants are due to several factors: (1) Both indigenous and transgenic strains prefer the simpler carbon sources normally available in wastewaters, which results in the target genes not being expressed; (2) Transgenic strains do not survive, or they cannot compete successfully with indigenous strains due to their too low metabolic activity; and (3) If the inoculated strains survive, the target genes cannot be transferred because the indigenous strains are not competent (Gallert & Winter, 2005).

9.1.5 Microbial Nexus in Effluent

Microbial diversity is significantly high in effluent. Common families are *Moraxellaceae*, *Steroidobacteraceae*, *Comamonadaceae*, *Anaerolineaceae*, and

Clostridiaceae, while *Weeksellaceae*, *Exiguobacteraceae*, and *Deinococcaceae* are significantly low (Bawiec et al., 2016; Wu et al., 2019). *Moraxellaceae* includes pseudomonads, human pathogens, and is the host of antibiotic resistance genes, while members of *Steroidobacteraceae* are gram-negative aerobes that reduce nitrate to nitric oxide or ammonia (Rossau et al., 1991; Stalder et al., 2019; Liu et al., 2019). *Aeromonas*, *Campylobacter*, *Enterobacter*, *Enterococcus*, *Escherichia*, *Klebsiella*, *Legionella*, *Micrococcus*, *Mycobacterium*, *Pseudomonas*, *Salmonella*, *Shigella*, and *Vibrio* species have also been recorded (Oliver, 2005). Total coliforms, thermotolerant coliforms (fecal coliforms previously), fecal streptococci, and enterococci have been considered indicator bacteria for effluent quality (Adams, 2017).

In effluent, indicator bacteria may become stressed or harmed by the presence of toxic substances such as heavy metal ions, solar radiation, or insufficient disinfection. These stressed bacteria may not be detected due to their inability to grow or form colonies under standard conditions, resulting in an inaccurate definition of water quality (Mcfeters et al., 1986; Bucklin et al., 1991). The abundances of mobile genes like antibiotic resistance genes and oxidative-stress-tolerance are significantly high (Lu et al., 2023).

9.2 CONVENTIONAL BIOLOGICAL TREATMENT

Water needs, climate change, and population growth have reduced the availability of freshwater resources. Freshwater pollution by hazardous microorganisms, heavy metals, and organic and inorganic contaminants also threaten freshwater resources (Yadav et al., 2017). Wastewater unbalances the elemental composition in the receiving water. To save water resources, wastewater must be treated and reused (Deng et al., 2019; Kumar et al., 2022). In this context, several treatment systems have been developed and among the well-studied systems are biological ones. The decomposition of organic matter is similar in a large proportion in anaerobic and aerobic bacteria. Aerobic treatment is usually faster than anaerobic in degradation rate. Aerobic treatment is effective with a small amount of organic pollutants although it is more costly, and yields more sludge residues for discarding. Constructing an anaerobic treatment system is expensive but the operating cost is lower than an aerobic system (Gallert & Winter, 2005; Kayser, 2005).

Beginning in 1910, continuous aeration was tested for wastewater treatment. Subsequently, the activated sludge was settled after a period of intermittent aeration, repeated floc settling, and decanting of the supernatant; however, there were many problems due to manual operation. An activated sludge plant included an aeration tank to mix the activated sludge with wastewater and oxygen and achieve adequate turbulence to prevent settling of the mixed liquor. The biomass was separated by settling from the treated wastewater in a final clarifier. Part of the collected sludge was pumped back into the aeration tank, while the excess was drawn out. Initially, aeration used diffused bubbles, then surface aerators were developed due to clogging problems in the ceramic diffusers. Several aerators were developed including a vertical shaft cone surface, cone surface, the horizontal axis brush, and the horizontal axis mammoth rotor. Membrane diffusers were developed around the 1970s, making diffused air ventilation more popular (Gallert & Winter, 2005).

Among the well-studied systems are biological ones including ponds and wetlands, where treatment is based on synergies between microorganisms (Abouhend et al., 2020; Ji et al., 2020; Kumar & Chandra, 2020; Hassan et al., 2022). Wetlands were one of the earliest biological treatment systems. Constructed wetlands improve water quality and create valuable wildlife habitats, while being competitive with other systems considering cost issues. *Phragmites australis*, *Typha angustifolia*, or similar plants planted on soil filters are applied to purify gray water from single households (Ji et al., 2021).

9.3 MICROBIAL FUEL CELLS (MFCS)

The MFC is a technology in which microorganisms directly convert biochemical substrates into electricity. MFCs are considered a promising technology for wastewater purification and energy production. This technology can improve sanitation in developing regions or countries (Ida & Mandal, 2022). Through redox reactions, MFCs process organic pollutants and produce electrons that generate current by transferring them in an electrical circuit (Vidhyeswari et al., 2022). MFCs treat wastewater using the biological degradation of organic matter to produce electricity. Wastewater treatment with MFCs has been applied for organic pollutants, heavy metals, and NH_3 removal, as well as recovery of high-value elements like silver (Ag) or chromium (Cr) (Munoz-Cupa et al., 2021).

9.3.1 A Look at the History and Current Status of MFCs

Michael C. Potter was the first developer of electricity from organic compounds in 1911 but generated low energy. What drew attention toward the MFC was the progress made by Barnett Cohen who produced 35 V by a series mode of connected cells in 1931. Subsequently, several investigations were conducted to understand the fundamentals of MFCs that led to effective performance.

Different types of MFC were constructed, different electrode components were tested, a mediator and catalyst were added to maximize the power produced, and the possibility of other applications was investigated (Ida & Mandal, 2022).

9.3.2 Principles of MFCs

Two chambers bonded by a membrane, or salt bridge, are the main components of MFCs. One chamber is anaerobic and biological, consisting of electrogenic bacteria (electron donors), an anode, and an anolyte, while the other is an abiotic, aerobic cathode chamber, consisting of a cathode and a catholyte. The electrogenic bacteria decompose organic compounds (wastewater) and generate protons and electrons. Electricity is generated through a redox reaction where protons and electrons are transferred to the cathode due to the potential difference at which they interact with electron acceptors and safe byproducts are produced. Protons are transported across an ion exchange membrane or salt bridge while electrons are transported through an external circuit. The best and most abundant electron acceptor is O_2, which has a high reducing potential and produces water as a byproduct (Saran et al., 2023; Khandaker et al., 2021).

Anodic reaction:

$$CH_3COO^- + 2H_2O \rightarrow 8e^- + 2CO_2 + 7\ H_2O$$

Cathodic reaction:

$$O_2 + 4e^- + 4H^+ \rightarrow 2H_2O$$

9.3.3 Components of MFCs

The anode, cathode chambers, and proton exchange membrane (PEM) are constructed with specific materials; these materials determine the performance of MFCs.

9.3.3.1 Anode

The anode chamber is anaerobic, consisting of an electrode, a substrate, and active microorganisms. The active microorganisms oxidize the substrate (organic matter) and produce both protons and electrons. Microorganisms that transfer electrons outside of their cells are known as exoelectrogens and are mostly bacteria. While the electrodes receive the produced electrons directly from the exoelectrogens or through mediators, the protons pass into the cathode chamber via the PEM. Electricity generation occurs through e^- and H^+ transfer depending on the charge difference between the anode and cathode chambers. Anode chamber efficiency in MFC cells is determined by exoelectrogens, biofilm development, system design, and electron mediators (Khandaker et al., 2021).

The active electrode must be biocompatible and electrically conductive with a high surface area and its material should have a low operating cost, chemical stability, corrosion resistance, be nontoxic, and also mechanically strong. The most used material for anode electrodes is carbon as a brush, rod, mesh, plate, paper, veil, felt, or reticulated vitreous carbon, granular activated carbon, granular graphite, or carbonaceous-based materials like carbonized cardboard and carbon cloth. Electrodes may also consist of metals such as gold, aluminum, stainless steel, copper, molybdenum, silver, nickel, and titanium (Ida & Mandal, 2022).

9.3.3.2 Cathode

The operating efficiency of the cathode depends on the cathode material, configuration, and the redox potential. The higher the redox potential, the easier it is to capture a proton and the more electricity is generated. Several electron acceptors are used for reduction such as oxygen permanganate or ferricyanide but the most widely used is oxygen. What makes oxygen better than other oxidants is that it has a high oxidation potential, is sustainable, and is easier to access. The electrode material in the cathode chamber is the same as that used for the anode but a catalyst can be integrated into the electrode to improve the reduction reaction (Khandaker et al., 2021). Many catalysts such as carbon-supported platinum or palladium-based materials are fused to a cathode electrode by spraying, drop casting, or a doctor's blade rolling technique (Ida & Mandal, 2022).

9.3.3.3 Membrane

In MFCs, the membranes or separators are also termed ion exchange membranes (IEM) or PEM and are semipermeable to allow the flow of protons and limit the transfer of the substrate from the anode chamber to the cathode chamber, while preventing the backflow of oxygen/hydrogen (Liu & Lee, 2021). The proton flow rate to the cathode is less than that of the electron and therefore, the membrane surface area must be greater than that of the electrode to accelerate proton transfer. The most effective H^+ transport membranes are Nafion and Ultrexmembranes and are therefore widely used in MFCs (Gil et al., 2003). The cost and cations like Ca^{2+}, K^+, Mg^{2+}, Na^+, and NH_4^+ are also important to consider when selecting a membrane (Munoz-Cupa et al., 2021).

Salt bridge, size-selective, and IEMs are three categories of PEM. The ionic groups affixed to the membrane matrix divide the IEMs into cation exchange membranes (CEM) or anion exchange membranes (AEM). Other separators like a microfiltration membrane (MFM), ultrafiltration membrane (UFM), bipolar membrane (BPM), salt bridge, glass fibers, and porous fabrics are relevant (Munoz-Cupa et al., 2021; Rahimnejad et al., 2015).

9.3.4 Types of MFC

Several factors such as shape, size, electrode, and MFC configuration are significant for efficient reactor design, so there are many types of MFC. Single or double-chambered MFCs are the most common models, with a combination of the two models also being applicable (Nawaz et al., 2022).

9.3.4.1 Single-chambered MFC

A single-chambered or single-chambered air cathode MFC has only a single chamber (Saran et al., 2023). In this chamber, the anode is embedded in the substrate and located close to the PEM to be separated from the cathode directly exposed to air (Fig. 9.1). Different designs of single-chambered MFCs such as cube, cylindrical, and bottle reactors have been constructed and investigated. In a single-chamber MFC, the internal resistance between the anode/membrane and cathode is related to the distance between them, the lower the distance, the lower the resistance. This kind of MFC is characterized by its simple design, low operating cost, and high energy output (Vidhyeswari et al., 2022).

9.3.4.2 Double-chamber MFC

A double-chamber or dual-chamber MFC consists of two compartments (Fig. 9.2); an anaerobic anode and aerobic cathode. Both chambers are joined externally through a wire and internally by a PEM to complete the circuit and limit the diffusion of

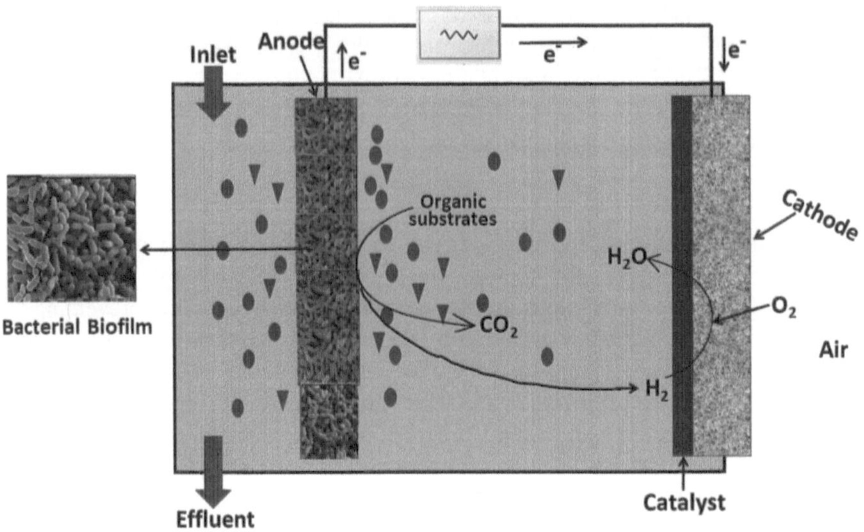

FIGURE 9.1 Single-chambered MFC.

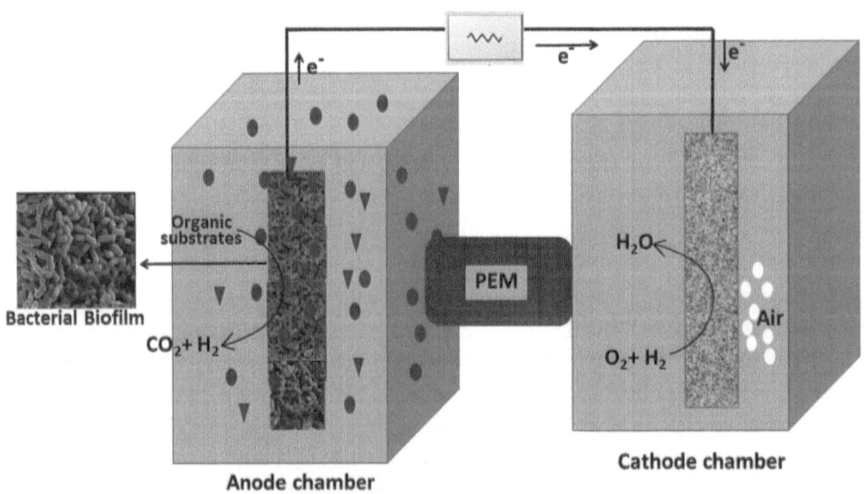

FIGURE 9.2 Double-chamber MFC.

oxygen to the anodic part (Nawaz et al., 2022). Various models of double-chamber MFCs such as cylindrical, rectangular, miniature shapes, and flat plates have been developed to overcome the high cost of ventilation and are widely applied (Munoz-Cupa et al., 2021).

The up-flow is a tubular-shaped model at the top, cathode compartment, then there are a glass wool film and glass beads, while the anode is at the base. The electrolytes of the two compartments are the same. The substrate flow starts at the anode compartment, then proceeds to the cathode, and finally, exits from the top. The main application of this design is wastewater treatment; however, the generated electricity is low due to the high energy required to pull the substrate (Saran et al., 2023).

9.3.4.4 Stacked MFC

Several MFCs are assembled in parallel or series to maximize the energy produced. In anode compartments, organic compounds are oxidized under anaerobic conditions, and there is an accumulation of protons into the anolyte which has a negative impact on the performance and operation time. By connecting a single-chambered to a double-chamber MFC in parallel, this accumulation will be reduced (Ida & Mandal, 2022).

9.3.4.5 Hybrid MFC System

Hybridization of MFCs to the conventional constructed wetland has recently been proposed for different applications. Because the constructed wetland consists of anaerobic and aerobic conditions and is effective in wastewater treatment, it can be combined with MFCs to enhance biodegradation and generate more energy (Vidhyeswari et al., 2022).

9.3.5 MFC Operation System

Wastewater is a valuable source of renewable energy (Tatinclaux et al., 2018). The most promising feature of an MFC is that instead of consuming energy, it produces energy by treating wastewater and successful operation depends on the appropriate feeding of substrates to the anode chamber and oxygen to the cathode. The surface area of the electrodes and the internal resistance control the efficiency of the MFC (Khandaker et al., 2021). Microorganisms in wastewater decompose organic matter for their metabolism yielding electrons, protons, and CO_2. The delivered electrons from the microbial reduction are transferred to the anode, while protons produce H_2 and OH^- in the cathode chamber. Electrons are transferred to the cathode chamber via an external circuit, received mostly by oxygen, and produce water (H_2O). The produced water in MFCs can be stored and then passed through tunnels to fall on turbines to generate electricity on a larger scale (Nawaz et al., 2022; Ucar et al., 2017).

9.3.6 Microbial Behavior in MFCs

In MFCs, microbes convert chemical energy into electrical energy, hence the name "microbial fuel cell." Organic matter in wastewater is oxidized by microbial metabolism to produce protons and electrons in various forms such as NAD^+, $FAD^{+,}$ and FMN that reduce the final electron acceptor, mostly oxygen (Vidhyeswari et al., 2022). The generation of electrons outside the cells is characteristic of exoelectrogens and there are two pathways to transfer the electrons to electrodes. The first pathway, known as direct electron transfer (DET), is through cytochromes or pili/nanowires, while the other pathway is through soluble chemicals acting as mediators. Microbes with the DET pathway usually develop a biofilm on the electrode, while mediators may be produced by microbes (redox-active proteins) or added to the system (Boas et al., 2022).

A few families were tested in MFC systems such as *Acetobacteraceae* (*Acidiphilium* sp.), *Aeromonadaceae* (*Aeromonas hydrophila*), *Brucellaceae* (*Ochrobactrum nthropic*), *Burkholderiaceae* (*Cupriavidus basilensis*), *Clostridiaceae* (*Clostridium beijerinckii* and *Clostridium butyricum*), *Comamonadaceae* (*Rhodoferax ferrireducens*), *Desulfobulbaceae* (*Desulfobulbus propionicus*), *Desulfovibrionaceae* (*Desulfovibrio desulfuricans*, *Desulfovibrio vulgaris*), *Desulfuromonaceae* (*Desulfuromonas acetoxidans*), *Enterobacteriaceae* (*Erwinia dissolvens*, *Escherichia coli*, *Klebsiella pneumonia*, and *Proteus vulgaris*), *Enterococcaceae* (*Enterococcus faecium*), *Geobacteraceae* (*Geobacter metallireducens* and *Geobacter sulfurreducens*), *Holophagaceae* (*Geothrix fermentans*), *Nitrobacteraceae* (*Rhodopseudomonas palustris*), *Peptococcaceae* (*Thermincola* sp.), *Pseudomonadaceae* (*Pseudomonas aeruginosa*), *Saccharomycetaceae* (*Hansenula anomala*), and *Shewanellaceae* (*Shewanella putrefaciens*, *Shewanella oneidensis*) (Vidhyeswari et al., 2022).

Active microorganisms in MFCs are of two types, electrogenic and nonelectrogenic. Electrogenic microorganisms are capable of transferring their electrons to the anode such as *Shewanella putrefaciens*, *Geobacter metallireducens*, *Clostridium butyricum*, *Desulfuromonas acetoxidans*, and *Acidiphilium* sp. Nonelectrogenics mediate electron transfers to anodes such as *Proteus mirabilis*, *Saccharomyces cerevisiae*, *Shewanella oneidensis*, *Streptococcus lactis*, and *Micrococcus luteus*.

9.3.7 Advantages and Disadvantages of MFCs

The use of MFCs to treat wastewater is a sustainable technology, the more waste degraded, the more electricity generated, COD levels are reduced by up to 50%, and the generated power reaches 420–460 mW/m^2 (Meylani et al., 2023). Although the generated power is low compared to the input or less to operate devices such as transmitters or sensors constantly, it can be improved. To increase the power generated, the surface area of the electrodes must be increased, the material of the electrodes must be properly selected, reactor design and configuration can be considered, and catalysts can be added. Any biological system needs proper temperatures, which is why MFCs do not operate at temperatures that are too low (Rahimnejad et al., 2015). Electrode materials, catalysts, increased surface area, and warming make the cost of MFCs high, but using MFCs in wastewater treatment can reduce the overall cost (Munoz-Cupa et al., 2021).

9.4 BIONANOTECHNOLOGICAL APPROACHES

Bionanotechnology is defined as the novel approaches to emerging advances in nanotechnology and biotechnology-based techniques. Nanotechnology has started to provide reliable solutions for wastewater purification. Therefore, there is an essential requirement to use advanced nanotechnological techniques for providing clean water to meet human needs worldwide. Recently, researchers have paid significant attention to the use of nanoparticles, nanomaterials, and microbes for the removal of pollutants from wastewater (Mandeep & Shukla, 2020). Nanoparticles of approximately 1–100 nm are of significant interest in the area of water purification due to their unique physicochemical properties that enable them to be applied in such modern applications (Taher et al., 2022; Kunduru et al., 2017; Zhang et al., 2016). Nanoparticles not only offer a larger specific surface area than bulk materials but also have high reactivity, strong sorption, and a small size (Bethi & Sonawane, 2018). Nanoparticles and nanomaterials can be synthesized through various routes comprised of chemical, physical, and biological pathways (Iravani et al., 2014). The nano-sized particles obtained by chemical and physical methods are toxic, expensive, and harmful to the environment. Green/biological synthesis of nano-sized particles is considered a promising alternative to chemical and physical methods (Zhang et al., 2022; Sayed & Saad, 2021). In the green synthesis process, nanoparticles can be obtained from plant extracts as well as microorganisms. Green synthesis technology is safe, low cost, and easy. The applicability of using green synthesized nanoparticles and nanomaterials in wastewater treatment is high due to their antimicrobial activity (Taher et al., 2022).

Metal oxide nanoparticles like gold, silver, iron, and titanium oxide, which are synthesized from biological routes, are common nanomaterials applied for the decontamination of polluted water (Bethi et al., 2016; Taiba & Tayyiba, 2021; Nikita et al., 2022). Metal nanoparticles can be used for the sensitive detection of toxic materials due to their surface plasmon resonance (Prasad & Aranda, 2018). Green synthesized metal nanoparticles have great advantages for the treatment of water contaminated by heavy metals, pathogenic microbes, and organic and inorganic solutes. The nanoparticles that have high adsorption and reaction capacities due to their small size are considered perfect candidates for water and wastewater decontamination. The main goal of bionanotechnology approaches is developing low-cost and environmentally friendly nano-sized particles to purify contaminated water. Nanoparticles and nanomaterials have unique functionalities that are required for toxic metal ions' removal.

The most important nano-sized particles that have interesting applications in water purification and wastewater treatment can be categorized as outlined in the following sections.

9.4.1 Metal Nanoparticles

Metal nanoparticles have active antimicrobial capacities and disinfecting mechanisms which enable them to be employed for the removal of different organic and inorganic contaminants from wastewater. The use of these nano-sized particles represents an environmentally friendly, energy-saving, and smart water purification solution. Metal nanoparticles such as silver and gold are some of the most researched candidates in nanotechnology for water treatment (Qian et al., 2013; Que et al., 2018). Silver nanoparticles (Ag NPs) are one of the best candidates for water purification purposes due to their high antimicrobial activity. Ag NPs release silver ions (Ag^+) in large quantities and attack bacterial cells, forming highly reactive oxygen species (ROS). Silver ions cause a large increase in membrane permeability and DNA and RNA damage which prevent DNA replication. Silver ions with UV radiation display photocatalytic activity which is effective in disinfection. Several studies utilized membranes impregnated with Ag NPs for water purification and disinfection systems. A recent study reported that silver nanoparticles, biosynthesized from *Streptomyces* sp. U13 (KP109813), were utilized to construct a wastewater disinfection filter to kill all coliform bacteria present (Taher et al., 2022). Another study revealed that biogenic Ag NPs produced by *Lactobacillus fermentum* can be used to remove viruses from drinking water, leading to a commercially available application for home water ultrapure systems (Maynard, 2007).

9.4.2 Metal Oxide Nanoparticles

Extensive techniques of adsorption are highly valued for removing dyes and toxic ions from industrial water. Metal oxide nano-sized particles including titanium oxide nanoparticles (TiO_2 NPs), iron oxide nanoparticles (Fe_3O_4 NPs), copper oxide nanoparticles (CuO NPs), zinc oxide nanoparticles (ZnO NPs), aluminum oxide nanoparticles (Al_2O_3 NPs), and manganese dioxide nanoparticles (MnO_2 NPs) can be used as adsorbents to remove metallic contaminants from water/wastewater sources (Nikita et al., 2022; Naseem & Durrani, 2021). These nanoparticles can be fabricated via green routes and can be utilized in different fields such as photocatalytic activity, adsorption, and antibacterial activity. Among these, ZnO NPs are considered the most widely used for wastewater treatment (Spoială et al., 2021; Dimapilis et al., 2018). Zinc oxide particles at the nanoscale show a high efficiency of ultraviolet absorption and photocatalytic activity. After ZnO NPs, CuO and TiO_2 nanoparticles are the most widely used metal oxides for the treatment of wastewater (Ighalo et al., 2021; Deshmukh et al., 2021). Titanium dioxide nanoparticles display excellent photocatalytic activity with ultraviolet radiation, and they can show this photocatalytic activity even in the presence of visible sunlight. Their antimicrobial activity is mainly due to the production of ROS. TiO_2 nanoparticles can be used in the disinfection process in water treatment because of their stability in water, nontoxicity, and low cost (Magaña-López et al., 2021). They can be used to construct filters by doping TiO_2 nanoparticles with various metals or coating reactors with thin films.

9.4.3 Magnetic Nanoparticles

Magnetic nanoparticles are considered suitable candidates for the process of wastewater treatment including adsorption, filtration, and photocatalytic activities. Their magnetic properties facilitate the removal of contaminants from wastewater in an efficient way. Water treatment processes based on magnetic nanoparticles are easy and unique. One study presented a novel low-cost magnetic nanomaterial used for heavy metal ions' removal from various types of water. This material consisted of iron oxide magnetic nanoparticles (Fe_3O_4 NPs) treated with humic acid (HA) to enhance material stability and efficiency in removing toxic heavy metals (Liu et al., 2008). Another study revealed that Fe_2O_3 nanoparticles could be used for fast removal of (Pb^{2+}), a typical metal ion present in industrial wastewater, with maximum removal efficiency during a very short time (Cheng et al., 2012). Magnetic nanoparticles were recommended as rapid, effective, and commercial for metal ions' removal from wastewater and adsorbing organic pollutants (Gutierrez et al., 2017).

9.5 MEMBRANE BIOREACTOR (MBR)

A membrane bioreactor (MBR) is one of the latest technologies for wastewater treatment. It has emerged to treat various wastewaters over the last century. Combining membrane filtration (microfiltration or ultrafiltration) with biological treatment is the method used to create MBR plants. This approach has several advantages over traditional methods like activated sludge (Al-Asheh et al., 2021).

9.5.1 A Look at the History and Current Status of MBRs

Early versions of MBRs, dating back to the 1960s, used a membrane filter fed by an aeration tank and operated in cross-flow mode. This construction allowed for process optimization (biological treatment and solids' separation); however, it was characterized by high energy consumption which was necessary to maintain sufficient cross-flow velocity. Such cross-flow velocity was required to control membrane fouling. In Japan in 1989, Yamamoto applied submerged membranes in a bioreactor (Rahman et al., 2023). Submerged systems, with the membrane in the bioreactor, resulted in a substantial decrease in energy requirements. By the end of 2008, 37 MBR plants with capacities greater than 5000 m^3/d were operational in Europe (Krzeminski et al., 2017). Currently (2023), there are over 800 commercial MBR applications in use. The global MBR market has an annual growth rate of 7.6% with large-scale (≥10,000 m^3/d) and super-large-scale (≥100,000 m^3/d) plants in operation (Xiao et al., 2019; Rahman et al., 2023).

9.5.2 MBR Design and Operation System

The MBR is considered a hybrid treatment system as it combines the biologically activated sludge process (aerobic, anaerobic) with membrane filters to separate the sludge produced by biological treatments (Vaishnavi et al., 2023).

9.5.2.1 MBR Design

The first part of the MBR plant is the bioreactor tank in which microorganisms aid in the degradation of biological wastes. The second part is the membrane unit. Depending on the location of the membrane unit, the MBR plant could be designed as a side-stream MBR or immersed MBR (Fig. 9.3). In the side-stream MBR, the membrane unit is outside the reactor, whereas, inside in the immersed MBR.

The operating procedure of the MBR consists of an organized process which includes: (1) Pretreatment of the wastewater by removing large debris and heavy particles; (2) Pretreated wastewater is passed through an activated sludge tank where microorganisms degrade water pollutants; (3) Membrane filtration, where the

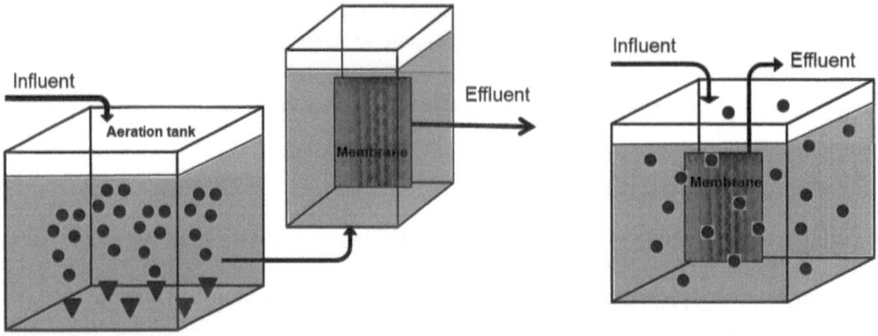

FIGURE 9.3 Membrane bioreactor (MBR) design: Side-stream MBR (right) and immersed MBR (left).

biologically treated water is pumped through a membrane unit; and finally (5) To kill any remaining microorganisms and reduce the risk of disease, the treated water is disinfected.

9.5.2.2 Pretreatment Process

The influent undergoes a pretreatment process to remove large debris and heavy particles that can block the pores in the membrane unit and reduce the flux. In addition, the pretreatment process is applied to control the pH of the influent. Extreme pH values in some industrial wastewaters could influence biological performance and damage membrane permeability, reducing the membrane's lifespan (Lin et al., 2013). Therefore, several wastewater pretreatment strategies (e.g., filtration, pH adjustment, ozone, and ultrasound treatment) are crucial to removing contaminants in the influent.

9.5.2.3 MBR Bioreactor Tank

The dynamics of biological communities in activated sludge are critical in pollutant degradation. Wide groups of microorganisms are found in the activated sludge of MBRs such as bacteria, protozoa, metazoa, algae, and fungi (Table 9.1). Nevertheless, bacteria are the common microorganisms existing in MBRs.

9.5.2.4 Microbial Behavior in MBRs

Microorganisms in MBRs are found in aggregated forms or as single cells. They maintain their growth through synthesis and respiration processes that use organic/inorganic pollutants in wastewater. Accordingly, metabolic products like extracellular polymeric substances excreted by living microorganisms and dead cells are produced (Deb et al., 2022). The accumulation of such microbial exudates and other colloids results in membrane fouling which in turn affects effluent quality. Solid retention time (SRT) is a significant parameter that characterizes activated sludge. For instance, MBR operation at a short SRT restricts the development of functional microorganisms including slow-growing nitrifiers. When the process operation is carried out at a longer SRT, the biomass undergoes extensive growth, which can lead to substrate

TABLE 9.1
Previous studies showing different microorganisms used in MBR

Microorganism	Reference
Ammonium-oxidizing and nitrite-oxidizing bacteria	Leyva-Díaz et al. (2020); Leyva-Díaz et al. (2015)
Serratia liquefaciens and *Aeromonas hydrophila* (bacteria)	Al-Malack (2006)
Shewanella baltica KB30 (bacteria)	Leyva-Díaz et al. (2017)
Azonexaceae, unclassified *Bacteroidetes*, unclassified *Pseudomonadaceae*, *Candidatus Cloacamonas*, *Fervidobacterium*, *Rectinema*, and *Zoogloea* (bacteria)	Ramadan et al. (2023)
Rhodobacter Sphaeroides strain 1.2174 (bacteria)	Lu et al. (2013)
Eucheuma spinosum (algae)	Sim et al. (2021)
Chlorella emersonii (algae)	Xu et al. (2014)
Neurospora crassa (fungi)	Luke and Burton (2001)
Phanerochaete chrysosporium (fungi)	Girijan and Kumar (2019)
Escherichia coli (bacteria), *Cryptosporidium* oocysts and *Giardia* cysts (protozoa)	Ugarte et al. (2022)

deficiency, increased aeration cost, and membrane fouling (Yu et al., 2018). For a high-quality effluent, biomass concentration, bioreactor volume, and factors influencing microbial reaction rate must be considered. Oxygen concentration is one of the limiting factors for microbial growth and metabolism in aerobic MBRs. It acts as an electron acceptor, yielding energy. In MBRs, the dissolved oxygen (DO) levels may favor some microbes while discouraging others. Another key factor for the MBR operation process is temperature. The growth of desirable microbes is facilitated by optimal temperatures, but inappropriate microbes may decline or diminish in number within the reactors. Microorganisms hydrolyze and degrade organic or inorganic substances enzymatically, and the enzyme activity is temperature sensitive. Some enzymes' activities (e.g., phosphatase and esterase) positively correlate with temperature, whereas others (e.g., glucosidase) can do well at a low temperature (Molina-Muñoz et al., 2010; Molina-Muñoz et al., 2007). Impaired enzyme activities result in less organic substance biodegradation, so a high concentration of waste is retained in the reactors. Therefore, it is fundamental to control environmental temperatures that impact microbial growth and activity in MBRs.

9.5.2.5 MBR Membrane Unit

Membranes of MBRs are usually fabricated from either polymeric materials or ceramic. For instance, ceramic membranes could be alumina, silicon carbide, titanium dioxide, or zirconia. They are flexible for cleaning and characterized by high resistance to fouling and chemicals. However, the high cost of fabrication frequently makes ceramic materials uneconomical; therefore, polymeric membranes such as polyethylsulphone, polyacrylonitrile, polysulphone, or polyvinylidene difluoride are widely applied.

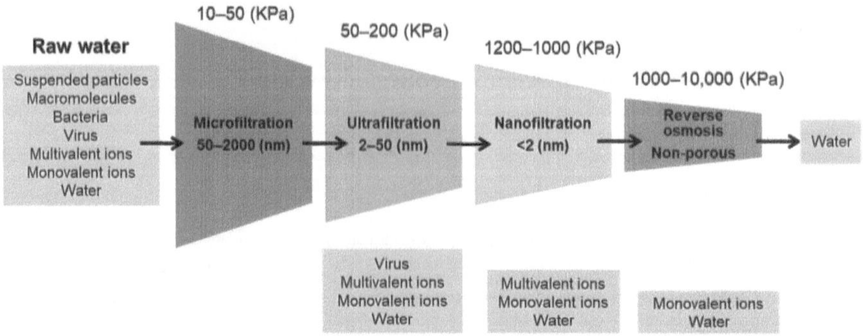

FIGURE 9.4 Various types of filtration systems in membrane bioreactors (MBR) showing pore sizes (nm) and operating pressure (KPa) of each membrane type.

Membrane units in MBRs could be also classified into two types according to the porosity of membrane filters; porous and nonporous (Shirazi et al., 2010) (Fig. 9.4). Porous membranes use filters (e.g., microfiltration, ultrafiltration, or nanofiltration) with various pore sizes to exclude particles. Nonporous membranes that separate monovalent salts and small molecules are based on the variation in solubility of the solute and the solvent within the membranes (e.g., reverse osmosis) (Paul et al., 2023).

9.5.2.6 MBR Membrane Fouling and Cleaning

The permeability of the membrane and its transmembrane pressure (TMP) are critical to its operation. TMP is the dynamic power for membrane filtration, whereas permeability indicates filtration performance. The primary challenge faced while executing membrane separation techniques is the accumulation of fouling and blockage. The fouling process consists of four steps (Fig. 9.5) (Al-Asheh et al., 2021; Rahman et al., 2023): (1) A blockage of the smallest membrane pores; followed by (2) the concealment of larger pores' interior surfaces; then (3) accumulation of particles that directly block the larger pores; and finally (4) formation of the cake layer. Microbial cells, debris, sludge clogs, colloids, solutes, and inorganics are all potential foulants in MBR membranes. Regeneration of the fouled membranes may be achieved physically, chemically, or biologically (Lin et al., 2013). Physical methods include membrane relaxation and back flushing. Chemical cleaning involves the use of agents such as sodium hypochlorite, sodium hydroxide, citric acid, hydrochloric acid (HCl), nitric acid, and EDTA. Biological cleaning is an environmentally friendly method that employs purified enzymes and surfactants capable of extracting foulants from membranes.

9.5.2.7 Operating Cost and Energy Consumption

The capital cost of MBRs includes the cost of building the plant in addition to the pipeline operating costs. Energy, chemicals, labor, and sludge disposal are all part of the operating costs (Rahman et al., 2023). The average energy requirement starting from the operation process to get a treated effluent was estimated at 0.4 and 2.3

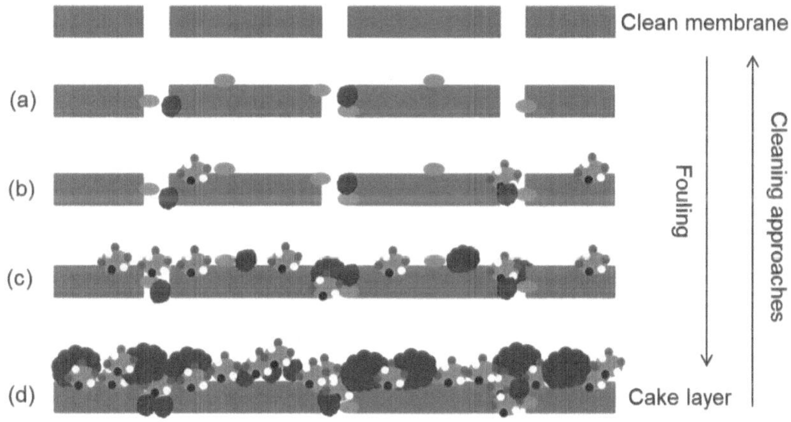

FIGURE 9.5 The fouling process and cleaning in the membrane bioreactor (MBR).

kWh/m3 (Krzeminski et al., 2017). In 2020, Arif and colleagues applied economic modeling to compare the cost of two conventional wastewater treatment plants with MBR (Arif et al., 2020). They found that the MBR had the highest cost among other processes for the same amount of treated effluent.

9.5.3 Advantages and Disadvantages of MBRs

The MBR technique has many advantages in wastewater treatment. Activated sludge, when combined with the membrane separation technique, generates a high-quality effluent due to the completely retained biomass and suspended solids' separation by the membrane. The MBR is a small bioreactor that does not include a sedimentation tank and a long SRT is accomplished within a compact area. On the other hand, the most challenging issue for MBRs is fouling and biofilm formation on the membrane. Chemical cleaning of the membrane could be a solution; however, it is a complex process and reduces membrane longevity. The cost of antifouling strategies in addition to the cost of the membrane is considered high.

9.6 MOVING BED BIOFILM REACTOR (MBBR)

A moving bed biofilm reactor (MBBR) is an innovative wastewater treatment reactor. It is a biological unit that operates similarly to activated sludge processes, with the exception that biofilm grows on small carrier elements suspended in continuous motion throughout the reactor's entire volume (Barwal & Chaudhary, 2014). Moreover, it does not need any sludge recirculation.

9.6.1 A Look at the History and Current Status of MBBRs

In the 1980s, the Norwegian company, Kaldnes Miljteknologi AS, developed the MBBR process (now AnoxKaldnes AS). The first MBBR facility opened in Lardnal,

Norway, in 1990. In 2014, around 50 countries had nearly 1200 operational MBBR plants (Biswas et al., 2014). MBBR technology has proven to be effective in the treatment of different wastewaters, including pharmaceutical wastewater, industrial wastewater, laundry wastewater, and dairy wastewater. New research has introduced a hybrid MBBR technology that merges both attached and suspended biomass to enhance the effectiveness of wastewater treatment.

9.6.2 MBBR Design and Operation System

An MBBR plant consists of a single reactor or is designed in series. Each MBBR has a length-to-width ratio (L:W) in the range of 0.5:1 to 1.5:1 (McQuarrie & Boltz, 2011). The reactor contains plastic biofilm carriers with a volume of up to 70% of the empty bed liquid volume. In aerobic tanks, the aeration system uniformly distributes plastic biofilm carriers, which increases the contact area between the biological membrane and oxygen as well as the efficiency of oxygen transfer. An agitator is used to increase water flow under anaerobic conditions to fully expose the biofilm and pollutants and achieve the goal of biological decomposition (McQuarrie & Boltz, 2011). Recent studies applied the combination of both anaerobic and aerobic tanks in municipal wastewater treatment (Fig. 9.6).

9.6.2.1 Biocarriers and Biofilms

Biocarriers are usually cylindrical, with a cross section on the inside and longitudinal fins on the outside. Although the traditional Kaldnes-5 (K5-biocarrier) is commonly used, a new biocarrier, Saddle-Chips has shown a high potential in MBBR plants (Tadda et al., 2021) (Fig. 9.7). Generally, biocarriers are fabricated from polyethylene, polypropylene, or polyethylene with a density of ~0.95 g cm^{-3}. The shape, density, and biofilm areas of the carriers all have an impact on the performance of MBBR processes enabling a stable biofilm during the process. For instance, low-density plastic carriers with a biofilm attachment can be stored in wastewater tanks without mixing. The range of the carrier's biofilm-specific areas is 300–1000 m^2/m^3. Because

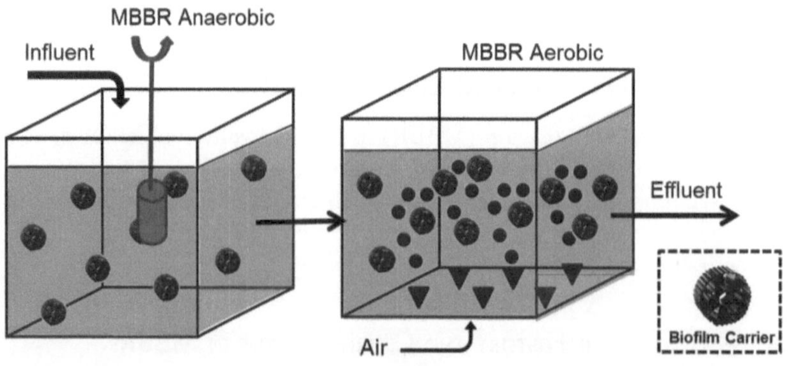

FIGURE 9.6 Moving bed biofilm reactor (MBBR) design combining aerobic and anaerobic treatment tanks.

FIGURE 9.7 Plastic biocarriers in a moving bed biofilm reactor (MBBR): Conventional K5-biocarrier (left) and saddle-chips (right).

of their cost-effectiveness and process advantages, carriers with biofilm-specific areas of 500–1000 m^2/m^3 are commonly used in full-scale MBBR plants. For maximum efficacy, porous carriers with surfaces containing spherical holes with diameters > 1 mm are preferred. They provide microenvironments that produce uneven biofilms and retain large amounts of biomass. The plastic biocarriers used in MBBRs have a life span of 10–30 years and do not need to be replaced or replenished regularly.

Biofilms are "aggregates of microorganisms in which cells are frequently embedded in a self-produced matrix of extracellular polymeric substances that are adherent to each other and/or a surface" (Flemming et al., 2016). In MBBRs, a mature biofilm is formed on the surface of biocarriers by a mechanism involving the adherence and proliferation of microbial cells. A biofilm may encompass a variety of microorganisms. Several studies have shown that bacterial communities belonging to *Proteobacteria*, *Actinobacteria*, *Bacteroidetes*, and *Firmicutes* form a high-quality biofilm on MBBR plastic carriers (Arif et al., 2020; Deng et al., 2016).

9.6.2.2 Microbial Behavior in MBBRs

Substrate removal by biofilm-forming microbes is essential for keeping viable microbial films. However, such a process depends on the biomass depth and substrate concentration. According to the bulk depth of biofilms, they are classified as: (1) Deep and thick biofilms, wherein the substrate concentration ultimately becomes zero within the biofilm; (2) Shallow biofilms, wherein the substrate concentration is higher than zero; and (3) Fully penetrated biofilms, wherein the concentration gradient of the substrate is negligible (Barwal & Chaudhary, 2014). Generally, biofilms with a thickness of < 100 µm are the most functional for substrate removal due to their high penetration capacity. The separation of a biofilm from the surface of the carrier is a process that takes place naturally wherein the biomass is released into the bulk liquid.

Such a separation process limits biofilm accumulation and therefore, influences the quality of substrate removal (Flemming et al., 2016).

The microbial reactions in wastewater treatment are biodegradation of organic matter, nitrification, and denitrification. The biodegradation of organic compounds is carried out by heterotrophs on the outer surface of the biofilm (Gupta et al., 2022). Nitrogen is eliminated by nitrification and denitrification processes. Firstly, ammonia-oxidizing bacteria produce nitrite from ammonia by oxidation as follows:

$$NH_3 + O_2 \rightarrow NO_2^- + 3H^+ + 2e^-$$

Next, nitrite is oxidized to nitrate by nitrite-oxidizing bacteria as follows:

$$NO_2^- + H_2O \rightarrow NO_3^- + 2H^+ + 2e^-$$

The nitrification process is carried out by the autotrophs in the biofilm's inner layer. After nitrification, denitrification is carried out by heterotrophs.

9.6.3 Advantages and Disadvantages of MBBRs

The MBBR is an appealing wastewater treatment technology due to improvements in both design and operation that have resulted in smaller footprints, significantly lower production of suspended solids, reliable generation of reusable water that meets high-quality standards, and minimal waste disposal. In MBBRs, hydraulic retention time (required time for efficient influent treatment) is short because of the combination of the moving carriers with the highly concentrated biofilm. On the other hand, most MBBRs have a filling ratio of 60–70% to keep carriers continuously mixed and the entire reactor homogeneous. Higher filling ratios necessitate more energy consumption to ensure adequate mixing of the suspended carriers. Moreover, it is required that experts in biological treatment monitor the factors influencing biofilms.

9.7 CONCLUSIONS

The effluent content of pollutants is significantly influenced by the treatment process. Effluent discharge unbalances the elemental composition in the surface water which leads to eutrophication. Differences in microbial community and elemental composition between surface water and discharged effluent must be minimized. Effluent discharging will raise the bacterial count in the receiving water; either pathogenic or purification-related bacteria. Physical, chemical, and biological treatment must be applied synergistically to deal with the diversity of pollutants in wastewater and to meet environmental legislation standards. To effectively control water pollution, new processes for wastewater treatment must be developed, add-on processes may also be required. Many benefits result from wastewater treatment including resource recovery, fertilizer production, agricultural irrigation, and industrial cooling. The improvement of treatment techniques comes from the knowledge of microbial interactions and their ecophysiology. Biostimulation of autochthonous water purification bacteria should be considered.

REFERENCES

Abouhend A.S., Milferstedt K., Hamelin J., Ansari A.A., Butler C., Carbajal-González B.I., and Park C. 2020. Growth progression of oxygenic photogranules and its impact on bioactivity for aeration-free wastewater treatment. *Environmental Science and Technology*, 54: 486–496.

Adams, V.D., 2017. *Water and Wastewater Examination Manual*. Routledge.

Al-Asheh, S., Bagheri, M. and Aidan, A. 2021. Membrane bioreactor for wastewater treatment: A review. *Case Studies in Chemical and Environmental Engineering* 4: 100109. https://doi.org/10.1016/j.cscee.2021.100109

Al-Malack, M.H. 2006. Determination of biokinetic coefficients of an immersed membrane bioreactor. *Journal of Membrane Science.* 271: 47–58. https://doi.org/10.1016/j.memsci.2005.07.008

Arif, A.U.A., Sorour, M.T. and Aly, S.A. 2020. Cost analysis of activated sludge and membrane bioreactor WWTPs using CapdetWorks simulation program: Case study of Tikrit WWTP (middle Iraq). *Alexandria Engineering Journal* 59: 4659–4667. https://doi.org/10.1016/j.aej.2020.08.023

Barwal, A. and Chaudhary, R. 2014. To study the performance of biocarriers in moving bed biofilm reactor (MBBR) technology and kinetics of biofilm for retrofitting the existing aerobic treatment systems: A review. *Reviews in Environmental Science and Bio/Technology* 13: 285–299. https://doi.org/10.1007/s11157-014-9333-7

Bawiec, A., Paweska, K., and Jarzab, A. 2016. Changes in the microbial composition of municipal wastewater treated in biological processes. *Journal of Ecological Engineering* 17: 41–46. https://doi.org/10.12911/22998993/63316.

Bethi, B. and Sonawane, S.H. 2018. "Nanomaterials and its application for clean environment." In *Nanomaterials for Green Energy*, edited by Bhanvase, B.A., Pawade, V.B., Dhoble, S.J., Sonawane, S.H., Ashokkumar, M. 385–409. https://doi.org/10.1016/B978-0-12-813731-4.00012-6

Bethi, B., Sonawane, S.H., Bhanvase, B.A., and Gumfekar, S.P. 2016. Nanomaterials-based advanced oxidation processes for wastewater treatment: A review. *Chemical Engineering and Processing – Process Intensification* 109: 178–189. https://doi.org/10.1016/j.cep.2016.08.016

Biswas, K., Taylor, M.W. and Turner, S.J. 2014. Successional development of biofilms in moving bed biofilm reactor (MBBR) systems treating municipal wastewater. *Applied Microbiology and Biotechnology* 98: 1429–1440. https://doi.org/10.1007/s00253-013-5082-8

Boas, J.V., Oliveira, V.B., Simões, M., and Pinto, A.M.F.R. 2022. Review on microbial fuel cells applications, developments and costs. *Journal of Environmental Management* 307: 114525. https://doi.org/10.1016/j.jenvman.2022.114525

Broughton, M. J., Thiele, J., Birch, E. J., and Cohen, A. 1998. Anaerobic batch digestion of sheep tallow. *Water Research* 32, 1323–1428.

Bryant, M. P. 1979. Microbial methane production: Theoretical aspects. *Journal of Animal Science* 48: 193–201.

Brynhildsen, L., and Rosswall, T. 1997. Effects of metals on the microbial mineralization of organic acids. *Water, Air, and Soil Pollution* 94: 45–57.

Bucklin K.E., Mcfeters G.A., and Amirtharajah A. 1991. Penetration of coliforms through municipal drinking water filters. *Water Research* 25: 1013.

Bueno E, Mesa S, Bedmar E. J., Richardson D. J., and Delgado M. J. 2012. Bacterial adaptation of respiration from oxic to microoxic and anoxic conditions: Redox control. *Antioxid Redox Signal* 16(8): 819–852. https://doi: 10.1089/ars.2011.4051

Chandra, R. and Kumar, V. 2015. Biotransformation and biodegradation of organophosphates and organohalides. In: Chandra, R. (Ed.) *Environmental Waste Management.* Boca Raton: CRC Press. DOI:10.1201/b19243-17.

Cheng, Z., Tan, A., Tao, Y., Shan, D., Ting, K.E., and Yin, X.J. 2012. Synthesis and characterization of iron oxide nanoparticles and applications in the removal of heavy metals from industrial wastewater. *International Journal of Photoenergy* 2012: 608298 https://doi.org/10.1155/2012/608298

Conrad, R. 1996. Soil microorganisms as controllers of the atmospheric trace gases (H_2, CO, CH_4, OCS, N_2O, and NO).*Microbiology Review* 60: 609–640.

Deb, A., Gurung, K., Rumky, J., Sillanpää, M., Mänttäri, M., and Kallioinen, M. 2022. Dynamics of microbial community and their effects on membrane fouling in an anoxic-oxic gravity-driven membrane bioreactor under varying solid retention time: A pilot-scale study. *Science of The Total Environment* 807: 150878. https://doi.org/10.1016/j.scitotenv.2021.150878

Deng, L., Guo, W., Ngo, H.H., Zhang, X., Wang, X.C., Zhang, Q., and Chen, R. 2016. New functional biocarriers for enhancing the performance of a hybrid moving bed biofilm reactor–membrane bioreactor system. *Bioresource Technology* 208: 87–93. https://doi.org/10.1016/j.biortech.2016.02.057

Deng, S., Yan, X., Zhu, Q., andLiao, C., 2019. The utilization of reclaimed water: Possible risks arising from waterborne contaminants. *Environmental Pollution.* 254: 113020. https://doi.org/10.1016/j.envpol.2019.113020.

Deshmukh, S.P., Dalavi, D.K., and Hunge, Y.M. 2021. Disinfection of water using supported titanium dioxide photocatalysts. *American Journal of Engineering and Applied Sciences* 13 (4): 819.826. https://doi.org/10.3844/ajeassp.2020.819.826

Dimapilis, E.A.S., Hsu, C., Mendoza, R.M.O. and Lu, M. 2018. Zinc oxide nanoparticles for water disinfection. *Sustainable Environment Research* 28 (2): 47–56. https://doi.org/10.1016/j.serj.2017.10.001

Flemming, H.C., Wingender, J., Szewzyk, U., Steinberg, P., Rice, S.A., and Kjelleberg, S. 2016. Biofilms: An emergent form of bacterial life. *Nature Reviews Microbiology* 14: 563–575. https://doi.org/10.1038/nrmicro.2016.94

Gallert C. and Winter J. 2005. Bacterial metabolism in wastewater treatment systems In: Jördening H. J. and Winter J. (eds.) *Environmental Biotechnology. Concepts and Applications.* Wiley-VCH Verlag GmbH & Co. KGaA, Weinheim.

Gil, G. C., Chang, I. S., Kim, B. H., Kim, M., Jang, J. K., Park, H. S., and Kim, H. J. 2003. Operational parameters affecting the performance of a mediator-less microbial fuel cell. *Biosensors and Bioelectronics* 18(4): 327–334. https://doi.org/10.1016/S0956-5663(02)00110-0

Girijan, S. and Kumar, M. 2019. Immobilized biomass systems: An approach for trace organics removal from wastewater and environmental remediation. *Current Opinion in Environmental Science & Health* 12: 18–29. https://doi.org/10.1016/j.coesh.2019.08.005

Gupta, B., Gupta, A.K., Ghosal, P.S., Lal, S., Saidulu, D., Srivastava, A., and Upadhyay, M. 2022. Recent advances in application of moving bed biofilm reactor for wastewater treatment: Insights into critical operational parameters, modifications, field-scale performance, and sustainable aspects. *Journal of Environmental Chemical Engineering* 10: 107742. https://doi.org/10.1016/j.jece.2022.107742

Gutierrez A.M., Dziubla T.D., and Hilt J.Z. 2017. Recent advances on iron oxide magnetic nanoparticles as sorbents of organic pollutants in water and wastewater treatment. *Reviews on Environmental Health* 32 (1-2): 111–117 https://doi.org/10.1515/reveh-2016-0063

Hassan, S., Sabreena, Khurshid, Z., Bhat, S.A., Kumar, V., Ameen, F., and Ganai, B.A., 2022. Marine Bacteria and omic approaches: A novel and potential repository for bioremediation assessment. *Journal of Applied Microbiology*. https://doi.org/10.1111/jam.15711

He P.J., Lü F., Shao L.M., Pan X.J., Lee, D.J., 2006. Enzymatic hydrolysis of polysaccharide-rich particulate organic waste. *Biotechnology and Bioengineering* 936: 1145–1151.

Ida, T.K., and Mandal, B. 2022. Microbial fuel cell design, application and performance: A review. *Materials Today: Proceedings* 76: 88–94. https://doi.org/10.1016/j.matpr.2022.10.131

Ighalo, J.O., Sagboye, P.A., Umenweke, G., Ajala, O.J., Omoarukhe, F.O., Adeyanju, C.A., and Ogunniyi, S. 2021. CuO nanoparticles (CuO NPs) for water treatment: A review of recent advances. *Environmental Nanotechnology, Monitoring & Management* 15: 100443. https://doi.org/10.1016/j.enmm.2021.100443

Iravani, S., Korbekandi, H., Mirmohammadi, S.V., and Zolfaghari, B. 2014. Synthesis of silver nanoparticles: chemical, physical and biological methods. *Research in Pharmaceutical Sciences* 9 (6): 385–406. PMID: 26339255; PMCID: PMC4326978.

Ji, B., Wang, S., Silva, M.R.U., Zhang, M. and Liu, Y., 2021. Microalgal-bacterial granular sludge for municipal wastewater treatment under simulated natural diel cycles: Performances-metabolic pathways-microbial community nexus. *Algal Research* 54: 102198.

Ji B., Zhang M., Gu J., Ma Y, and Liu Y. 2020. A self-sustaining synergetic microalgal-bacterial granular sludge process towards energy-efficient and environmentally sustainable municipal wastewater treatment. *Water Research* 179: 115884.

Kayser R. 2005. Activated sludge process In: Jördening H. J. and Winter J. (eds). Environmental Biotechnology. Concepts and Applications. Wiley-VCH Verlag GmbH & Co. KGaA, Weinheim.

Khandaker, S., Das, S., Hossain, M.T., Islam, A., Miah, M.R., and Awual, M.R. 2021. Sustainable approach for wastewater treatment using microbial fuel cells and green energy generation – A comprehensive review. *Journal of Molecular Liquids* 344: 117795. https://doi.org/10.1016/j.molliq.2021.117795

Krzeminski, P., Leverette, L., Malamis, S. and Katsou, E. 2017. Membrane bioreactors – A review on recent developments in energy reduction, fouling control, novel configurations, LCA and market prospects. *Journal of Membrane Science* 527: 207–227. https://doi.org/10.1016/j.memsci.2016.12.010

Kunduru, K.R., Nazarkovsky, M., Farah, S., Pawar, R.P., Basu, A., and Domb, A.J. 2017. "Nanotechnology for water purification: Applications of nanotechnology methods in wastewater treatment." In: Alexandru Mihai Grumezescu (ed). *Water Purification*, 33–74. Academic Press. https://doi.org/10.1016/B978-0-12-804300-4.00002-2

Kumar, V., Agrawal, S., Bhat, S.A., Américo-Pinheiro, J.H.P., Shahi, S.K., and Kumar, S., 2022. Environmental impact, health hazards, and plant-microbes synergism in remediation of emerging contaminants. *Cleaner Chemical Engineering* 2: 100030. https://doi.org/10.1016/j.clce.2022.100030

Kumar, V., 2018. Mechanism of microbial heavy metal accumulation from polluted environment and bioremediation. In: Sharma, D. and Saharan, B.S. (eds.) *Microbial Fuel factories*. Boca Raton: CRC Press

Leyva-Díaz, J. C., González-Martínez, A., González-López, J., Muñío, M.M. and Poyatos, J.M. 2015. Kinetic modeling and microbiological study of two-step nitrification in a membrane bioreactor and hybrid moving bed biofilm reactor–membrane bioreactor for wastewater treatment. *Chemical Engineering Journal* 259: 692–702. https://doi.org/10.1016/j.cej.2014.07.136

Leyva-Díaz, J.C., Muñío, M.M., Fenice, M. and Poyatos, J.M. 2020. Respirometric method for kinetic modeling of ammonium-oxidizing and nitrite-oxidizing bacteria in a membrane bioreactor. *AIChE Journal* 66: e16271. https://doi.org/10.1002/aic.16271

Leyva-Díaz, J.C., Poyatos, J.M., Barghini, P., Gorrasi, S. and Fenice, M. 2017. Kinetic modeling of *Shewanella baltica* KB30 growth on different substrates through respirometry. *Microbial Cell Factories* 16: 189. https://doi.org/10.1186/s12934-017-0805-7

Lin, H., Peng, W., Zhang, M., Chen, J., Hong, H., and Zhang, Y. 2013. A review on anaerobic membrane bioreactors: Applications, membrane fouling and future perspectives. *Desalination* 314: 169–188. https://doi.org/10.1016/j.desal.2013.01.019

Liu J., Zhao Z., and Jiang G. 2008. Coating Fe_3O_4 magnetic nanoparticles with humic acid for high efficient removal of heavy metals in water. *Environmental Science & Technology* 42 (18): 6949–6954. https://doi.org/10.1021/es800924c

Liu, Q., Liu, H.-C., Zhou, Y.-G., and Xin, Y.-H., 2019. *Stenotrophobium rhamnosiphilum* gen. nov., sp. nov., isolated from a glacier, proposal of *Steroidobacteraceae* fam. nov. in Nevskiales and emended description of the family *Nevskiaceae*. *International Journal of Systematic and Evolutionary Microbiology*. 69: 1404–1410. https://doi.org/10.1099/ijsem.0.003327.

Liu, S.H., and Lee, K.Y. 2021. Optimization preparation of composite membranes as proton exchange membrane for gaseous acetone fed microbial fuel cells. *Journal of Power Sources* 509: 230368. https://doi.org/10.1016/j.jpowsour.2021.230368

Lloyd, J. R., 2003. Microbial reduction of metals and radionuclides, *FEMS Microbiology Reviews*. 27: 411–425.

Lovley, D. R. and Coates, J. D., 1997. Bioremediation of metal contamination, *Current Opinion in Biotechnology*. 8: 285–289.

Lu, H., Zhang, G., Dai, X., Schideman, L., Zhang, Y., Li, B., and Wang, H. 2013. A novel wastewater treatment and biomass cultivation system combining photosynthetic bacteria and membrane bioreactor technology. *Desalination* 322: 176–181. https://doi.org/10.1016/j.desal.2013.05.007

Lu Q., Zhao R., Li Q., Ma Y., Chen J., Yu Q., Zhao D., and An S. 2023. Elemental composition and microbial community differences between wastewater treatment plant effluent and local natural surface water: A Zhengzhou city study. *Journal of Environmental Management* 325: 116398. https://doi.org/10.1016/j.jenvman.2022.116398

Luke, A.K. and Burton, S.G. 2001. A novel application for *Neurospora crassa*: Progress from batch culture to a membrane bioreactor for the bioremediation of phenols. *Enzyme and Microbial Technology* 29: 348–356. https://doi.org/10.1016/S0141-0229(01)00390-8

Magaña-López, R., Zaragoza-Sánchez, P.I., Jiménez-Cisneros, B.E., and Chávez-Mejía, A.C. 2021. The use of TiO_2 as a disinfectant in water sanitation applications. *Water* 13 (12): 1641. https://doi.org/10.3390/w13121641

Mandeep and Shukla P. 2020. Microbial nanotechnology for bioremediation of industrial wastewater. *Frontiers in Microbiology* 11: 590631. https://doi.org/10.3389/fmicb.2020.590631

Maynard, A.D. 2007. Nanotechnologies: Overview and issues In Simeonova, P.P., Opopol, N., and Luster, M.I., (eds). *Nanotechnology – Toxicological Issues and Environmental Safety and Environmental Safety*. NATO Science for Peace and Security Series. Springer, Dordrecht. https://doi.org/10.1007/978-1-4020-6076-2_1

Mcfeters G.A., Kippin J.S., and Lechevallier M.W. 1986. Injured coliforms in drinking water. *Applied and Environmental Microbiology*. 51: 1.

McInerney, M. J., 1988. Anaerobic hydrolysis and fermentation of fats and proteins. In: Zehnder, A. J. B. (ed.) *Biology of Anaerobic Microorganisms*. 373–415.Wiley, New York.

McQuarrie, J.P. and Boltz, J.P. 2011. Moving bed biofilm reactor technology: process applications, design, and performance. *Water Environment Research* 83: 560–575. https://doi.org/10.2175/106143010x12851009156286

Meylani, V., Surahman, E., Fudholi, A., Almalki, W.H., Ilyas, N., and Sayyed, R.Z. 2023. Biodiversity in microbial fuel cells: Review of a promising technology for wastewater treatment. *Journal of Environmental Chemical Engineering* 11 (2): 109503. https://doi.org/10.1016/j.jece.2023.109503

Molina-Muñoz, M., Poyatos, J.M., Rodelas, B., Pozo C., Manzanera, M., Hontoria, E., and Gonzalez-Lopez, J. 2010. Microbial enzymatic activities in a pilot-scale MBR experimental plant under different working conditions. *Bioresource Technology* 101: 696–704. https://doi.org/10.1016/j.biortech.2009.08.071

Molina-Muñoz, M., Poyatos, J.M., Vílchez, R., Hontoria, E., Rodelas, B., and González-López, J. 2007. Effect of the concentration of suspended solids on the enzymatic activities and biodiversity of a submerged membrane bioreactor for aerobic treatment of domestic wastewater. *Applied Microbiology and Biotechnology* 73: 1441–1451. https://doi.org/10.1007/s00253-006-0594-0

Munoz-Cupa, C., Hu, Y., Xu, C., and Bassi, A. 2021. An overview of microbial fuel cell usage in wastewater treatment, resource recovery and energy production. *Science of the Total Environment* 754: 142429. https://doi.org/10.1016/j.scitotenv.2020.142429

Naseem, T. and Durrani, T. 2021. The role of some important metal oxide nanoparticles for wastewater and antibacterial applications: A review, *Environmental Chemistry and Ecotoxicology* 3: 59–75. https://doi.org/10.1016/j.enceco.2020.12.001

Nawaz, A., ul Haq, I., Qaisar, K., Gunes, B., Raja, S.I., Mohyuddin, K., and Amin, H. 2022. Microbial fuel cells: Insight into simultaneous wastewater treatment and bioelectricity generation. *Process Safety and Environmental Protection* 161: 357–373. https://doi.org/10.1016/j.psep.2022.03.039

Nikita, G., Susmita, D., Goutam, B., and Prabir, K.H. 2022. Review on some metal oxide nanoparticles as effective adsorbent in wastewater treatment. *Water Science & Technology* 85 (12): 3370–3395. https://doi.org/10.2166/wst.2022.153

Oliver, J.D. 2005. The viable but nonculturable state in bacteria. *Journal of Microbiology.* 43: 93.

Paul, A., Dasgupta, D., Hazra, S., Chakraborty, A., Haghighi, M., and Chakraborty, N. 2023. "Membrane bioreactor: A potential stratagem for wastewater treatment." InNadda, A.K., Banerjee, P., Sharma, S. and Nguyen-Tri (eds.) *Membranes for Water Treatment and Remediation*, 133–155, Springer Nature Singapore. https://doi:10.1007/978-981-19-9176-9_6.

Prasad, R. and Aranda E. 2018. *Approaches in Bioremediation:* The *New Era of Environmental Microbiology and Nanobiotechnology*. Springer Nature Switzerland AG. https://doi.org/10.1007/978-3-030-02369-0

Prosser, J.I., Head, I.M., and Stein, L.Y. (2014). The family nitrosomonadaceae. In: Rosenberg, E., DeLong, E.F., Lory, S., Stackebrandt, E., and Thompson, F. (eds.) *The Prokaryotes*. Springer, Berlin, Heidelberg. https://doi.org/10.1007/978-3-642-30197-1_372.

Qian, H., Pretzer, L.A., Velazquez, J.C., Zhao, Z. and Wong, M.S. 2013. Gold nanoparticles for cleaning contaminated water. *Journal of Chemical Technology and Biotechnology* 88: 735–741. https://doi.org/10.1002/jctb.4030

Que, Z.G., Torres, J.G.T., Vidal, H.P., Rocha, M.A.L., Pérez, J.C.A., López, I.C., Romero, D.C., Reyna, A., Sosa, J., Pavón, A. and Hernández, J. 2018. "Application of Silver Nanoparticles for Water Treatment." In Khan Maaz (ed.) *Silver Nanoparticles*,. http://dx.doi.org/10.5772/intechopen.74675

Rahimnejad, M., Adhami, A., Darvari, S., Zirepour, A., and Oh, S.E. 2015. Microbial fuel cell as new technology for bioelectricity generation: A review. *Alexandria Engineering Journal* 54 (3): 745–756. https://doi.org/10.1016/j.aej.2015.03.031

Rahman, T.U., Roy, H., Islam, M.R., Tahmid, M., Fariha, A., Mazumder, A., Tasnim, N., Pervez, M.N., Cai, Y., Naddeo, V., and Islam, M.S. 2023. The advancement in aembrane bioreactor (MBR) technology toward sustainable industrial wastewater management. *Membranes* 13: 181. https://doi.org/10.3390/membranes13020181

Ramadan, L., Deeb, R., Sawaya, C., El Khoury, C., Wazne, M., and Harb, M. 2023. Anaerobic membrane bioreactor-based treatment of poultry slaughterhouse wastewater: Microbial community adaptation and antibiotic resistance gene profiles. *Biochemical Engineering Journal* 192: 108847. https://doi.org/10.1016/j.bej.2023.108847

Rossau, R., Van Landschoot, A., Gillis, M., and De Ley, J., 1991. Taxonomy of *Moraxellaceae* fam. Nov., a new bacterial family to accommodate the genera *Moraxella, acinetobacter*, and *psychrobacter* and related organisms. *International Journal of Systematic and Evolutionary Microbiology*. 41:310–319. https://doi.org/10.1099/00207713-41-2-310.

Saran, C., Purchase, D., Saratale, G.D., Saratale, R.G., Romanholo Ferreira, L.F., Bilal, M., and Bharagava, R.N. 2023. Microbial fuel cell: A green eco-friendly agent for tannery wastewater treatment and simultaneous bioelectricity/power generation. *Chemosphere* 312 (P1): 137072. https://doi.org/10.1016/j.chemosphere.2022.137072

Sayed, R. and Saad, H. 2021. Gold nanoparticles: Green synthesis, characterization and biological activities. *Egyptian Journal of Chemistry* 64 (12): 2–3. https://doi.org/10.21608/ejchem.2021.68846.3553

Schlegel, H.G., and Zaborosch, C. 1993. *General Microbiology*. Cambridge University Press.

Shirazi, S., Lin, C.-J. and Chen, D. 2010. Inorganic fouling of pressure-driven membrane processes — A critical review. *Desalination* 250: 236–248. https://doi.org/10.1016/j.desal.2009.02.056

Sharma, U., Sharma, S., Rana, V.S., Rana, N., Kumar, V., Sharma, S., Qadri, H., Kumar, V., and Bhat, S.A. 2023. Assessment of microplastics pollution on soil health and eco-toxicological risk in horticulture. Soil Systems 7(1), 7. https://doi.org/10.3390/soilsystems7010007

Sim, Y., Jung, J., Baik, J., Park, J., Kumar, G., Banu, J.R., and Kim, S. 2021. Dynamic membrane bioreactor for high rate continuous biohydrogen production from algal biomass. *Bioresource Technology* 340: 125562. https://doi.org/10.1016/j.biortech.2021.125562

Spoială, A., Ilie, C.I., Trușcă, R.D., Oprea, O.C., Surdu, V.A., Vasile, B.Ș., Ficai A., Ficai, D., Andronescu, E., and Dițu, LM. 2021. Zinc oxide nanoparticles for water purification. *Materials* 14: 4747. https://doi.org/10.3390/ma14164747

Stalder, T., Press, M.O., Sullivan, S., Liachko, I., Top, E.M., 2019. Linking the resistome and plasmidome to the microbiome. *ISME Journal*. 13: 2437–2446. https://doi.org/ 10.1038/s41396-019-0446-4.

Straub, K. L., Benz, M., Schink, B., Widdel, F., 1996. Anaerobic, nitrate-dependent microbial oxidation of ferrous iron. *Applied and Environmental Microbiology*. 62: 1458–1460.

Tadda, M.A., Altaf, R., Gouda, M., Rout, P.R., Shitu, A., Ye, Z., Zhu, S., and Liu, D. 2021. Impact of Saddle-Chips biocarrier on treating mariculture wastewater by moving bed biofilm reactor (MBBR): Mechanism and kinetic study. *Journal of Environmental Chemical Engineering* 9: 106710. https://doi.org/10.1016/j.jece.2021.106710

Taher, H.S., Sayed, R, Loutfi, A., and Abdulla, H. 2022. Construction of a domestic wastewater disinfection filter from biosynthesized and commercial nanosilver: a comparative study. *Annals of Microbiology* 72: 31. https://doi.org/10.1186/s13213-022-01688-2

Taiba, N., and Tayyiba, D. 2021. The role of some important metal oxide nanoparticles for wastewater and antibacterial applications: A review. *Environmental Chemistry and Ecotoxicology* 3: 59–75. http://dx.doi.org/10.1016/j.enceco.2020.12.001

Tatinclaux, M., Gregoire, K., Leininger, A., Biffinger, J. C., Tender, L., Ramirez, M., and Kjellerup, B.V. 2018. Electricity generation from wastewater using a floating air cathode microbial fuel cell. *Water-Energy Nexus* 1 (2): 97–103. https://doi.org/10.1016/j.wen.2018.09.001

Ucar, D., Zhang, Y., and Angelidaki, I. 2017. An overview of electron acceptors in microbial fuel cells. *Frontiers in Microbiology* 8: 1–14. https://doi.org/10.3389/fmicb.2017.00643

Ugarte, P., Ramo, A., Quílez, J., Bordes, M., Mestre, S., Sánchez, E., Peña J. Á., and Menéndez, M. 2022. Low-cost ceramic membrane bioreactor: Effect of backwashing, relaxation and aeration on fouling. Protozoa and bacteria removal. *Chemosphere* 306: 135587. https://doi.org/10.1016/j.chemosphere.2022.135587

Unz, R. F. and Shuttleworth, K. L., 1996. Microbial mobilization and immobilization of heavy metals, *Current Opinion in Biotechnology* 7: 307–310.

Vaishnavi, M., Gopinath, K.P. and Ghodke, P.K. 2023. Membrane bioreactor (MBR) technologies for treatment of tannery waste water and biogas production. In: Shah, M.P. (ed.) *Biorefinery for Water and Wastewater Treatment*, 217–247. Springer International Publishing, 2023. https://doi:10.1007/978-3-031-20822-5_11.

Van Limbergen, H., Top, E. M., and Verstraete, W. 1998. Bioaugmentation inactivated sludge: Current features and future perspectives, *Applied Microbiology and Biotechnology* 50: 16–23.

Vidhyeswari, D., Surendhar, A., and Bhuvaneshwari, S. 2022. General aspects and novel PEMss in microbial fuel cell technology: A review. *Chemosphere* 309 (P1): 136454. https://doi.org/10.1016/j.chemosphere.2022.136454

Winter, J., Hilpert, H., and Schmitz, H., 1992. Treatment of animal manures and wastes for ultimate disposal: Review, *Asian-Australasian Journal of Animal Sciences* 5: 199–215.

Wu, L., Ning, D., Zhang, B., Li, Y., Zhang, P., Shan, X., Zhang, Q., Brown, M.R., Li, Z., Van Nostrand, J.D., Ling, F., Xiao, N., Zhang, Y., Vierheilig, J., Wells, G.F., Yang, Y., Deng, Y., Tu, Q., Wang, A., Zhang, T., He, Z., Keller, J., Nielsen, P.H., Alvarez, P.J.J., Criddle, C.S., Wagner, M., Tiedje, J.M., He, Q., Curtis, T.P., Stahl, D.A., Alvarez-Cohen, L., Rittmann, B.E., Wen, X., and Zhou, J. 2019. Global diversity and biogeography of bacterial communities in wastewater treatment plants. *Nature Microbiology* 4: 1183–1195. https://doi.org/10.1038/s41564-019-0426-5.

Xiao, K., Liang, S., Wang, X., Chen, C. and Huang, X. 2019. Current state and challenges of full-scale membrane bioreactor applications: A critical review. *Bioresource Technology* 271: 473–481. https://doi.org/10.1016/j.biortech.2018.09.061

Xu, M., Bernards, M. and Hu, Z. 2014. Algae-facilitated chemical phosphorus removal during high-density *Chlorella emersonii* cultivation in a membrane bioreactor. *Bioresource Technology* 153: 383–387. https://doi.org/10.1016/j.biortech.2013.12.026

Yadav A., Chowdhary P., Kaithwas G., and Bharagava R.N., 2017. Toxic metals in the environment: Threats on ecosystem and bioremediation approaches. In *Handbook of Metal-Microbe Interactions and Bioremediation*. CRC Press. 128–141.

Yu, L., Yang, Y., Yang, B., Li, Z., Zhang, X., Hou Y., Lei, L., and Zhang, D. 2018. Effects of solids retention time on the performance and microbial community structures in membrane bioreactors treating synthetic oil refinery wastewater. *Chemical Engineering Journal* 344: 462 468. https://doi.org/10.1016/j.cej.2018.03.073

Zhang, K., Zhang, J., Li, J., Zheng Z., and Sun, M. 2022. Green synthesis of nanoparticles: Current developments and limitations. *Environmental Technology & Innovation* 26: 102336. https://doi.org/10.1016/j.eti.2022.102336

Zhang, Y., Wu, B., Xu, H., Liu, H., Wang, M., He, Y., and Pan, B. 2016. Nanomaterials-enabled water and wastewater treatment. *Nano Impact* 3: 22–39. https://doi.org/10.1016/j.impact.2016.09.004

10 Microbial Nexus for the Circular Economy and Climate Change

Recent Advances in Bioelectrochemical Systems and Future Opportunities

Dolores Hidalgo, Francisco Corona, and Jesús M. Martín-Marroquín

10.1 INTRODUCTION

Freshwater scarcity is a pressing global issue, as it is essential to human survival. The world's population is growing rapidly, leading to increased demand and wastewater discharge, and depleting freshwater supplies in developing nations (Shindhal et al., 2021; Kumar et al., 2021; Kayastha et al., 2022). In response to concerns about the use of fossil fuels, there is an increasing demand for alternative energy sources and unconventional energy extraction methods (Varjani et al., 2021). Industrialization has had a significant impact on less developed countries' economies, resulting in substantial health and economic losses due to pollution of air, water, and soil (Chen et al., 2021; Pandey et al., 2021; Gaur et al., 2021; Patel et al., 2021). Industrial processes produce effluents as a byproduct, and treating industrial wastewater consumes significant energy. Therefore, wastewater treatment and water pollution reduction have become critical pathways for freshwater reuse (Varjani, 2022; Kumar et al., 2022a, 2022b).

The utilization of bioelectrochemical systems (BESs) has recently increased, providing a novel method of generating bioelectricity by desalinating seawater and promoting organic matter oxidation through the use of bioelectrically active bacteria (Wilberforce et al., 2021; Yadav et al., 2022). The evolution of BESs has the potential to revolutionize the utilization of wastewater resources. Recently, there has been a surge in research activities on bioelectrochemical cells due to the urgent need for sustainable and environmentally friendly energy generation. This has led to the development of several types of BESs (Ramanaiah et al., 2023; Zhang et al., 2020). These systems have emerged as effective solutions for treating wastewater and reducing waste since they are designed to perform various actions as a function of the type of

 DOI: 10.1201/9781003441069-10

microbial and bioelectrochemical systems used. As an example, microbial fuel cells (MFCs) are capable of recovering bioelectricity, while microbial electrolysis cells (MECs) can produce biohydrogen. Microbial electrosynthesis systems (MESs) can be used for generating high-value compounds, and microbial desalination cells (MDCs) are effective for removing salts from water. Furthermore, BESs can be combined with other devices to enhance particular redox reactions, such as specialized enzyme catalysts including enzymatic fuel cells (EFCs) or microbial solar cells (MSCs). MFCs, MECs, and MDCs are the most commonly utilized types among them (Wilberforce et al., 2021). MFCs are the most popular, and they have been extensively studied for their ability to simultaneously generate electricity and treat wastewater. These systems have numerous applications including biosensors, wastewater desalination, and water treatment (Meena et al., 2019). Researchers worldwide have explored a wide range of substrates suitable for MFCs, primarily to enhance electricity production (Obileke et al., 2021). The substrates used in these studies have varied significantly, from simple single organic molecules to complex substrates. MFCs have been successfully fed with industrial wastewater, which contains both complex and simple organic pollutants (Saravanan et al., 2022). In MFCs the anode biofilm is a complex mixture of fermentative and electroactive microorganisms. Electricity can be generated from the degradation of simple organic compounds through the use of electroactive microorganisms (Naaz et al., 2023). Simultaneously treating wastewater and generating electricity or chemicals is a common application of MECs (Wilberforce et al., 2021). In the past decade, this kind of BES has been significantly researched for generating hydrogen from waste sources (Koo & Jung, 2022). MECs use biodegradable organic materials as an electron source, and are formulated to surpass the fermentative barrier that typically limits hydrogen production in fermentation-based processes. Unlike other hydrogen production methods, MECs can utilize several substrates (Bora et al., 2022). Although the technology shows promise, there are still challenges that need to be addressed before it can be commercialized, particularly in optimizing the production of hydrogen gas and reducing overall costs, which are critical steps toward eventual commercialization.

The development of low-cost desalination technologies has resulted in the increasing use of MDCs (Guijala et al., 2022; Jatoi et al., 2022; Yang et al., 2023). These modified MFCs enable the in situ desalination of water and have the potential to generate potable water from seawater, making them an effective technology for low-energy desalination using bioelectrochemical energy obtained from wastewater (Ghasemi et al., 2022). Moreover, MDCs also aid in environmental preservation by treating brackish water (Zahid et al., 2022). MDC technology has the added advantage of generating bioelectricity, which makes it an attractive option for remote areas where grid systems are neither feasible nor environmentally sustainable (Esmaeilion et al., 2021). MDCs can promote seawater desalination and remove organic materials from waste effluents without requiring additional energy, unlike traditional desalination methods (Shinde et al., 2018). MDCs function by introducing a third chamber between the anodic and the cathodic chambers and utilizing the potential difference originated by bacteria for water desalination, similar to the principle of water electrodialysis that requires an external electricity source (Tawalbeh et al., 2020). The discovery of MDC technology has triggered numerous research projects involving microbiology and electrochemistry, resulting in significant research progress in recent years to enhance the functional and structural characteristics of MDCs to improve desalination efficiency and prolong the technology's scope (Imoro et al., 2021).

Biotechnology has made progress in the development of BESs, which require two electrodes, namely the cathode and anode, for the occurrence of oxidation and reduction reactions. While some BESs have been reported without membranes, typically, the electrodes are positioned in separate compartments that are divided by a membrane (Rohbohm et al., 2022). BESs can utilize various types of membranes to enhance treatment efficiency, including ultrafiltration (UF), nanofiltration (NF), microfiltration (MF), reverse and forward osmosis membranes, ion exchange membranes, and gas permeable membranes (Ramanaiah et al., 2023). BESs have the capability to eliminate contaminants like suspended solids, viruses, bacteria, and protozoa, and generate 10–20% more energy, complementing other treatment operations. The low biomass production in BESs is also an advantage as it can achieve high levels of total suspended solids (higher than 99.0%) and COD (92%) removal, saving costs of sludge disposal (Ren et al., 2014). However, the use of membrane processes in BESs has raised some environmental concerns, including limited availability and high operation costs.

It can be concluded that bioelectrochemical systems provide a technological solution to collect energy during the treatment process of wastewater, which presents a significant opportunity for reducing the energy consumption involved in wastewater treatment. Consequently, this technology has gained significant attention and has been the subject of extensive research over the past two decades. Although the technology is not yet fully developed, it has now reached applications of BESs, with a special focus on sustainable wastewater treatment.

10.2 BIOELECTROCHEMICAL TECHNOLOGIES FOR WASTEWATER

The use of green technologies that utilize wastewater for resource restoration or power generation has grown in popularity, and BESs are often considered the most suitable option (Ramanaiah et al., 2023). BESs are hybrid biosystems that depend on both electrochemistry and microbiology to produce bioelectricity or value-added biochemicals via redox reactions catalyzed by microorganisms (Li et al., 2018a; Ramanaiah et al., 2021). Electrochemical devices produce electricity through the movement of electrons from the anode to the cathode, either autonomously or with the assistance of external power (Li et al., 2018a). BESs are highly efficient in breaking down various organics simultaneously, thanks to their capability to use a variety of biodegradable electron donors (Tavker & Kumar, 2023). BESs come in different reactor designs, sizes, microbial populations, substrates, etc., depending on the system requirements (Li et al., 2018a; Ramanaiah et al., 2021). However, MECs, MFCs, MDCs, MESs, EFCs, and MSCs, are the common types of BESs, according to Li et al. (2018a), as shown in Figure 10.1. These systems have undergone significant improvements in terms of their mechanisms, component enhancements, and applications, including industrial and household wastewater treatment (El Ichi-Ribault et al., 2018). Despite

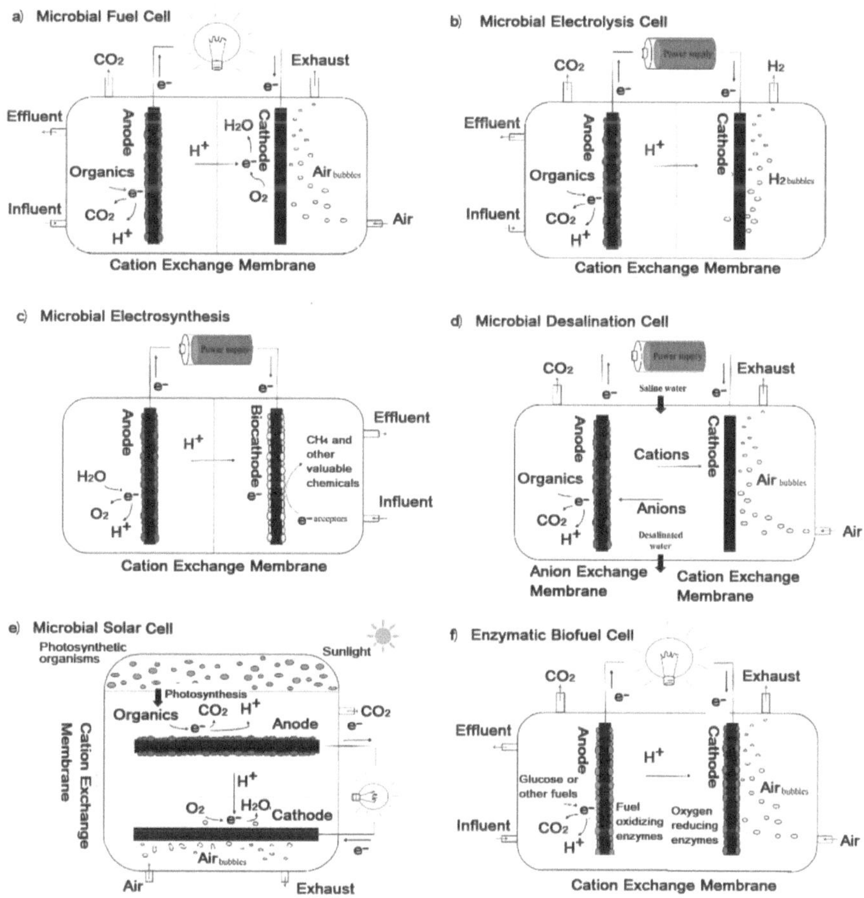

FIGURE 10.1 Different types of bioelectrochemical systems.

Source: Adapted from Li et al. (2018a).

these advances BESs' industrial applications have not received sufficient attention. The next part of this section will explore the latest developments in BES research on wastewater-related applications.

10.2.1 Microbial Fuel Cells

Figure 10.1a illustrates MFCs, which are the most commonly studied BES for practical applications. MFCs utilize microorganisms, especially bacteria, to produce electricity in a bioelectrochemical system. By utilizing electrochemically active microorganisms in the anodic chamber, bioenergy is directly generated through the degradation of various compounds, which results in the generation of bioelectricity (Li & Chen, 2018; Ramanaiah et al., 2021). Exoelectrogenic microorganisms consume organic compounds and produce electrons outside of the cell on the anode, which functions as a donor of electrons. The potential gradient created leads to the flow of electrons to the cathode from the anode over an external circuit, which results in electric current production that can be stored using batteries or used immediately (Li et al., 2018a). In contrast to traditional two-chamber MFCs, single-compartment MFCs have enhanced the feasibility of this technology for wastewater treatment and energy production (Logan et al., 2015). Numerous studies have demonstrated that MFCs can generate significant electric power using residual streams, for example, industrial waste (Pant et al., 2012), municipal effluents (Park et al., 2017), swine wastewater (Min et al., 2005), landfill leachate (Li & Chen, 2018), and other sources, according to several researchers (Ramanaiah et al., 2023). However, two primary factors have limited the commercial-scale utilization of MFCs, namely the relatively low power output of electrodes and their cost. Despite these challenges, MFCs have been extensively researched due to their potential to manage diverse agro-food wastewater and waste while simultaneously producing sustainable electricity for on-site usage, such as on farms (Ramanaiah et al., 2021; Pant et al., 2010). Table 10.1 illustrates the variability of electricity production according to the type and quality of feedstock and the type of MFC used.

The use of high-loaded effluents and waste as sources for MFCs has seen considerable advances. However, co-bioprocesses such as methane production and ammonification tend to limit the bioelectricity generation from these materials. These processes consume electrons, lowering the power generation and coulombic efficiency of MFCs. Moreover, the majority of exoelectrogens are adversely affected by high concentrations of ammonia, which can be toxic to their powering MFCs. As such, future research efforts should focus on reducing the negative impacts of ammonification and methanogenesis on MFCs, to increase energy recovery from agro-waste and wastewater. Researchers have recently shown interest in microbial soil cells and plant microbial fuel cells (PMFCs) for their capacity to produce electric power from rhizodeposits released by plant roots. In some studies, the MFC anode has been embedded in plant-growing soil to harvest rhizodeposits and other soil compounds for power production. However, further research is necessary to advance the understanding of the fundamental principles and practical uses of PMFCs. It is crucial to develop mechanistic models to optimize system design and operation requirements (Kuleshova et al., 2022; Kabutey et al., 2019; Rusyn et al., 2021).

10.2.2 Microbial Electrolysis Cells

MECs, which are derived from MFCs and illustrated in Figure 10.1b, have the ability to produce hydrogen, methane, and other valuable compounds using indirect energy and storage, as noted by Radhika et al. (2022). Theoretical studies suggest that MECs have the capability to produce biochemical products using external electric power, even in the presence of redox variations on the electrodes. It is recommended to maintain the cathode in an anoxic or anaerobic state for maximum efficiency. MECs can produce hydrogen as a product, which is increasingly in demand as a green energy source, as reported by Park et al. (2022). Researchers have shown interest in utilizing a variety of carbon-rich wastewater with the potential to be used as a feedstock for

TABLE 10.1
Electric power generation and carbon organic demand (COD) removal in MFCs using wastewater

MFC type	Wastewater	Power generation (mW/m²)	Original COD (mg/L)	COD Removal efficiency (%)
Single chamber	Starch processing wastewater	239.4	4852	98
Single chamber	Brewery wastewater	205	1600	98
Single chamber	White wine lees wastewater	262	6400	90
Single chamber	Beer brewery wastewater	205	2240	87
Single chamber	Swine wastewater	261	8320	80
Single chamber	Swine wastewater	205	1820	69
Single chamber	Winery wastewater	317	2200	65
Single chamber (upflow tubular)	Animal carcass wastewater	219	11180	51
Two chamber	Cereal wastewater	81	595	95
Two chamber	Cheese whey	1800	-	93
Two chamber	Acidogenic food, waste leachate	432	1000	87
Two chamber (upflow)	Chocolatery wastewater	98	-	70
Two chamber	Dairy wastewater	123.5	-	64
Two chamber	Distillery wastewater	31.5	-	61
Two chamber (continuous upflow)	Retting wastewater (Coconut husk)	254	2690	32
Two chamber	Palm oil mill effluent sludge	451	2680	3
Two chamber	Fermented apple juice	78	3501	-
Two chamber	Yogurt wastewater	54	8169	-
Two chamber (anoxic/oxic)	Saline sea food wastewater	162	-	-
Two chamber	Wheat straw hydrolysate	148	-	-
Earthen pot-MFC	Rice mill wastewater	2	2250	96.5
Membrane electrode assembly-MFC	Molasses wastewater	16	1500	96

Source: Adapted from Ramanaiah et al. (2023).

MEC-mediated hydrogen generation, which is noticeably more efficient than other BESs (Dinesh et al., 2018). MEC-based hydrogen production has made significant progress in recent years. According to a report by Lu and Ren (2016), H_2 production rates can increase from less than 0.2 m^3H_2/m^3 reactor/day to as high as 60 m^3H_2/m^3 reactor/day, while efficiency can rise from ≤50% to almost 100%.

TABLE 10.2
Hydrogen production in MECs from different types of wastewater

MEC type	Wastewater	Max. H_2 production rate (m^3/m^3/day)	Original COD (mg/L)	COD removal efficiency (%)
Single chamber	Soybean oil refinery	0.133	2900	95.8
Single chamber	Potato wastewater	0.74	7700	79
Single chamber	Glycerol, starch, and milk	0.94	-	74
Single chamber	Milk	0.086	1000	73.5
Single chamber	Leachate and dairy wastewater	15	24000	73
Single chamber	Fermentation effluent	2.11	3200	60
Single chamber	Fermentation effluent	4.55	8840	51
Single chamber	Winery	0.17	2200	47
Single chamber	Swine wastewater	1.0	2000	29
Single chamber	Table olive oil brine processing	130 Nml CH_4/g COD rem	-	29
Single chamber	Molasses	10.72	2000	-
Two chamber	Cheese whey: diluted and fermented	0.5	15.26	82
Two chamber	Switchgrass wastewater	4.3	2000	48
Two chamber	Animal urine	32.0	1360	46
Two chamber	Swine	0.061	1298	52

Source: Adapted from Ramanaiah et al. (2023).

Several types of agro-food residual streams, such as dairy manure, pig wastewater, and digestate, have been utilized in nano- and microscale energy conversion systems to produce biohydrogen (Table 10.2). Compared to other hydrogen generation processes, for example, photobiological processes and anaerobic digestion (AD), MECs have exhibited superior results due to their ability to surpass thermodynamic restrictions (Saratale et al., 2022. Moreover, MEC systems can be powered by conventional energy sources as well as renewable sources like MFCs and nano- and small-scale hydroelectric plants (Kadier et al., 2022). Fermentation facilities, such as biogas digesters, are frequently utilized to decrease agricultural waste while extracting valuable biogases such as bio-CH_4 and bio-H_2. The combination of anaerobic digestion with MECs intensifies the processes since AD can break down complex carbohydrates into simple and easily degradable molecules using various microbial populations (Shtepa et al., 2023). Additionally, CO_2 synthesized during fermentation can be recirculated for pH control (Saratale et al., 2022. It is possible to design MECs to generate hydrogen with high purity using fermentation-produced substrates under various operating conditions (Chaurasia & Mondal, 2021). The incorporation of MECs into existing systems facilitates the conversion into CO_2 of complex carbon-rich organic matter, a feat that cannot be achieved by fermentation operations alone.

Agricultural wastes such as sugar beet, plant leaves, maize stover, and corn stalks, which were previously considered refractory, have been utilized in such integrated systems to generate hydrogen, resulting in an average production rate increase of 148% (with a range of 42–538%) and a yield increase of 225% (with a range of 2–400%) (Lu & Ren, 2016). Balancing the fermentation and hydrogen production rates is crucial for favoring the overall system behavior. Tailored feeding techniques and inventive sequential procedures can be employed to achieve this (Dhar et al., 2015; Khan et al., 2018; Ramanaiah et al., 2023). Despite the extensive research conducted on MECs, pilot-scale MECs are yet to be constructed, probably because of monetary issues resulting from the expensive nature of materials for electrodes and the necessity for an external power source (Aiken et al., 2019).

10.2.3 Microbial Electrosynthesis

Figure 10.1c illustrates that MESs have the potential to generate valuable biochemicals including, acetic acid, butyric acid, ethanol, and biodiesel, via the process of biocathode reduction, which involves reducing CO_2 and other products (Sadhukhan et al., 2016). The potential value-added products that MESs can yield are reliant on the types of redox mediators, electron acceptors, and biocatalysts utilized. In addition, the oxidation reactions that take place at the anode in MESs have the potential to produce biofuels, for example, bioethanol (Ghangrekar et al., 2022). It is crucial to note that MESs and MECs share similarities, where external energy sources aid in driving the reduction reactions that occur at the cathode, including the conversion of H^+ to H_2. Due to the broad range of feedstocks and possible specificity inherent to MESs, it is anticipated that in the field of biological energy production, their significance is expected to grow considerably in the future (Roy et al., 2022; Thulluru et al., 2023; Zhao et al., 2023). Sadhukhan et al. (2016) also suggest that organic-rich waste products from agro-food and wastewater sources can be utilized in MESs, allowing for the transformation of agricultural and forestry waste, like wood, leaves, grass, corn stalks, immature cereal, and stover, into a diverse range of products, including syngas, biohydrogen, biomethane, formate, ethylene, methanol, dimethyl ether, urea, succinic acid, and others (Ramanaiah et al., 2023). Additionally, applying appropriate potentials to MESs can trigger redox reactions at the electrodes, which can lead to the generation of high-value biological compounds and fuels such as ethanol and diesel.

Agricultural waste and wastewater are potential resources that can be utilized for the production of sustainable organic feed for energy production. However, the composition of lignocellulosic biomass is complex, consisting of polysaccharides, such as hemicellulose and cellulose, that are surrounded by lignin, making them resistant to microbial oxidation, and make the conversion process more complicated. Pretreatment of agricultural waste streams can overcome this obstacle, allowing the recovery and processing of cellulose and hemicellulose into various chemicals of commercial interest, such as L-arabinose, xylite, furan resin, and nylon (Sadhukhan et al., 2016). Enzymatic hydrolysis of the media plays an important role in regulating the efficiency of substrate biodegradation. The process of converting lignocellulosic biomass into useful organic compounds occurs naturally in three steps: hydrolysis, acidogenesis, and methanogenesis, with acetogenesis being the optimal stage for generating electrons at the anode in bioelectrochemical systems (Li & Chen, 2018). By using electrons produced from organics, carbon dioxide can be reduced to create organic compounds like formate. Technologies for metal recovery, including MESs or MECs, have been developed recently to extract metals from polluted sites, wastewater streams, and waste material in an eco-friendly and less disruptive way (Bajracharya et al., 2016). Wang et al. (2022) state that BES technologies are more efficient and cost-effective than traditional metal recovery methods that rely on chemical precipitation. Moreover, BESs can be integrated with other food processing and agricultural practices within the industry to selectively synthesize value-added products (Sadhukhan et al., 2016). While still in its infancy, this technology is considered a promising platform for the transformation of organic matter into fuels and biochemicals.

10.2.4 Microbial Desalination Cell

Sevda et al. (2015) transformed an MFC into an MDC (illustrated in Fig. 10.1d) to reduce the energy demand required by desalination techniques such as electrodialysis, reverse osmosis, and mechanical vapor compression. Between the anode and cathode electrodes lies the desalination chamber which is equipped with cation and anion exchange membranes (CEM and AEM, respectively), which act as a barrier between the electrodes. The movement of anions and cations across the AEM and CEM into the anode and cathode electrodes, respectively, due to the potential difference between them, results in the desalination of saltwater. The organic matter in the anode electrode generates electrical energy, which can be utilized or accumulated. By using an electron acceptor, ferricyanide, and an electron donor such as acetate, a desalination effectiveness of up to 90% can be achieved at a maximum power density of 30 W/m^3, according to Sevda et al. (2015). Moreover, follow-up studies indicate that using sparged air as an electron acceptor and household wastewater as an electron donor can decrease salinity by as much as 60% (Luo et al., 2012). Energy-efficient technology, MDCs, have surfaced as a viable option that can address several pressing issues, including brackish water and saltwater desalination, wastewater and groundwater treatment, remediation, energy conversion, and biochemical manufacturing (Danaeifar et al., 2023; Guijala et al., 2022; Jatoi et al., 2022). Around 70% of the world's freshwater consumption is due to agricultural activities (Wu et al., 2022), with the problem of water scarcity becoming more severe as a result of population growth and global warming. Although desalination techniques, such as electrodialysis, reverse osmosis, and thermal desalination, have been implemented to provide freshwater to regions lacking freshwater sources, they are not environmentally sustainable due to their energy-intensive character (Sevda et al., 2015). In contrast, the development of low-cost desalination techniques using MDCs provides significant environmental, energy, and financial profits (Al-Mamun et al., 2018). However, the desalination rate attainable currently with MDCs is limited, primarily due to their use for desalinating or pre-desalinating brackish water, which can cause long-term drawbacks, for instance, the accumulation of harmful species of ions and expansion of soil salinity. The high osmotic potentials caused by elevated soil salinity resulting from expansion can impede a plant's capacity to take up water (Ping et al., 2015). As a consequence, it is crucial to keep the irrigation water with salinity levels below 500 g total dissolved solids (TDS) per liter to minimize damage to crop development and soil properties (Misaghi et al., 2017). Currently, modern MDCs can desalinate brackish water to an industry-standard level of 500 mg/L TDS. According to Jacobson et al. (2011), an upflow MDC that operates continuously can remove more than 99% of sodium chloride from a salt solution. Ping et al. (2015) observed that MDCs reduced the TDS levels of saltwater to 100 mg/L, comparable to drinking water. The desalination rates achieved with MDCs are considered satisfactory (Patel et al., 2021). Zhang and He (2015) conducted experiments on MDCs with a capacity of 105 L of water for desalinating wastewater and observed a salt removal rate of 4–9 g TDS/L d. Future research should focus on scaling up MDCs for higher freshwater yields. Alternatively, integrating MDCs into reverse osmosis plants to lower energy use is a feasible approach until MDCs can achieve higher efficiencies (ElMekawy et al., 2014).

10.2.5 Microbial Solar Cell

The integration of photosynthesis and exoelectrogens enables the retrieval of in situ bioelectricity through MSCs, as illustrated in Figure 10.1e. These systems use photosynthetic organisms to convert sunlight into chemical energy that is subsequently preserved as organic substances (Beauzamy et al., 2020a,b). This stored energy is then utilized to generate electricity through MFCs, where the transfer of electrons to the cathode from the anode plays a crucial role (Beauzamy et al., 2020a). The most common modes of transmission include phototrophic biofilm dispersal, rhizodeposition from higher plants, and pumping for relocation from photobioreactors (Jadhav et al., 2021). Among different forms of MSCs, PMFCs that incorporate higher plants have been found to be most effective in power generation, producing up to 3.2 W/m^2 (Mateo et al., 2014; Strik et al., 2011). Power output in PMFCs is proportional to the presence of rhizodeposits for anodic oxidation processes (Bajracharya et al., 2016; Li et al., 2018a). Unlike traditional solar cells, MSCs also produce various high-value-added compounds alongside energy production (Wang et al., 2014). As the population of microorganisms involved in both photosynthetic and electrochemical processes continues to increase, MSCs have a longer lifespan and need lower maintenance, making them self-repairing systems (Strik et al., 2011).

Bioelectrochemical systems, including MFCs and EFCs, are being explored as potential power sources for wireless sensor networks (WSNs) used in remote agricultural monitoring (Jadhav et al., 2021). While WSN technology has revolutionized precision and productivity in agriculture, its dependence on batteries or solar energy makes it challenging to maintain in remote areas subject to variable weather (Brunelli et al., 2018; Osorio-de-la-Rosa et al., 2020; Thakur et al., 2019). To overcome these limitations, BESs incorporating MFCs and EFCs are being developed as a long-term and sustainable power source for WSNs, with capacitators utilized as a means of storing the energy generated (Bajracharya et al., 2016; Brunelli et al., 2018).

Specifically, sediment MFCs (SMFCs), a type of MFC, have been utilized to power wireless sensors and telemetry devices for remote data transmission (Kiran & Gaur, 2013). SMFCs have demonstrated the capability of producing up to 2.1 V of voltage. The combination of these capacitors with a DC-DC converter makes them suitable for powering commercially available electronic circuits (Shantaram et al., 2005). However, limited potential generation and intermittent power recovery in some SMFCs require an energy management system to supply consistent and stable energy to biosensors (Donovan et al., 2008). Integration of bioelectrochemical systems with sensors in agriculture has the potential to power various wireless devices collecting data such as irrigation management information, weather statistics, insecticides, availability of plant nutrients, O_2 concentration, and soil pH (Sartori & Brunelli, 2016). With improved electrode performance and system efficiency, BESs have the potential to become hybrid-microbial electrochemical supplemental technologies for WSNs, transforming remote agricultural monitoring.

10.2.6 Enzymatic Fuel Cell

Electrochemical cells in EFCs generate electrical energy with the help of specific oxidoreductase enzymes acting as catalysts, as illustrated in Figure 10.1f. The enzymes facilitate the conversion of energy present in a waste substrate containing carbon into valuable electrical energy. This process involves the occurrence of reduction reactions at the cathode and substrate degradation processes at the anode (Barelli et al., 2019). Previously, enzymes with selective properties were used as biocatalysts within the electrolyte suspension in EFCs, but the low robustness and efficiency of the system made it no more advantageous than the in-bulk technique (Beilke et al., 2009). However, the recent advancement of stabilizing and immobilizing enzymes onto electrode surfaces (Feng et al., 2022; Rasmussen et al., 2016) has sparked interest in EFCs. Scientists have developed various enzyme immobilization approaches, such as cross-linking redox hydrogels, sandwiching between encapsulation modalities, and coatings of compatible materials (Liu et al, 2021; Meyer et al., 2022). Researchers have extensively studied biocatalysts derived from thermophilic microorganisms to improve the stability of EFCs, given their capacity for thermal stability even when operating at high temperatures (Shi et al., 2022; Xiao et al., 2019). EFCs have been widely used in implantable structures, portable devices, and biosensors, since their rapid development in recent years (Marks et al., 2022; Nishaa et al., 2022). Moreover, the feasibility of EFCs for energy conversion or storage has drawn increased attention (Suvvala & Kumar, 2023).

As indicated above, BESs such as EFCs, MSCs, MDCs, MESs, MECs, and MFCs are extensively researched and applied due to their environmental sustainability, easy operation, suitability, and economic potential. Table 10.3 provides a summary of the valuable products obtained by various BESs, together with their corresponding pros and cons.

TABLE 10.3
Pros and cons of different BESs

BES	Action	Pros	Cons
MFD	Converts waste and wastewater to energy	Low treatment cost Nutrient recovery Low sludge yield	Expensive catalyst Less power production Moderate COD removal
MEC	Produces biohydrogen, methane, and other value-added biochemical compounds	Removal of complex organic and inorganic pollutants Easily modifiable to produce required end product	External power supply required Low valuable production rate
MES	Produces valuable biochemicals like ethanol, acetate, butanol, biofuels, and biomethane	Carbon sequestration High-value and multi-carbon organic compounds recovered from industrial applications Metal recovery Wide variety of applications	External voltage supply required Lower production rate of chemicals with longer carbon chains
MDC	Performs saltwater and brackish water desalination while generating bioelectricity	Desalination and simultaneous waste and wastewater treatment Does not rely on external power supply unlike other concurrent desalination technology	Increase in anolyte salinity decreases bacterial removal efficiency Membrane fouling Inorganic elements in wastewater adversely impact the efficiency of MDC Scaling
MSC	Uses sunlight and self-repairing systems to generate bioelectricity and treat waste and wastewater	Self-repairing systems Waste and wastewater treatment Use of wireless sensor network (WSN) technology solutions in agricultural operations	Employing solar energy sources is inefficient, costly, and highly dependent on the weather
EFC	Performs water treatment while generating bioelectricity	Can be employed in biosensors and portable devices for efficient and portable energy generation Energy storage and conversion applications can help store and utilize energy more effectively Waste and wastewater treatment can help mitigate environmental pollution	Loss of enzyme activity upon immobilization Cost of immobilization process Mass transfer limitations

Source: Adapted from Ramanaiah et al. (2023).

10.2.7 Microorganisms used in Bioelectrochemical Systems

Microorganisms are widely distributed and have diverse metabolic capabilities, making them highly suitable for use in bioelectrochemical systems. In 1911, it was discovered that *E. coli* had electroactive properties, but the process by which exoelectrons are delivered was not understood until over a century later. Lovley et al. (1987) found, in the late 1980s, that bacteria obtained from river sediment demonstrated the ability to utilize different electron acceptors in anaerobic conditions. Further discoveries of *Shewanella* and *Geobacter*, and their ability to reduce metals without oxygen, stimulated curiosity in exoelectron transfer systems among other microorganisms. Subsequently, the capacity of this microbiota to produce electricity in BESs has been extensively studied.

Bioelectrochemical systems rely on the activity of microorganisms to generate an electrical current or produce chemicals from organic matter. These microorganisms, also known as electroactive bacteria, are able to move electrons from organic compounds to electrodes, producing an electric current in the process. They play a crucial role in the functioning of BESs (Thapa et al., 2022). There are several types of electroactive bacteria that are commonly used in bioelectrochemical systems, including *Geobacter*, *Shewanella*, and *Desulfobacter* (Ren et al., 2022). A diverse array of complex substrates, including paddy soil, sediment, and wastewater, have been used as significant sources of electrogenic microbial communities. *Geobacter*, in particular, has been extensively studied due to its ability to form conductive filaments known as pili, which enable it to transfer electrons to electrodes (Guo et al., 2023). *Shewanella* is another electroactive bacterium that can produce electricity in BESs and is known for its ability to reduce heavy metals (Dundas & Keitz, 2022). Apart from bacteria, there are other microorganisms that can be used in BESs, such as archaea and fungi. Methanogens, a type of archaea, have been shown to produce methane in bioelectrochemical systems, while fungi can be used to produce biofuels or other high-value products (Afsharian & Rahimnejad, 2022). The selection of microorganisms used in BESs is a function of several factors, including the type of feedstock, the desired product, and the environmental conditions. Researchers are continually exploring new microorganisms and developing strategies to optimize their activity in BESs for more efficient and sustainable wastewater treatment and resource recovery, some of which are shown in Table 10.4.

Exoelectrogens, including *Shewanella*, *Geobacter*, and *Desulfobacter*, are a type of bacteria that by oxidizing organic substances and transferring electrons to an external electron acceptor, such as O_2, can produce electricity (Yan et al., 2020; Zhu et al., 2022). Bacterial electrochemical systems commonly employ exoelectrogens to treat wastewater and generate electricity. In the anode chamber, organic materials are oxidized by bacteria, producing H^+ and CO_2 while directing electrons to the cathode via an external electrical circuit, which generates a current. External electron acceptors, typically O_2, in the cathode chamber perform reduction reactions to produce pure water. The potential difference between the cathodic and anodic chambers drives anions such as SO_4^{2-} and Cl^- from the saline water across an AEM and into the anode chamber (Huang et al., 2020). The survival and growth of exoelectrogens in BESs are critical for their effectiveness, and various parameters such as medium composition,

TABLE 10.4
Typical microorganisms found in BESs

		Reported mechanism	
Kingdom	**Strain**	**Anode**	**Cathode**
Bacteria	*Geobacter sulfureducens*	Direct electron transfer: Outer Membrane Cytochromes (OMCs) system (c-, d-types cytochromes) Nanowires: flexible filament	Direct electron transfer
	Shewanella oneidensis	Direct electron transfer: Mtr pathway Mediated electron transfer: soluble extracellular shuttles (riboflavin) Direct nanowires: flexible filament	Mediated electron transfer: methyl viologen used
	Clostridium ljungdahlii *Acetobacterium woodii*	-	Hypothetical direct Electron transfer: Rnfcomplex and soluble electron-bifurcating complexes
	Sporomusa ovata	-	H_2-mediated electron transfer
Fungi	*Saccharomyces cerevisiae*	Mediated electron transfer	Mediated electron transfer: neutral red and 7α-hydroxylase used
Algae	*Chlorella vulgaris*	-	Not determined

Source: Adapted from Thapa et al. (2022).

solute concentration, pH, or temperature, are used to find differences among these bacteria (Iannaci et al., 2020; Guang et al., 2020). Using appropriate exoelectrogen(s) in BESs can improve their efficiency, particularly in generating electricity (Huang et al., 2020; Wang et al., 2021). Exoelectrogens can be obtained as mixed cultures or pure cultures from anaerobic waste from various sources, including industry, households, sediments, effluent, and agricultural land, and utilized in BESs (Vyas et al., 2022). Due to its ease of accessibility, rich organic matter, and diverse microbial communities, wastewater is frequently employed as a microbial source in the anode chamber of BESs. Anode characteristics significantly affect the exoelectrogenic content of wastewater. Dominant microbial communities in anaerobic sludge/activated sludge inoculums include *Geobacter*, *Chloroflex* sp., *Firmicutesand*, *Bacteriodetes*, and *Proteobacteria* (Zhao et al., 2017). Mixed cultures are typically used due to the advantages of having various species with different metabolisms (Shah et al., 2019; Amanze et al., 2022). Although dominant species produce most of the electrons in the anode via organic matter oxidation, other microorganisms are capable of generating electrons through respiration, according to Pachapur et al. (2019). Mixed cultures have the potential to improve power generation, desalination, and wastewater treatment by utilizing fewer electron transfer mechanisms compared to pure cultures. This can lead to increased efficiency and performance in these processes (Jafary et al., 2018). The design of structural elements and operational conditions can modify the electrogenic content of multi-species substrates.

A BES offers a variety of possibilities including bioelectricity production, wastewater treatment, hydrogen gas generation, and biosensors by utilizing electrode-microorganism interactions (Lekshmi et al., 2023). In anaerobic conditions, the anode of MFCs enables the electroactive microorganisms to oxidize organic compounds, which generates electricity by allowing free electrons to travel toward the cathode through an external circuit. Some fungi, particularly yeast, have an electrogenic function that involves electron transfer reactions via cytochrome-c (Samukaite et al., 2023). Recently, BESs have increasingly employed the cathodic reaction approach due to its cost-effectiveness, the ability to use nonmetal catalysts, the potential for biochemical production by cathodic biocatalysts, and the appearance of novel endo-electrogenic microorganisms. This method utilizes biocatalysts like acetogenic bacteria to reduce carbon dioxide at the cathode and acquire value-added multi-carbon chemicals in microbial electrosynthesis (MES), which has been intensively researched in recent years (Roy et al., 2022). Electrotrophic microorganisms located in the cathode of bioelectrochemical systems are able to transform CO_2 into multichain organic compounds (alcohols, volatile fatty acids, and acetate). This is in contrast to electrogenesis, which primarily involves the production of electricity through the transfer of electrons (Song et al., 2022). Furthermore, microalgae and fungi are capable of catalyzing the same reaction at the cathode. The performance and output of BESs vary depending on operational conditions, electrode materials, reactor configuration, and the source of inoculum. Although mixed consortia are commonly used due to their diverse and rich population variety, the system's performance and reproducibility are affected by the fluctuation of microbial communities over time, as well as changes in the nutritional composition of the media. Researchers have discovered various electroactive microbial communities over time that could be harnessed for different applications in BESs (Zhou et al., 2022).

10.3 FACTORS AFFECTING THE PERFORMANCE OF BESS

BESs consist of several components including a cathode and an anode electrode, ion exchange membranes, substrates, and microbial communities. The anode serves as a site of substrate breakdown via bacterial fermentation, resulting in the production of electrons and ions. The microbial population facilitates the transfer of electrons to the anode. Within BESs, anaerobic conditions are maintained to enable optimal electricity generation, while oxygen is allowed to flow through the cathode electrode. A potential difference between the cathode and anode electrodes is created by the transport of ions across a solution to the cathode, resulting in an interface across an ion-selective membrane (Abubakari et al., 2019; Guang et al., 2020). BESs' performance is affected by several variables such as voltage, pH, electrode material, bacteria used, catholyte, and membranes (Abbaspour et al., 2021; Liang et al., 2021). Electrode materials in BESs should exhibit high electron flow, chemical stability, biocompatibility, mass transfer, porosity, surface area, scalability, and mechanical strength to perform optimally. Carbon electrodes are commonly used in BESs due to their high potential for generating electricity through the transfer of electrons (Bejjanki et al., 2021). The presence of an AEM inhibits the transfer of protons to the cathode from the anode, which creates a significant potential gradient (Abubakari et al., 2019). A low pH level in the anolyte in BESs could hinder the activity of exoelectrogens. To overcome this issue, recirculation of electrolytes and also buffering are practices employed to keep the pH level constant (Koomson et al., 2021; Kaleekkal et al., 2021). Recirculation of electrolytes is an effective way to achieve efficient desalination, remove pH imbalances, and increase power. Qu et al. (2012) found that recirculation of electrolytes significantly reduced the salinity of salt solutions. The use of a membrane as a separator in BESs can pose challenges, such as reduced proton permeability, back diffusion of oxygen, substrate loss, pH gradient, biofouling, and internal resistance. These issues affect the maintenance, design, efficiency, and scalability of membrane-based BESs.

There are two commonly used ion exchange membranes (IEMs) in microbial cells: CEMs and anion exchange membranes (AEMs). The formation of a microbial biofilm on the anode chamber side of the AEM can compromise the structural integrity of the IEM and increase the internal resistance, leading to decreased efficiency of the microbial cell system (Muhammad et al., 2020; Liu et al., 2019). However, the use of coatings, nanomaterials, and other materials in the AEM can reduce biofilm formation and enhance its structural stability, thereby improving the effectiveness and extending the lifespan of the microbial cell system (Guang et al., 2020). IEMs have a significant impact on the behavior of microbial cells as a consequence of their high resistance, which affects the power output and desalination rate (Moruno et al., 2018). In terms of cathode materials, carbon paper, carbon cloth, and graphite can significantly affect power output (Hubenova, 2018). While air cathodes are commonly used, they have some limitations, such as slow redox kinetics and high costs linked with the use of metal catalysts for the process of oxidation, according to Xu et al. (2020). The utilization of microbiota as electron receptors in biocathodes presents a promising alternative, as it eliminates the need for metals for O_2 reduction. This makes biocathodes a cost-effective option (Prakash et al., 2022).

The performance of BESs is highly dependent on the strain of bacteria used, and careful selection of microbes is critical for optimal operation. Mixed-culture bacteria can also influence the cell's power output. Ma and Hou (2019) found that certain species of nitrifying and denitrifying bacteria and algae can act as biological catalysts. For desalinating brackish water and seawater in MDCs, Abbaspour et al. (2021) observed that potassium ferricyanide was more effective than air cathodes, and Febriana et al. (2020) were able to attain a desalination efficiency exceeding 93% through the utilization of an air cathode/Ferro-ferricyanide redox in MDCs. Compared to oxygen, the redox couple of Fe and Fe^{2+} can remove up to 98% of salt from seawater. In MDCs, different concentrations of sodium percarbonate (Na_2CO^3) catholyte ranging from 0 to 0.2M have been used to remove salt, with higher concentrations resulting in a greater salt recovery (Jaroo et al., 2021). However, back diffusion may occur during the desalination cycle at 0.2M due to the high concentration of electrolyte. The percarbonate catholyte can serve as an antifoulant cathode. Other alternative catholyte choices have been investigated in MDCs, such as phosphate buffer solution and acidified water. Li et al. (2020) found that hypochlorite is a potent cathodic electron acceptor with a high redox potential, resulting in increased desalination behavior compared to other electron acceptors.

The appropriate selection of bacterial species is crucial for optimal performance in BESs. If the anode chamber contains an exceptionally large amount of organic substrates, this can harm the growth of exoelectrogenic microbes, resulting in a 12% reduction in power output at 500 mM Cl concentration. However, the generation of power can be improved by increasing the concentration of KCl, with enhancements observed up to 300 mM (Wang et al., 2020; Guang et al., 2020). The impact of salt level also depends on the type of exoelectrogens used, with decreased activity of *Geobactersulfur* and increased activity of *Pelobacter propionicus*, when NaCl concentration increases (Oh & Logan, 2006). Hence, a particular level of salinity can only be appropriate for a specific type of exoelectrogen, and individual strains of certain exoelectrogens might be less efficient in terms of desalination. Mixed cultures have been shown to improve desalination in MDCs (Lefebvre et al., 2012). BESs can also be affected by temperature sensitivity in terms of lifespan and efficiency. The lack of real knowledge regarding microbes under nonconventional BES environments hinders the scaling up of this technology (Tavker & Kumar, 2023).

In summary, the performance of BESs can be influenced by several factors that may impede their efficiency and effectiveness. These factors include the microbial community present in the system, comprising the type, abundance, and metabolic activities of microorganisms, which may impact the rate of electron transfer, thereby leading to higher power output. Another key factor is the electrode material used, where different materials possess unique properties that may affect their ability to donate or accept electrons. For example, carbon-based electrodes are frequently utilized as they offer more surface area favoring the transfer of electrons. Operational conditions of the BES, e.g., temperature, nutrient availability, and pH can also have a bearing on the performance of the system. The optimal conditions for microbial growth and metabolism change as a consequence of the microorganisms present, and hence, it is critical to regulate these operational conditions for optimal performance. The design of the system, including the electrode geometry and spacing, distance between electrodes, and system configuration, can also impact the rate of electron transfer and overall performance. Lastly, the amount and type of organic matter fed into the system can affect the system's performance, where an excessive organic load can inhibit microbial activity and reduce performance. In conclusion, an array of factors can influence BES performance, and understanding and managing these factors are critical to developing and operating efficient BES systems.

10.4 APPLICATIONS OF MICROBIAL CELLS: A FOCUS ON MFCS

MFCs are one of the most promising technologies that can address two of society's most pressing challenges: the energy crisis and the treatment and purification of wastewater. MFCs use a bioelectrochemical process to convert chemical energy into electrical energy by introducing microorganisms into the cell. The microorganisms consume organic matter (substrate) to produce a current of electrons through metabolic action (Ye et al., 2019; Callegari et al., 2018). The MFC typically consists of a cathode and an anode, with the electron current generated by organic matter degradation and transferred from the anode to the cathode. The electrodes can be made of various materials and shapes, such as graphite, titanium, or platinum, and brush, cylindrical, sheet, or wire. The electron movement generates electrical energy and a movement of ions to the cathode chamber from the anode chamber. Two-chamber MFCs have a membrane, usually made of a polymeric material that serves as a pathway for ions between the chambers (Yan et al., 2018). MFCs have the potential for various applications, such as pollutant removal, nutrient recovery, or high-value product generation, together with electricity generation. MFCs are distinct from other power generation systems because they can work properly at ambient temperatures and have a low carbon footprint, making them suitable for remote areas where basic electricity is required.

10.4.1 Electric Energy Generation

As previously mentioned, the main aim of MFCs in wastewater treatment is to obtain electrical energy by taking advantage of the biological action of microorganisms when degrading organic matter. However, like any other technology developed, obtaining electrical energy from MFCs depends on a series of factors or operating parameters. The parameters that have a greater influence on the process are type or material of electrodes, electron exchange system, pH, and temperature. The material of the appliance is a crucial issue both in the performance of the process and in its economic profitability. The electrodes should be made of a conductive, compact material (to avoid disintegration), that is, they have good mechanical properties and are non-corrosive (to avoid electrode degradation and contamination of the liquid solutions in each chamber). Therefore, a priori, the best candidates would be metallic electrodes (made of materials such as titanium or platinum) (Watanabe, 2008). With metallic electrodes, higher yields can be obtained than with graphite electrodes, however, they have a much higher cost, which can jeopardize the profitability of the process and its scalability. Consequently, the most common is to use combinations of graphite electrodes, although lately, metal-doped graphite electrodes have also begun to be used, which improves the performance of the process more than if simple graphite electrodes were used, at a lower cost than metallic electrodes (Muthukumar et al., 2019). Regarding the morphology of the electrode, studies carried out (Haavisto et al., 2019; Shahid et al., 2021; Ye et al., 2019; Ye et al., 2020), show that the performance of the process increases as the available surface on the electrode also increases.

The electron exchange system is the solution that is applied between the electrodes of an MFC to facilitate the exchange of protons and cations from the anode to the cathode to compensate for the circulation of electrons that is being produced by the generation of organic energy. In general, the most widely used system is a porous polymeric membrane that is placed between the two electrodes (cation exchange polymer). Although different materials such as UltrexTM or ceramics have been tested, the material that gives the best results is NafionTM (Flimban et al., 2020). There are various grades of Nafion available, such as Nafion 117, 115, and 112 which differ in thickness (0.178, 0.127, and 0.051 mm, respectively), as well as their hydration and permeation characteristics. Nafion is more permeable with lower resistance than a traditional salt bridge (NaCl). A follow-up of the fouling of the membrane must be carried out, since during the bioelectrochemical reaction depositions are made on the membrane (especially on the anode side), producing a blockage of the pores, hindering the diffusion and transport of protons and ions through the membrane and therefore a decrease in electrical production. The pH is another of the important factors in the process, and on the one hand, it influences the activity of the bacteria that degrade organic matter and produce electrical energy and on the other hand, it affects the circulation of electrons (columbic efficiency). In both cases, it has been shown that the operation with neutral pH values (between 6–8) favors both the correct development of the biochemical and electrical parts (Halim et al., 2021). Another important aspect to consider is that if excessive electrical current is produced in the MFC, it will lead to a large accumulation, thereby increasing the acidity in the anode and, therefore, inhibiting microbial activity (Mohan et al., 2011).

Lastly, temperature is another of the parameters with the greatest degree of influence in MFC technology. As in any process in the field of chemical engineering in which a chemical reaction comes into play, temperature influences the thermodynamics and kinetics of the reaction, and the phenomena of matter and energy transport. On the other hand, as it is also a biological process, the temperature will affect the activity of the microorganisms responsible for degrading organic matter and generating electrical energy. In studies carried out in this regard (Tee et al., 2018; Wang et al., 2018) it has been shown that the best temperatures to operate with MFCs are slightly higher than room temperature (between 25–35 °C). Above these temperatures, although the reaction continues to be activated equally or better, there is a proliferation of different populations of bacteria on the anode that can compete and compromise the correct degradation of organic matter.

10.4.2 Pollutants' Removal

One of the most important aspects of MFCs is the possibility of using them not only to obtain energy but also to obtain clean water. In recent times, the scientific community has delved into the contribution of MFCs to bioremediation and elimination of typical pollutants present in wastewater (both industrial and urban), such as potentially toxic elements, xenobiotic compounds, or even organic contaminants (insecticides, antibiotics, pesticides, etc.). This is achieved by reducing the different polluting species, thanks to the capture of electrons from the cathode by the microorganisms inoculated in the MFC chamber. In this way, more environmentally friendly effluents are finally obtained that can be discharged into rivers (complying with current legislation) or reclaimed for different uses (irrigation, cleaning, etc.). It is important to highlight that apart from the elimination of both inorganic and organic compounds in the water treatment process, MFCs produce important technical and environmental advantages (beyond electricity generation and therefore the energy self-sustainability of the process) compared to systems from typical water treatments such as wastewater (such as aerobic treatments). The main of these advantages resides in the minimization of sludge generation (one of the great environmental problems of traditional technologies). On the other hand, studies have found that not only is effluent pollution reduced, but also that MFCs are capable of decreasing gaseous pollutants, thus reducing not only the water footprint but also the emissions of gaseous polluting compounds.

One of the last benefits necessary to highlight concerning the use of MFCs in wastewater treatment is the flexibility and adaptability of the technology to work with other innovative technologies in the removal of pollutants. Authors such as Vinayak et al. (2021) have worked on wastewater treatment with combined photocatalytic technology and MFCs with quite promising prospects. In addition, studies have been reported (Elshobary et al., 2021), in which microalgae have been added to the microbiological treatment of wastewater, with an undoubted synergistic effect, not only for technical aspects but also for improving profitability. These are subjects that will be dealt with in a later section of this work (value-added products' generation). With all this, there are various works in which COD removal yields close to 90% (Toczyłowska-Mamińska et al., 2018) or even higher (Marassi et al., 2020) are achieved. In the elimination of pesticides through bioelectrochemical processes, studies have been carried out (Zhang et al., 2019) in which, although a total elimination of the contaminant was not achieved, it was reduced by a significant percentage (79% elimination of Fipronil in an operation time of 12 h). In the work itself, fish were incubated in the treated effluent and no four-day death was recorded. Regarding the treatment of antibiotics, Long et al. (2021) studied tetracycline degradation in a single-chamber MFC and a Fenton coupled system and dual-chamber MFC in H form. The removal yield for the single-chamber system only reached 74% while for the latter system, it increased up to 99%. The increase was due to the generation of OH· and O^{2-}· radicals by the Fenton process. In addition, in the single-chamber MFC reactor, there was a decrease in electrical energy production due to the inactivation of cell microorganisms by antibiotics. The latter was avoided in the second process. Therefore, the viability of the MFC as a technology for the treatment of wastewater pollutants is clear.

10.4.3 Value-added Product Generation

It is clear that wastewater is a raw material with the potential to obtain high-value-added products from the different components it contains. This is in such a manner that the trend is that shortly wastewater treatment plants (WWTPs) will move to a circular economy model, transforming into biofactories or biorefineries. In this model, obtaining value-added products is a fundamental stage to ensure the economic and environmental sustainability of the process. Therefore, there is an important trend in scientific research into innovative technologies and processes that allow high-value products to be obtained in wastewater treatment processes. One such trend is the bioelectrochemical technology of MFCs. The type of final product obtained will depend on the composition of the wastewater as well as the type of process carried out inside the MFCs (Mohan & Chandrasekhar, 2011; Gambino et al., 2021). Thus, through bioelectrochemistry, it is possible to obtain valuable short-chain hydrocarbons (biomethane) (Ning et al., 2021), even alcohols such as butanol (Zaybak et al., 2013), or hexanol (Vassilev et al., 2018), or other derivatives such as those obtained by Chandrasekhar et al. (2015) (acetate, propionate, butyrate, ethanol, etc.). In addition to the aforementioned hydrocarbons, it is also possible to perform the extraction of nutrients (phosphorus and nitrogen) from wastewater to obtain biofertilizers (Adesra et al., 2021; Sabin et al., 2022) and to obtain hydrogen gas (Kadier et al., 2018; 2022).

The most common way to obtain biomethane from bioelectrochemical processes is using the hydrogenotrophic methanogenesis pathway (Kadier et al., 2018). In this route, as methane is generated through the action of the bacteria inoculated in the cell, the production of hydrogen is reduced. Therefore, through this technology a predominant route of methane is achieved, reaching productions of 0.055 mol CH_4/g VSS·day. Regarding the recovery of nutrients, the strategy focuses mainly on the recovery of nitrogen and phosphorus contained in the wastewater as NH_4^+ and PO_3^{2-}, respectively. The higher the organic matter content of the wastewater, the greater the recovery potential. This is because the potential for bacterial growth and therefore, degradation of organic matter, will be greater and the transport of electrons between the electrodes of the MFC will increase and consequently, the flow of NH_4^+ and PO_3^{2-} ions from the anode to the cathode will be higher (to compensate for electric charge transport). In this sense, in works such as those collected by Paucar and Sato (2021), recovery yields of 70% for N and 80% for P have been achieved. An enhanced approach to nutrient recovery involves harnessing the growth of microalgae. By incorporating microalgae into the microbial population utilized in MFCs, a superior final product can be obtained, thereby increasing its value compared to the conventional process (Wu et al., 2021). This innovative method not only improves nutrient retrieval but also yields a high-value end product derived from microalgae. Finally, it is worth noting the possibility of obtaining products such as bioethanol or medium-chain fatty acids through bioelectrochemical technologies. In these cases, an electrode is used where the biological reaction (cathode) and hydrogen and electron facilitators take place. Even so, it has been possible to obtain bioethanol from acetic acid (Rosenbaum et al., 2011) or even achieve a cathodic recovery of butanol and ethanol (Steinbusch et al., 2010).

10.5 PRACTICAL LIMITATIONS AND PROSPECTS OF BESS

BESs have numerous applications, including bioelectricity generation and clean energy production, as well as wastewater treatment and environmental remediation. Research has shown that microbes can be used to convert chemical substrates into energy, making them ideal for use in MFCs, which have the potential for commercialization. Additionally, microbial action can produce biohydrogen from renewable substrates like acetate, propionate, or wastewater by fermenting processes. BESs are also well-suited for treating wastewater and can be used to produce biochemicals, bioenergy, and bioproducts, as well as for valorizing biomass wastes, remediating heavy metal-polluted soils and water, and conducting bioassays and biosensors. In particular, MFC-based biosensors have shown promise in detecting metabolic processes and toxicity.

The use of BESs shows great potential for both wastewater treatment and bioenergy production, but there are still some challenges that must be addressed before these systems can reach their full potential. While there has been considerable progress made in laboratory settings, it is important to evaluate the performance of BESs in real-world conditions and at larger scales to ensure successful commercialization. The low efficiency and production rates remain major concerns that have hindered the widespread adoption of BESs technologies. There are several limitations that impede the practical application of BESs. The high investment cost is the first of these limitations, mainly as a consequence of the costly electrodes, catalysts, current collectors, and membranes. Additionally, the power output yields are low, and these systems need to operate under ideal conditions to maintain the activity of microorganisms. To reduce the cost of electrodes, carbon-based electrodes obtained from carbon waste need to be synthesized and investigated. Novel cell configurations are necessary to reduce capital and maintenance costs while improving system efficiency. An approach that could simplify the replacement and maintenance process is employing multiple anodes and cathodes and not a single large-size electrode. However, the cost and fouling associated with the membrane are major challenges in the application and further development of BESs. Fouling adversely affects the behavior of both MECs and MFCs. To circumvent these problems, it is advisable to use membraneless BESs. However, membraneless systems are prone to oxygen and substrate crossover, which ultimately reduces microorganism activity, power generation, and columbic efficiency. Additionally, the formation of a biofilm on the cathode surface is a common problem faced by membraneless MFCs, which leads to a decrease in the rate at which the oxygen reduction reaction occurs on the surface of the cathode.

To fully realize the potential of BESs, it is crucial to move from laboratory-scale experiments to pilot-scale and ultimately full-scale applications. This requires a thorough understanding of the metabolic pathways of microorganisms involved in the process, which can be used to enrich their growth and enhance electron transfer to the anode surface from the cell interior. In addition, simple designs that avoid operational complexity and reduce capital and maintenance costs should be prioritized. BESs have applications in various fields, including hazardous material remediation, wastewater treatment, and chemical production. Hybrid systems that combine BESs with other technologies, such as water desalination, should be investigated to increase overall efficiency. Lastly, the application of nanotechnology in the development of high-performance membranes and electrodes that reduce biofouling and promote biofilm formation represent a crucial area for future research and development endeavors.

ACKNOWLEDGMENT

The authors gratefully acknowledge support of this work by the European Union's Horizon 2020 and Horizon Europe Research and Innovation Programs under grant agreement No 101000752 (WALNUT project) and under grant agreement No 63249400 (CRONUS project), respectively.

REFERENCES

Abbaspour, M., Habibi, A., Javid, A. H., & Hassani, A. H. (2021). Investigation on bio-cathode MDC performance for desalination and waste water treatment. *Environment and Water Engineering*, https://doi.org/10.22034/jewe.2021.273836.1540

Abubakari, Z. I., Mensah, M., Buamah, R., & Abaidoo, R. C. (2019). Assessment of the electricity generation, desalination and wastewater treatment capacity of a plant microbial desalination cell (PMDC). *International Journal of Energy and Water Resources*, 3, 213–218.

Adesra, A., Srivastava, V. K., & Varjani, S. (2021). Valorization of dairy wastes: integrative approaches for value added products. *Indian Journal of Microbiology*, 61(3), 270–278.

Afsharian, Y. P., & Rahimnejad, M. (2022). Functional dynamics of microbial communities in bioelectrochemical systems: The importance of eco-electrogenic treatment of complex substrates. *Current Opinion in Electrochemistry*, 31, 100816.

Aiken, D. C., Curtis, T. P., & Heidrich, E. S. (2019). Avenues to the financial viability of microbial electrolysis cells [MEC] for domestic wastewater treatment and hydrogen production. *International Journal of Hydrogen Energy*, 44(5), 2426–2434.

Al-Mamun, A., Ahmad, W., Baawain, M. S., Khadem, M., & Dhar, B. R. (2018). A review of microbial desalination cell technology: configurations, optimization and applications. *Journal of Cleaner Production*, 183, 458–480.

Amanze, C., Zheng, X., Man, M., Yu, Z., Ai, C., Wu, X., ... & Zeng, W. (2022). Recovery of heavy metals from industrial wastewater using bioelectrochemical system inoculated with novel Castellaniella species. *Environmental Research*, 205, 112467.

Bajracharya, S., Sharma, M., Mohanakrishna, G., Benneton, X. D., Strik, D. P., Sarma, P. M., & Pant, D. (2016). An overview on emerging bioelectrochemical systems (BESs): Technology for sustainable electricity, waste remediation, resource recovery, chemical production and beyond. *Renewable Energy*, 98, 153–170.

Barelli, L., Bidini, G., Calzoni, E., Cesaretti, A., Di Michele, A., Emiliani, C., ... & Sisani, E. (2019, December). Enzymatic fuel cell technology for energy production from bio-sources. In AIP Conference Proceedings (Vol. 2191, No. 1, p. 020014). AIP Publishing LLC.

Beauzamy, L., Delacotte, J., Bailleul, B., Tanaka, K., Nakanishi, S., Wollman, F. A., & Lemaître, F. (2020a). Mediator-microorganism interaction in microbial solar cell: A fluo-electrochemical insight. *Analytical Chemistry*, 92(11), 7532–7539.

Beauzamy, L., Lemaître, F., & Derr, J. (2020b). Underlying mechanisms in microbial solar cells: How modeling can help. *Sustainable Energy & Fuels*, 4(12), 6004–6010.

Beilke, M. C., Klotzbach, T. L., Treu, B. L., Sokic-Lazic, D., Wildrick, J., Amend, E. R., ... & Minteer, S. D. (2009). Enzymatic biofuel cells. In Micro Fuel Cells (pp. 179–241). Academic Press.

Bejjanki, D., Muthukumar, K., Radhakrishnan, T. K., Alagarsamy, A., Pugazhendhi, A., & Mohamed, S. N. (2021). Simultaneous bioelectricity generation and water desalination using Oscillatoria sp. as biocatalyst in photosynthetic microbial desalination cell. *Science of The Total Environment*, 754, 142215.

Bora, A., Mohanrasu, K., Swetha, T. A., Ananthi, V., Sindhu, R., Chi, N. T. L., ... & Mathimani, T. (2022). Microbial electrolysis cell (MEC): Reactor configurations, recent advances and strategies in biohydrogen production. *Fuel*, 328, 125269.

Brunelli, D., Tosato, P., & Rossi, M. (2018). Flora monitoring with a plant-microbial fuel cell. In *Applications in Electronics Pervading Industry, Environment and Society: APPLEPIES 2016* 5 (pp. 41–48). Springer International Publishing.

Callegari, A., Cecconet, D., Molognoni, D., & Capodaglio, A. G. (2018). Sustainable processing of dairy wastewater: Long-term pilot application of a bio-electrochemical system. *Journal of Cleaner Production*, 189, 563–569.

Chandrasekhar, K., Amulya, K., & Mohan, S. V. (2015). Solid phase bio-electrofermentation of food waste to harvest value-added products associated with waste remediation. *Waste Management*, 45, 57–65.

Chandrasekhar, K., Velvizhi, G., & Mohan, S. V. (2021). Bio-electrocatalytic remediation of hydrocarbons contaminated soil with integrated natural attenuation and chemical oxidant. *Chemosphere*, 280, 130649.

Chaurasia, A. K., & Mondal, P. (2021). Hydrogen production from waste and renewable resources. In *Hydrogen Fuel Cell Technology for Stationary Applications* (pp. 22–46). IGI Global.

Chen, C. Y., Kuo, E. W., Nagarajan, D., Dong, C. D., Lee, D. J., Varjani, S., ... & Chang, J. S. (2021). Semi-batch cultivation of Chlorella sorokiniana AK-1 with dual carriers for the effective treatment of full strength piggery wastewater treatment. *Bioresource Technology*, 326, 124773.

Danaeifar, M., Ocheje, O. M., & Mazlomi, M. A. (2023). Exploitation of renewable energy sources for water desalination using biological tools. *Environmental Science and Pollution Research*, 30(12), 32193–32213.

Dhar, B. R., Elbeshbishy, E., Hafez, H., & Lee, H. S. (2015). Hydrogen production from sugar beet juice using an integrated biohydrogen process of dark fermentation and microbial electrolysis cell. *Bioresource Technology*, 198, 223–230.

Dinesh, G. K., Chauhan, R., & Chakma, S. (2018). Influence and strategies for enhanced biohydrogen production from food waste. *Renewable and Sustainable Energy Reviews*, 92, 807–822.

Donovan, C., Dewan, A., Heo, D., & Beyenal, H. (2008). Batteryless, wireless sensor powered by a sediment microbial fuel cell. *Environmental Science & Technology*, 42(22), 8591–8596.

Dundas, C. M., & Keitz, B. K. (2022). Tapping the potential of gram-positive bacteria for bioelectrochemical applications. *Trends in Biotechnology* 41(3), 273–275.

El Ichi-Ribault, S., Alcaraz, J. P., Boucher, F., Boutaud, B., Dalmolin, R., Boutonnat, J., ... & Martin, D. K. (2018). Remote wireless control of an enzymatic biofuel cell implanted in a rabbit for 2 months. *Electrochimica Acta*, 269, 360–366.

ElMekawy, A., Hegab, H. M., & Pant, D. (2014). The near-future integration of microbial desalination cells with reverse osmosis technology. *Energy & Environmental Science*, 7(12), 3921–3933.

Elshobary, M. E., Zabed, H. M., Yun, J., Zhang, G., & Qi, X. (2021). Recent insights into microalgae-assisted microbial fuel cells for generating sustainable bioelectricity. *International Journal of Hydrogen Energy*, 46(4), 3135–3159.

Esmaeilion, F., Ahmadi, A., Hoseinzadeh, S., Aliehyaei, M., Makkeh, S. A., & Astiaso Garcia, D. (2021). Renewable energy desalination; A sustainable approach for water scarcity in arid lands. *International Journal of Sustainable Engineering*, 14(6), 1916–1942.

Febriana, T. A., Khairiza, M. R., Maulina, R., Utami, T. S., Arbianti, R., & Hermansyah, H. (2020). Concentration optimization of sodium percarbonate as buffering catholyte on stacked microbial desalination cell by utilizing tofu wastewater as a substrate. *Engineering Journal*, 24(4), 217–228.

Feng, Y., Xu, Y., Liu, S., Wu, D., Su, Z., Chen, G., ... & Li, G. (2022). Recent advances in enzyme immobilization based on novel porous framework materials and its applications in biosensing. *Coordination Chemistry Reviews*, 459, 214414.

Flimban, S. G., Hassan, S. H., Rahman, M. M., & Oh, S. E. (2020). The effect of Nafion membrane fouling on the power generation of a microbial fuel cell. *International Journal of Hydrogen Energy*, 45(25), 13643–13651.

Gambino, E., Chandrasekhar, K., & Nastro, R. A. (2021). SMFC as a tool for the removal of hydrocarbons and metals in the marine environment: A concise research update. *Environmental Science and Pollution Research*, 28(24), 30436–30451.

Gaur, V. K., Sharma, P., Gaur, P., Varjani, S., Ngo, H. H., Guo, W., ... & Singhania, R. R. (2021). Sustainable mitigation of heavy metals from effluents: Toxicity and fate with recent technological advancements. *Bioengineered*, 12(1), 7297–7313.

Ghangrekar, M. M., Das, S., & Das, S. (2022). Biofuel cell: Existing formats, production level, constraints, and potential uses. In *Handbook of Biofuels* (pp. 531–550). Academic Press.

Ghasemi, M., Sedighi, M., & Usefi, M. M. B. (2022). A comprehensive review on membranes in microbial desalination cells; Processes, utilization, and challenges. *International Journal of Energy Research*, 46(11), 14716–14739.

Guang, L., Koomson, D. A., Jingyu, H., Ewusi-Mensah, D., & Miwornunyuie, N. (2020). Performance of exoelectrogenic bacteria used in microbial desalination cell technology. *International Journal of Environmental Research and Public Health*, 17(3), 1121.

Gujjala, L. K. S., Dutta, D., Sharma, P., Kundu, D., Vo, D. V. N., & Kumar, S. (2022). A state-of-the-art review on microbial desalination cells. *Chemosphere*, 288, 132386.

Guo, W., Ying, X., Zhao, N., Yu, S., Zhang, X., Feng, H., ... & Yu, H. (2023). Interspecies electron transfer between Geobacter and denitrifying bacteria for nitrogen removal in bioelectrochemical system. *Chemical Engineering Journal*, 455, 139821.

Haavisto, J., Dessì, P., Chatterjee, P., Honkanen, M., Noori, M. T., Kokko, M., ... & Puhakka, J. A. (2019). Effects of anode materials on electricity production from xylose and treatability of TMP wastewater in an up-flow microbial fuel cell. *Chemical Engineering Journal*, 372, 141–150.

Halim, M., Rahman, M., Ibrahim, M., Kundu, R., & Biswas, B. K. (2021). Effect of anolyte pH on the performance of a dual-chambered microbial fuel cell operated with different biomass feed. *Journal of Chemistry*, 2021.

Huang, X., Guida, S., Jefferson, B., & Soares, A. (2020). Economic evaluation of ion-exchange processes for nutrient removal and recovery from municipal wastewater. *NPJ Clean Water*, 3(1), 7.

Hubenova, Y.V. (2018). Yeast-based biofuel cells. In *Encyclopedia of Interfacial Chemistry: Surface Science and Electrochemistry*, pp. 537–550 https://doi.org/10.1016/B978-0-12-409547-2.13467-5.

Iannaci, A., Myles, A., Flinois, T., Behan, J. A., Barrière, F., Scanlan, E. M., & Colavita, P. E. (2020). Tailored glycosylated anode surfaces: Addressing the exoelectrogen bacterial community via functional layers for microbial fuel cell applications. *Bioelectrochemistry*, 136, 107621.

Imoro, A. Z., Mensah, M., & Buamah, R. (2021). Developments in the microbial desalination cell technology: A review. *Water-Energy Nexus*, 4, 76–87.

Jacobson, K. S., Drew, D. M., & He, Z. (2011). Efficient salt removal in a continuously operated upflow microbial desalination cell with an air cathode. *Bioresource Technology*, 102(1), 376–380.

Jadhav, D. A., Ghosal, D., Chendake, A. D., Pandit, S., & Sajana, T. K. (2021). Plant microbial fuel cell as a biomass conversion technology for sustainable development. *Catalysis for Clean Energy and Environmental Sustainability: Biomass Conversion and Green Chemistry*, 1, 135–147.

Jafary, T., Daud, W. R. W., Aljlil, S. A., Ismail, A. F., Al-Mamun, A., Baawain, M. S., & Ghasemi, M. (2018). Simultaneous organics, sulphate and salt removal in a microbial desalination cell with an insight into microbial communities. *Desalination*, 445, 204–212.

Jaroo, S. S., Jumaah, G. F., & Abbas, T. R. (2021). The catholyte effects on the microbial desalination cell performance of desalination and power generation. *Journal of Engineering*, 27(7), 53–65.

Jatoi, A. S., Hashmi, Z., Mazari, S. A., Mubarak, N. M., Karri, R. R., Ramesh, S., & Rezakazemi, M. (2022). A comprehensive review of microbial desalination cells for present and future challenges. *Desalination*, 535, 115808.

Kabutey, F. T., Zhao, Q., Wei, L., Ding, J., Antwi, P., Quashie, F. K., & Wang, W. (2019). An overview of plant microbial fuel cells (PMFCs): Configurations and applications. *Renewable and Sustainable Energy Reviews*, 110, 402–414.

Kadier, A., Kalil, M. S., Chandrasekhar, K., Mohanakrishna, G., Saratale, G. D., Saratale, R. G., ... & Sivagurunathan, P. (2018). Surpassing the current limitations of high purity H_2 production in microbial electrolysis cell (MECs): Strategies for inhibiting growth of methanogens. *Bioelectrochemistry*, 119, 211–219.

Kadier, A., Singh, R., Song, D., Ghanbari, F., Zaidi, N. S., Aryanti, P. T. P., ... & Ma, P. C. (2022). A novel pico-hydro power (PHP)-Microbial electrolysis cell (MEC) coupled system for sustainable hydrogen production during palm oil mill effluent (POME) wastewater treatment. *International Journal of Hydrogen Energy*, 48(55), 21066–21087.

Kaleekkal, N. J., Mural, P. K. S., Vigneswaran, S., Ghosh, U. (Eds.), (2021). *Sustainable Technologies for Water and Wastewater Treatment*. CRC Press, Boca Raton.

Kayastha, V., Patel, J., Kathrani, N., Varjani, S., Bilal, M., Show, P. L., ... & Bui, X. T. (2022). New Insights in factors affecting ground water quality with focus on health risk assessment and remediation techniques. *Environmental Research*, 212, 113171.

Khan, M. A., Ngo, H. H., Guo, W., Liu, Y., Zhang, X., Guo, J., ... & Wang, J. (2018). Biohydrogen production from anaerobic digestion and its potential as renewable energy. *Renewable Energy*, 129, 754–768.

Kiran, V., & Gaur, B. (2013). Microbial fuel cell: technology for harvesting energy from biomass. *Reviews in Chemical Engineering*, 29(4), 189–203.

Koo, B., & Jung, S. P. (2022). Trends and perspectives of microbial electrolysis cell technology for ultimate green hydrogen production. *Journal of Korean Society of Environmental Engineers*, 44(10), 383–396.

Koomson, D. A., Huang, J., Li, G., Miwornunyuie, N., Ewusi-Mensah, D., Darkwah, W. K., & Opoku, P. A. (2021). Comparative studies of recirculatory microbial desalination cell–microbial electrolysis cell coupled systems. *Membranes*, 11(9), 661.

Kuleshova, T., Rao, A., Bhadra, S., Garlapati, V. K., Sharma, S., Kaushik, A., ... & Sevda, S. (2022). Plant microbial fuel cells as an innovative, versatile agro-technology for green energy generation combined with wastewater treatment and food production. *Biomass and Bioenergy*, 167, 106629.

Kumar, V., Agrawal, S., Shahi, S.K., Motghare, A., Singh, S., Ramamurthy, P.C., 2022a. Bioremediation potential of newly isolated *Bacillus albus* strain VKDS9 for decolourization and detoxification of biomethanated distillery effluent and its metabolites characterization for environmental sustainability. *Environmental Technology & Innovation* 26, 102260. https://doi.org/10.1016/j.eti.2021.102260

Kumar, V., Ameen, F., Islam, M.A., Agrawal, S., Motghare, A., Dey, A., Shah, M.P., Américo-Pinheiro, J.H.P., Singh, S., Ramamurthy, P.C., 2022b. Evaluation of cytotoxicity and genotoxicity effects of refractory pollutants of untreated and biomethanated distillery effluent using *Allium cepa. Environmental Pollution* 300, 118975. https://doi.org/10.1016/j.envpol.2022.118975

Kumar, V., Shahi, S.K., Ferreira, L.F.R., Bilal, M., Biswas, J.K., Bulgariu, L., 2021. Detection and characterization of refractory organic and inorganic pollutants discharged in biomethanated distillery effluent and their phytotoxicity, cytotoxicity, and genotoxicity assessment using *Phaseolus aureus* L. and *Allium cepa* L. *Environmental Research* 201, 111551. https://doi.org/10.1016/j.envres.2021.111551

Lefebvre, O., Tan, Z., Kharkwal, S., & Ng, H. Y. (2012). Effect of increasing anodic NaCl concentration on microbial fuel cell performance. *Bioresource Technology*, 112, 336–340.

Lekshmi, G. S., Bazaka, K., Ramakrishna, S., & Kumaravel, V. (2023). Microbial electrosynthesis: carbonaceous electrode materials for CO_2 conversion. *Materials Horizons*, 10, 292–312.

Li, H., Xiong, J., Zhang, G., Liang, A., Long, J., Xiao, T., ... & Zhang, H. (2020). Enhanced thallium (I) removal from wastewater using hypochlorite oxidation coupled with magnetite-based biochar adsorption. *Science of The Total Environment*, 698, 134166.

Li, S., & Chen, G. (2018). Factors affecting the effectiveness of bioelectrochemical system applications: Data synthesis and meta-analysis. *Batteries*, 4(3), 34.

Li, S., Chen, G., & Anandhi, A. (2018a). Applications of emerging bioelectrochemical technologies in agricultural systems: A current review. *Energies*, 11(11), 2951.

Li, S., Barreto, V., Li, R., Chen, G., & Hsieh, Y. P. (2018b). Nitrogen retention of biochar derived from different feedstocks at variable pyrolysis temperatures. *Journal of Analytical and Applied Pyrolysis*, 133, 136–146.

Liang, D., Li, C., He, W., Li, Z., & Feng, Y. (2021). Response of exoelectrogens centered consortium to nitrate on collaborative metabolism, microbial community, and spatial structure. *Chemical Engineering Journal*, 426, 130975.

Liu, S., Bilal, M., Rizwan, K., Gul, I., Rasheed, T., & Iqbal, H. M. (2021). Smart chemistry of enzyme immobilization using various support matrices–A review. *International Journal of Biological Macromolecules*, 190, 396–408.

Liu, Z., Xiang, P., Duan, Z., Fu, Z., Zhang, L., & Zhang, Z. (2019). Electricity generation, salinity, COD removal and anodic biofilm microbial community vary with different anode CODs in a microbial desalination cell for high-salinity mustard tuber wastewater treatment. *RSC Advances*, 9(43), 25189–25198.

Logan, B. E., Wallack, M. J., Kim, K. Y., He, W., Feng, Y., & Saikaly, P. E. (2015). Assessment of microbial fuel cell configurations and power densities. *Environmental Science & Technology Letters*, 2(8), 206–214.

Long, S., Zhao, L., Chen, J., Kim, J., Huang, C. H., & Pavlostathis, S. G. (2021). Tetracycline inhibition and transformation in microbial fuel cell systems: Performance, transformation intermediates, and microbial community structure. *Bioresource Technology*, 322, 124534.

Lovley, D.R., Stolz, J.F., Nord, G.L., & Phillips, E.J.P., 1987. Anaerobic production of magnetite by a dissimilatory iron-reducing microorganism. *Nature,* 330, 252–254.

Lu, L., & Ren, Z. J. (2016). Microbial electrolysis cells for waste biorefinery: A state of the art review. *Bioresource Technology*, 215, 254–264.

Luo, H., Xu, P., & Ren, Z. (2012). Long-term performance and characterization of microbial desalination cells in treating domestic wastewater. *Bioresource Technology*, 120, 187–193.

Ma, C. Y., & Hou, C. H. (2019). Enhancing the water desalination and electricity generation of a microbial desalination cell with a three-dimensional macroporous carbon nanotube-chitosan sponge anode. *Science of The Total Environment*, 675, 41–50.

Marassi, R. J., Queiroz, L. G., Silva, D. C., Dos Santos, F. S., Silva, G. C., & de Paiva, T. C. (2020). Long-term performance and acute toxicity assessment of scaled-up air–cathode microbial fuel cell fed by dairy wastewater. *Bioprocess and Biosystems Engineering*, 43, 1561–1571.

Marks, A., Griggs, S., Gasparini, N., & Moser, M. (2022). Organic electrochemical transistors: an emerging technology for biosensing. *Advanced Materials Interfaces*, 9(6), 2102039.

Mateo, S., Del Campo, A. G., Cañizares, P., Lobato, J., Rodrigo, M. A., & Fernandez, F. J. (2014). Bioelectricity generation in a self-sustainable microbial solar cell. *Bioresource Technology*, 159, 451–454.

Meena, R. A. A., Kannah, R. Y., Sindhu, J., Ragavi, J., Kumar, G., Gunasekaran, M., & Banu, J. R. (2019). Trends and resource recovery in biological wastewater treatment system. *Bioresource Technology Reports*, 7, 100235.

Meyer, J., Meyer, L. E., & Kara, S. (2022). Enzyme immobilization in hydrogels: A perfect liaison for efficient and sustainable biocatalysis. *Engineering in Life Sciences*, 22(3-4), 165–177.

Min, B., Kim, J., Oh, S., Regan, J. M., & Logan, B. E. (2005). Electricity generation from swine wastewater using microbial fuel cells. *Water Research*, 39(20), 4961–4968.

Misaghi, F., Delgosha, F., Razzaghmanesh, M., & Myers, B. (2017). Introducing a water quality index for assessing water for irrigation purposes: A case study of the Ghezel Ozan River. *Science of the Total Environment*, 589, 107–116.

Mohan, S. V., & Chandrasekhar, K. (2011). Solid phase microbial fuel cell (SMFC) for harnessing bioelectricity from composite food waste fermentation: influence of electrode assembly and buffering capacity. *Bioresource Technology*, 102(14), 7077–7085.

Moruno, F. L., Rubio, J. E., Atanassov, P., Cerrato, J. M., Arges, C. G., & Santoro, C. (2018). Microbial desalination cell with sulfonated sodium poly (ether ether ketone) as cation exchange membranes for enhancing power generation and salt reduction. *Bioelectrochemistry*, 121, 176–184.

Muhammad, M. H., Idris, A. L., Fan, X., Guo, Y., Yu, Y., Jin, X., ... & Huang, T. (2020). Beyond risk: Bacterial biofilms and their regulating approaches. *Frontiers in Microbiology*, 11, 928.

Muthukumar, H., Mohammed, S. N., Chandrasekaran, N., Sekar, A. D., Pugazhendhi, A., & Matheswaran, M. (2019). Effect of iron doped Zinc oxide nanoparticles coating in the anode on current generation in microbial electrochemical cells. *International Journal of Hydrogen Energy*, 44(4), 2407–2416.

Naaz, T., Kumar, A., Vempaty, A., Singhal, N., Pandit, S., Gautam, P., & Jung, S. P. (2023). Recent advances in biological approaches towards anode biofilm engineering for improvement of extracellular electron transfer in microbial fuel cells. *Environmental Engineering Research*, 28(5), 220666.

Ning, X., Lin, R., O'Shea, R., Wall, D., Deng, C., Wu, B., & Murphy, J. D. (2021). Emerging bioelectrochemical technologies for biogas production and upgrading in cascading circular bioenergy systems. *Iscience*, 24(9), 102998.

Nishaa, V., Spoorthi, B. V., Soumya, B. T., Meda, U. S., & Desai, V. S. (2022). Powering implantable medical devices with biological fuel cells. *ECS Transactions*, 107(1), 19197.

Obileke, K., Onyeaka, H., Meyer, E. L., & Nwokolo, N. (2021). Microbial fuel cells, a renewable energy technology for bio-electricity generation: A mini-review. *Electrochemistry Communications*, 125, 107003.

Oh, S. E., & Logan, B. E. (2006). Proton exchange membrane and electrode surface areas as factors that affect power generation in microbial fuel cells. *Applied Microbiology and Biotechnology*, 70, 162–169.

Osorio-de-la-Rosa, E., Vazquez-Castillo, J., Castillo-Atoche, A., Heredia-Lozano, J., Castillo-Atoche, A., Becerra-Nunez, G., & Barbosa, R. (2020). Arrays of plant microbial fuel cells for implementing self-sustainable wireless sensor networks. *IEEE Sensors Journal*, 21(2), 1965–1974.

Pachapur, V. L., Kutty, P., Pachapur, P., Brar, S. K., Le Bihan, Y., Galvez-Cloutier, R., & Buelna, G. (2019). Seed pretreatment for increased hydrogen production using mixed-culture systems with advantages over pure-culture systems. *Energies*, 12(3), 530.

Pandey, A. K., Gaur, V. K., Udayan, A., Varjani, S., Kim, S. H., & Wong, J. W. (2021). Biocatalytic remediation of industrial pollutants for environmental sustainability: research needs and opportunities. *Chemosphere*, 272, 129936.

Pant, D., Van Bogaert, G., Diels, L., & Vanbroekhoven, K. (2010). A review of the substrates used in microbial fuel cells (MFCs) for sustainable energy production. *Bioresource Technology*, 101(6), 1533–1543.

Pant, D., Singh, A., Van Bogaert, G., Olsen, S. I., Nigam, P. S., Diels, L., & Vanbroekhoven, K. (2012). Bioelectrochemical systems (BES) for sustainable energy production and product recovery from organic wastes and industrial wastewaters. *RSC Advances*, 2(4), 1248–1263.

Park, S. G., Rajesh, P. P., Sim, Y. U., Jadhav, D. A., Noori, M. T., Kim, D. H., ... & Chae, K. J. (2022). Addressing scale-up challenges and enhancement in performance of hydrogen-producing microbial electrolysis cell through electrode modifications. *Energy Reports*, 8, 2726–2746.

Park, Y., Park, S., Kim, J. R., Kim, H. S., Kim, B. G., Yu, J., & Lee, T. (2017). Effect of gradual transition of substrate on performance of flat-panel air-cathode microbial fuel cells to treat domestic wastewater. *Bioresource Technology*, 226, 158–163.

Patel, M., Patel, S. S., Kumar, P., Mondal, D. P., Singh, B., Khan, M. A., & Singh, S. (2021). Advancements in spontaneous microbial desalination technology for sustainable water purification and simultaneous power generation: a review. *Journal of Environmental Management*, 297, 113374.

Patel, G. B., Shah, K. R., Shindhal, T., Rakholiya, P., & Varjani, S. (2021). Process parameter studies by central composite design of response surface methodology for lipase activity of newly obtained Actinomycete. *Environmental Technology & Innovation*, 23, 101724.

Paucar, N. E., & Sato, C. (2021). Microbial fuel cell for energy production, nutrient removal and recovery from wastewater: A review. *Processes*, 9(8), 1318.

Ping, Q., Huang, Z., Dosoretz, C., & He, Z. (2015). Integrated experimental investigation and mathematical modeling of brackish water desalination and wastewater treatment in microbial desalination cells. *Water Research*, 77, 13–23.

Prakash, S., Ponnusamy, K., & Naina Mohamed, S. (2022). An insight on Biocathode Microbial Desalination Cell: Current challenges and prospects. *International Journal of Energy Research*, 46(7), 8546–8559.

Qu, Y., Feng, Y., Wang, X., Liu, J., Lv, J., He, W., & Logan, B. E. (2012). Simultaneous water desalination and electricity generation in a microbial desalination cell with electrolyte recirculation for pH control. *Bioresource Technology*, 106, 89–94.

Radhika, D., Shivakumar, A., Kasai, D. R., Koutavarapu, R., & Peera, S. G. (2022). Microbial electrolysis cell as a diverse technology: Overview of prospective applications, advancements, and challenges. *Energies*, 15(7), 2611.

Ramanaiah, S.V., Chandrasekhar, K., Cordas, C.M., & Potoroko, I., (2023). Bioelectrochemical systems (BESs) for agro-food waste and wastewater treatment, and sustainable bioenergy-A review. *Environmental Pollution*, 325, 121432.

Ramanaiah, S. V., Cordas, C. M., Matias, S. C., Reddy, M. V., Leitao, J. H., & Fonseca, L. P. (2021). Bioelectricity generation using long-term operated biocathode: RFLP based microbial diversity analysis. *Biotechnology Reports*, 32, e00693.

Rasmussen, M., Abdellaoui, S., & Minteer, S. D. (2016). Enzymatic biofuel cells: 30 years of critical advancements. *Biosensors and Bioelectronics*, 76, 91–102.

Ren, L., Ahn, Y., & Logan, B.E., 2014. A two-stage microbial fuel cell and anaerobic fluidized bed membrane bioreactor (MFC-AFMBR) system for effective domestic wastewater treatment. *Environmental Science and Technology*, 48, 4199–4206.-

Ren, Z., Ji, G., Liu, H., Yang, M., Xu, S., Ye, M., & Lichtfouse, E. (2022). Accelerated start-up and improved performance of wastewater microbial fuel cells in four circuit modes: Role of anodic potential. *Journal of Power Sources*, 535, 231403.

Rohbohm, N., Sun, T., Blasco-Gomez, R., Byrne, J. M., Kappler, A., & Angenent, L. T. (2022). Carbon oxidation with sacrificial anodes to inhibit O_2 evolution in membrane-less bioelectrochemical systems for microbial electrosynthesis. *bioRxiv*, 2022-08.

Rosenbaum, M., Aulenta, F., Villano, M., & Angenent, L. T. (2011). Cathodes as electron donors for microbial metabolism: Which extracellular electron transfer mechanisms are involved?. *Bioresource Technology*, 102(1), 324–333.

Roy, M., Aryal, N., Zhang, Y., Patil, S. A., & Pant, D. (2022). Technological progress and readiness level of microbial electrosynthesis and electrofermentation for carbon dioxide and organic wastes valorization. *Current Opinion in Green and Sustainable Chemistry*, 35, 100605.

Rusyn, I. B., Medvediev, O. V., & Valko, B. T. (2021). Enhancement of bioelectric parameters of multi-electrode plant–microbial fuel cells by combining of serial and parallel connection. *International Journal of Environmental Science and Technology*, 18(6), 1323–1334.

Sabin, J. M., Leverenz, H., & Bischel, H. N. (2022). Microbial fuel cell treatment energy-offset for fertilizer production from human urine. *Chemosphere*, 294, 133594.

Sadhukhan, J., Lloyd, J. R., Scott, K., Premier, G. C., Eileen, H. Y., Curtis, T., & Head, I. M. (2016). A critical review of integration analysis of microbial electrosynthesis (MES) systems with waste biorefineries for the production of biofuel and chemical from reuse of CO2. *Renewable and Sustainable Energy Reviews*, 56, 116–132.

Samukaite, U., Zukauskas, S., Ratautaite, V., Vilkiene, M., Mockeviciene, I., Liustrovaite, V., ... & Ramanavicius, A. (2022). Assessment of cytochrome c and chlorophyll a as natural redox mediators for enzymatic biofuel cells powered by glucose. *Energies*, 15(18), 6838.

Saratale, G. D., Banu, J. R., Nastro, R. A., Kadier, A., Ashokkumar, V., Lay, C. H., ... & Chandrasekhar, K. (2022). Bioelectrochemical systems in aid of sustainable biorefineries for the production of value-added products and resource recovery from wastewater: A critical review and future perspectives. *Bioresource Technology*, 359, 127435.

Saravanan, A., Kumar, P. S., Srinivasan, S., Jeevanantham, S., Kamalesh, R., & Karishma, S. (2022). Sustainable strategy on microbial fuel cell to treat the wastewater for the production of green energy. *Chemosphere*, 290, 133295.

Sartori, D., & Brunelli, D. (2016, April). A smart sensor for precision agriculture powered by microbial fuel cells. In *2016 IEEE Sensors Applications Symposium (SAS)* (pp. 1–6). IEEE.

Sevda, S., Yuan, H., He, Z., & Abu-Reesh, I. M. (2015). Microbial desalination cells as a versatile technology: Functions, optimization and prospective. *Desalination*, 371, 9–17.

Shah, S., Venkatramanan, V., & Prasad, R. (2019). Microbial fuel cell: Sustainable green technology for bioelectricity generation and wastewater treatment. In Shah, S., Venkatramanan, V., Prasad, R. (eds). *Sustainable Green Technologies for Environmental Management* (pp. 199–218). Springer, Singapore.

Shahid, K., Ramasamy, D. L., Haapasaari, S., Sillanpää, M., & Pihlajamäki, A. (2021). Stainless steel and carbon brushes as high-performance anodes for energy production and nutrient recovery using the microbial nutrient recovery system. *Energy*, 233, 121213.

Shantaram, A., Beyenal, H., Veluchamy, R. R. A., & Lewandowski, Z. (2005). Wireless sensors powered by microbial fuel cells. *Environmental Science & Technology*, 39(13), 5037–5042.

Shi, P., Wu, R., Wang, J., Ma, C., Li, Z., & Zhu, Z. (2022). Biomass sugar-powered enzymatic fuel cells based on a synthetic enzymatic pathway. *Bioelectrochemistry*, 144, 108008.

Shinde, O. A., Bansal, A., Banerjee, A., & Sarkar, S. (2018). Bioremediation of steel plant wastewater and enhanced electricity generation in microbial desalination cell. *Water Science and Technology*, 77(8), 2101–2112.

Shindhal, T., Rakholiya, P., Varjani, S., Pandey, A., Ngo, H. H., Guo, W., ... & Taherzadeh, M. J. (2021). A critical review on advances in the practices and perspectives for the treatment of dye industry wastewater. *Bioengineered*, 12(1), 70–87.

Shtepa, V., Balintova, M., Shykunets, A., Chernysh, Y., Chubur, V., Plyatsuk, L., & Junakova, N. (2023). Intensification of waste valorization techniques for biogas production on the example of clarias gariepinus droppings. *Fermentation*, 9(3), 225.

Song, Y. E., Mohamed, A., Kim, C., Kim, M., Li, S., Sundstrom, E., ... & Kim, J. R. (2022). Biofilm matrix and artificial mediator for efficient electron transport in CO_2 microbial electrosynthesis. *Chemical Engineering Journal*, 427, 131885.

Steinbusch, K. J., Hamelers, H. V., Schaap, J. D., Kampman, C., & Buisman, C. J. (2010). Bioelectrochemical ethanol production through mediated acetate reduction by mixed cultures. *Environmental Science & Technology*, 44(1), 513–517.

Strik, D. P., Timmers, R. A., Helder, M., Steinbusch, K. J., Hamelers, H. V., & Buisman, C. J. (2011). Microbial solar cells: applying photosynthetic and electrochemically active organisms. *Trends in Biotechnology*, 29(1), 41–49.

Suvvala, J., & Kumar, K. S. (2023). Implementation of EFC charging station by multiport converter with integration of RES. *Energies*, 16(3), 1521.

Tavker, N., & Kumar, N. (2023). Bioelectrochemical systems: Understanding the basics and overcoming the challenges. In *Development in Wastewater Treatment Research and Processes* (pp. 79–98). Elsevier.

Tawalbeh, M., Al-Othman, A., Singh, K., Douba, I., Kabakebji, D., & Alkasrawi, M. (2020). Microbial desalination cells for water purification and power generation: A critical review. *Energy*, 209, 118493.

Tee, P. F., Abdullah, M. O., Tan, I. A., Amin, M. A., Nolasco-Hipolito, C., & Bujang, K. (2018). Bio-energy generation in an affordable, single-chamber microbial fuel cell integrated with adsorption hybrid system: Effects of temperature and comparison study. *Environmental Technology*, 39(8), 1081–1088.

Thakur, D., Kumar, Y., Kumar, A., & Singh, P. K. (2019). Applicability of wireless sensor networks in precision agriculture: A review. *Wireless Personal Communications*, 107, 471–512.

Thapa, B. S., Kim, T., Pandit, S., Song, Y. E., Afsharian, Y. P., Rahimnejad, M., ... & Oh, S. E. (2022). Overview of electroactive microorganisms and electron transfer mechanisms in microbial electrochemistry. *Bioresource Technology*, 347, 126579.

Thulluru, L. P., Ghangrekar, M. M., & Chowdhury, S. (2023). Progress and perspectives on microbial electrosynthesis for valorisation of CO_2 into value-added products. *Journal of Environmental Management*, 332, 117323.

Toczyłowska-Mamińska, R., Szymona, K., & Kloch, M. (2018). Bioelectricity production from wood hydrothermal-treatment wastewater: Enhanced power generation in MFC-fed mixed wastewaters. *Science of The Total Environment*, 634, 586–594.

Varjani, S. (2022). Prospective review on bioelectrochemical systems for wastewater treatment: Achievements, hindrances and role in sustainable environment. *Science of The Total Environment*, 841, 156691.

Varjani, S., Pandey, A., & Upasani, V. N. (2021). Petroleum sludge polluted soil remediation: Integrated approach involving novel bacterial consortium and nutrient application. *Science of The Total Environment*, 763, 142934.

Vassilev, I., Hernandez, P. A., Batlle-Vilanova, P., Freguia, S., Krömer, J. O., Keller, J., ... & Virdis, B. (2018). Microbial electrosynthesis of isobutyric, butyric, caproic acids, and corresponding alcohols from carbon dioxide. *ACS Sustainable Chemistry & Engineering*, 6(7), 8485–8493.

Vinayak, V., Khan, M. J., Varjani, S., Saratale, G. D., Saratale, R. G., & Bhatia, S. K. (2021). Microbial fuel cells for remediation of environmental pollutants and value addition: Special focus on coupling diatom microbial fuel cells with photocatalytic and photoelectric fuel cells. *Journal of Biotechnology*, 338, 5–19.

Vyas, S., Prajapati, P., Shah, A. V., Srivastava, V. K., & Varjani, S. (2022). Opportunities and knowledge gaps in biochemical interventions for mining of resources from solid waste: A special focus on anaerobic digestion. *Fuel*, 311, 122625.

Wang, S., Adekunle, A., & Raghavan, V. (2022). Bioelectrochemical systems-based metal removal and recovery from wastewater and polluted soil: Key factors, development, and perspective. *Journal of Environmental Management*, 317, 115333.

Wang, X. C., Jiang, P., Yang, L., Van Fan, Y., Klemeš, J. J., & Wang, Y. (2021). Extended water-energy nexus contribution to environmentally-related sustainable development goals. *Renewable and Sustainable Energy Reviews*, 150, 111485.

Wang, X., Kimbrel, E. A., Ijichi, K., Paul, D., Lazorchak, A. S., Chu, J., ... & Xu, R. H. (2014). Human ESC-derived MSCs outperform bone marrow MSCs in the treatment of an EAE model of multiple sclerosis. *Stem cell Reports*, 3(1), 115–130.

Wang, Y., Xu, A., Cui, T., Zhang, J., Yu, H., Han, W., ... & Wang, L. (2020). Construction and application of a 1-liter upflow-stacked microbial desalination cell. *Chemosphere*, 248, 126028.

Wang, S., Zhao, J., Liu, S., Zhao, R., & Hu, B. (2018). Effect of temperature on nitrogen removal and electricity generation of a dual-chamber microbial fuel cell. *Water, Air, & Soil Pollution*, 229, 1–13.

Wilberforce, T., Sayed, E. T., Abdelkareem, M. A., Elsaid, K., & Olabi, A. G. (2021). Value added products from wastewater using bioelectrochemical systems: Current trends and perspectives. *Journal of Water Process Engineering*, 39, 101737.

Wu, J. Y., Lay, C. H., Chia, S. R., Chew, K. W., Show, P. L., Hsieh, P. H., & Chen, C. C. (2021). Economic potential of bioremediation using immobilized microalgae-based microbial fuel cells. *Clean Technologies and Environmental Policy*, 23(8), 2251–2264.

Wu, B., Tian, F., Zhang, M., Piao, S., Zeng, H., Zhu, W., ... & Lu, Y. (2022). Quantifying global agricultural water appropriation with data derived from earth observations. *Journal of Cleaner Production*, 358, 131891.

Xiao, X., Xia, H. Q., Wu, R., Bai, L., Yan, L., Magner, E., ... & Liu, A. (2019). Tackling the challenges of enzymatic (bio) fuel cells. *Chemical Reviews*, 119(16), 9509–9558.

Xu, C., Lu, J., Zhao, Z., Lu, X., Zhang, Y., Cheng, M., & Zhang, J. (2020). Simultaneous bioelectricity generation, desalination, organics degradation, and nitrogen removal in air–cathode microbial desalination cells. *SN Applied Sciences*, 2, 1–11.

Yadav, R. K., Das, S., & Patil, S. A. (2022). Are integrated bioelectrochemical technologies feasible for wastewater management?. *Trends in Biotechnology*. https://doi.org/10.1016/j.tibtech.2022.09.001

Yan, X., Lee, H. S., Li, N., & Wang, X. (2020). The micro-niche of exoelectrogens influences bioelectricity generation in bioelectrochemical systems. *Renewable and Sustainable Energy Reviews*, 134, 110184.

Yan, T., Ye, Y., Ma, H., Zhang, Y., Guo, W., Du, B., ... & Ngo, H. H. (2018). A critical review on membrane hybrid system for nutrient recovery from wastewater. *Chemical Engineering Journal*, 348, 143–156.

Yang, Z., Li, Y., Zhan, Z., Song, Y., Zhang, L., Jin, Y., ... & Chen, F. (2023). Enhanced power generation, organics removal and water desalination in a microbial desalination cell (MDC) with flow electrodes. *Science of The Total Environment*, 858, 159914.

Ye, Y., Ngo, H. H., Guo, W., Chang, S. W., Nguyen, D. D., Liu, Y., ... & Wang, J. (2019). Effect of organic loading rate on the recovery of nutrients and energy in a dual-chamber microbial fuel cell. *Bioresource Technology*, 281, 367–373.

Ye, Y., Ngo, H. H., Guo, W., Chang, S. W., Nguyen, D. D., Zhang, X., ... & Liu, Y. (2020). Impacts of hydraulic retention time on a continuous flow mode dual-chamber microbial fuel cell for recovering nutrients from municipal wastewater. *Science of the Total Environment*, 734, 139220.

Zahid, M., Savla, N., Pandit, S., Thakur, V. K., Jung, S. P., Gupta, P. K., ... & Marsili, E. (2022). Microbial desalination cell: Desalination through conserving energy. *Desalination*, 521, 115381.

Zaybak, Z., Pisciotta, J. M., Tokash, J. C., & Logan, B. E. (2013). Enhanced start-up of anaerobic facultatively autotrophic biocathodes in bioelectrochemical systems. *Journal of Biotechnology*, 168(4), 478–485.

Zhang, F., & He, Z. (2015). Scaling up microbial desalination cell system with a post-aerobic process for simultaneous wastewater treatment and seawater desalination. *Desalination*, 360, 28–34.

Zhang, S., Yang, Y. L., Lu, J., Zuo, X. J., Yang, X. L., & Song, H. L. (2020). A review of bioelectrochemical systems for antibiotic removal: Efficient antibiotic removal and dissemination of antibiotic resistance genes. *Journal of Water Process Engineering*, 37, 101421.

Zhang, Q., Zhang, L., Li, Z., Zhang, L., & Li, D. (2019). Enhancement of fipronil degradation with eliminating its toxicity in a microbial fuel cell and the catabolic versatility of anodic biofilm. *Bioresource Technology*, 290, 121723.

Zhao, J., Ma, H., Wu, W., Bacar, M. A., Wang, Q., Gao, M., ... & Qian, D. (2023). Conversion of liquor brewing wastewater into medium chain fatty acids by microbial electrosynthesis: Effect of cathode potential and CO2 supply. *Fuel*, 332, 126046.

Zhao, Q., Yu, H., Zhang, W., Kabutey, F. T., Jiang, J., Zhang, Y., ... & Ding, J. (2017). Microbial fuel cell with high content solid wastes as substrates: A review. *Frontiers of Environmental Science & Engineering*, 11, 1–17.

Zhou, L., Wu, Y., Zhang, S., Li, Y., Gao, Y., Zhang, W., ... & Sun, S. (2022). Recent development in microbial electrochemical technologies: Biofilm formation, regulation, and application in water pollution prevention and control. *Journal of Water Process Engineering*, 49, 103135.

Zhu, T. T., Cheng, Z. H., Yu, S. S., Li, W. W., Liu, D. F., & Yu, H. Q. (2022). Unexpected role of electron-transfer hub in direct degradation of pollutants by exoelectrogenic bacteria. *Environmental Microbiology*, 24(4), 1838–1848.

11 Microbial Technologies
Effectiveness and Feasibility Viewpoint

Shubhra Singh and Douglas J. H. Shyu

11.1 INTRODUCTION

11.1.1 Overview of Microbial Technologies in Wastewater Treatment

Microbial technologies have revolutionized the field of wastewater treatment by providing efficient and cost-effective methods for removing pollutants and contaminants from sewage and industrial wastewater. These technologies employ microorganisms, such as bacteria, fungi, and protozoa, to degrade organic matter and remove nitrogen and phosphorus. Activated sludge processes, anaerobic digestion, biofiltration, membrane bioreactors, and sequencing batch reactors are commonly used microbial technologies in wastewater treatment (Fig. 11.1). Each technology comprises a unique set of advantages and disadvantages. For instance, anaerobic digestion produces biogas that can be used as a renewable energy source, while membrane bioreactors produce high-quality effluent and save space. Microbial technologies are highly effective in treating wastewater, and have a crucial role in ensuring the safe exit of wastewater into various water systems and also soil systems (Singh et al., 2016; Anekwe et al., 2016; Nancharaiah & Sarvajith, 2019).

11.1.2 Importance of Microbial Technologies for Sustainable Wastewater Treatment

Microbial technologies are essential for sustainable wastewater treatment, providing efficient and eco-friendly methods for removing pollutants and contaminants from sewage and industrial wastewater. These technologies use microorganisms to degrade organic substances and remove the primary pollutants in wastewater, such as nitrogen and phosphorus (Kumar et al., 2022a, 2022b). By harnessing the power of microorganisms, these technologies can achieve high levels of wastewater treatment without relying on expensive and energy-intensive processes such as developing advanced biosensors or bioelectrochemical systems in addition to microbial fuel cells (Ramírez-Vargas et al., 2018; Yi et al., 2020). Moreover, microbial technologies promote sustainability by reducing the environmental and conservational impact of wastewater treatment. For example, anaerobic digestion produces biogas, a renewable bioenergy; reducing reliance on fossil fuels.

DOI: 10.1201/9781003441069-11

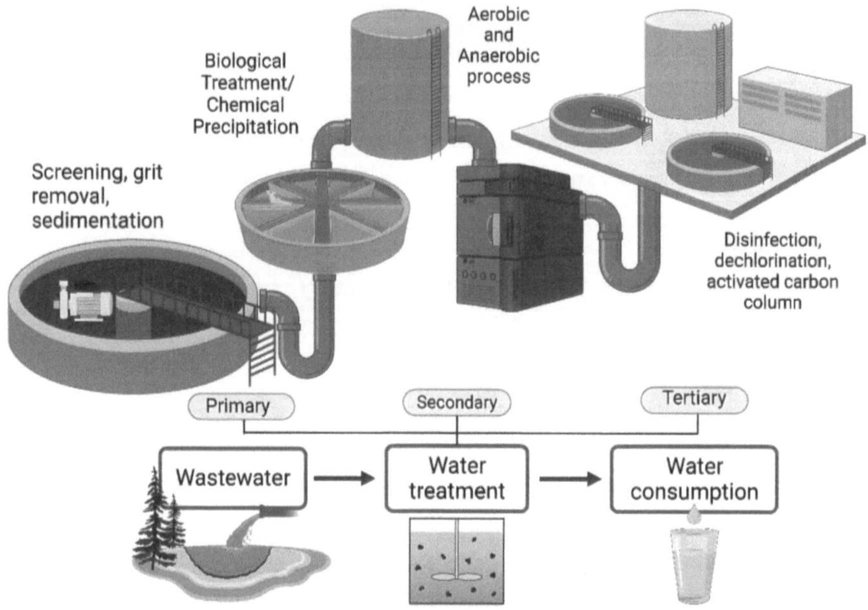

FIGURE 11.1 Diagram with all the typical processes in wastewater treatment (Created with BioRender.com).

Similarly, biofiltration and sequencing batch reactors can remove pollutants from wastewater without generating significant amounts of sludge, reducing the need for disposal and lowering carbon emissions (Tshemese et al., 2023). Microbial technologies are also cost-effective and easy to implement, making them accessible to developed and developing countries. This accessibility is critical for achieving sustainability goals, such as the United Nations Sustainable Development Goals (UN SDGs). One of their aims is to provide the safe utilization of clean water and sanitation for all. Therefore, microbial technologies are crucial for sustainable wastewater treatment. They provide efficient, eco-friendly, and cost-effective methods for removing pollutants and contaminants from wastewater, while improving access to clean water and sanitation (Carlsen & Bruggemann, 2022).

11.1.3 A Brief History of Microbial Technologies in Wastewater Treatment

Proper use of microorganisms in wastewater treatment can be traced back to ancient civilizations such as the Egyptians, Greeks, and Romans, who used biological methods to treat wastewater. However, it was not until the early twentieth century that the scientific understanding of microbial processes led to the advancement of modern wastewater treatment technologies. In the 1910s, the activated sludge process was developed, which used microorganisms to treat sewage by breaking down organic matter. The process became widely adopted in the 1930s and remained the most commonly used treatment technology. In the 1950s, anaerobic digestion was introduced to treat industrial wastewater and produce biogas. The 1970s and 1980s were the age for developing advanced technologies such as membrane bioreactors. These used membrane filtration combined with biological processes to create high-quality cleaned water and sequencing batch reactors, which combined several treatment stages in a single tank. Today, microbial technologies continue to evolve, with ongoing research focused on improving treatment efficiency, reducing energy consumption, and developing new technologies that can address emerging contaminants such as pharmaceuticals and microplastics. In summary, the history of microbial technologies in wastewater treatment spans centuries. Still, it was not until the twentieth century that the scientific understanding of microbial processes led to the development of modern wastewater treatment technologies (Lofrano & Brown, 2010; Yadav et al., 2020; Angelakis et al., 2023).

11.2 MICROBIAL DIVERSITY AND FUNCTIONALITY IN WASTEWATER TREATMENT

11.2.1 Microbial Taxonomy and Phylogeny

Microbial taxonomy and phylogeny play a crucial role in wastewater treatment by providing insights into microorganisms' diversity, function, and ecology. Microbial taxonomy is the science of identifying, classifying, and naming microorganisms based on their morphological, biochemical, and genetic characteristics. In wastewater treatment, microbial taxonomy identifies the microorganisms in treatment processes to understand their roles in removing pollutants and contaminants from wastewater (Das & Dash, 2019).

Microbial phylogeny, on the other hand, is the study of the evolutionary relationships among microorganisms based on their genetic sequences. Phylogenetic analysis can provide insights into the evolutionary history of microorganisms and their functional capabilities (Konstantinidis & Tiedje, 2007). The use of molecular techniques such as DNA sequencing, metagenomics, next-generation sequencing, and other computational tools has revolutionized microbial taxonomy and phylogeny to a more accurate and reproducible classification (Kumar et al., 2020, 2021). These techniques allow for identifying and characterizing microorganisms that are difficult or impossible to culture using traditional microbiological methods. They also enable the study of microbial communities rather than focusing on the individual microorganism. For example, data analysis through whole genome sequencing brings on more scientific approaches for classification. The computational analysis, such as average amino acid identity (AAI), average nucleotide identity (ANI), and *in silico* genome-to-genome distance hybridization (GGDH), leads to more precise information about microbial diversification (Hugenholtz et al., 2016; Mahato et al., 2017). Another viable method for microbial taxonomic and phylogenetic identification includes using a 16S rRNA gene sequence that provides a targeted approach through primer designing for different taxa. In addition, 23S rRNA gene sequences are coupled with 16S rRNA for better identification of species with distinguished taxonomic relationships, for example, the *Bartonella* and *Rickettsia* citrate synthase genes. Researchers can develop targeted interventions to improve treatment efficiency and understand the impact of environmental factors such as temperature, pH, and nutrients on microbial communities. Therefore, microbial taxonomy and phylogeny are essential tools in wastewater treatment research, providing insights into the diverse functions and ecological relations of microorganisms involved in treatment processes (Zeaiter et al., 2002; Fournier et al., 2003; Romalde et al., 2019). Flow cytometry can also be used for species delineation to count and identify microorganisms in wastewater samples. This technique involves using lasers and fluorescent dyes to detect and quantify microorganisms. Further quantification is done using a real-time PCR machine (Gonzalez & Saiz-Jimenez, 2005; Tindall et al., 2010).

The use of genomic data has revolutionized the field of microbial taxonomy and phylogeny and has opened new avenues for species delineation and understanding inter- and intra-species relationships (Wani et al., 2022). In addition, concepts such as AAI, generalized genomic signatures' (GGSs), and *in silico* GGDH are commonly used to compare and analyze genomic data. Karlin genomic signatures are based on statistical measures that reflect the nucleotide composition of genomes. These signatures can be used to compare the genomic similarity between different organisms and to identify potential evolutionary relationships among microbial species from wastewater. AAI measures the average amino acid identity between two distinct genomes. This method can determine the relatedness of two genomes and estimate the evolutionary distance between them. *In silico* GGDH calculates the distance between two genomes based on the number of shared orthologous genes. This method can infer the phylogenetic relationships between different organisms and identify potential new species (Thompson et al., 2013; Loman & Pallen, 2015). These genomic comparison concepts have several advantages over the conventional polyphasic approach, including higher resolution and accuracy, faster analysis times, and the ability to identify novel species. However, it is essential to note that the traditional polyphasic approach is still valuable and should be used with genomic data to understand microbial diversity and taxonomy comprehensively.

11.2.2 Microbial Metabolism and Functionality

Microbial metabolism and functionality are critical to wastewater treatment, as microorganisms play a crucial role in removing pollutants and contaminants from sewage and industrial wastewater. Microbial metabolism refers to the biochemical processes that microorganisms use to transform organic and inorganic compounds in wastewater. These processes are dependent upon the availability of oxygen as aerobic metabolism uses oxygen to break down organic matter and convert it into carbon dioxide, water, and energy through microorganisms. In anaerobic metabolism, microorganisms use electron acceptors other than oxygen, such as nitrates, sulfates, and carbon dioxide, to break down organic matter and produce methane or hydrogen gas. Microbial functionality refers to the roles that microorganisms play in wastewater treatment processes. Different types of microorganisms perform different functions in the treatment process as *Nitrosomonas* and *Nitrobacter* convert ammonia to nitrate during nitrification, while denitrifying bacteria convert nitrate to nitrogen gas during denitrification. Fungi and protozoa also play important roles in breaking down complex organic compounds and removing suspended solids from wastewater. Understanding microbial metabolism and functionality is critical to optimizing wastewater treatment processes. By adjusting the conditions of the treatment system, such as the temperature, pH, and nutrient availability, researchers can encourage the growth and activity of specific microorganisms to achieve targeted treatment goals. For example, optimizing the conditions for nitrifying bacteria can improve the removal of nitrogen from wastewater (Fig. 11.2). In summary, microbial metabolism and functionality are critical to wastewater treatment, as microorganisms play a crucial role in removing pollutants and contaminants from wastewater. By understanding the biochemical processes that microorganisms use to transform organic and inorganic compounds and the roles that different types of microorganisms play in the treatment process, researchers can optimize treatment processes to achieve targeted treatment goals (Wagner et al., 2002).

Meanwhile, the anammox (anaerobic ammonium oxidation) method, a promising approach, can be integrated with the nitrogen removal method conventionally used in wastewater treatment systems and tends to reduce operational costs and

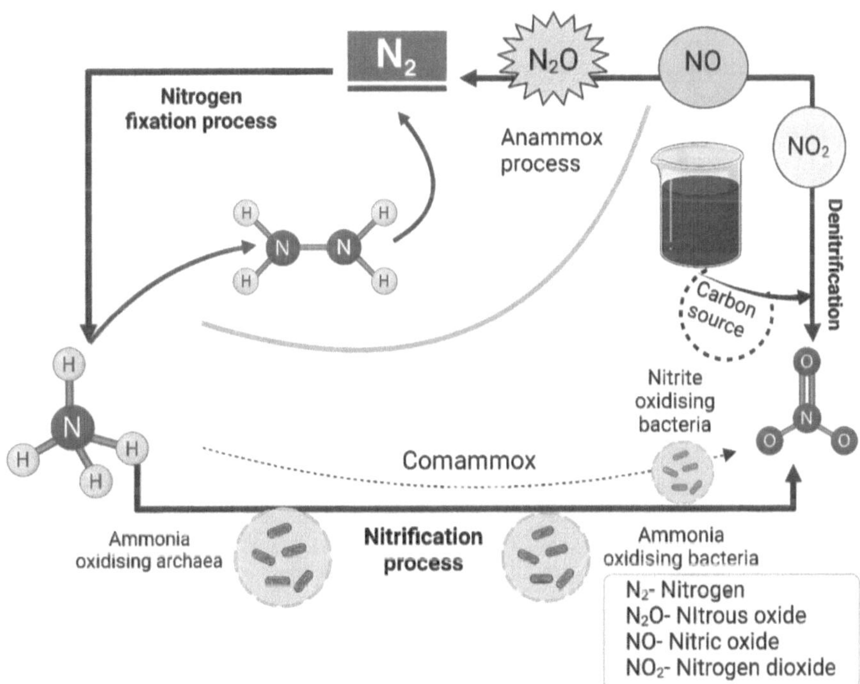

FIGURE 11.2 Schematic of the nitrogen cycle including anammox-mediated nitrogen transformation (Modified from Du et al., 2019 and created with BioRender.com).

environmental impacts. Formerly, traditional nitrogen removal typically involved nitrification. The ammonia gets oxidized to nitrite and then to nitrate, followed by a series of steps to convert nitrate to nitrogen gas as denitrification. These processes are energy-intensive steps, require oxygen and organic carbon as electron donors, and produce greenhouse gases (Xiao et al., 2021). On the contrary, anammox emerges as an energy-saving process where ammonia is directly converted to nitrogen gas. It uses nitrite as an electron acceptor without needing oxygen or organic carbon. This process can significantly reduce conventional nitrogen removal's operational costs and environmental impacts. Integrating anammox with traditional nitrogen removal involves adding anammox bacteria to the existing wastewater treatment system. This can be done by adding a separate anammox reactor or creating anaerobic conditions within the current aerobic reactor. One of the key challenges in integrating anammox is system stability, and anammox bacteria must be able to compete with other microorganisms for nutrients. This can be achieved by regulating the pH, temperature, and nutrient concentrations in the system and selecting appropriate anammox bacteria. Overall, integrating anammox with conventional nitrogen removal methods can potentially improve the efficiency of wastewater treatment systems. However, it is an area of ongoing research and development (Al-Hazmi et al., 2022; Fofana et al., 2022).

11.2.3 Microbial Interactions and Coexistence

Microbial interactions and coexistence are essential for effectively forming complex communities, where they interact and coexist through various mechanisms such as competition, cooperation, and predation. Competition is a common mechanism in microbial interactions, where microorganisms compete for resources such as nutrients and space. For example, ammonia-oxidizing bacteria compete with other microorganisms for oxygen and ammonium ions during nitrification. In contrast, denitrifying bacteria compete with other microorganisms for organic carbon sources (Xiao et al., 2021).

Cooperation is another mechanism in microbial interactions, where microorganisms work together to achieve a common goal. For example, some microorganisms, such as syntrophic bacteria, cooperate with methanogens to convert complex organic compounds into methane and carbon dioxide during anaerobic digestion. Predation is a less common mechanism in microbial interactions where microorganisms prey on other microorganisms. For example, some protozoa consume bacteria and other microorganisms, playing a role in controlling the growth of microbial populations together with understanding microbial interactions and coexistence for optimizing wastewater treatment processes. By manipulating the conditions of the treatment system, researchers can encourage the growth and activity of specific microorganisms to achieve targeted treatment goals. For example, by controlling the dissolved oxygen concentration, it is possible to promote the development of nitrifying bacteria over denitrifying bacteria, thus improving the removal of nitrogen from wastewater. In summary, microbial interactions and coexistence are essential for effective functioning wastewater treatment processes. Understanding the mechanisms of competition, cooperation, and predation in microbial communities can help optimize treatment processes to achieve targeted treatment goals (Damore & Gore, 2013; Lechón-Alonso et al., 2021).

11.2.4 Microbial Communities and Microbiomes

Wastewater treatment with recycling and reuse is important for maintaining global health and sustaining the increasing world population, expected to reach 10 billion in 2050 (Maquet, 2020). Microbial communities and microbiomes are crucial as microorganisms work together to remove pollutants and contaminants from sewage and industrial wastewater. Microbial communities in wastewater treatment systems are complex and diverse, consisting of bacteria, fungi, archaea, and protozoa. In addition to interacting with each other, these microorganisms interact with the environment. They perform different functions, such as organic matter degradation, nitrification, denitrification, and phosphorus removal. The microbiome is the collection of microorganisms and their genetic material in a specific environment, such as a wastewater treatment system. Recent advances in molecular biology techniques, such as DNA sequencing and metagenomics, have allowed for identifying and characterizing microbial communities and their functional capabilities in wastewater treatment (Gude, 2015).

Understanding the microbiome of wastewater treatment systems is critical to optimizing treatment processes. By analyzing the genetic material of microorganisms in the system, researchers can identify the presence of specific microorganisms and their functional capabilities. This information can then be used to manipulate the conditions of the treatment system to encourage the growth and activity of specific microorganisms to achieve targeted treatment goals. For example, by analyzing the microbiome of a wastewater treatment system, researchers can identify the presence of microorganisms that are responsible for removing specific pollutants, such as nitrogen or phosphorus. They can then manipulate the treatment system's conditions, such as the concentration of dissolved oxygen, the pH values, and the nutrient availability, to promote the growth and activity of these microorganisms, thus improving the removal of the target pollutants. Recent advances such as microbiome-based machine learning have allowed a better understanding of predicting the functional characteristics required to design the desired wastewater treatment systems (Fig. 11.3). They enhance the stability of the systems and signal a warning against process failure to achieve targeted treatment goals (Cai et al., 2021). This proactive approach of machine learning to utilize microbial indicators would identify and assess the potential drawbacks of the microbiome used in wastewater systems. However, specific problems need to be addressed, such as microbiome management requiring high-throughput experimentation to monitor the performance of the system. Exploring models to predict system functions needs more research and development. Further, combining metagenomic and metatranscriptomic tools can help study the genetic basis and expression profile of a rapidly evolving microbiome (Ju et al., 2019).

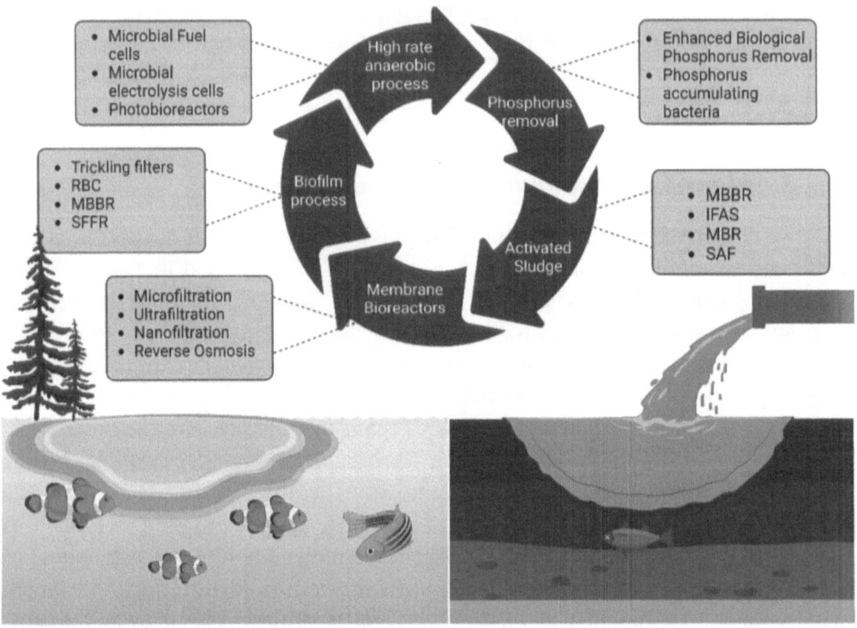

FIGURE 11.3 Combined effects of conventional and modern methods on the wastewater microbiome (Created with BioRender.com).

11.3 MICROBIAL PROCESSES IN WASTEWATER TREATMENT

11.3.1 Aerobic and Anaerobic Microbial Processes for Organic Matter Removal

Aerobic and anaerobic microbial processes are two common methods for organic matter removal in wastewater treatment. Each method has its advantages and limitations, and they can be used together in different configurations to achieve efficient organic matter removal. Aerobic microbial processes involve using oxygen to support the growth of aerobic bacteria that decompose organic substances into carbon dioxide, water, and other derived biomass. This process is known as aerobic digestion or oxidation and requires energy to supply oxygen, and it can produce large amounts of carbon dioxide, contributing to greenhouse gas emissions. On the other hand, anaerobic microbial processes involve microorganisms that do not require oxygen to decompose organic matter and can convert organic matter into biogas, which can be used as an energy source. Anaerobic digestion is highly efficient in removing organic matter and can produce a valuable byproduct. However, it is slower than aerobic digestion and requires careful monitoring of pH and temperature to ensure optimal conditions for the anaerobic bacteria. To achieve efficient organic matter removal, wastewater treatment plants can combine aerobic and anaerobic processes in different configurations (Goli et al., 2019). One common configuration is using an anaerobic digester to convert organic matter into biogas, followed by an aerobic reactor to remove any remaining organic matter and nutrients. This configuration allows for both organic matter removal and energy recovery. Another configuration is the use of a sequencing batch reactors (SBRs), which alternates between aerobic and anaerobic conditions to achieve efficient organic matter removal (Mahvi, 2008; Dutta & Sarkar, 2015). Recently, for disinfection purposes, metallic and metal oxide nanomaterials have been used for wastewater treatment and environmental remediation (Yang et al., 2013).

11.3.2 Microbial Nitrification and Denitrification Processes to Remove Nitrogen

Two microbial processes, nitrification and denitrification, are used as biological wastewater treatment to remove nitrogen. Nitrification involves the use of ammonia, which is then oxidized to nitrate. On the other hand, denitrification involves the use of nitrate, which is then reduced to nitrogen gas. Nitrification is typically carried out by two kinds of bacteria: ammonia-oxidizing bacteria (AOB), which convert ammonia to nitrite, and nitrite-oxidizing bacteria (NOB), which convert nitrite to nitrate. This process requires oxygen and is typically carried out in an aerobic reactor. Nitrification can be affected by temperature, pH, and also toxic substances that affect microbial activities. If not removed from the wastewater, it can produce nitrate, contributing to eutrophication.

In contrast, denitrification is carried out by a group of bacteria that use nitrates as electron acceptors and are converted, forming nitrogen gas. The process is typically carried out under anaerobic conditions and requires a source of organic carbon as an electron donor. Denitrification can be performed in a separate denitrification reactor or the same reactor as nitrification in a Modified Ludzack-Ettinger (MLE) process. Both nitrification and denitrification processes can be combined into a single process known as biological nitrogen removal for wastewater treatment. This process involves alternating conditions between aerobic and anaerobic status to facilitate nitrification and denitrification. This can be achieved in a variety of configurations, such as the use of anoxic reactors or the use of SBRs. Recent advances in microbial processes used 16S rRNA genes for nitrogen removal to verify microbial diversity in wastewater treatment. They can be combined in different configurations for efficient and effective nitrogen removal. The choice of configuration depends on the specific requirements and constraints of the wastewater treatment plant, such as energy and resource availability, treatment goals, and regulatory requirements (Peng & Zhu, 2006; Chai et al., 2019).

Advances in microbial genomics have analyzed microbial communities, including ammonia-oxidizers, anammox, comammox, and denitrifiers. The high-throughput sequencing targeted specific genes, such as ammonia monooxygenase genes and nitrous oxide reductase genes involved in nitrifying and denitrifying processes, to effectively accelerate the whole process. The structure and composition of microbial genetic communities have indicated that differences in oxygen levels were not the main drivers for the composition of the microbial community. The number of communities is ubiquitous in all types of wastewater treatment processes. Additionally, soil bacteria are important inputs during treatments for proper operation and maintenance (Wigginton et al., 2020).

11.3.3 Phosphorus Removal Microbial Processes

Wastewater treatment plants use various microbial processes to remove phosphorus from wastewater. For example, enhanced biological phosphorus removal (EBPR) is described as a biological process involving specific bacteria to remove phosphorus from wastewater. During this process, bacteria consume organic matter in the wastewater and store excess phosphorus in their cells (Yuan et al., 2012). The bacteria are then removed from the wastewater as sludge, which can be further treated or disposed of. Highly selected microorganisms perform EBPR under aerobic conditions, such as phosphorus-accumulating organisms (PAOs). However, actual conditioning or PAOs occur under anaerobic conditions by exposing them to volatile fatty acids (VFAs). Therefore, PAOs store food as VFAs under anaerobic conditions, and further processing of stored material occurs under aerobic conditions. The EBPR process is considered successful in the influent wastewater quality and the amount of VFAs available in proportion to the amount of phosphorus to be removed (Stokholm-Bjerregaard et al., 2017; Muduli et al., 2021).

Another one is phosphorus-accumulating biofilters; a type of fixed-film biological process that can be used to remove phosphorus from wastewater. These biofilters are typically constructed of a porous material, such as sand or plastic; which provides a surface area for bacterial growth. As the wastewater flows through the biofilter, bacteria consume organic matter and store excess phosphorus in their cells. The biofilter media can be periodically removed, replaced, or cleaned to maintain effectiveness (Lin et al., 2019; Wang et al., 2022).

On a small scale, chemical precipitation using ion exchange is widely used to remove phosphorus from wastewater. In this process, a coagulant, such as aluminum hydroxide or ferric chloride, is added to increase the selectivity in wastewater to form a precipitate that concentrates phosphorus (Seo et al., 2013). Microorganisms can also be added to the wastewater to enhance the precipitation process, such as *Tetrasphaera*-like bacteria in sludge systems. One of the important biological processes, EBNR, is a combination of biological and chemical processes that can be used to remove phosphorus using PAOs under regulated conditions (Nguyen et al., 2011). This process typically involves three stages: anoxic, aerobic, and anaerobic. During the anoxic stage, bacteria consume nitrates instead of oxygen, which allows for the removal of phosphorus in the aerobic stage. The anaerobic stage promotes the growth of phosphorus-accumulating bacteria. Overall, the effectiveness of these microbial processes for phosphorus removal in wastewater treatment can vary depending on the specific conditions and phosphorus uptake carried out during the process (Bunce et al., 2018). In addition, understanding biochemical requirements is necessary to increase the production of PAOs for EBPR, including carbon, glycogen, and electron acceptors. *Accumulibacter*, a PAO, is considered to be the basis of most models. The complete genome provides detailed information about the gene expression and metabolic processes that are required for effective EBPR systems. Other known PAOs, such as *Enterobacter* sp., *Paracoccus* sp., and *Pseudomonas* sp., were found at small amounts from phosphorus removal wastewater treatment plants (Li et al., 2003; López-Vázquez et al., 2008).

11.3.4 Microbial Processes for Emerging Contaminant Removal

Emerging contaminants refer to substances that have recently been discovered or have become a concern due to their environmental presence. They have a potential impact on the ecosystem and human health. These contaminants include pharmaceutical compounds, healthcare products, pesticides, household products, and various chemicals not typically monitored in water or soil. Microbial processes can effectively remove emerging contaminants from the environment (Shah, 2021). Many emerging contaminants can be degraded by microorganisms in the environment. For example, bacteria can break down antibiotics and other pharmaceuticals, while fungi can degrade pesticides. Biodegradation can occur naturally but can also be enhanced through bioremediation strategies. These include adding nutrients to stimulate microbial growth or introducing specific strains of bacteria or fungi (Sidhu et al., 2017). Anaerobic treatment involves using microorganisms that do not require oxygen to break down contaminants. This process can effectively remove emerging contaminants resistant to other treatment methods, such as some pharmaceuticals. Anaerobic treatment can be used with other methods, such as aerobic treatment or bioreactors. Overall, microbial processes can be a promising approach for removing emerging contaminants from the environment. However, the effectiveness of these processes can vary depending on the specific contaminants being targeted and the environmental conditions. Therefore, careful evaluation and monitoring are important to ensure these methods are effective and sustainable (Niehaus et al., 2019).

11.4 MICROBIAL TECHNOLOGIES FOR WASTEWATER TREATMENT

11.4.1 Activated Sludge and the Variations of Attached-growth Processes

Activated sludge is a widely used wastewater treatment process that relies on microorganisms to remove pollutants from wastewater. However, variations of this process have been developed to improve its efficiency and effectiveness. Attached-growth processes, such as integrated fixed-film activated sludge (IFAS), moving-bed biofilm reactor (MBBR), membrane bioreactor (MBR), and submerged aerated filter (SAF), involve adding a substrate or media to facilitate the growth of microorganisms. These processes provide a sufficient surface area for microorganism growth, allow for a relatively large amount of biomass concentration in the treatment tank, and can provide a high-quality effluent. However, the effectiveness of these processes can vary according to the wastewater conditions and the adopted treatment plant. Overall, attached-growth processes can be effective variations of the activated sludge process for wastewater treatment, and their implementation can lead to improved wastewater treatment efficiency and reduced environmental impact (Rendueles & Ghigo, 2015; Sharma et al., 2019).

MBBR involves adding plastic media to the treatment tank to create a surface for microbial growth. The microorganisms consume the organic substances and pollutants in the wastewater flowing via the tank. The plastic media provides a sufficient surface area and a large biomass concentration in the treatment tank for microorganism growth (Biase et al., 2019). MBR combines a membrane filtration system with the activated sludge process and involves adding a membrane filter to the treatment tank to separate the treated wastewater from the activated sludge. The membrane filter allows for a high biomass concentration in the treatment tank and can provide high-quality effluent (Goswami et al., 2018). IFAS is a process that combines the activated sludge process with an attached-growth system. This process involves adding plastic media to the activated sludge tank to facilitate microorganism growth. The purpose of using plastic media is to provide the same functions for microorganism growth as those for MBBR (Biase et al., 2019). SAF is a process that involves adding a filter medium to the treatment tank to promote the growth of microorganisms. The filter medium provides a large surface area for microbial growth and allows for a high biomass concentration in the treatment tank. The microorganisms consume the organic substances and pollutants as the wastewater flows through the filter medium (Melin et al., 2006; Waqas et al., 2020).

11.4.2 Biofilm Processes and the Variations of Attached-growth Processes

Biofilm processes are a type of wastewater treatment process involving microorganisms' growth on a surface or substrate. Biofilm processes are commonly used for the removal of pollutants from wastewater and involve the growth of microorganisms on surfaces in a matrix of extracellular polymeric substances (EPS). This matrix provides protection and support for the microorganisms and allows for their attachment to various substrates (Biase et al., 2019; Hamoda & Al-Sharekh, 2000).

Attached-growth processes are also commonly used in wastewater treatment. These processes involve using a substrate or media to promote the growth of microorganisms. Attached-growth processes are like biofilm processes but typically use a higher surface area-to-volume ratio and a shorter hydraulic retention time. There are several variations of attached-growth processes, including:

a. Biological trickling filters (BTFs): The wastewater flows through a bed of porous media, such as stones or plastic material, where microorganisms are attached to the surface of the media. Microorganisms remove organic matter from the water as the wastewater trickles down through the media. A previous study used BTFs with different media, such as rubber, polystyrene, plastic, and stones, to assess the better BTF systems for removing organic matter. The results showed that using different filters can produce as much effluent as carbonaceous matter (biological oxygen demand ((BOD)) and remove fecal coliform contaminants at a high rate of 93.4% at 25–35 °C—showing the potential for BTF systems to be scaled up for large-scale wastewater treatment applications in developing nations. Further investigation is required to determine whether such BTFs can effectively nitrify at high nitrogen and BOD loading rates (Naz et al., 2015).

b. Rotating biological contactor (RBC): In this process, wastewater flows over a series of rotating disks or drums, partially submerged in the wastewater. The microorganisms attached to the surface of the disks or drums remove organic matter from the wastewater as they rotate. The RBC has a limited success rate in eliminating biological phosphorus, with 70% removal. However, the energy costs are far less than trickling filters. The RBC process can remove 99% of fecal coliforms found in wastewater such as hybrid activated sludge (HYBACS), demonstrating new combinations of suspended growth and fixed film on rotating media with efficient denitrification and increased organic matter removal (Khomela, 2018). Recently, novel RBC reactors used algae and microbial fuel cells as systems for direct energy generation for higher efficiency. The scaling-up considerations of RBC technology remain challenging for future applications; however, simplicity, adaptability, low maintenance, and high volumetric activity indicate that this method will still be used for the treatment of wastewater (Hassard et al., 2015).
c. Submerged fixed-film reactor (SFFR): In this process, microorganisms attach to the surface of fixed media, such as a plastic or ceramic tubesettler that is submerged in the wastewater. The microorganisms remove organic matter from the wastewater flowing through the tank. For example, hospital wastewater was treated with SFFR with a tubesettler arranged in series, and the efficiency was 75% chemical oxygen demand (COD), 67% BOD, and 67% phosphate (Khan et al., 2022).

11.4.3 Membrane Bioreactors and the Variations of Membrane Processes

MBRs use membrane filtration and biological treatment as a wastewater treatment technology. In MBRs, microorganisms are used to biodegrade organic matter in wastewater, while a membrane barrier separates the treated water from the biomass. MBRs are becoming increasingly popular due to their ability to produce high-quality treated water, having a small footprint, and low sludge production. They are commonly used for municipal and industrial wastewater treatment and water reuse. The membrane process in MBRs can vary depending on the desired treatment level and the feed water quality. There are four main types of membrane processes: microfiltration (MF), nanofiltration (NF), reverse osmosis (RO), and ultrafiltration (UF). Among them, the pore size of MF membranes is the largest of all the membrane processes and can typically remove suspended solids and bacteria from wastewater.

The pore size of MF membranes generally ranges from 0.1 to 10 microns. UF membranes have a smaller pore size than MF membranes and can remove smaller particles, such as viruses and proteins, from wastewater. The UF membrane pore size is typically between 0.001–0.1 microns. NF membranes have an even smaller pore size than UF membranes and can remove divalent calcium and magnesium ions from water. The pore size of NF membranes is typically in the 0.001–0.01 microns range. RO membranes have the smallest pore size of all the membrane processes and can remove dissolved salts and other contaminants from water. The RO membrane pore size is typically from 0.0001 to 0.001 microns. In MBRs, UF or a combination of UF and MF membranes are generally used to filter wastewater. The choice of membrane process depends on the feed water quality and the desired treatment level. UF membranes are commonly used in MBRs because they can effectively remove particles and microorganisms from wastewater. UF membranes are also more resistant to fouling than MF membranes, leading to longer membrane life and lower maintenance costs.

In some cases, a combination of UF and MF membranes may provide a more efficient treatment. The MF membranes can be used as a pretreatment step to remove large particles and protect the UF membranes from fouling. The membrane process used in an effective MBR's wastewater biological treatment technology can vary depending on the feed water quality and the desired treatment level. UF membranes are commonly used in MBRs due to their ability to remove particles and microorganisms from wastewater effectively. In contrast, MF membranes can be used as a pretreatment step to protect the UF membranes from fouling (Melin et al., 2006; Goswami et al., 2018).

11.4.4 High-rate Anaerobic Processes with Electrochemical Systems for Organics and Energy Recovery

High-rate anaerobic processes with electrochemical systems are a promising technology for treating organic wastewater and recovering energy. These systems combine anaerobic digestion with electrochemical reactions to achieve high levels of organic removal and energy recovery. The anaerobic digestion process is the break down of organic substances by microorganisms in the absence of oxygen. This process generates biogas primarily composed of carbon dioxide and methane. Biogas can be used as an energy source, either for electricity generation or as a fuel for heating or transportation. Electrochemical systems use microorganisms to catalyze electrochemical reactions that generate electrical current or produce hydrogen gas, for example, microbial electrolysis cells (MECs) and microbial fuel cells (MFCs). MECs and MFCs can be integrated with anaerobic digestion to improve organic removal and increase energy recovery (Oh et al., 2010; Kamali et al., 2023). In MFCs, bacteria oxidize organic substances, and the electrons are transferred to an anode electrode, creating a flow of electrons that can be harnessed as electrical current. In MECs, the opposite reaction occurs, where an electrical current is used to drive the reduction of organic matter and generate hydrogen gas at a cathode electrode. Integrating MFCs or MECs with anaerobic digestion can increase organic matter degradation and biogas production rate while generating electrical or hydrogen energy (Gude, 2016; Wang et al., 2019). This approach has several advantages over conventional anaerobic digestion systems, including higher organic removal efficiency, increased energy recovery, and reduced sludge production. High-rate anaerobic processes with electrochemical systems can potentially improve the sustainability of wastewater treatment and energy production. These systems can be used in municipal and industrial wastewater and agricultural waste treatment. Further research is needed to optimize these systems' performance and evaluate their economic viability (Khandaker et al., 2021; Nguyen & Babel, 2022).

11.4.5 Algal-based Strategies for Resource Recovery and Wastewater Treatment

Algal-based strategies are becoming increasingly popular due to their ability to remove nutrients and organic matter from wastewater and produce valuable biomass that can be used for various purposes. Algae are photosynthetic organisms that can proliferate in nutrient-rich wastewater, capturing carbon dioxide and producing oxygen. They can be used to remove nitrogen and phosphorus, a major source of water pollution from wastewater, if not properly treated. Algae can also remove organic matter from wastewater, reducing the demand for oxygen in downstream treatment processes. In addition to their wastewater treatment capabilities, algae can also be used to produce valuable biomass that can be used for various purposes. Algae can be used as a food source for humans or animals, as a feedstock for biofuels, or as a source of high-value compounds, for example, pigments or omega-3 fatty acids. Several algal-based strategies have been developed for resource recovery and wastewater treatment (Gude, 2016). One approach is to use open ponds or raceways to cultivate algae, which can be harvested and processed for biomass production. Another approach is to use photobioreactors, closed systems that allow for precise control of environmental conditions and higher biomass productivity. Sometimes, algal-based systems can be integrated with other wastewater treatment processes to enhance nutrient and organic matter removal. For example, algal-based systems can be combined with anaerobic digestion to produce biogas and nutrient-rich digested products. The resulting digestate can be used as a fertilizer for agricultural utilization. There are also emerging technologies that combine algal-based systems with other innovative processes, such as membrane filtration or electrochemical systems, to further improve resource recovery efficiency and wastewater treatment. Overall, algal-based strategies for resource recovery and wastewater treatment offer several advantages, including high nutrient and organic matter removal efficiency, biomass production for various applications, and potential for integration with other treatment processes. Further research is needed to optimize these systems' performance and evaluate their economic viability for widespread adoption (Hamoda & Al-Sharekh, 2000; Rendueles & Ghigo, 2015).

11.5 FEASIBILITY AND IMPLEMENTATION OF MICROBIAL TECHNOLOGIES IN WASTEWATER TREATMENT

11.5.1 Design and Optimization of Microbial Technologies for Wastewater Treatment

This involves the selection and manipulation of microorganisms to efficiently and effectively remove contaminants from wastewater. Microbial technologies can be used to remove various pollutants, including organic matter, nitrogen, phosphorus, and pathogens (Rivas et al., 2008; Li & Yu, 2016). The design and optimization of microbial technologies for wastewater treatment involves several key steps:

a. Selection of microbial communities: The selection of microbial communities depends on the wastewater to be treated, for example, the type of wastewater and its characteristics. The sources of selected microbial communities can be from natural environments, such as soil and water, or commercial products.
b. Process design: This involves selecting the most appropriate treatment process, such as activated sludge, biofilm reactors, or membrane bioreactors. The process design also determines process parameters, for example, hydraulic retention time (HRT), solids' retention time (SRT), aeration rate, dissolved oxygen, and biomass concentration.
c. Operational optimization: Once the treatment process is established, operational optimization involves adjusting process parameters to achieve optimal treatment performance. Operational optimization can include adapting aeration rates, changing biomass concentration, or modifying influent characteristics.
d. Microbial manipulation: Microbial manipulation involves the introduction of microorganisms to enhance treatment performance. It can include the addition of specific microorganisms, such as denitrifying bacteria or PAOs, or microbial community modifications through changes in operating conditions.
e. Monitoring and control: Monitoring and control of the treatment process involves regularly measuring process parameters, such as pH, dissolved oxygen, and biomass concentration, to ensure that the treatment process is functioning correctly. Control strategies can be implemented to adjust process parameters in response to changes in influent characteristics or process performance.

Optimizing microbial technologies for wastewater treatment can be achieved using mathematical modeling and simulation tools. These tools can predict treatment performance and identify the most effective operating conditions and process parameters. Therefore, designing and optimizing microbial technologies for wastewater treatment involve selecting appropriate microbial communities, process design and optimization, microbial manipulation, and monitoring and control. The use of mathematical modeling and simulation tools can aid in the optimization of these processes. The goal is to efficiently and effectively remove contaminants from wastewater while minimizing energy and resource consumption (Morgenroth et al., 2002; Kudryavtsev et al., 2019).

11.5.2 Operation and Maintenance of Microbial Technologies for Wastewater Treatment

The successful operation and maintenance of microbial technologies for wastewater treatment are critical to efficiently and effectively remove contaminants from wastewater. The operation and maintenance of microbial technologies requires careful attention to several key factors, including process monitoring, equipment maintenance, and microbial management (Quansheng et al., 2017; Priya et al., 2021).

a. Process monitoring: Process monitoring involves regular testing of wastewater and process parameters to ensure the treatment process operates effectively. Commonly monitored process parameters include pH, temperature, dissolved oxygen, and nutrient levels. Regular monitoring helps identify any process upsets or issues affecting treatment performance.
b. Equipment maintenance: Equipment maintenance is an essential part of the operation and maintenance of microbial technologies. It includes regularly inspecting and cleaning equipment, such as pumps, aerators, and valves, to ensure they function correctly. Any necessary repairs or replacements should be performed promptly to prevent downtime and maintain treatment performance.
c. Microbial management: Microbial management involves maintaining the health and activity of the microbial community within the treatment system. It includes monitoring biomass levels, maintaining proper aeration and mixing, and avoiding shock loads that may disrupt the microbial community. Additionally, regular cleaning and maintenance of aeration systems, membranes, and other components can help prevent the buildup of biofilms or further microbial growth that may affect treatment performance.
d. Safety and compliance: Safety is a critical aspect of operating and maintaining microbial technologies for wastewater treatment. Proper safety procedures should be implemented to protect personnel and ensure compliance with applicable regulations. This includes properly handling and disposing of wastewater, chemicals, and other materials used in the treatment process.
e. Staff training: Proper staff training is essential for successfully operating and maintaining microbial technologies for wastewater treatment, and staff should be trained on operating procedures, maintenance practices, and safety protocols. Additionally, ongoing training and education can help the team stay up to date on the latest advances and best practices in wastewater treatment.

11.5.3 Performance Evaluation and Monitoring of Microbial Technologies

Performance evaluation and monitoring of microbial technologies for wastewater treatment are essential to ensure that the treatment system is operating efficiently and effectively. Performance evaluation involves measuring and analyzing various parameters related to the treatment process, while monitoring involves regularly tracking these parameters over time. The following are some key parameters that can be monitored and evaluated to assess the performance of microbial technologies for wastewater treatment. Monitoring the influent wastewater characteristics, such as pH, temperature, organic matter, nitrogen, and phosphorus, is critical to understanding the treatment process and identifying potential issues. The process parameters, such as HRT, SRT, aeration rate, dissolved oxygen, and biomass concentration, should be monitored to ensure the treatment process is optimally functioning. The effluent quality parameters, such as BOD, total nitrogen (TN), total phosphorus (TP), and total suspended solids (TSS), can be used for the assessment of the effectiveness of the treatment process. This will ensure that the effluent meets regulatory standards (Lorenzo et al., 2020).

Microbial activity can be assessed by measuring oxygen uptake rate (OUR) and specific oxygen uptake rate (SOUR) parameters which provide information about the activity and health of the microbial community. Additionally, monitoring energy and resource consumption, such as electricity and chemical usage, can help identify optimization and cost-reduction opportunities. Establishing a monitoring schedule and a system for recording and analyzing data is important to ensure accuracy and consistency. This can be done manually or using automated monitoring systems, for example, the supervisory control and data acquisition (SCADA) system. In addition to regular monitoring, performance evaluations can be conducted periodically to assess the overall performance of the treatment system. This can involve conducting bench-scale or pilot-scale testing to simulate different operating conditions or to evaluate the effectiveness of various treatment processes or technologies. Modeling and simulation can also be used to predict treatment performance and optimize the treatment process. Overall, performance evaluation and monitoring microbial technologies for wastewater treatment are critical to maintaining optimal treatment performance, meeting regulatory requirements, and identifying optimization and cost-reduction opportunities. Regular monitoring and performance evaluations can help identify issues early and ensure the treatment system functions efficiently and effectively (Jones et al., 2012).

11.6 STUDIES AND PRACTICES OF MICROBIAL TECHNOLOGIES FOR WASTEWATER TREATMENT

11.6.1 Microbial Technologies of Urban and Rural Areas' Wastewater Treatment

Microbial technologies have been widely studied and applied for wastewater treatment in urban and rural areas. However, the application and optimization of microbial technologies may change depending on the conditions of treatment sites, such as wastewater composition, population density, available resources, and regulations. Urban areas typically generate large volumes of wastewater with high organic loads and nutrient concentrations. As a result, microbial technologies such as activated sludge, MBRs, and SBRs have been commonly used for treating municipal wastewater in urban areas. These technologies can achieve high removal efficiencies for organic substances, nitrogen, and phosphorus and produce standard effluent to be safely discharged or reused.

On the other hand, rural areas often face challenges such as limited access to infrastructure, low population density, and lack of financial resources, making wastewater treatment difficult. Decentralized treatment systems such as anaerobic digesters, constructed wetlands, and soil infiltration systems have been widely used in rural areas. These technologies have low energy and maintenance costs and can effectively treat wastewater while producing valuable byproducts such as biogas and fertilizers. There has been a growing interest in combining microbial technologies with other sustainable practices such as resource recovery, water reuse, and circular economy approaches in recent years. Algal-based systems, for example, can be used for nutrient removal and bioenergy production while producing valuable biomass for animal feed or fertilizer. In addition, energy and resources in wastewater can be recovered by using microbial electrochemical technologies (METs) to convert organic substances into electricity or hydrogen. To successfully apply microbial technologies for wastewater treatment in both urban and rural areas, it is important to consider the site's specific conditions and requirements and optimize the treatment process based on local conditions. Regular monitoring and performance evaluation are critical to ensuring the treatment system operates efficiently and effectively (Wang et al., 2011; Michael et al., 2013).

11.6.2 Microbial Technologies for Industrial Wastewater Treatment

Industrial wastewater is typically more complex and variable than municipal wastewater, as it contains various organic and inorganic pollutants that are challenging to treat. Microbial technologies offer an effective and sustainable solution for industrial wastewater treatment through the activity of microorganisms to degrade and remove pollutants. Below are selected potential microbial technologies used for the treatment of industrial wastewater (Pandey et al., 2017):

a. Anaerobic digestion: The process by which anaerobic microorganisms break down organic substances in the absence of oxygen. This technology is commonly used to treat high-strength industrial wastewater, such as that generated in the food and beverage, chemical, and pharmaceutical industries. Anaerobic digestion can result in highly efficient organic substance removal while producing biogas that can be used for energy generation.
b. Aerobic treatment: On the contrary, this is a microbial process that breaks down organic substances in the presence of oxygen. This technology effectively treats wastewater with high organic loads and is commonly used in the pulp, paper, textile, and food and beverage industries. Aerobic treatment systems include activated sludge, MBRs, and SBRs.
c. Biofiltration: Biofiltration is a technology that uses solid media-attached microorganisms to treat wastewater. The solid media provides a sufficient surface area for the growth of microorganisms. While the wastewater is passed through the bed, the microorganisms degrade and remove pollutants. Biofiltration is commonly used in the petrochemical and pharmaceutical industries.
d. METs: METs are a group of technologies that use electrochemical reactions to promote microbial activity for wastewater treatment. METs can produce electricity, hydrogen, or other valuable products while simultaneously

treating wastewater. This technology is still in the development stage but has the potential to be widely used in the future.

In conclusion, microbial technologies offer a sustainable and effective solution for industrial wastewater treatment. By harnessing the natural ability of microorganisms, these technologies can remove pollutants from wastewater and produce valuable byproducts such as biogas and electricity. The choice of technology depends on the specific wastewater characteristics, site conditions, and regulatory requirements.

11.6.3 Microbial Technologies for Hospital Wastewater Treatment

Hospital wastewater comprises a complex mixture of domestic, medical, and chemical disinfectants, making it a challenging wastewater stream to treat. Microbial technologies offer an effective and sustainable solution for hospital wastewater treatment by utilizing the natural ability of microorganisms to degrade and remove pollutants. Activated sludge is a commonly used biological treatment technology that effectively treats hospital wastewater. The wastewater is mixed with a population of microorganisms in a tank, which degrades and removes organic substances, nitrogen, and phosphorus from the wastewater. The treated effluent can then be discharged to the environment or reused for nonpotable purposes. Other methods are MBRs combining membrane filtration with biological treatment to make effluent of a standard quality. MBRs are commonly used in hospital wastewater treatment due to their ability to remove contaminants such as bacteria, viruses, and suspended solids. The treated effluent can be harmlessly discharged or reused for nonpotable purposes.

Additionally, constructed wetlands are a natural treatment technology that uses microorganisms, plants, and soil to treat wastewater. Hospital wastewater is treated as it passes through a bed of gravel and plants, which absorb nutrients and remove pollutants. This can effectively treat hospital wastewater while providing benefits such as wildlife habitats and aesthetic value. Ozonation is a chemical treatment technology that effectively removes pathogens and other pollutants from hospital wastewater. The wastewater is treated with ozone, which breaks down organic matter and disinfects the wastewater. Ozonation can be combined with other microbial technologies to effectively treat hospital wastewater (Kontominas et al., 2021; Kamali et al., 2023).

METs are a group of technologies that use electrochemical reactions to promote microbial activity for wastewater treatment. This can produce electricity, hydrogen, or other valuable products while simultaneously treating wastewater. This technology is still in the development stage but can potentially be used in hospital wastewater treatment. These technologies offer a sustainable and effective solution for hospital wastewater treatment. The choice of technology depends on the specific wastewater characteristics, site conditions, and regulatory requirements. Integrating different microbial technologies and chemical treatments can achieve effective treatment results for hospital wastewater, producing standard-quality effluent that can be discharged without harm or reused for nonpotable purposes (Ramírez-Vargas et al., 2018).

11.6.4 Microbial Technologies for Aquatic Ecosystem Wastewater Treatment

Aquatic ecosystems, such as lakes, rivers, and estuaries, are essential resources for human society and the environment. Nevertheless, wastewater discharge from various sources can cause their severe pollution and ecological degradation. Microbial technologies offer a sustainable and effective solution for aquatic ecosystem wastewater treatment by utilizing the natural ability of microorganisms to degrade and remove pollutants. Bioremediation is a natural treatment technology that uses microorganisms to degrade and remove pollutants from wastewater. It can be applied in situ, where the contaminated water is treated in its natural location, or ex situ, where the contaminated water is transported to a treatment plant. Bioremediation effectively treats a wide range of contaminants, including organic and inorganic pollutants (Muduli et al., 2021).

Algal-based treatment is a technology that uses microalgae to remove pollutants from wastewater. Microalgae are photosynthetic microorganisms that can multiply and consume nutrients from wastewater. The treated wastewater can then be utilized for irrigation, aquaculture, or discharged into the environment. Algal-based treatment offers a sustainable solution for aquatic ecosystem wastewater treatment, producing valuable byproducts such as biofuels and fertilizer. Biofilm-based treatment is a technology that uses microbial communities attached to a solid surface to treat wastewater (Hu & Li, 2021). Biofilms are formed by microorganisms that attach to the surfaces and produce an EPS matrix. The EPS matrix protects the microorganisms and enhances their ability to degrade and remove pollutants from wastewater. Biofilm-based treatment can be applied in both fixed-film and moving-bed systems. Constructed wetlands are a natural treatment technology that use a combination of microorganisms, plants, and soil to treat wastewater. The wastewater is treated as it passes through a bed of gravel and plants, which absorb nutrients and remove pollutants. Constructed wetlands can effectively treat aquatic ecosystem wastewater while providing benefits such as wildlife habitats and aesthetic value (Vymazal, 2010; Wu et al., 2015). METs are another group of technologies that use electrochemical reactions to promote microbial activity for wastewater treatment. METs can produce electricity, hydrogen, or other valuable products while simultaneously treating wastewater. This technology is still in the development stage but can potentially be used in aquatic ecosystem wastewater treatment in the future (Li & Yu, 2016).

11.6.5 Microbial Technologies for Decentralization and Small-scale Wastewater Treatment

Decentralized and small-scale wastewater treatment systems are gaining popularity due to their numerous advantages over traditional centralized systems. These systems are cost-effective, flexible, and energy-efficient, ideal for treating wastewater in rural areas, small communities, and individual households. Microbial technologies play a crucial role in these systems by providing efficient and sustainable solutions for treating wastewater on a small scale.

Anaerobic digestion is a common microbial technology in decentralized and small-scale wastewater treatment systems that takes place without oxygen. It involves using microorganisms to break down the organic substances into biogas and nutrient-rich effluent. Anaerobic digestion is particularly suitable for treating high-strength organic wastewater from households and food processing plants. The produced biogas is a source of renewable energy for heating or electricity generation. In small-scale systems, anaerobic digestion can be achieved using straightforward technologies such as septic tanks or biodigesters (Yang et al., 2013).

Constructed wetlands are another microbial technology used in decentralized wastewater treatment systems on a smaller scale by the use of a combination of physical, chemical, and biological processes. The wetlands are designed to treat wastewater from individual households or small communities, and can remove organic substances, nutrients, and pathogenic microorganisms and produce reusable effluent. Constructed wetlands are low cost and low maintenance, making them ideal for small-scale applications (Kamali et al., 2023).

Bioelectrochemical systems (BES) use microorganisms to generate electricity or other valuable products from wastewater. BES can be used in decentralized wastewater treatment systems to produce renewable energy while simultaneously treating wastewater. BES technology is still in the development stage, but it has the potential to be used in wastewater treatment systems on a small scale in the future (Chin et al., 2022).

11.6.6 Microbial Technologies for Water Reuse and Energy Recovery

Microbial technologies are increasingly being used in water reuse and energy recovery applications. These technologies use microorganisms to remove or degrade pollutants and produce renewable energy sources from the organic substances that are present in the wastewater. The RO membrane is the most common microbial technology used in water reuse applications. RO membranes use a physical barrier to remove contaminants, such as salts, from wastewater. However, these membranes can become fouled with organic matter, reducing efficiency. Microbial technologies, such as biofouling control and biofilm control, are used to reduce membrane fouling and improve the capability of RO membranes to overcome this issue (Bui et al., 2019).

Another microbial technology used in water reuse applications is biological treatments that use microorganisms to break down organic substances and remove nutrients from wastewater. This process produces an effluent that can be reused for nonpotable applications, such as irrigation, industrial processes, and toilet flushing. Biological treatment systems can be operated in a small footprint and are energy-efficient, making them ideal for water reuse applications (Jones et al., 2012).

Anaerobic digestion is another microbial technology used for water reuse and energy recovery. In this process, microorganisms break down organic substances in the wastewater to produce biogas and an effluent rich in nutrients. Biogas can be used as a renewable energy source for heating or electricity generation, while nutrient-rich effluent can be reused for nonpotable applications.

MECs are another microbial technology used for energy recovery from wastewater. In MECs, microorganisms oxidize organic substances and release electrons, which are then transferred to an electrode to generate electricity. This process can simultaneously treat high-strength wastewater and produce renewable energy. Microbial technologies offer a range of solutions for water reuse and energy recovery applications, and these technologies can be adapted to suit different wastewater types and treatment needs. Microbial technologies in water reuse and energy recovery applications can help reduce the pressure on freshwater resources and provide sustainable solutions for energy production (Kamali et al., 2023).

11.7 FUTURE DIRECTIONS AND CHALLENGES OF MICROBIAL TECHNOLOGIES IN WASTEWATER TREATMENT

11.7.1 Emerging Microbial Technologies for Wastewater Treatment

Emerging microbial technology for the treatment of wastewater refers to innovative and cutting-edge methods being developed and tested to improve the efficiency and effectiveness of wastewater treatment. Some of these emerging technologies include:

a. MFCs: MFCs use microorganisms to convert the chemical energy in organic substances into electrical energy. The microorganisms break down the organic matter, producing electrons that can be harvested to generate electricity. MFCs have the potential to provide an environmentally sustainable and cost-effective solution for treating wastewater and producing electricity (Waqas et al., 2020).
b. Membrane distillation: This technology uses a hydrophobic membrane that is permeable to water vapor but not liquid water to separate wastewater from clean water. The wastewater is heated, and the water vapor passes through the membrane, leaving behind the contaminants. This emerging technology is energy-efficient and has the potential to produce high-quality water for reuse (Chin et al., 2022).
c. Electrocoagulation: This technology uses an electrical current to destabilize and coagulate contaminants in wastewater. The coagulated particles can then be removed through sedimentation or filtration. Electrocoagulation has the potential to be an effective and efficient method for treating a variety of wastewater types (Krishnan et al., 2023).
d. Microbial desalination cells: This technology uses microorganisms to remove salt from seawater or brackish water while simultaneously producing electricity. The microorganisms generate a proton gradient that drives salt ions through a membrane, leaving behind fresh water. This emerging technology can potentially be an energy-efficient and cost-effective solution for produ cing freshwater from seawater.
e. Anammox: Anammox is a type of microbial technology that uses specific microorganisms to remove nitrogen from wastewater without needing oxygen or carbon sources. This technology is efficient and cost-effective for treating wastewater with high nitrogen concentrations (Al-Hazmi et al., 2022).

11.7.2 Efficient Wastewater Treatment with Microbial Communities and Microbiome Technologies

Efficient wastewater treatment can be achieved using microbial communities and microbiome technologies. Microbial communities are complex collections of microorganisms that work together to break down organic substances and remove pollutants from wastewater. Microbiome technologies use advanced molecular techniques to analyze and manipulate microbial communities to optimize their performance. The following are some ways microbial communities and microbiome technologies can be used to achieve efficient wastewater treatment. The first is bioaugmentation, which is the process of adding specific microbial cultures to a wastewater treatment system to enhance its performance. The added cultures can help degrade specific pollutants or improve the efficiency of the treatment process. This approach can be particularly useful in treating wastewater that contains high levels of recalcitrant pollutants. Microbiome analysis involves using advanced molecular techniques to identify and analyze the microorganisms present in a wastewater treatment system. This approach can help identify key microbial players and determine how they interact with each other and the wastewater. This information can be used to optimize the performance of the system and develop more efficient treatment strategies. Biofilm-based treatments are communities of microorganisms that attach to a surface and form a complex matrix. Biofilm-based treatment systems use biofilms to remove pollutants from wastewater. This approach can be particularly useful in treating wastewater that contains high levels of organic matter or nutrients (Paddock et al., 2020).

11.7.3 Regulatory and Social Barriers to the Adoption of Microbial Technologies

Despite the potential benefits of microbial technologies for wastewater treatment, several regulatory and social barriers can hinder their widespread adoption. Microbial technologies often face regulatory obstacles that make it difficult for them to be adopted on a large scale. For example, there may be strict regulations governing the use of genetically modified microorganisms, which can limit their use in wastewater treatment. Additionally, regulatory agencies may hesitate to approve new technologies that have not been extensively tested, which can slow adoption. There may be concerns about the safety of using microorganisms to treat wastewater, particularly if the microorganisms are genetically modified.

Additionally, some people may prefer traditional treatment methods and resist change. Microbial technologies can be more expensive to implement than traditional treatment methods, particularly if they require specialized equipment or extensive testing. This can make it difficult for smaller municipalities or organizations with limited resources to adopt these technologies. Microbial technologies may require specialized knowledge and training, which can be a barrier for some organizations. There may be a shortage of trained professionals who can design, implement, and maintain these systems. Regarding infrastructure, some microbial technologies may require specific infrastructure or modifications to existing infrastructure, which can be costly and time-consuming. For example, anaerobic digestion systems may require additional storage capacity for biogas, or membrane bioreactors may need extra space for the membrane filtration system. Overall, various regulatory and social barriers can hinder the adoption of microbial technologies for wastewater treatment. Collaboration between regulatory agencies and industry professionals will be required to address these barriers, promote the benefits of these technologies, and address concerns (Capodaglio, 2020; Robbins et al., 2022).

11.8 CONCLUSIONS AND RECOMMENDATIONS

Microbial technologies have been increasingly recognized as innovative and sustainable solutions for the treatment of wastewater. They exhibit the potential to offer significant advantages compared to traditional treatment methods, such as reduced energy consumption, increased efficiency, and the production of valuable byproducts such as renewable energy.

One of the key advantages of microbial technologies is their ability to treat a wide range of pollutants. Microorganisms naturally occur in wastewater and can break down organic matter, remove nutrients, and reduce pathogens. By harnessing the power of these microorganisms using microbial technologies, wastewater treatment plants can achieve higher levels of treatment and produce higher-quality effluent. Microbial technologies have also shown great potential in reducing the environmental influence resulting from wastewater treatment. For example, anaerobic digestion systems can produce biogas, a renewable energy source, as a byproduct of the treatment process. This can help reduce a wastewater treatment plant's carbon footprint and promote sustainability.

However, several barriers exist to adopting microbial technologies for wastewater treatment. As mentioned earlier, regulatory and social obstacles can limit their widespread adoption. Furthermore, cost is also a significant factor, particularly for smaller municipalities or organizations with limited resources. It is crucial to continue investing in research and development to optimize the performance and reduce the costs of microbial technologies to address these challenges. This can include developing more efficient and cost-effective processes for cultivating and maintaining microbial communities and investigating the potential of new microbial species or strains for wastewater treatment. Another key strategy is promoting education and training programs that prepare professionals to work with microbial technologies in wastewater treatment. This can help develop a skilled workforce that can design, implement, and maintain these systems, which is critical for long-term success. In conclusion, microbial technologies offer a promising pathway for achieving more sustainable and effective wastewater treatment. By addressing the challenges and barriers to adoption and promoting research, education, and collaboration, we can accelerate the adoption of these technologies and create a more sustainable future for our water resources.

REFERENCES

Al-Hazmi, Hussein E., Gamal K. Hassan, Mojtaba Maktabifard, Dominika Grubba, Joanna Majtacz, and Jacek Mąkinia. 2022. "Integrating Conventional Nitrogen Removal with Anammox in Wastewater Treatment Systems: Microbial Metabolism, Sustainability and Challenges." *Environmental Research* 215. https://doi.org/10.1016/j.envres.2022.114432.

Anekwe, Ifeanyi Michael Smarte, Jeremiah Adedeji, Stephen Okiemute Akpasi, and Sammy Lewis Kiambi. 2016. "Available Technologies for Wastewater Treatment." *Intech* 11 (tourism): 13. www.intechopen.com/books/advanced-biometric-technologies/liveness-detection-in-biometrics.

Angelakis, Andreas N., Andrea G. Capodaglio, and Emmanuel G. Dialynas. 2023. "Wastewater Management: From Ancient Greece to Modern Times and Future." *Water (Switzerland)* 15 (1): 1–26. https://doi.org/10.3390/w15010043.

Biase, Alessandro di, Maciej S. Kowalski, Tanner R. Devlin, and Jan A. Oleszkiewicz. 2019. "Moving Bed Biofilm Reactor Technology in Municipal Wastewater Treatment: A Review." *Journal of Environmental Management* 247: 849–66. https://doi.org/10.1016/j.jenvman.2019.06.053.

Bui, Xuan-Thanh, Chart Chiemchaisri, Takahiro Fujioka, and Suita Varjani. 2019. *Water and Wastewater Treatment Technologies*. https://doi.org/10.1007/978-981-13-3259-3_18.

Bunce, Joshua T., Edmond Ndam, Irina D. Ofiteru, Andrew Moore, and David W. Graham. 2018. "A Review of Phosphorus Removal Technologies and Their Applicability to Small-Scale Domestic Wastewater Treatment Systems." *Frontiers in Environmental Science* 6: 1–15. https://doi.org/10.3389/fenvs.2018.00008.

Cai, Wenfang, Fei Long, Yunhai Wang, Hong Liu, and Kun Guo. 2021. "Enhancement of Microbiome Management by Machine Learning for Biological Wastewater Treatment." *Microbial Biotechnology* 14 (1): 59–62. https://doi.org/10.1111/1751-7915.13707.

Capodaglio, Andrea G. 2020. "Fit-for-Purpose Urban Wastewater Reuse: Analysis of Issues and Available Technologies for Sustainable Multiple Barrier Approaches." *Critical Reviews in Environmental Science and Technology* 51 (15): 1–48. https://doi.org/10.1080/10643389.2020.1763231.

Carlsen, Lars, and Rainer Bruggemann. 2022. "The 17 United Nations' Sustainable Development Goals: A Status by 2020." *International Journal of Sustainable Development and World Ecology* 29 (3): 219–29. https://doi.org/10.1080/13504509.2021.1948456.

Chai, Hongxiang, Yu Xiang, Rong Chen, Zhiyu Shao, Li Gu, Li Li, and Qiang He. 2019. "Enhanced Simultaneous Nitrification and Denitrification in Treating Low Carbon-to-Nitrogen Ratio Wastewater: Treatment Performance and Nitrogen Removal Pathway." *Bioresource Technology* 280: 51–58. https://doi.org/10.1016/j.biortech.2019.02.022.

Chin, Min Yee, Zhen Xin Phuang, Kok Sin Woon, Marlia M. Hanafiah, Zhen Zhang, and Xiaoming Liu. 2022. "Life Cycle Assessment of Bioelectrochemical and Integrated Microbial Fuel Cell Systems for Sustainable Wastewater Treatment and Resource Recovery." *Journal of Environmental Management* 320: 115778. https://doi.org/10.1016/j.jenvman.2022.115778.

Damore, James A., and Jeff Gore. 2013. "Understanding Microbial Cooperation." *Journal of Theoretical Biology* 299 (1): 31–41. https://doi.org/10.1038/nature08365.Reconstructing.

Das, Surajit, and Hirak Ranjan Dash. 2019. *Microbial Diversity in the Genomic Era*. Academic Press. https://doi.org/10.1016/C2017-0-01759-7

Du, Rui, Shenbin Cao, Yongzhen Peng, Hanyu Zhang, and Shuying Wang. 2019. "Combined Partial Denitrification (PD)-Anammox: A Method for High Nitrate Wastewater

Treatment." *Environment International* 126: 707–16. https://doi.org/10.1016/j.envint.2019.03.007.

Dutta, Aparna, and Sudipta Sarkar. 2015. "Sequencing Batch Reactor for Wastewater Treatment: Recent Advances." *Current Pollution Reports* 1 (3): 177–90. https://doi.org/10.1007/s40726-015-0016-y.

Fofana, Rahil, Michael Parsons, Chenghua Long, Kartik Chandran, Kimberly Jones, Stephanie Klaus, Bob Trovato, Chris Wilson, Haydee De Clippeleir, and Charles Bott. 2022. "Full-Scale Transition from Denitrification to Partial Denitrification–Anammox (PdNA) in Deep-Bed Filters: Operational Strategies for and Benefits of PdNA Implementation." *Water Environment Research* 94 (5). https://doi.org/https://doi.org/10.1002/wer.10727.

Fournier, Pierre Edouard, J. Stephen Dumler, Gilbert Greub, Jianzhi Zhang, Yimin Wu, and Didier Raoult. 2003. "Gene Sequence-Based Criteria for Identification of New Rickettsia Isolates and Description of Rickettsia Heilongjiangensis Sp. Nov." *Journal of Clinical Microbiology* 41 (12): 5456–65. https://doi.org/10.1128/JCM.41.12.5456-5465.2003.

Gude, Veera Gnaneswa. 2015. "A New Perspective on Microbiome and Resource Management in Wastewater Systems." *Journal of Biotechnology & Biomaterials* 05 (02). https://doi.org/10.4172/2155-952x.1000184.

Goli, Amin, Ahmad Shamiri, Susan Khosroyar, Amirreza Talaiekhozani, Reza Sanaye, and Kourosh Azizi. 2019. "A Review on Different Aerobic and Anaerobic Treatment Methods in Dairy Industry Wastewater." *Journal of Environmental Treatment Techniques* 7 (1): 113–41.

Gonzalez, Juan M., and Cesareo Saiz-Jimenez. 2005. "A Simple Fluorimetric Method for the Estimation of DNA-DNA Relatedness between Closely Related Microorganisms by Thermal Denaturation Temperatures." *Extremophiles* 9 (1): 75–79. https://doi.org/10.1007/s00792-004-0417-0.

Goswami, Lalit, R. Vinoth Kumar, Siddhartha Narayan Borah, N. Arul Manikandan, Kannan Pakshirajan, and G. Pugazhenthi. 2018. "Membrane Bioreactor and Integrated Membrane Bioreactor Systems for Micropollutant Removal from Wastewater: A Review." *Journal of Water Process Engineering* 26: 314–28. https://doi.org/10.1016/j.jwpe.2018.10.024.

Gude, Veera Gnaneswar. 2016. "Wastewater Treatment in Microbial Fuel Cells – An Overview." *Journal of Cleaner Production* 122: 287–307. https://doi.org/10.1016/j.jclepro.2016.02.022.

Hamoda, Mohamed F., and Hamed A. Al-Sharekh. 2000. "Performance of a Combined Biofilm-Suspended Growth System for Wastewater Treatment." *Water Science and Technology* 41 (1): 167–75. https://doi.org/10.2166/wst.2000.0026.

Hassard, Francis, Jeremy Biddle, Elise Cartmell, Bruce Jefferson, Sean Tyrrel, and Tom Stephenson. 2015. "Rotating Biological Contactors for Wastewater Treatment – A Review." *Process Safety and Environmental Protection* 94 (C): 285–306. https://doi.org/10.1016/j.psep.2014.07.003.

Hu, Ming, and Lei Li. 2021. "Treatment Technology of Microbial Landscape Aquatic Plants for Water Pollution." *Advances in Materials Science and Engineering* 2021. https://doi.org/10.1155/2021/4409913.

Hugenholtz, Philip, Adam Skarshewski, and Donovan H. Parks. 2016. "Genome-Based Microbial Taxonomy Coming of Age." *Cold Spring Harbor Perspectives in Biology* 8 (6): 1–12. https://doi.org/10.1101/cshperspect.a018085.

Jones, Kim D., Naga Yadavalli, Anand K. Karre, and Jan Paca. 2012. "Microbial Monitoring and Performance Evaluation for H 2S Biological Air Emissions Control at a Wastewater Lift Station in South Texas, USA." *Journal of Environmental Science and Health – Part*

A Toxic/Hazardous Substances and Environmental Engineering 47 (7): 949–63. https://doi.org/10.1080/10934529.2012.667294.

Ju, Feng, Karin Beck, Xiaole Yin, Andreas Maccagnan, Christa S. McArdell, Heinz P. Singer, David R. Johnson, Tong Zhang, and Helmut Bürgmann. 2019. "Wastewater Treatment Plant Resistomes Are Shaped by Bacterial Composition, Genetic Exchange, and Upregulated Expression in the Effluent Microbiomes." *ISME Journal* 13 (2): 346–60. https://doi.org/10.1038/s41396-018-0277-8.

Kamali, Mohammadreza, Yutong Guo, Tejraj M. Aminabhavi, Rouzbeh Abbassi, Raf Dewil, and Lise Appels. 2023. "Pathway towards the Commercialization of Sustainable Microbial Fuel Cell-Based Wastewater Treatment Technologies." *Renewable and Sustainable Energy Reviews* 173: 113095. https://doi.org/10.1016/j.rser.2022.113095.

Khan, Nadeem A., Awais Bokhari, Muhammad Mubashir, Jiří Jaromír Klemeš, Rachida El Morabet, Roohul Abad Khan, Majed Alsubih, et al. 2022. "Treatment of Hospital Wastewater with Submerged Aerobic Fixed Film Reactor Coupled with Tube-Settler." *Chemosphere* 286. https://doi.org/10.1016/j.chemosphere.2021.131838.

Khandaker, Shahjalal, Sudipto Das, Md Tofazzal Hossain, Aminul Islam, Mohammad Raza Miah, and Md Rabiul Awual. 2021. "Sustainable Approach for Wastewater Treatment Using Microbial Fuel Cells and Green Energy Generation – A Comprehensive Review." *Journal of Molecular Liquids* 344: 117795. https://doi.org/10.1016/j.molliq.2021.117795.

Khomela, Emmanuel Marang. 2018. "Innovative and Efficient Waste Water Technologies for Tsakane Waste Water Treatment Works," 106–10. 2018 IMESA Conference Port Elizabeth Conference 2018 - Institute of Municipal Engineering of Southern Africa (imesa.org.za))(PAPER #9 paper-9-page-106-110.pdf (imesa.org.za)

Konstantinidis, Konstantinos T., and James M. Tiedje. 2007. "Prokaryotic Taxonomy and Phylogeny in the Genomic Era: Advancements and Challenges Ahead." *Current Opinion in Microbiology* 10 (5): 504–9. https://doi.org/10.1016/j.mib.2007.08.006.

Kontominas, Michael G., Anastasia V. Badeka, Ioanna S. Kosma, and Cosmas I. Nathanailides. 2021. "Innovative Seafood Preservation Technologies: Recent Developments." *Animals* 11 (1): 1–40. https://doi.org/10.3390/ani11010092.

Krishnan, Radhakrishnan Yedhu, Sivasubramanian Manikandan, Ramasamy Subbaiya, Natchimuthu Karmegam, Woong Kim, and Muthusamy Govarthanan. 2023. "Recent Approaches and Advanced Wastewater Treatment Technologies for Mitigating Emerging Microplastics Contamination – A Critical Review." *Science of the Total Environment* 858 (October 2022): 159681. https://doi.org/10.1016/j.scitotenv.2022.159681.

Kudryavtsev, V. A., S. L. Rychkov, and A. V. Shatrov. 2019. "The Mathematical Modeling and Monitoring for Process of a Biological Sewage Treatment." *IOP Conference Series: Earth and Environmental Science* 321 (1). https://doi.org/10.1088/1755-1315/321/1/012013.

Kumar, V., Agrawal, S., Shahi, S. K., Motghare, A., Singh, S., and Ramamurthy, P. C. 2022a. "Bioremediation Potential of Newly Isolated *Bacillus albus* Strain VKDS9 for Decolourization and Detoxification of Biomethanated Distillery Effluent and Its Metabolites Characterization for Environmental Sustainability." *Environmental Technology & Innovation* 26, 102260. https://doi.org/10.1016/j.eti.2021.102260

Kumar, V., Bilal, M., and Ferreira, L. F. R. 2022b. "Editorial: Recent Trends in Integrated Wastewater Treatment for Sustainable Development." *Frontiers in Microbiology* 13, 846503. https://doi.org/10.3389/fmicb.2022.846503

Kumar, V., Thakur, I.S., Singh, A. K., and Shah, M. P. 2020. "Application of Metagenomics in Remediation of Contaminated Sites and Environmental Restoration." In: Shah M.,

Rodriguez-Couto S., and Sengor SS (Eds.), *Emerging Technologies in Environmental Bioremediation*. Elsevier. https://doi.org/10.1016/B978-0-12-819860-5.00008-0.

Kumar, V., Singh, K., Shah, M. P., Singh, A. K, Kumar, A., and Kumar, Y., 2021. "Application of Omics Technologies for Microbial Community Structure and Function Analysis in Contaminated Environment." In Shah, M. P., Sarkar, A., and Mandal, S. (Eds.), *Wastewater Treatment: Cutting Edge Molecular Tools, Techniques & Applied Aspects in Waste Water Treatment*. Elsevier. https://doi.org/10.1016/B978-0-12-821925-6.00013-7

Lechón-Alonso, Pablo, Tom Clegg, Jacob Cook, Thomas P. Smith, and Samraat Pawar. 2021. "The Role of Competition versus Cooperation in Microbial Community Coalescence." *PLoS Computational Biology* 17 (11): 1–15. https://doi.org/10.1371/journal.pcbi.1009584.

Li, Jun, Xin Hui Xing, and Bao Zhen Wang. 2003. "Characteristics of Phosphorus Removal from Wastewater by Biofilm Sequencing Batch Reactor (SBR)." *Biochemical Engineering Journal* 16 (3): 279–85. https://doi.org/10.1016/S1369-703X(03)00071-8.

Li, Wen Wei, and Han Qing Yu. 2016. "Advances in Energy-Producing Anaerobic Biotechnologies for Municipal Wastewater Treatment." *Engineering* 2 (4): 438–46. https://doi.org/10.1016/J.ENG.2016.04.017.

Lin, Ziyuan, Yingmu Wang, Wei Huang, Jiale Wang, Li Chen, Jian Zhou, and Qiang He. 2019. "Single-Stage Denitrifying Phosphorus Removal Biofilter Utilizing Intracellular Carbon Source for Advanced Nutrient Removal and Phosphorus Recovery." *Bioresource Technology* 277: 27–36. https://doi.org/10.1016/j.biortech.2019.01.025.

Lofrano, Giusy, and Jeanette Brown. 2010. "Wastewater Management through the Ages: A History of Mankind." *Science of the Total Environment* 408 (22): 5254–64. https://doi.org/10.1016/j.scitotenv.2010.07.062.

Loman, Nicholas J., and Mark J. Pallen. 2015. "Twenty Years of Bacterial Genome Sequencing." *Nature Reviews Microbiology* 13 (12): 787–94. https://doi.org/10.1038/nrmicro3565.

López-Vázquez, Carlos M., Christine M. Hooijmans, Damir Brdjanovic, Huub J. Gijzen, and Mark C.M. van Loosdrecht. 2008. "Factors Affecting the Microbial Populations at Full-Scale Enhanced Biological Phosphorus Removal (EBPR) Wastewater Treatment Plants in The Netherlands." *Water Research* 42 (10–11): 2349–60. https://doi.org/10.1016/j.watres.2008.01.001.

Lorenzo, Fernando, Maria Sanz-Puig, Ramón Bertó, and Enrique Orihuel. 2020. "Assessment of Performance of Two Rapid Methods for On-Site Control of Microbial and Biofilm Contamination." *Applied Sciences* 10 (3). https://doi.org/10.3390/app10030744.

Mahato, Nitish Kumar, Vipin Gupta, Priya Singh, Rashmi Kumari, Helianthous Verma, Charu Tripathi, Pooja Rani, et al. 2017. "Microbial Taxonomy in the Era of OMICS: Application of DNA Sequences, Computational Tools and Techniques." *Antonie van Leeuwenhoek, International Journal of General and Molecular Microbiology* 110 (10): 1357–71. https://doi.org/10.1007/s10482-017-0928-1.

Mahvi, A. H. 2008. "Sequencing Batch Reactor: A Promising Technology in Wastewater Treatment." *Iranian Journal of Environmental Science and Engineering* 5 (2): 79–90.

Maquet, C. 2020. "Wastewater Reuse: A Solution with a Future." *Field Actions Science Report* 2020: 64–69.

Melin, T., B. Jefferson, D. Bixio, C. Thoeye, W. De Wilde, J. De Koning, J. van der Graaf, and T. Wintgens. 2006. "Membrane Bioreactor Technology for Wastewater Treatment and Reuse." *Desalination* 187 (1–3): 271–82. https://doi.org/10.1016/j.desal.2005.04.086.

Michael, I., L. Rizzo, C. S. McArdell, C. M. Manaia, C. Merlin, T. Schwartz, C. Dagot, and D. Fatta-Kassinos. 2013. "Urban Wastewater Treatment Plants as Hotspots for the Release of Antibiotics in the Environment: A Review." *Water Research* 47 (3): 957–95. https://doi.org/10.1016/j.watres.2012.11.027.

Morgenroth, E., E. Arvin, and P. Vanrolleghem. 2002. "The Use of Mathematical Models in Teaching Wastewater Treatment Engineering." *Water Science and Technology* 45 (6): 229–33. https://doi.org/10.2166/wst.2002.0110.

Muduli, Monali, Vasavdutta Sonpal, Krutika Trivedi, Soumya Haldar, Madhava Anil Kumar, and Sanak Ray. 2021. *Enhanced Biological Phosphate Removal Process for Wastewater Treatment: A Sustainable Approach. Wastewater Treatment Reactors: Microbial Community Structure*. BV. https://doi.org/10.1016/B978-0-12-823991-9.00012-5.

Nancharaiah, Yarlagadda V., and Manjunath Sarvajith. 2019. "Aerobic Granular Sludge Process: A Fast Growing Biological Treatment for Sustainable Wastewater Treatment." *Current Opinion in Environmental Science and Health* 12: 57–65. https://doi.org/10.1016/j.coesh.2019.09.011.

Naz, Iffat, Devendra P. Saroj, Sadia Mumtaz, Naeem Ali, and Safia Ahmed. 2015. "Assessment of Biological Trickling Filter Systems with Various Packing Materials for Improved Wastewater Treatment." *Environmental Technology* 36 (4): 424–34. https://doi.org/10.1080/09593330.2014.951400.

Nguyen, Hien Thi Thu, Vang Quy Le, Aviaja Anna Hansen, Jeppe Lund Nielsen, and Per Halkjær Nielsen. 2011. "High Diversity and Abundance of Putative Polyphosphate-Accumulating Tetrasphaera-Related Bacteria in Activated Sludge Systems." *FEMS Microbiology Ecology* 76 (2): 256–67. https://doi.org/10.1111/j.1574-6941.2011.01049.x.

Nguyen, Hoang Dung, and Sandhya Babel. 2022. "Insights on Microbial Fuel Cells for Sustainable Biological Nitrogen Removal from Wastewater: A Review." *Environmental Research* 204 (PB): 112095. https://doi.org/10.1016/j.envres.2021.112095.

Niehaus, Lori, Ian Boland, Minghao Liu, Kevin Chen, David Fu, Catherine Henckel, Kaitlin Chaung, et al. 2019. "Microbial Coexistence through Chemical-Mediated Interactions." *Nature Communications* 10 (1). https://doi.org/10.1038/s41467-019-10062-x.

Oh, Sung T., Jung Rae Kim, Giuliano C. Premier, Tae Ho Lee, Changwon Kim, and William T. Sloan. 2010. "Sustainable Wastewater Treatment: How Might Microbial Fuel Cells Contribute." *Biotechnology Advances* 28 (6): 871–81. https://doi.org/10.1016/j.biotechadv.2010.07.008.

Paddock, Matthew B., Jesús Dionisio Fernández-Bayo, and Jean S. VanderGheynst. 2020. "The Effect of the Microalgae-Bacteria Microbiome on Wastewater Treatment and Biomass Production." *Applied Microbiology and Biotechnology* 104 (2): 893–905. https://doi.org/10.1007/s00253-019-10246-x.

Pandey, Kritika, Brajesh Singh, Ashutosh Kumar Pandey, Ishrat Jahan Badruddin, Srinath Pandey, Ved Kumar Mishra, and Prashant Ankur Jain. 2017. "Application of Microbial Enzymes in Industrial Waste Water Treatment." *International Journal of Current Microbiology and Applied Sciences* 6 (8): 1243–54. https://doi.org/10.20546/ijcmas.2017.608.151.

Peng, Yongzhen, and Guibing Zhu. 2006. "Biological Nitrogen Removal with Nitrification and Denitrification via Nitrite Pathway." *Applied Microbiology and Biotechnology* 73 (1): 15–26. https://doi.org/10.1007/s00253-006-0534-z.

Priya, A. K., Rekha Pachaiappan, P. Senthil Kumar, A. A. Jalil, Dai Viet N. Vo, and Saravanan Rajendran. 2021. "The War Using Microbes: A Sustainable Approach for Wastewater Management." *Environmental Pollution* 275: 116598. https://doi.org/10.1016/j.envpol.2021.116598.

Quansheng, Dong, Chen Qixing, Li Zehui, and Zeng Shouwei. 2017. "Application of Microbial Technology in Wastewater Treatment." *Progress in Applied Microbiology*, 3–6.

Ramírez-Vargas, Carlos A., Amanda Prado, Carlos A. Arias, Pedro N. Carvalho, Abraham Esteve-Núñez, and Hans Brix. 2018. "Microbial Electrochemical Technologies for Wastewater Treatment: Principles and Evolution from Microbial Fuel Cells to Bioelectrochemical-Based Constructed Wetlands." *Water* 10 (9): 1–29. https://doi.org/10.3390/w10091128.

Rendueles, Olaya, and Jean-marc Ghigo. 2015. "Mechanisms of Competition in Biofilm Communities." *Microbiology Spectrum* 3 (3): 0009. https://doi.org/10.1128/microbiolspec.MB-0009-2014.f1.

Rivas, A., I. Irizar, and E. Ayesa. 2008. "Model-Based Optimisation of Wastewater Treatment Plants Design." *Environmental Modelling and Software* 23 (4): 435–50. https://doi.org/10.1016/j.envsoft.2007.06.009.

Robbins, Cristian A., Xuewei Du, Thomas H. Bradley, Jason C. Quinn, Todd M. Bandhauer, Steven A. Conrad, Kenneth H. Carlson, and Tiezheng Tong. 2022. "Beyond Treatment Technology: Understanding Motivations and Barriers for Wastewater Treatment and Reuse in Unconventional Energy Production." *Resources, Conservation and Recycling* 177: 106011. https://doi.org/10.1016/j.resconrec.2021.106011.

Romalde, Jesús L., Sabela Balboa, and Antonio Ventosa. 2019. "Editorial: Microbial Taxonomy, Phylogeny and Biodiversity." *Frontiers in Microbiology* 10: 1–2. https://doi.org/10.3389/fmicb.2019.01324.

Seo, Young Ik, Ki Ho Hong, Se Hoon Kim, Duk Chang, Kyu Hwan Lee, and Young Do Kim. 2013. "Phosphorus Removal from Wastewater by Ionic Exchange Using a Surface-Modified Al Alloy Filter." *Journal of Industrial and Engineering Chemistry* 19 (3): 744–47. https://doi.org/10.1016/j.jiec.2012.11.008.

Shah, Maulin P. 2021. *Removal of Emerging Contaminants Through Microbial Processes. Removal of Emerging Contaminants Through Microbial Processes*. https://doi.org/10.1007/978-981-15-5901-3.

Sharma, Sumit, Saurabh Jyoti Sarma, and Joo Hwa Tay. 2019. *Aerobic Granulation in Wastewater Treatment: A General Overview. Microbial Wastewater Treatment*. Elsevier Inc. https://doi.org/10.1016/B978-0-12-816809-7.00004-X.

Sidhu, Chandni, Surendra Vikram, and Anil Kumar Pinnaka. 2017. "Unraveling the Microbial Interactions and Metabolic Potentials in Pre- and Post-Treated Sludge from a Wastewater Treatment Plant Using Metagenomic Studies." *Frontiers in Microbiology* 8: 1–10. https://doi.org/10.3389/fmicb.2017.01382.

Singh, Jay Shankar, Sumit Koushal, Arun Kumar, Shobhit R. Vimal, and Vijai K. Gupta. 2016. "Book Review: Microbial Inoculants in Sustainable Agricultural Productivity-Vol. II: Functional Application." *Frontiers in Microbiology* 7. https://doi.org/10.3389/fmicb.2016.02105.

Stokholm-Bjerregaard, Mikkel, Simon J. McIlroy, Marta Nierychlo, Søren M. Karst, Mads Albertsen, and Per H. Nielsen. 2017. "A Critical Assessment of the Microorganisms Proposed to Be Important to Enhanced Biological Phosphorus Removal in Full-Scale Wastewater Treatment Systems." *Frontiers in Microbiology* 8: 1–18. https://doi.org/10.3389/fmicb.2017.00718.

Thompson, Cristiane C., Luciane Chimetto, Robert A. Edwards, Jean Swings, Erko Stackebrandt, and Fabiano L. Thompson. 2013. "Microbial Genomic Taxonomy." *BMC Genomics* 14 (1). https://doi.org/10.1186/1471-2164-14-913.

Tindall, Brian J., R. Rosselló-Móra, H. J. Busse, W. Ludwig, and P. Kämpfer. 2010. "Notes on the Characterization of Prokaryote Strains for Taxonomic Purposes." *International Journal of Systematic and Evolutionary Microbiology* 60 (1): 249–66. https://doi.org/10.1099/ijs.0.016949-0.

Tshemese, Zikhona, Nirmala Deenadayalu, Linda Zikhona Linganiso, and Maggie Chetty. 2023. "An Overview of Biogas Production from Anaerobic Digestion and the Possibility of Using Sugarcane Wastewater and Municipal Solid Waste in a South African Context." *Applied System Innovation* 6 (1): 13. https://doi.org/10.3390/asi6010013.

Vymazal, Jan. 2010. "Constructed Wetlands for Wastewater Treatment." *Water* 2 (3): 530–49. https://doi.org/10.3390/w2030530.

Wagner, Michael, Alexander Loy, Regina Nogueira, Ulrike Purkhold, Natuschka Lee, and Holger Daims. 2002. "Microbial Community Composition and Function in Wastewater Treatment Plants." *Antonie van Leeuwenhoek, International Journal of General and Molecular Microbiology* 81 (1–4): 665–80. https://doi.org/10.1023/A:1020586312170.

Wang, Jicun, Shuai Zhao, Apurva Kakade, Saurabh Kulshreshtha, Xiangkai Li, and Pu Liu. 2019. "A Review on Microbial Electrocatalysis Systems Coupled with Membrane Bioreactor to Improve Wastewater Treatment." *Microorganisms* 7 (10). https://doi.org/10.3390/microorganisms7100372.

Wang, Jixiang, Zejiao Li, Qian Wang, Zhongfang Lei, Tian Yuan, Kazuya Shimizu, Zhenya Zhang, Yasuhisa Adachi, Duu-Jong Lee, and Rongzhi Chen. 2022. "Achieving Stably Enhanced Biological Phosphorus Removal from Aerobic Granular Sludge System via Phosphorus Rich Liquid Extraction during Anaerobic Period." *Bioresource Technology* 346 (1): 126–439.

Wang, Longmian, Feihong Guo, Zheng Zheng, Xingzhang Luo, and Jibiao Zhang. 2011. "Enhancement of Rural Domestic Sewage Treatment Performance, and Assessment of Microbial Community Diversity and Structure Using Tower Vermifiltration." *Bioresource Technology* 102 (20): 9462–70. https://doi.org/10.1016/j.biortech.2011.07.085.

Wani, A.K., Akhtar, N., Naqash, N., Chopra, C., Singh, R., Kumar, V., Kumar, S. Mulla, S.I., and Américo-Pinheiro, J.H.P. 2022. "Bioprospecting Culturable and Unculturable Microbial Consortia Through Metagenomics for Bioremediation." *Cleaner Chemical Engineering* 2, 100017. https://doi.org/10.1016/j.clce.2022.100017

Waqas, Sharjeel, Muhammad Roil Bilad, Zakaria Man, Yusuf Wibisono, Juhana Jaafar, Teuku Meurah Indra Mahlia, Asim Laeeq Khan, and Muhammad Aslam. 2020. "Recent Progress in Integrated Fixed-Film Activated Sludge Process for Wastewater Treatment: A Review." *Journal of Environmental Management* 268: 110718. https://doi.org/10.1016/j.jenvman.2020.110718.

Wigginton, Sara K., Elizabeth Q. Brannon, Patrick J. Kearns, Brittany V. Lancellotti, Alissa Cox, Serena Moseman-Valtierra, George W. Loomis, and Jose A. Amador. 2020. "Nitrifying and Denitrifying Microbial Communities in Centralized and Decentralized Biological Nitrogen Removing Wastewater Treatment Systems." *Water* 12 (6). https://doi.org/10.3390/w12061688.

Wu, Haiming, Jian Zhang, Huu Hao Ngo, Wenshan Guo, Zhen Hu, Shuang Liang, Jinlin Fan, and Hai Liu. 2015. "A Review on the Sustainability of Constructed Wetlands for Wastewater Treatment: Design and Operation." *Bioresource Technology* 175: 594–601. https://doi.org/10.1016/j.biortech.2014.10.068.

Xiao, Rui, Bing Jie Ni, Sitong Liu, and Huijie Lu. 2021. "Impacts of Organics on the Microbial Ecology of Wastewater Anammox Processes: Recent Advances and Meta-Analysis." *Water Research* 191: 116817. https://doi.org/10.1016/j.watres.2021.116817.

Yadav, Bhoomika, Ashutosh K. Pandey, Lalit R. Kumar, Rajwinder Kaur, Sravan K. Yellapu, Balasubramanian Sellamuthu, R. D. Tyagi, and Patrick Drogui. 2020. *Introduction to Wastewater Microbiology: Special Emphasis on Hospital Wastewater*. Vol. 21. Elsevier. http://journal.um-surabaya.ac.id/index.php/JKM/article/view/2203.

Yang, Yu, Chiqian Zhang, and Zhiqiang Hu. 2013. "Impact of Metallic and Metal Oxide Nanoparticles on Wastewater Treatment and Anaerobic Digestion." *Environmental Sciences: Processes and Impacts* 15 (1): 39–48. https://doi.org/10.1039/c2em30655g.

Yi, Huan, Minfang Li, Xiuqin Huo, Guangming Zeng, Cui Lai, Danlian Huang, Ziwen An, et al. 2020. "Recent Development of Advanced Biotechnology for Wastewater Treatment." *Critical Reviews in Biotechnology* 40 (1): 99–118. https://doi.org/10.1080/07388551.2019.1682964.

Yuan, Zhiguo, Steven Pratt, and Damien J. Batstone. 2012. "Phosphorus Recovery from Wastewater through Microbial Processes." *Current Opinion in Biotechnology* 23 (6): 878–83. https://doi.org/10.1016/j.copbio.2012.08.001.

Zeaiter, Zaher, Zhongxing Liang, and Didier Raoult. 2002. "Genetic Classification and Differentiation of Bartonella Species Based on Comparison of Partial FtsZ Gene Sequences." *Journal of Clinical Microbiology* 40 (10): 3641–47. https://doi.org/10.1128/JCM.40.10.3641-3647.2002.

12 Linear to Circular Economy for Sustainable Wastewater Treatment

Sehliselo Ndlovu and Anil Kumar

12.1 INTRODUCTION

Demand for water resources has rapidly accelerated due to the rise in global population and economy. It is reported that approximately 36% of the global population is residing in water-stressed areas (HLPW, 2018), which is further expected to rise to greater than 50% by the year 2050 (Rodriguez et al., 2020). The competing water needs are increasing pressure on sharing of water resources. In this context, wastewater can be a treasure that can help overcome the water crisis, scarcity, and security since 380 trillion L (380 billion m^3) of wastewater is being generated per annum across the globe. Furthermore, projections suggest a 51% increase in wastewater generation over current volumes (Qadir et al., 2020). Wastewater treatment (WWT) and reuse will further lower dependency on natural water resources, prevent their contamination with human and industrial waste as well as contribute to the "United Nations Sustainable Development Goals (SDGs)."

Traditionally, WWT follows a linear approach, whereby wastewater treatment plants (WWTPs) collect and process (remove pathogens and other contaminants) wastewater for disposal, just to meet the restriction limits set by regulatory authorities (Somoza-Tornos et al., 2019). However, along with WWT for environmental reasons, attention is currently being paid to resource recovery as wastewater contains plenty of energy, nutrients, and other valuable products, under the umbrella of a circular economy (CE). Reciprocal to a linear economy approach, i.e., take–consume–dispose, a CE with a regenerating spirit is necessary to progressively decouple the expansion of the wastewater sector from the consumption of finite resources. Harnessing resources from wastewater is an old concept that has been implemented successfully in many European countries to harness nitrogen and phosphate, prior to the development of WWT facilities. Currently, the same strategy is being revisited following advanced and sustainable techniques to address the resource crisis. Numerous methods including chemical, biological, physical, and hybrid have been developed to treat as well as valorize energy, nutrients, and value-added products embedded in wastewater (Kundu et al., 2022).

Wastewater treatment plants are central to our day-to-day life and capable of handling and processing significant wastewater volumes. The wastewater coming to WWTPs can be recycled and used for potable and nonpotable use. Generally, the

DOI: 10.1201/9781003441069-12

recycled water is used for nonpotable applications such as agricultural irrigation, toilet flushing, car washing, fisheries, aquaculture, fire protection, etc. However, in a few countries, wastewater is being utilized for potable use following high-level purification and treatment. Furthermore, several WWTPs are easily recovering energy and nutrients (discussed in this chapter). Recently, phosphorus recycling using struvite crystallization processes (NuReSys, Pearl, and AirPrex) has been employed on a large scale (Hukari et al., 2016). Another major byproduct of wastewater treatment is sewage sludge, which requires treatment before being disposed of. The sludge produced can be stabilized to biosolids, a nutrient and energy-rich product that can be utilized as a fertilizer or soil conditioner, or raw material for renewable energy production. The commonly employed technology for the treatment of sewage sludge is anaerobic digestion. In the US, WWTPs digest sewage sludge to produce biogas (Shen et al., 2015). While contemplating the recovery of resources, it is important to look beyond the boundaries and consider a larger context for WWT. This is because the quantities of resources a single plant can provide may be insufficient to be accepted as a resource factory, for example, production of bioplastics. Thus, there is a need for integrating wastewater management into the whole urban system when it comes to the recovery of resources. Integration can prove to be a more comprehensive strategy for wastewater management and potentially provide an opportunity for cost-economic solutions (Solon et al., 2019).

This chapter discusses various aspects of wastewater treatment using a linear and CE approach. Since resource recovery and maximizing their use is central to CE, the major resources embedded in wastewater are identified and technologies used to recover are presented. Further, WWTPs successfully recovering key resources and attaining sustainability and self-sufficiency are discussed with examples. Moreover, the possibility of reusing treated water for potable and nonpotable uses by economies is also discussed.

12.2 FROM A LINEAR TO A CIRCULAR ECONOMY MODEL

Most industries and processes that use water, for example, mining, manufacturing, agriculture, municipality, and even domestic use have been epitomized by an approach that focuses more on maximizing production and profitability rather than process efficiency. As a result, such an approach tends to lead to the generation of large volumes of wastewater. While this was viewed in the past as an unavoidable side effect of a modernizing society, the huge volumes of generated wastewater have, however, had an impact on the environment causing significant pollution on land and in some cases of freshwater reservoirs such as dams and rivers. This is especially so in areas where there is weak policing and monitoring of environmental regulations and legislation. Such an approach to the utilization of fresh water and the discard of the generated waste is associated with the linear economy model.

The linear economy model, with its focus on the concept of production, use, and discard assumes an infinite availability of the resources which the world's population needs to survive. However, as the demand for technological advancements in modernizing society increases, so does the demand for resources such as water and energy. Although water covers 71% of the Earth's surface, only 3% of that water is considered fresh, with the rest, i.e., 97%, found in oceans, which is thus too salty to drink or utilize in most residential and industrial applications. Recent events such as extreme climate changes, decreasing amounts of rainfall, geopolitical instabilities, etc. make it more difficult to access fresh water. These events have further shown that a continuous, unrestricted, and irresponsible consumption and utilization of water resources cannot be done without consideration for the developmental needs of current and future generations. Thus, it has become crucial to safeguard the uninterrupted supply of resources in a sustainable, economically viable way for the world to continue developing. This has led to an increased awareness of how resources are utilized, consumed, and discarded, thereby, supporting a push for sustainable development that is underpinned by the CE model.

Sustainable development concedes that industrial and economic growth must continue into the distant future, but in a responsible manner and at rates that allow sustenance for the generations of tomorrow (Brundtland, 1987). Because of its relative impact on boosting the economic potential of various industry sectors and protecting the environment, through approaches such as industrial ecology, industrial symbiosis, and cleaner production, it is therefore, not surprising that the CE model can be placed at the heart of sustainable development. A CE model (Fig. 12.1)

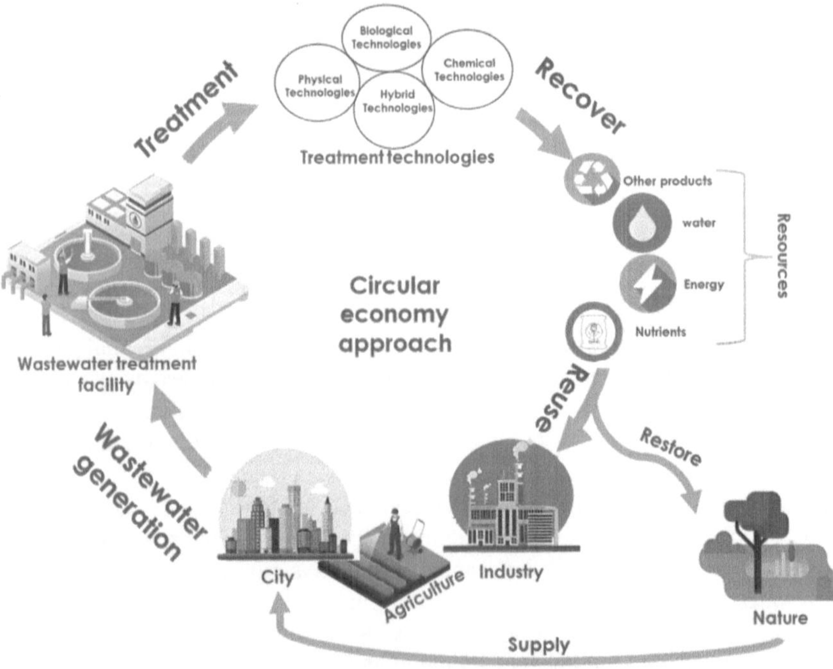

FIGURE 12.1 The circular economy approach for the possible harnessing of resources and treatment of wastewater.

Source: Adapted from Rodriguez et al. (2020); Delgado Martin et al. (2021); Morais et al. (2021).

is defined as one that extends the usage of resources for as long as possible, harnessing the maximum possible value from them while they are in use, and then extracting and revitalizing products and materials at the end of each service life as well as pushing the resources back into the economy to build economic, natural, and social capital (Kirchherr et al., 2017). It describes an economy in which, by design, waste and pollution do not exist and is thus based on principles such as: eliminating or reducing waste, recovering, circulating products and materials, and finally, regenerating nature. Based on the above definitions, wastewater can, therefore, be viewed as a valued resource rather than just waste. In this context, the circular water economy can be viewed as a model that provides sustainable production and consumption with the reclamation and reuse of water, and subsequent resource and product recovery.

According to Kirchherr et al. (2017), the shift from a linear to a CE can help to maintain high-quality resources and product value as well as reduce waste generation. In the CE approach for wastewater, opportunities are explored for wastewater to be reused and processed to generate resources and byproducts (water, energy, nutrients, etc.) that could be used in different applications and sectors. This model has thus, become a key discussion point for most industries because of the opportunities for reduced demand on primary resources. The CE concept, therefore, has been noted as having the advantage of not only providing a boost to the primary production industries but also helping to further de-risk these operations by providing alternative and/or secondary supply chains of materials and resources. In addition, it can also help achieve a "zero-waste society" by improving resource efficiency, minimizing pollution, and lowering emissions and hence, the adverse effects on the environment (Kirchherr et al., 2017).

The CE concept favors sustainable production, reasonable consumption, and efficient waste management, which all address the "United Nations Sustainable Development Goals (SDGs)." In the context of wastewater treatment (WWT), the CE concept, therefore, holds promise for accomplishing several SDGs, including SDGs 6 on clean water, 7 on energy, 8 on economic growth, 11 on sustainable cities, 12 on responsible consumption and production, 14 on oceans, and 15 on life on land. For example, CE activities involving small-scale WWT and purification, water reclamation and reuse, nutrient recovery, and biogas systems as highlighted in Table 12.1 can aid in expanding access to safe potable water and adequate sanitation, reduce pollution, and improve water quality which speaks to SDG6. At the same time, renewable energy approaches, such as small-scale biomass technologies and biofuels, and energy (heat) recovery highlighted in Table 12.1 all contribute to SDG7, while the act of decoupling economic activity from primary resources addresses SDG12. It is therefore imperative that if the world is to align with the goals of sustainable development, the CE model and its practices should be at the core of wastewater management and treatment. This model can be easily supported by designing for low or zero waste through the implementation of existing technologies and practices for the treatment of wastewater leading to product (nutrients) and resource (energy and water) recovery for sustainable economic growth and development.

TABLE 12.1
Commonly employed technologies to extract resources from wastewater

Technology	Description	Type of resource recovered	Reference
Anaerobic digestion	• Eco-friendly and cost-economic technology for sludge stabilization, degradation of biodegradable material, and production of biogas from wastewater • Organic matter of wastewater contains chemical energy which is then anaerobically digested to produce biogas (50–70% methane) • The methane produced (need to remove nonmethane components of biogas) can be fed into gas engines to generate electricity and thermal energy • Anaerobic digestion is a highly recommended approach for on-site energy recovery • The process of anaerobic digestion comprises four phases, i.e., hydrolysis, acidogenesis, acetogenesis, and methanogenesis	Biogas (mainly), hydrogen, electricity, and thermal energy	Elalami et al. (2019), Zarei (2020)
Heat exchanger	• This is a device that enables thermal energy (internal) transfer between two fluids while also preventing the mixing of these fluids • Requires no external energy • Types of heat exchangers used in wastewater heat recovery applications are plate and frame, double-pipe parallel flow, shell-and-tube, and double-pipe counterflow heat exchangers	Thermal energy	Nagpal et al. (2021)
Heat pump technology	• Uses electricity and reverse refrigeration to transport thermal energy from one place to another • Requires low-temperature energy followed by its conversion to usable heating energy through mechanical work • In WWTPs, wastewater is a low-temperature source of heat for the heat pumps • The types of heat pumps and their applications in wastewater heat recovery are thoroughly described by Hepbasli et al. (2014)	Thermal energy	Hepbasli et al. (2014), Grassi (2017)

TABLE 12.1 (Continued)
Commonly employed technologies to extract resources from wastewater

Technology	Description	Type of resource recovered	Reference
Gasification	• Biomass heating under limited oxygen conditions results in gasification and syngas (a combustion fuel) production • Using chemical or biological processes, syngas can be converted to liquid fuels • Using Fischer-Tropsch synthesis or through the Mobil methanol to gasoline process, syngas can be used to generate synthetic petroleum	Fuel (Syngas)	Burton et al. (2009)
Combustion	• Biomass/syngas combustion under excess oxygen supply leads to complete oxidation and the production of hot flue gases • The hot flue gases are generally used to generate steam to drive turbines to produce electricity (with 30% efficiency) • Combined heat and power (CHP) can boost the efficiency to 80% if the heat energy is also captured	Electricity	Burton et al. (2009)
Crystallization	• A technique based on solid-liquid separation where solute from liquid solution gets crystallized into a crystalline phase • Does not require supplementary materials like adsorbents, membrane catalysts, oxidation or reducing agents, ion-exchange resins, etc. • Can recover salts as well as water because it produces pure products from impure solution • Different crystallization techniques are applied for resource recovery from wastewater such as cooling, evaporation, drowning-out reaction, and membrane distillation crystallization	Nutrients	Lu et al. (2017)
Ammonia stripping	• A simple desorption technique to reduce ammonium content in wastewater • In the process, caustic or lime is added to the wastewater so that it reaches a pH of 10.8–11.5 standard units resulting in the formation of ammonia gas from ammonium hydroxide ions	Nutrients	EPA (2000)

(continued)

TABLE 12.1 (Continued)
Commonly employed technologies to extract resources from wastewater

Technology	Description	Type of resource recovered	Reference
Microalgae-based technologies	• A sustainable and economic approach within the context of a CE • Photobioreactor (open system and closed, or hybrid photobioreactors) is a key factor to consider for high efficiency and cost-effectivity • Most frequently used are open pond reactors for wastewater treatment • Nutrient recovery process uses the following steps: pretreatment, nutrient uptake and biomass generation, harvesting of biomass, and transforming biomass into end product • Before the final step, dewatering or drying is needed. The final step includes thermochemical (hydrothermal carbonization, pyrolysis, etc.) or biochemical processes (anaerobic digestion, transesterification, fermentation, etc.) depending upon the type of product required • In addition, algal biomass can be utilized for energy production, livestock, and agriculture use	Nutrients, biodiesel, bioethanol, biofertilizer, and feed (proteinaceous)	Alkarawi et al. (2018), Santos and Pires (2018), Robles et al. (2020)
Photosynthetic bacteria based (PSB) technology	• PSBs are classified as purple sulfur and nonsulfur bacteria, gliding filamentous green sulfur bacteria • In wastewater treatment, purple nonsulfur bacteria are mostly applied due to their versatile metabolism • The nutrient recovery process using PSB includes pretreatment, growth, separation, and downstream processing • Pre-fermentation is needed since PSB can only use micro-organic molecules like sugar, alcohol, and organic acids • Then, growth in a configured reactor and assimilation of C, N, and P • Most research inoculated PSB straight into the wastewater	Nutrients	Lu et al. (2019), Robles et al. (2020)

TABLE 12.1 (Continued)
Commonly employed technologies to extract resources from wastewater

Technology	Description	Type of resource recovered	Reference
	• The most frequently applied technology is photobioreactor technology • Commonly used separation processes are coagulation and centrifugation		
Microbial fuel cells (MFCs)	• A type of device that operates on electrochemical energy conversion • Microbes produce substances that are electrochemically active and thereafter react on the anode to form a biofilm which functions as a biocatalyst to generate electricity • MFC contains two chambers, i.e., an anodic (anaerobic part; maintained by nitrogen gas) and a cathodic chamber (aerobic; contains high potential electron acceptors, mainly O_2) separated by a proton exchange membrane • Wastewater is a suitable substrate because it contains a high organic load • A variety of electro-active microbes transform chemical energy contained in wastewater into electricity	Bioelectricity	Zhou et al. (2018), Nawaz et al. (2022), Tsekouras et al. (2022)
Electrodialysis (ED)	• Membrane-dependent separation method in which ions are selectively transferred via ion-exchange membranes under the influence of an electric field • Two streams with distinct concentrations can be successfully produced in separate compartments by alternately partitioning the anion- and cation-exchange membranes • Cost-effective • Desalinating municipal wastewater using ED has been proven cost-economic, thus enabling the reuse of water for gardening or irrigation, etc. • When coupled to forward osmosis, it can be effectively employed to recover high-quality water and to control the salinity build-up	Nutrients, salts, acid and bases, and metals	Zhang et al. (2013), Bohra et al. (2022)

(continued)

TABLE 12.1 (Continued)
Commonly employed technologies to extract resources from wastewater

Technology	Description	Type of resource recovered	Reference
Membrane filtration	• Separates two phases (soluble ions and solids) by selectively controlling the passage of components crossing via the membrane • The reverse osmosis technique is very efficient among all membrane filtration processes (99.5% removal of small waste particles) • Coupling membrane technologies with other techniques will enhance the removal/reclamation efficacy, e.g., membrane bioreactor	Nutrients	Xie et al. (2016)
Anaerobic membrane bioreactors (AnMBRs)	• Process where anaerobic digestion is integrated with membrane filtration • An MBR-mediated treatment biologically converts organics contained in wastewater to biogas (rich in methane) • Anaerobic treatment transforms nutrients into their chemical forms, i.e., phosphate and ammonia, thus can be precipitated for subsequent nutrient recovery	Energy and nutrient	Song et al. (2018)
Advanced oxidation processes (AOPs)	• Treats hazardous material of wastewater • AOPs work on the principle of hydroxyl radical formation that are highly reactive oxidants and are capable of degrading nondegradable pollutants, protozoa, bacteria, etc. • AOPs that have been applied widely include photocatalysis (UV/TiO_2) ultraviolet irradiation, ozonation, and sonolysis	Water for reuse	(Zhang and Huang, 2020)

12.3 RESOURCES IN WASTEWATER

In the recent past, with a paradigm shift from a linear to a CE, wastewater has begun to be seen as a resource rather than a waste stream. The potentially valued resources present in municipal wastewater include water, energy, nutrients, and other products (Fig. 12.1). Water is the most vital resource recoverable from WWT, and its reuse is crucial to a CE. Globally, 380 billion m^3 of wastewater is generated per year, which is anticipated to rise to 470 billion m^3 and 574 billion m^3 by 2030 and 2050, respectively (Qadir et al., 2020). These projections indicate that surplus amounts of wastewater will be available in the coming years, providing a prospect of alleviating water scarcity through WWT. According to Jones et al. (2021), of total global wastewater production, only 65% is collected and 52% is treated. Reclamation and reuse of water from WWTPs can effectively lower the freshwater consumption of a particular city or area. Presently, the intentional reuse of treated wastewater is projected to be 40.7 billion m^3 (Jones et al., 2021).

A significant portion of chemical energy that can be transformed into heat or electrical energy, i.e., in the form of organic matter, is contained in wastewater. The energy contained in wastewater is five times the amount needed in the treatment process of wastewater. According to estimates, 53.2 billion m^3 of methane (CH_4) with caloric values of 1908 MJ energy is embedded in 380 billion m^3 of wastewater. This energy is reported as being enough to supply electricity to approximately 158 million households, considering the average need of a household is 3350 kWh (Qadir et al., 2020).

Qadir et al. (2020) estimated the nutrient (nitrogen (N), phosphorus (P), and potassium (K)) potential of wastewater on the waste of 53 global wastewater datasets. These researchers reported that 16.6, 3.0, and 6.3 million metric tons (MMT) of N, P, and K are entrenched in the global wastewater generated annually. The extraction of nutrients entrenched in wastewater might generate $13.6 billion in revenue, globally (calculated based on the monthly average value of the year 2018) (World Bank, 2019; Qadir et al., 2020). Other products extractable from wastewater include cellulosic fibers, volatile fatty acids (VFAs), single-cell proteins (SCPs), extracellular polymeric substances, alginates, etc. However, data about the recovery route of these products remain scant.

12.3.1 Technologies for Resource Recovery from Wastewater

Management of wastewater plays a key role in the sustainable development of a city. Previously, wastewater was treated before release into the environment with the aim of protecting people from health risks. However, the increasing shortage in resource availability and water scarcity, a paradigm shift to a circular use of resources, and strict environmental regulations have focused attention toward the development of WWT technologies that can recover resources as well as enable water reuse. While scientists have widely researched and developed wastewater recycling and resource recovery technologies over the past few years, large-scale usage, however, remains limited. The most prevalent resource recovery from large-scale WWTPs is recovering energy as biogas or electricity produced from sludge. Among the two observed clusters of energy, i.e., biogas (1.3 to 2.9 MJ/m^3) and electricity (0.14 to 0.97 MJ/m^3), the electricity range has been less due to the low transition efficiency of electricity generators (Diaz-Elsayed et al., 2019). Further, the efficiency of energy recovery varies with technology, for example, approximately 6.95 MJ/m^3 of energy is retrievable from the organic matter of a typical municipal wastewater plant. On the other hand, the thermal energy that can be recovered through a heat pump is 25.20 MJ/m^3 although this also depends on ambient and wastewater temperature (McCarty et al., 2011). Moreover, it has been reported that for thermal energy, the on-site recovery approach is more suitable since it lowers the heat loss during the transfer of wastewater (Diaz-Elsayed et al., 2019). To valorize nutrients contained in wastewater, several technologies including chemical, biological, physical, and physiochemical have been developed (Meena et al., 2019). Compared to other methods, biological nutrient removal (BNR) holds advantages such as cost-effectiveness, low energy demand, less sludge production, and high removal efficiency. Further, while conventional activated sludge (CAS) treatment is the most widely employed technology for domestic and industrial WWT, it is limited by high energy demand, high capital cost, and offensive odor in comparison to BNR. In addition to the above, the bioelectrochemical approach is another promising technique applied to recover nutrients and generate bioelectricity from wastewater. Although removing personal care products and pharmaceuticals is a challenge in WWT, technologies like advanced oxidation processes, reverse osmosis, nanofiltration, and activated carbon have shown significant removal efficiency of selected contaminants. Some of the commonly employed resource recovery technologies to extract resources from wastewater are summarized in Table 12.1.

12.3.2 Harnessing Resources from Wastewater for a Circular Economy

The notion of re-source, re-make, and re-think to transform waste into value-added resources is becoming increasingly attractive. Previously, wastewater was not viewed as a source of immense benefits or use, it was just treated and disposed of, but now it is viewed as a valued resource in support of a CE (Puchongkawarin et al., 2015; Rao et al., 2017). Water supply is not the only rationale behind WWT, but also resource recovery and value addition, which will help WWTPs gain energy self-sufficiency and contribute to revenue generation. The key resources recovered given a CE from WWTPs are described below.

12.3.2.1 Energy Production

12.3.2.1.1 Biofuels

Although WWT consumes a significant portion of energy, it can be a major source of energy, since it contains chemical energy five to ten times more than that needed in CAS process operations (Shen et al., 2015; Qadir et al., 2020). This can help achieve energy-neutral WWTP or net energy-positive WWTPs. The energy embedded in wastewater can be harvested in the form of thermal energy, electrical energy, and biofuels, especially biogas (Bertanza et al., 2018). This is a cheap and clean fuel that has been successfully produced from wastewater (primary and secondary sewage sludge). Biogas is a mix of methane (50–70%), carbon dioxide (30–50%), and traces of gases like hydrogen sulfide (0.0001%), nitrogen (1–2%), hydrogen (5–10%), and water vapors (0.5%) (Burton et al., 2009; Zarei, 2020; Kundu et al., 2022). Biogas is produced by anaerobically digesting the sewage sludge, a byproduct formed during WWT (Shen et al., 2015). Zhen et al. (2017) documented that biogas produced through anaerobic digestion in WWTPs has tremendous energy potential (65% CH_4). They estimated approximately 40% less consumption of energy in WWTPs operating with anaerobic digestion than in WWTPs operating without sludge digestion. Moreover, the development of combined heat and power (CHP) systems has enabled biogas transformation into electricity at WWTPs (Bennett, 2007). It has been reported that biogas produced from wastewater in Sweden is mostly used for heat production (35%) followed by its use in electricity (26%), vehicle use (14%), torching (13%), and delivered to local gas networks (12%) (Jonasson & Ulf Jeppsson, 2007).

Co-digesting sewage sludge with other wastes (mainly organic) and WWTPs' thermal hydrolysis technologies is recommended to further improve the efficiency of anaerobic digestion (Hagos et al., 2017). Sewage sludge digestion along with other organic waste has been a general practice in Europe. In addition, 216 WWTPs were estimated to operate following the co-digestion approach in the USA. Shen et al. (2015) reported that the "East Bay Municipal Utility District (EBMUD)" WWT facility situated in Oakland, California, North America, was the first energy-neutral WWTP. The EBMUD wastewater treatment facility uses mixed organic waste (pretreated) to co-digest with sewage sludge in a thermophilic digester, enhancing the biogas yield by approximately 70% approx.

A further example of co-digestion includes the "Des Moines Metropolitan Wastewater Reclamation Authority (DMMWRA) WWTP," which treats wastewater from the Des Moines metropolitan area of Iowa, US. The plant operates on a co-digestion approach, whereby mixed feedstock containing high-strength waste [(HSW) biodiesel processing waste, whey, food processing waste, etc.] and approximately 42% of fat, oil, and grease is digested with municipal primary and secondary sludge. It was reported that DMMWRA-WWTP sells 40–50% of its biogas to Cargill, making an annual revenue of $460,000–800,000 (Shen et al., 2015). Further, in 2010, the plant was upgraded to accommodate a higher organic loading rate and produce more biogas energy. To further expand its power generation, i.e., from 1.8 MW to 4.8 MW, DMMWRA constructed two cogeneration units of 1.5 MW capacity. To utilize an excess of biogas, DMMWRA also proposed establishing a biogas-CNG (compressed natural gas) fueling station. Other examples include Austria's Strass WWTP, a net-energy-positive plant with an average flow rate of 6 MGD. Over two decades, the plant has implemented numerous process optimization strategies to attain self-sufficiency of 100% (Wett et al., 2007; Crawford, 2010). Similarly, WWTP in Prague, the Czech Republic increased biogas production from 15 to 23.5 KWh/(PE. year) using different approaches, and claimed 100% energy self-sufficiency (Jenicek et al., 2012, 2013).

As stated above, pretreatment using thermal hydrolysis technologies has the potential to boost biogas production. A few technologies including Cambi™ (thermal hydrolysis), BTA® process (hydromechanical screw-mill), and Exelys™ (thermal hydrolysis) have been successfully employed in large-scale WWTPs in the USA and Europe. For example, the Csepel WWTP is an advanced wastewater treatment plant in Budapest, Hungary that utilizes a BNR process to treat a usual flow of 93 MGD. Recently, the plant has incorporated a thermal hydrolysis system (Exelys™) and an additional digester operating at mesophilic temperature, thus creating a distinct Exelys™–Digestion–Lysis–Digestion configuration to boost biogas production and thereby, power generation. Using the new design, the self-sufficiency in terms of power was improved to 65% compared to the old design where power generation could only meet 49% of the demand from a biogas-fueled CHP facility (Gurieff et al., 2012). In view of the CE, biogas production from wastewater using anaerobic digestion is progressing toward an economical, practical, and sustainable solution.

Other than biogas, wastewater has been explored to produce bioethanol, biodiesel, and biohydrogen. The first step in bioethanol production is the fermentation of sludge by a variety of microorganisms followed by its purification using distillation. The distilled ethanol contains approximately 5% water and needs to be dehydrated to obtain high-grade bioethanol (Kundu et al., 2022). Similarly, biodiesel can also be produced from materials like waste sludge. The lipid content in wastewater sludge was reported as 5–20% (w/w dry sludge), which is equivalent to the lipid content of plant seeds (Chen et al., 2018). In addition, wastewater sludge is enriched with carbon, phosphorus, and nitrogen and hence, can serve as a suitable medium to cultivate microorganisms. Oleaginous microorganisms (capable of accumulating > 20% of their dry weight as oil) such as microalgae can be cultivated in large ponds of wastewater to accumulate lipids. The microalgae are harvested, dried, and subjected to oil extraction from algal biomass followed by transesterification of oil to produce biodiesel. However, several bottlenecks such as process parameters including temperature, pH, and hydraulic retention time for attaining the required concentration of VFAs and elimination of microbial contamination make biodiesel production from wastewater economically unfeasible (Turon et al., 2016; Puyol et al., 2017).

(Bio)hydrogen, which has a very high calorific value (~112 KJ/g), is another clean and reliable alternative fuel that can be generated from wastewater. Different processes like electro-hydrolysis and biological processes, i.e., autotrophic (biophotolysis) and heterotrophic (dark and photo fermentation) are rated as eco-friendly methods of biohydrogen production (Dincer & Acar, 2015). However, the only process that serves the double goal of energy recovery and WWT is dark fermentation (Puyol et al., 2017). Sharma et al. (2020) reported dark fermentation as being a key biological process to produce hydrogen from organic substrates. The organic material present in wastewater mainly contains cellulose and hemicellulose. The production of biohydrogen requires anaerobic bacteria during the fermentation of hemicelluloses and cellulose. Producing biohydrogen by dark fermentation is limited by its low yield as well as various other factors described by Puyol et al. (2017). However, the potential of dark fermentation to produce VFA-rich effluent makes it suitable for combination with other energy production processes. The biological processes integrated with dark fermentation involve microbial electrolysis cells (H_2), photofermentation (H_2), bioplastics (polyhydroxyalkanoates), microalgae cultivation (value-added products), and anaerobic digestion (biogas) (Muyzer & Stams, 2008; Ghimire et al., 2015; Roy & Das, 2016; Turon et al., 2016). Among these processes, integrated dark fermentation and anaerobic digestion are ideally recommended to generate hydrogen-rich biogas with enhanced thermal efficacy and reduced pollutant emission (Moreno et al., 2012). Microalgae extracted from wastewater treatment can also generate biohydrogen through dark fermentation or photolysis of waster (direct and indirect), resulting in the production of hydrogen as well as the synthesis of different VFAs (Banu et al., 2021; Arora et al., 2022). However, poor yield because of light, enzymatic sensitivity to O_2, and CO_2 fixation efficacy in the commercial setup poses a significant challenge.

12.3.2.1.2 Thermal Energy

Domestic wastewater resulting from the use of hot water for household activities such as cleaning and bathing holds a portion of its heat energy as it streams from the origin to the WWTP. Although this energy can be retrieved either at the WWTPs, at its source of origin, or along the cesspool line (Cipolla & Maglionico, 2014), it is still considered low-quality energy (Zarei, 2020). However, recovering heat energy from WWTPs attracts more attention given the volumetric flow associated, which in turn can generate a considerable amount of recoverable heat energy. To extract heat energy contained in digester sludge, filtrate, or effluents, heat exchangers or heat pumps are used at WWTPs. In the USA, Japan, China, South Korea, and Europe, heat pumps utilizing wastewater have been applied recently (Neczaj & Grosser, 2018). According to Hepbasli et al. (2014), globally, the wastewater source heat pumps that are in use number > 500, and the energy produced is higher than that produced using chemical energy. Zhao et al. (2010) reported that by using wastewater as a source, heat pumps generate thermal energy 3.5-fold higher compared to the energy (electrical) utilized in their processing. Based on the scale of energy recoveries, the retrieved energy can either be utilized on-site for sludge drying/office heating, or it can be circulated through district heating machinery (Tillman et al., 1998; Funamizu et al., 2001). For example, in Japan, thermal energy derived from wastewater and used for large-scale heating and cooling systems has significantly reduced energy consumption. Taking advantage of the thermal energy of wastewater, the municipal Government of Osaka is saving approximately 20–30% of energy (Shareefdeen et al., 2016). Further, to avoid a mismatch between demand and supply concerning time and location, thermal storage facilities like aquifers could be installed (Van der Hoek et al., 2016).

12.3.2.1.3 Electrical Energy

Fuel cells have recently been applied as a renewable source of energy that transforms chemical energy into electric energy. Among biological fuel cells, a special subcategory is microbial fuel cells (MFCs) which use microorganisms (as biocatalysts) for simultaneous electricity generation and WWT. MFCs transform chemical energy present in organic matter into electrical energy through bioelectrochemical reactions (Chen et al., 2019; Meena et al., 2019; Tang et al., 2019). The microorganisms used in MFCs are chosen depending on their capability of transporting electrons to the anode from the substrate. The microorganisms known to produce electrons such as *Shewanella*, *Geobacter*, and *Pseudomonas* play an important role in electricity generation and WWT through MFCs. Other bacterial strains such as those isolated from activated sludge, wastewater, domestic wastewater, and mixed microbial cultures from diverse habitats have been applied widely for WWT (Munoz-Cupa et al., 2021). The use of a mixture of microbial cultures where they grow as biofilms, improves electricity generation in MFCs. The formation of an effective biofilm and generation of electrons warrant continuous chemical oxygen demand (COD) elimination and voltage output. A variety of MFCs have been constructed such as MFCs with magnesium ammonium phosphate precipitation from urine (Zang et al., 2012), coupled with a sequencing batch reactor (Wang et al., 2014), fed with continuous flow (Wang et al., 2013), dual-chamber MFCs with an aerated catholyte on the cathode, and ferricyanide generating electricity by simultaneous removal of COD (Meena et al., 2019). In addition, a microbial electrochemical snorkel (MES) also called a short-circuited MFC has recently been found to be effective in reducing organic matter from wastewater. The construction of a MES involves the direct coupling of the bacterial anode to the cathode and assures the maximum feasible electrochemical processing rate (Hoareau et al., 2019).

12.3.2.2 Value-added Product Generation

12.3.2.2.1 Single-cell Proteins

Single-cell proteins (SCPs) are dead and dried microorganisms (biomass) that contain a high protein content and are also edible. The protein content in SCPs ranges between 60–80% of the entire dry cell weight (wt.), while essential amino acids, carbohydrates, minerals, and fats contribute to the rest of the dry cell weight. SCPs can be produced and harvested from wastewater using algae, fungi, yeast, and bacteria to match the protein demand as well as to reduce cost of the conventional production process (Puyol et al., 2017; Sharif et al., 2021). SCPs have been widely accepted and used for feed (animal) and food (human) purposes because of the high protein demand and a shortage of protein stemming from plants and animals. However, because of the anticipated limitation of resources, the cost will be impacted in the future. Thus, combining SCPs with nitrogen-rich solid or liquid waste streams such as wastewater might reduce the cost. In this context, algal ponds can successfully treat wastewater and the constituents removed can be utilized in SCP production (Zittelli et al., 2013). Further, nitrogen recovery via microbial resynthesis from waste streams has shown proven potential and can enable a circular biobased economy. A complete (100%) recovery as microbial protein is achievable using this process which can then be applied as an organic fertilizer, fodder, or food for humans (Matassa et al., 2015; Hülsen et al., 2016a; Puyol et al., 2017; Kundu et al., 2022).

To resynthesize the considerable number of proteins from wastewater constituents, large-scale actions are needed. Protein resynthesis in microalgae is dependent not only on wastewater load and retention time but also on nitrogen-removing potential, which has been documented to range between 36–87% for open high-rate algal ponds (Shoener et al., 2014). In case of ample carbon supply and sun energy, nitrogen availability is limited for the synthesis of proteins. Further, if wastewater is treated unsterile, it is expected to impact typical microalgal SCP composition (Maazouzi et al., 2008). So far, the cost of current investment significantly outweighs the cost of economically viable wastewater treatment using microalgae in closed photobioreactors (Posten, 2009). However, in closed photobioreactors, the growth (anaerobic) of purple phototrophic bacteria (PPB) as SCPs seems promising as removal of nitrogen is non-destructive with little loss, i.e., 10%, with the remaining (90%) being assimilated in biomass (Hülsen et al., 2016b). In a treatment system, up to 90% of PPB dominance can occur which would potentially enhance SCP composition (Hülsen et al., 2016b).

In the case of fungi, yeast-derived SCPs have been utilized as human and animal feed additives for a century. However, depending on species, fungal SCPs are limited by their elevated mycotoxins and nucleic acid levels that need to be eliminated before potential application as feed additives. The same applies to bacteria because of endo and exotoxins as well as nucleic acid, however, PPB does not produce toxins and the level of nucleic acid present is also comparable to algae; hence, it being reported as suitable for feed additives (Ponsano et al., 2004; Puyol et al., 2017).

12.3.2.2.2 Bioplastics

Recently, bioplastics such as polyhydroxyalkanoates (PHAs) have emerged as a preferable substitute to plastics derived from petroleum. These bioplastics are produced from renewable-biomass-based resources (Snell & Peoples, 2009; Możejko-Ciesielska & Kiewisz, 2016). Bioplastics are eco-friendly and have numerous applications including agriculture, biomedical use, consumable and food packaging, pharmaceuticals, and construction (Reddy et al., 2003; Ivanov et al., 2015). Several microorganisms including those present in sewage, marine environments, and soil synthesize PHA naturally, and are stored inside the cells as a reservoir of energy. Conventional bioplastic generation is restricted by the high-cost requirements of vast extraction and purification strategies. Producing bioplastics from waste streams has recently gained considerable attention. The wastewater arising from different streams such as municipal wastewater, agro-industrial wastewater (for example, oil mill, paper mill, dairy, and molasses,), spent glycerol, and sludge has been used to produce bioplastics. The synthesis of PHA can be coupled to organic and nutrient removal processes of WWTPs. For example, in a pilot-scale study over 22 months, Morgan-Sagastume et al. (2015) integrated the services of municipal wastewater and sludge management to evaluate PHA production at Brussels north WWTPs. Using feast–famine selection while treating the rapidly biodegradable COD from wastewater influent (average COD removal 70%), activated sludge with PHA accumulation potential (PAP) was generated. The full-scale WWTP sludge was subjected to batch fermentation at different temperatures resulting in the production of consistent VFA composition with optimum efficacy at 42 °C. These researchers were successful in producing biomass, i.e., 0.5 gPHA/gVSS within accessible carbon flow at municipal waste management facilities. Similarly, Basset et al. (2016) developed a novel approach to municipal wastewater treatment where nitritation/denitritation was coupled to the selection of PHA storing biomass under an oxic/anoxic, feast–famine regime. Their results reported 93% ammonia removal and reaching 98% removal of nitrite under optimal conditions. Further, the PHA accumulation was 10.6 wt.%. Studies revealed that a mixed culture in sludge can conserve approximately 20–60% PHA of sludge dry wt., which is comparatively less than that stored in pure culture (90%). However, research is continuing to enhance PHA production from such mixed cultures (Salehizadeh & Van Loosdrecht, 2004; Puyol et al., 2017). According to Bengtsson et al. (2017), the PHARIO project in the Netherlands is presently investigating PHA synthesis in full-scale municipal WWTPs from a mixed culture, however, no linkage has been established between PAP and storage.

12.3.2.2.3 Cellulose Fiber

Cellulose is another valuable resource recognized in WWTPs, which primarily comes from the use of toilet paper. Additionally, wastewater arising from the kitchen, pulp, paper mills, and dye industries also contains a significant fraction of cellulose. Ruiken et al. (2013) reported that approximately 40% of suspended solids (SS) and 25–30% of the COD in WWTPs account for cellulose fiber. Knowing that primary sludge holds a significant quantity of cellulose (approximately 20%, based on SS), Honda et al. (2002) introduced the notion of extracting cellulose from sludge for use as a resource. They proposed a simple method, i.e., high temperature and dilute acid hydrolysis to recover cellulose. It was reported that primary clarifiers can retain 68% of cellulose and condense it in sludge followed by its recovery before biological treatment to avoid cellulose degradation. The research of Ruiken et al. (2013) again attracted interest in recovering cellulose from wastewater by replacing the primary clarifier with a rotating belt filter (RBF) method and highlighted the advantages (Cipolletta et al., 2019). Based on the cellulose content in wastewater influent, RBF can recover approximately 80% of cellulose (about 37–76 mg/L-influent) from the total contained influent (Liu et al., 2022). Sources reported that in America and Europe about 700 RBF have been applied and installed. For instance, an RBF installed (after coarse screen) in Geestmerambacht WWTP of the Netherland could recover 400 Kg (approximately) cellulose/day, which is subsequently marketed under the brand name Recell® and utilized in a range of applications (Global Recycling, 2021; Liu et al., 2022).

Other than these, materials like alginate are also being recovered from WWTPs to promote the CE concept. For example, as part of the LIFE Waste2NeoAlginate project, Dutch regional water authorities are constructing new plants to generate alginate-like substances from sludge. In this project, a process will be developed to extract Kaumera Nereda® Gum, an alginate-like material, from sludge granules at WWTPs. Alginate has several uses for example, in paper production, plaster molds for medicine, textiles, and food stabilizers (European Commission, 2019; CEWP, 2022).

12.3.2.3 Nutrient Recovery

Over the years, different wastewater streams including industrial, agricultural, and municipal sewage have generated vast quantities of nutrient-rich wastewater. Discharge of wastewater rich in nutrients is most likely received by natural water bodies, enhancing the likelihood of eutrophication and thus, jeopardizing aquatic life (Saliu & Oladoja, 2021; Kundu et al., 2022). However, the same nutrients, if recovered from wastewater, can benefit the agriculture industry and other relevant chemical industries. According to estimates, a total of 25.9 MMT can be recovered from wastewater globally, which can satisfy the demand for global fertilizer nutrients by 13.4%.

12.3.2.3.1 Nitrogen

Urine and feces are the primary sources of nitrogen in WWTP influent, where 91% (approx.) of nitrogen loading comes from urine (Vu et al., 2022). Additionally, wastewater coming from dishwashers and food particles from kitchen sinks are other sources of nitrogen (Khajvand et al., 2022). The nitrogen species present in wastewater include ammonia (60%), organic nitrogen (40% nitrite and nitrate < 1%). The concentration of nitrogen in unprocessed wastewater ranges from 20 to 85 mg/L (Lusk et al., 2017). It is anticipated that the amount of nitrogen present in wastewater can compensate for 14.4% of the global demand for nitrogen as a fertilizer nutrient, if recovered completely. Several technologies including chemical and biological, have been applied to recover nitrogen from wastewater as a potential resource. For example, Chen et al. (2015) developed a "microbial nutrient recovery cell (MNRC)" for concurrent recovery of nutrients and purification of water. Their results reported that > 96% of NH_4^+–N was recovered using MNRC. The authors further reported that this is a self-driven approach whereby an electric field is produced utilizing the in situ energy of wastewater. In another attempt, Chen et al. (2017) used an advanced MNRC containing a membrane stack configuration for nutrient recovery and water purification. The recovery solution contained 260% and 150% of total nitrogen and NH_4^+–N, respectively. Ammonium was recovered as struvite during crystallization. Wu and Modin (2013) conducted the recovery of ammonium from rejected water in a bioelectrochemical reactor. Their study reported a 79% recovery of ammonium from the rejected water. Furthermore, the authors reported that ammonium recovery can be combined with hydrogen production. Since urine is the main source of N in wastewater, it is expected that separating urine at the source could be a feasible alternative to N recovery (Maurer et al., 2006; Pronk et al., 2006; Wilsenach et al., 2007; Lienert & Larsen, 2010; Meena et al., 2019). Xu et al. (2017) performed laboratory-scale combinatorial processes of precipitation and air stripping to extract nutrients (N, P, K) from urine separated at source. It was reported that ammonium recovery of > 70% was attained by optimization of factors such as increasing airflow and higher dosages of P and Mg. Nevertheless, the development of nitrogen recovery methods from wastewater for large-scale applications is still ongoing. Bio-drying sludge is one such method where under thermophilic conditions (65–75 °C) sludge is treated by a two-step forced aeration process resulting in a 73% reduction of the entire sludge. Ammonia produced from the microbial decomposition of organic sludge is then extracted from vented air in an acid gas scrubbing unit as an ammonium sulfate solution (7.3 Kton/year). The recovered ammonia can be sold as a substitute for synthetic fertilizers (Winkler et al., 2013). Recently, anaerobic membrane bioreactor (AnMBR) technology has gained more attention because of several advantages associated with it (Zhou et al., 2023). Several wastewater treatments have used AnMBR technology, for example, Grossman et al. (2021) recovered 77% of nitrogen from wastewater using this method.

12.3.2.3.2 Phosphate

Primary contributors of P in municipal wastewater are fecal matter, industrial effluents, and household detergents containing phosphorus. The concentrations of P in wastewater vary between 0.5–10 mg/L and exist as organic and inorganic phosphate (orthophosphate and polyphosphate) (Ramasahayam et al., 2014; Lusk et al., 2017; Vu et al., 2022). According to Egle et al. (2016), phosphate present in municipal wastewater can substitute 40–50% (theoretically) of phosphate fertilizer used yearly in agriculture. From WWTPs, it is mostly recovered as a phosphate-rich mineral namely struvite ($NH_4MgPO_4 \cdot 6H_2O$). A few technologies to recover phosphorus including EuPhoRe, Nuresys, Wastrip, Crystallization, and Airpex have been applied widely in central Europe (CEWP, 2022). In southern Europe and the Netherlands, biological removal of phosphorus is prevalent. The northern countries prefer parallel precipitation due to the CE use of ferric sulfate from industry which is the most cost-effective option.

Although struvite crystallization to recover P from wastewater has been applied commercially, the bioavailability of P in resulting struvite might be undesirable for direct agriculture use (Ye et al., 2017). The amorphous Ca-P and Mg-P are the alternate phosphate recovery products that have fertilizer potential. In the forward osmosis separation, when seawater is employed to draw the solution, Mg^{2+} and Ca^{2+} are transported with reverse flux to recover phosphorus with no extra precipitating agents. The product extracted from the hybrid system always comprises amorphous Mg-P and Ca-P and has high solubility and good bioavailability, hence, can be employed as fertilizer for soil (Lei et al., 2021; Nakarmi et al., 2022; Zheng et al., 2022).

According to Cai et al. (2018), dosing ferrous iron in the traditional sludge process can aid in granule formation by enhancing phosphorus bioavailability and ultimately phosphorus recovery. Given the presence of iron content in wastewater influents, several WWTPs in the Netherlands have attained approximately 20–50% removal of the influent phosphates (Wilfert et al., 2016). In addition to iron's presence in influent, the addition of a small amount of iron externally will precipitate as vivianite with phosphate. Similar to struvite, vivianite precipitation is another alternative way to increase phosphate extraction. Despite its use in several industrial processes (like oil paintings, water colors, murals, and the electronics' industry) (Yuan et al., 2021), vivianite may not be directly applicable to agricultural operations (Wilfert et al., 2018). Egle et al. (2016) performed a comparative assessment of technologies used to recover P by taking technological, environmental, and economic aspects into consideration. Their study reported that low-cost phosphorus recovery is achievable and in certain cases, economic benefits from phosphorus recovery can be earned if the soluble phosphorus is extracted from digested sewage sludge or digester supernatant. Although the rate of recovery is less, i.e., $< 25\%$ of phosphorus in raw wastewater, according to the authors, a higher percentage of phosphorus recovery (70–90%) is possible under specified conditions with little extra cost if sewage sludge is incinerated. Alternatively, fermentative putative polyphosphate-accumulating microorganisms (PAO) such as *Thiothix caldifontis*, *Candidatus Accumulibacter phosphatis*, and *Tetrasphaera* sp. are exploited to recover phosphorus from WWTPs. *Thiothix caldifontis* grow by oxidizing sulfide present in the influent and have a competitive advantage over other PAOs. Thus, the bacteria will produce a high level of biomass per unit of VFAs and have excess capacity for phosphorus storage (Rubio-Rincón et al., 2017; Bunce et al., 2018). Other than bacteria, microalgae such as *Scenedesmus* sp. and *Chlorella* sp. also carry out the uptake of phosphorus in the natural environment and have been proven as efficient in phosphorus removal from wastewater. Additionally, algal biofilm attained a 98% removal rate of phosphorus from real wastewater in a photobioreactor (Sukačová et al., 2015). Although phosphorus recovery into value-added fertilizer/

amendments from wastewater is promising and sustainable, there remains a need to educate society to make phosphorus recovery and its application acceptable and supported.

12.3.2.3.3 Potassium

The main source of potassium in municipal wastewater is urine, i.e., approximately 80–90% (Larsen & Gujer, 1996). This illustrates that potassium recovery from urine would be more viable than from municipal wastewater which is highly diluted. Xu et al. (2011) reported an effective method, i.e., magnesium potassium phosphate hexahydrate (struvite-K, $MgKPO_4 \cdot 6H_2O$) precipitation for concurrent recovery of K and P from urine. The resultant precipitates are valued slow-release fertilizers like magnesium ammonium phosphate hexahydrate. Thus, a complete recovery of nutrients, i.e., N, P, and K is conceivable from urine separated at source, through a combination of struvite-K precipitation and ammonia recovery route (Xu et al., 2017).

12.3.2.4 Reusing Treated Wastewater

With increasing water scarcity and advances in water purification technologies, large volumes of wastewater are being reclaimed and reused globally for a variety of potable and nonpotable purposes. However, wastewater that is subjected to a very high-level purification can only be utilized for potable use (both direct and indirect). In indirect use, treated water is released to environmental water sources such as rivers, reservoirs, aquifers, or lakes, and is subsequently retreated as a routine water supply process. On the other hand, direct use involves the treated water being supplied directly to a potable water facility for delivery. Countries such as the USA, Australia, Namibia, the United Kingdom, South Africa, Belgium, and Singapore have implemented drinking water reuse initiatives in their cities (Tortajada, 2020).

Globally, Windhoek, Namibia is the first and most successful example of direct potable reuse, where 25% of the city's drinking water arises from wastewater (Verstraete & Vlaeminck, 2011). Singapore is another example where ~40% of daily water needs are fulfilled by reused water (NEWater) which is further expected to increase to ~55% by 2060 (PUB, 2017). California is the leading state in the USA that is reusing reclaimed water for indirect potable use. The Orange Country Groundwater Replenishment System is the best-known (locally and internationally) potable reuse project in California, which provides potable reused water for nearly 850000 people (WHO, 2017; SWRCB, 2018). In Europe, EU policies mainly refer to nonpotable use with an emphasis on agricultural irrigation, while leaving the decision-making to member states regarding potable reuse of water. In Belgium, the Torrelle plant was supplying safe potable water in nearby areas (to ~60,000 people), in 2012. Further, the water from the plant was also used to prevent seawater intrusion by artificially recharging the Saint-Andre's dune aquifer (Van Houtte & Verbauwhede, 2012). The Langford Recycling project produces water to be used for indirect potable water supply. The plant supplies up to 70% of flow during drought periods, only when the Chelmer River flow is low (Janbakhsh, 2012). In Australia, only one project has been executed successfully whereas two were suspended because of public health issues and opposition by the public. Further, developing nations such as Kuwait, Mexico,

India, and Brazil have structured or planned schemes for reusing treated wastewater for potable use in some of their cities (Tortajada, 2020).

On the other hand, nonpotable use of treated wastewater includes agricultural irrigation, cooling water systems of powerplants, fisheries, algal cultivation, car washing, fire protection, road cleaning, sanitation infrastructure, toilet flushing, industrial operations, etc. (Neczaj & Grosser, 2018; Al-Ghouti et al., 2019; Meena et al., 2019). Globally, the agricultural sector dominates in the reuse of treated wastewater (Bauer & Wagner 2022). It is recommended that nonfood crops be irrigated using secondary treatment effluents and food crops via effluents from tertiary treatment (Neczaj & Grosser, 2018). The use of treated wastewater in industries is applicable mainly for cooling systems or production processes. However, outside production facilities, the application of treated wastewater within an industrial park is suggested as an efficient approach (Bauer & Wagner, 2022). Countries like Japan, Germany, Canada, and the USA have successfully applied recycled wastewater for industrial use. For example, in Tokyo, sand-filtered secondary effluent from Mikawashima WWTP was applied experimentally in a nearby paper mill for manufacturing paper, thus initiating wastewater reuse in Tokyo in 1951 (Maeda et al., 1996). The reuse of recycled water for irrigating green spaces, fire protection, toilet flushing, etc. is practiced in urban contexts (Bauer & Wagner, 2022). For instance, in Florida (Altamonte Springs) and California (Livermore) the fire hydrants connected to treated wastewater were 75 and 50, respectively. Similarly, in St. Petersburg, Florida, 308 hydrants were allied to over 460 km of reclaimed water supply conduits (Asano et al., 2007). The use of recycled wastewater in flushing toilets is largely practiced in Europe, Hong Kong, Australia, and Japan. In Japan, investigations related to wastewater reuse for toilet flushing started in 1964, and by 1996, approximately 2100 buildings had an operational on-site toilet cleansing water reuse system (Suzuki et al., 2002). Thus, from the preceding discussion, it appears that municipal wastewater should be perceived as a replenishing water resource for potable and nonpotable use, necessary for sustainable water management driven by a CE.

12.4 CONCLUSION

While water is considered a renewable resource, only 3% of the total water present on Earth is fresh water. Even in terms of freshwater, only 0.3% is available in liquid form on the surface. Because this amount is also being used at a much larger rate than it can be replenished, fresh water can be regarded as a potentially scarce resource. It is such global challenges that are pushing the world to focus more on sustainable development-driven initiatives. This is because the world has come to realize that resources such as water cannot continue being consumed and utilized in an irresponsible linear use and discard manner. As such, sustainably-driven initiatives have become synonymous with the CE model of doing things, which, focuses on ensuring maximum utilization of resources through the application of practices that keep resources in the loop as long as possible. Such practices include reuse, recycling, and value recovery from resources that have come to the end of life. Thus, where water utilization is concerned, wastewater treatment has become essential not only to minimize the demand on freshwater resources but also recovering value to sustain the economic development of communities.

This chapter has highlighted some of the resource values that can be recovered from wastewater through the application of different treatment technologies. This includes energy production (biofuel, thermal, and electrical energy), value-added products (SCPs, bioplastics, cellulose fiber), nutrients (nitrogen, phosphate, and potassium), and most importantly, water. All these products can be pushed back into the economy thus ensuring a closed loop system that reflects a near zero-waste concept, and not only reducing the negative impact on the environment but also reducing the demand for natural fresh resources of water. The implementation of such wastewater treatment practices will therefore go a long way in ensuring that resources to support the needs and lives of the current and future world generations are sustained.

REFERENCES

Al-Ghouti, M.A. *et al.* (2019) 'Produced water characteristics, treatment and reuse: A review', *Journal of Water Process Engineering*, 28, pp. 222–239.

Alkarawi, M.A., Caldwell, G.S. and Lee, J.G., (2018) Continuous harvesting of microalgae biomass using foam flotation. *Algal Research*, 36, pp.125–138.

Arora, A., Nandal, P. and Chaudhary, A. (2022) 'Critical evaluation of novel applications of aquatic weed Azolla as a sustainable feedstock for deriving bioenergy and feed supplement', *Environmental Reviews* [Preprint].

Asano, T., Burton, F. and Leverenz, H. (2007) *Water reuse: Issues, technologies, and applications*. McGraw-Hill Education.

Banu, J.R. *et al.* (2021) 'Integrated biorefinery routes of biohydrogen: Possible utilization of acidogenic fermentative effluent', *Bioresource Technology*, 319, p. 124241.

Basset, N. *et al.* (2016) 'Integrating the selection of PHA storing biomass and nitrogen removal via nitrite in the main wastewater treatment line', *Bioresource Technology*, 200, pp. 820–829.

Bauer, S. and Wagner, M. (2022) 'Possibilities and challenges of wastewater reuse—Planning aspects and realized examples', *Water*, 14(10), p. 1619.

Bengtsson, S. *et al.* (2017) *PHARIO: Stepping stone to a sustainable value chain for PHA bioplastic using municipal activated sludge*. Stichting Toegepast Onderzoek Waterbeheer Amersfoort, The Netherlands.

Bennett, A. (2007) 'Energy efficiency: Wastewater treatment and energy production', *Filtration & Separation*, 44(10), pp. 16–19.

Bertanza, G., Canato, M. and Laera, G. (2018) 'Towards energy self-sufficiency and integral material recovery in waste water treatment plants: Assessment of upgrading options', *Journal of Cleaner Production*, 170, pp. 1206–1218.

Bohra, V. et al. (2022) Energy and resources recovery from wastewater treatment systems. In *Clean Energy and Resource Recovery* (pp. 17–36). Elsevier.

Brundtland, G.H. (1987) 'What is sustainable development,' *Our Common Future*, 8(9).

Bunce, J.T. *et al.* (2018) 'A review of phosphorus removal technologies and their applicability to small-scale domestic wastewater treatment systems', *Frontiers in Environmental Science*, 6, p. 8.

Burton, S. *et al.* (2009) *Energy from Wastewater-A Feasibility Study*, WRC, South Africa, pp. 20399–09.

Cai, W. *et al.* (2018) 'Synthesis of peanut shell based magnetic activated carbon with excellent adsorption performance towards electroplating wastewater', *Chemical Engineering Research and Design*, 140, pp. 23–32.

CEWP. (2022) 'Circular economy of water – Waste water treatment'. China Europe Water Platform. Available at: www.cewp.eu/sites/default/files/2021-05/CEWP_Whitepaper_Circular%20Economy%20of%20Water%20in%20Waste%20Water%20Treatment_final.pdf (Accessed: 4 November 2023).

Chen, J. *et al.* (2018) 'Economic assessment of biodiesel production from wastewater sludge', *Bioresource Technology*, 253, pp. 41–48.

Chen, S. *et al.* (2019) 'Strategies for optimizing the power output of microbial fuel cells: transitioning from fundamental studies to practical implementation', *Applied Energy*, 233, pp. 15–28.

Chen, X. *et al.* (2015) 'Novel self-driven microbial nutrient recovery cell with simultaneous wastewater purification', *Scientific Reports*, 5(1), pp. 1–10.

Chen, X. *et al.* (2017) 'Self-sustaining advanced wastewater purification and simultaneous in situ nutrient recovery in a novel bioelectrochemical system', *Chemical Engineering Journal*, 330, pp. 692–697.

Cipolla, S.S. and Maglionico, M. (2014) 'Heat recovery from urban wastewater: analysis of the variability of flow rate and temperature in the sewer of Bologna, Italy', *Energy Procedia*, 45, pp. 288–297.

Cipolletta, G. *et al.* (2019) 'Toilet paper recovery from municipal wastewater and application in building sector', in *IOP Conference Series: Earth and Environmental Science*, IOP Publishing, p. 012024.

Crawford, G.V. (2010) 'Best practices for sustainable wastewater treatment: initial case study incorporating European experience and evaluation tool concept'. Iwa Publishing. ISBN electronic: 9781780407937 https://doi.org/10.2166/9781780407937

Delgado Martin, A. *et al.* (2021) 'Water in Circular Economy and Resilience (WICER): Position Paper Final'. World Bank Group. United States of America. Retrieved from https://policycommons.net/artifacts/1815919/water-in-circular-economy-and-resilience-wicer/2552412/ on 29 Mar 2024. CID: 20.500.12592/pkn10z.

Diaz-Elsayed, N. *et al.* (2019) 'Wastewater-based resource recovery technologies across scale: A review', *Resources, Conservation and Recycling*, 145, pp. 94–112.

Dincer, I. and Acar, C. (2015) 'Review and evaluation of hydrogen production methods for better sustainability', *International Journal of Hydrogen Energy*, 40(34), pp. 11094–11111.

Egle, L. *et al.* (2016) 'Phosphorus recovery from municipal wastewater: An integrated comparative technological, environmental and economic assessment of P recovery technologies', *Science of the Total Environment*, 571, pp. 522–542.

Elalami, D., Carrere, H., Monlau, F., Abdelouahdi, K., Oukarroum, A. and Barakat, A., 2019. Pretreatment and co-digestion of wastewater sludge for biogas production: Recent research advances and trends. *Renewable and Sustainable Energy Reviews, 114*, p.109287.

EPA. 2000. Wastewater Technology Fact Sheet Ammonia Stripping. USEPA EPA 832-F-00-019. www3.epa.gov/npdes/pubs/ammonia_stripping.pdf

European Commission. (2019) 'Extracting value from sewage sludge. European Commission, 19 Mar 2019.' Available at: https://ec.europa.eu/newsroom/eismea/newsletter-archives/14620 (Accessed: 4 January 2023).

Funamizu, N. *et al.* (2001) 'Reuse of heat energy in wastewater: implementation examples in Japan', *Water Science and Technology*, 43(10), pp. 277–285.

Ghimire, A. *et al.* (2015) 'A review on dark fermentative biohydrogen production from organic biomass: process parameters and use of by-products', *Applied Energy*, 144, pp. 73–95.

Global Recycling. (2021) 'Recovery of cellulose from waste water.', *Recovery of Cellulose from Waste Water.* Available at: https://global-recycling.info/archives/1665 (Accessed: 4 January 2023).

Grassi, W. (2017) *Heat pumps: fundamentals and applications.* Springer. https://doi.org/10.1007/978-3-319-62199-9

Grossman, A.D. *et al.* (2021) 'Advanced near-zero waste treatment of food processing wastewater with water, carbon, and nutrient recovery', *Science of the Total Environment*, 779, p. 146373.

Gurieff, N. *et al.* (2012) 'Moving towards an energy neutral WWTP–The positive impact of ExelysTM continuous thermal hydrolysis in achieving this goal', *Water Practice and Technology*, 7(2), pp 1–8.

Hagos, K. *et al.* (2017) 'Anaerobic co-digestion process for biogas production: Progress, challenges and perspectives', *Renewable and Sustainable Energy Reviews*, 76, pp. 1485–1496.

Hepbasli, A. *et al.* (2014) 'A key review of wastewater source heat pump (WWSHP) systems', *Energy Conversion and Management*, 88, pp. 700–722.

HLPW. (2018) 'Making every drop count. An agenda for water action: High level panel on water outcome document.' United Nations and World Bank Group. Available at: https://reliefweb.int/sites/reliefweb.int/ files/resources/17825HLPW_Outcome.pdf. (Accessed: 19 April 2023).

Hoareau, M., Erable, B. and Bergel, A. (2019) 'Microbial electrochemical snorkels (MESs): A budding technology for multiple applications. A mini review', *Electrochemistry Communications*, 104, p. 106473.

Honda, S., Miyata, N. and Iwahori, K. (2002) 'Recovery of biomass cellulose from waste sewage sludge', *Journal of Material Cycles and Waste Management*, 4, pp. 46–50.

Hukari, S., Hermann, L. and Nättorp, A. (2016) 'From wastewater to fertilisers—Technical overview and critical review of European legislation governing phosphorus recycling', *Science of the Total Environment*, 542, pp. 1127–1135.

Hülsen, T., Barry, E.M., Lu, Y., Puyol, D., Keller, J., *et al.* (2016a) 'Domestic wastewater treatment with purple phototrophic bacteria using a novel continuous photo anaerobic membrane bioreactor', *Water Research*, 100, pp. 486–495.

Hülsen, T., Barry, E.M., Lu, Y., Puyol, D. and Batstone, D.J. (2016b) 'Low temperature treatment of domestic wastewater by purple phototrophic bacteria: Performance, activity, and community.', *Water Research*, 100, pp. 537–545.

Ivanov, V. *et al.* (2015) 'Production and applications of crude polyhydroxyalkanoate-containing bioplastic from the organic fraction of municipal solid waste', *International Journal of Environmental Science and Technology*, 12, pp. 725–738.

Janbakhsh, A. (2012) *Langford Recycling Scheme: United Kingdom-Langford. In 2012 Guidelines for Water Reuse*, E114–E115. EPA. Available at: www3.epa.gov/region1/npdes/merrimackstation/pdfs/ar/AR-1530.pdf (Accessed: 4 December 2023).

Jenicek, P. *et al.* (2012) 'Potentials and limits of anaerobic digestion of sewage sludge: energy self-sufficient municipal wastewater treatment plant?', *Water Science and Technology*, 66(6), pp. 1277–1281.

Jenicek, P. *et al.* (2013) 'Energy self-sufficient sewage wastewater treatment plants: Is optimized anaerobic sludge digestion the key?', *Water Science and Technology*, 68(8), pp. 1739–1744.

Jonasson, M. and Ulf Jeppsson, I. (2007) 'Energy Benchmark for wastewater treatment processes—a comparison between Sweden and Austria'. *Lund University, Sweden.*

CODEN:LUTEDX/(TEIE-5247)/1-74/(2007). chrome-extension://efaidnbmnnnibpcajpcglclefindmkaj/https://www.iea.lth.se/publications/MS-Theses/Full%20document/5247_full_document.pdf

Jones, E.R. *et al.* (2021) 'Country-level and gridded estimates of wastewater production, collection, treatment and reuse', *Earth System Science Data*, 13(2), pp. 237–254.

Khajvand, M. *et al.* (2022) 'Management of greywater: environmental impact, treatment, resource recovery, water recycling, and decentralization', *Water Science & Technology*, 86(5), pp. 909–937.

Kirchherr, J., Reike, D. and Hekkert, M., 2017. Conceptualizing the circular economy: An analysis of 114 definitions. *Resources, Conservation and Recycling*, 127, pp. 221–232.

Kundu, D. *et al.* (2022) 'Valorization of wastewater: A paradigm shift towards circular bioeconomy and sustainability', *Science of The Total Environment*, 848, p. 157709.

Larsen, T.A. and Gujer, W. (1996) 'Separate management of anthropogenic nutrient solutions (human urine)', *Water Science and Technology*, 34(3–4), pp. 87–94.

Lei, Y. *et al.* (2021) 'Electrochemical recovery of phosphorus from wastewater using tubular stainless-steel cathode for a scalable long-term operation', *Water Research*, 199, p. 117199.

Lienert, J. and Larsen, T.A. (2010) 'High acceptance of urine source separation in seven European countries: A review', *Environmental Science & Technology*, 44(2), pp. 556–566.

Liu, R. *et al.* (2022) 'Review on the fate and recovery of cellulose in wastewater treatment', *Resources, Conservation and Recycling*, 184, p. 106354.

Lu, H. *et al.* (2017). Crystallization techniques in wastewater treatment: An overview of applications. *Chemosphere*, 173, pp.474–484.

Lu, H., Zhang, G., Zheng, Z., Meng, F., Du, T. and He, S., (2019). Bio-conversion of photosynthetic bacteria from non-toxic wastewater to realize wastewater treatment and bioresource recovery: a review. *Bioresource Technology*, 278, pp.383–399

Lusk, M.G. *et al.* (2017) 'A review of the fate and transport of nitrogen, phosphorus, pathogens, and trace organic chemicals in septic systems', *Critical Reviews in Environmental Science and Technology*, 47(7), pp. 455–541.

Maazouzi, C. *et al.* (2008) 'Midsummer heat wave effects on lacustrine plankton: variation of assemblage structure and fatty acid composition', *Journal of Thermal Biology*, 33(5), pp. 287–296.

Maeda, M. *et al.* (1996) 'Area-wide use of reclaimed water in Tokyo, Japan', *Water Science and Technology*, 33(10–11), pp. 51–57.

Matassa, S. *et al.* (2015) 'Can direct conversion of used nitrogen to new feed and protein help feed the world?', *Environmental Science & Technology*, 49(9), pp. 5247–5254.

Maurer, M., Pronk, W. and Larsen, T. (2006) 'Treatment processes for source-separated urine', *Water Research*, 40(17), pp. 3151–3166.

McCarty, P.L., Bae, J. and Kim, J. (2011) 'Domestic wastewater treatment as a net energy producer–can this be achieved?', *Environmental Science & Technology*, 45(17), pp. 7100–7106.

Meena, R.A.A. *et al.* (2019) 'Trends and resource recovery in biological wastewater treatment system', *Bioresource Technology Reports*, 7, p. 100235.

Morais, E.G. *et al.* (2021) 'Microalgal systems for wastewater treatment: Technological trends and challenges towards waste recovery', *Energies*, 14(23), p. 8112.

Moreno, F. *et al.* (2012) 'Efficiency and emissions in a vehicle spark ignition engine fueled with hydrogen and methane blends', *International Journal of Hydrogen Energy*, 37(15), pp. 11495–11503.

Morgan-Sagastume, F. *et al.* (2015) 'Integrated production of polyhydroxyalkanoates (PHAs) with municipal wastewater and sludge treatment at pilot scale', *Bioresource Technology*, 181, pp. 78–89.

Możejko-Ciesielska, J. and Kiewisz, R. (2016) 'Bacterial polyhydroxyalkanoates: Still fabulous?', *Microbiological Research*, 192, pp. 271–282.

Munoz-Cupa, C. *et al.* (2021) 'An overview of microbial fuel cell usage in wastewater treatment, resource recovery and energy production', *Science of the Total Environment*, 754, p. 142429.

Muyzer, G. and Stams, A.J. (2008) 'The ecology and biotechnology of sulphate-reducing bacteria', *Nature Reviews Microbiology*, 6(6), pp. 441–454.

Nagpal, H., Spriet, J., Murali, M.K. and McNabola, A., (2021) 'Heat recovery from wastewater—A review of available resource', *Water*, *13*(9), p.1274.

Nakarmi, A. *et al.* (2022) 'Applications of conventional and advanced technologies for phosphorus remediation from contaminated water', in *Green Functionalized Nanomaterials for Environmental Applications*. Elsevier, pp. 181–213.

Nawaz, A., ul Haq, I., Qaisar, K., Gunes, B., Raja, S.I., Mohyuddin, K. and Amin, H. (2022) Microbial fuel cells: Insight into simultaneous wastewater treatment and bioelectricity generation. *Process Safety and Environmental Protection*, 161, pp. 357-373.

Neczaj, E. and Grosser, A. (2018) 'Circular economy in wastewater treatment plant–Challenges and barriers', in. *Proceedings*, MDPI, p. 614.

Ponsano, E.H.G. *et al.* (2004) 'Performance and color of broilers fed diets containing Rhodocyclus gelatinosus biomass', *Brazilian Journal of Poultry Science*, 6, pp. 237–242.

Posten, C. (2009) 'Design principles of photo-Bioreactors for cultivation of microalgae', *Engineering in Life Sciences*, 9(3), pp. 165–177.

Pronk, W. *et al.* (2006) 'Nanofiltration for the separation of pharmaceuticals from nutrients in source-separated urine', *Water Research*, 40(7), pp. 1405–1412.

PUB. (2017) 'Annual Report 2016/17'. Public Utilities Board (PUB). Available at: www.pub.gov.sg/Documents/annualreport2017.pdf (Accessed: 4 December 2023).

Puchongkawarin, C. *et al.* (2015) 'Optimization-based methodology for the development of wastewater facilities for energy and nutrient recovery', *Chemosphere*, 140, pp. 150–158.

Puyol, D. *et al.* (2017) 'Resource recovery from wastewater by biological technologies: Opportunities, challenges, and prospects', *Frontiers in Microbiology*, 7, p. 2106.

Qadir, M. *et al.* (2020) 'Global and regional potential of wastewater as a water, nutrient and energy source', in *Natural Resources Forum*, Wiley Online Library, pp. 40–51.

Ramasahayam, S.K. *et al.* (2014) 'A comprehensive review of phosphorus removal technologies and processes', *Journal of Macromolecular Science, Part A*, 51(6), pp. 538–545.

Rao, K.C. *et al.* (2017) 'Resource recovery and reuse as an incentive for a more viable sanitation service chain', *Water Alternatives*, 10(2), p. 493.

Reddy, C., Ghai, R. and Kalia, V. (2003) 'Polyhydroxyalkanoates: An overview', *Bioresource Technology*, 87(2), pp. 137–146.

Robles, Á., et al. (2020) Microalgae-bacteria consortia in high-rate ponds for treating urban wastewater: Elucidating the key state indicators under dynamic conditions. *Journal of Environmental Management*, *261*, p.110244.

Rodriguez, D.J. *et al.* (2020) *From Waste to Resource: Shifting paradigms for smarter wastewater interventions in Latin America and the Caribbean*. World Bank.

Roy, S. and Das, D. (2016) 'Biohythane production from organic wastes: present state of art', *Environmental Science and Pollution Research*, 23, pp. 9391–9410.

Rubio-Rincón, F. *et al.* (2017) 'Long-term effects of sulphide on the enhanced biological removal of phosphorus: the symbiotic role of *Thiothrix caldifontis*', *Water Research*, 116, pp. 53–64.

Ruiken, C. *et al.* (2013) 'Sieving wastewater–Cellulose recovery, economic and energy evaluation', *Water Research*, 47(1), pp. 43–48.

Salehizadeh, H. and Van Loosdrecht, M. (2004) 'Production of polyhydroxyalkanoates by mixed culture: Recent trends and biotechnological importance', *Biotechnology Advances*, 22(3), pp. 261–279.

Saliu, T. and Oladoja, N. (2021) 'Nutrient recovery from wastewater and reuse in agriculture: A review', *Environmental Chemistry Letters*, 19(3), pp. 2299–2316.

Santos, F.M. and Pires, J.C. (2018) 'Nutrient recovery from wastewaters by microalgae and its potential application as bio-char'. *Bioresource Technology*, 267, pp. 725–731.

Shareefdeen, Z., Elkamel, A. and Kandhro, S. (2016) 'Modern water reuse technologies: Membrane bioreactors', in *Urban Water Reuse Handbook*. CRC Press, pp. 419–428.

Sharif, M. *et al.* (2021) 'Single cell protein: Sources, mechanism of production, nutritional value and its uses in aquaculture nutrition', *Aquaculture*, 531, p. 735885.

Sharma, S. *et al.* (2020) 'Waste-to-energy nexus for circular economy and environmental protection: Recent trends in hydrogen energy', *Science of the Total Environment*, 713, p. 136633.

Shen, Y. *et al.* (2015) 'An overview of biogas production and utilization at full-scale wastewater treatment plants (WWTPs) in the United States: Challenges and opportunities towards energy-neutral WWTPs', *Renewable and Sustainable Energy Reviews*, 50, pp. 346–362.

Shoener, B. *et al.* (2014) 'Energy positive domestic wastewater treatment: The roles of anaerobic and phototrophic technologies', *Environmental Science: Processes & Impacts*, 16(6), pp. 1204–1222.

Snell, K.D. and Peoples, O.P. (2009) 'PHA bioplastic: A value-added coproduct for biomass biorefineries', *Biofuels, Bioproducts and Biorefining: Innovation for a Sustainable Economy*, 3(4), pp. 456–467.

Solon, K. *et al.* (2019) 'Resource recovery and wastewater treatment modelling', *Environmental Science: Water Research & Technology*, 5(4), pp. 631–642.

Somoza-Tornos, A. *et al.* (2019) 'A circular economy approach to the design of a water network targeting the use of regenerated water', in *Computer Aided Chemical Engineering*. Elsevier, pp. 119–124.

Song, X. *et al.* (2018) Resource recovery from wastewater by anaerobic membrane bioreactors: Opportunities and challenges. *Bioresource Technology*, 270, pp. 669–677.

Sukačová, K., Trtílek, M. and Rataj, T. (2015) Phosphorus removal using a microalgal biofilm in a new biofilm photobioreactor for tertiary wastewater treatment. *Water Research*, 71, pp. 55–63.

Suzuki, Y. *et al.* (2002) 'Large-area and on-site water reuse in Japan', in. *International Seminar on Sustainable Development of Water Resource Pluralizing Technology, held at Korean National Institute of Environmental Research*, Citeseer.

SWRCB. (2018) 'Adopting the proposed regulations for surface water augmentation using recycled water', *State Water Resources Control Board Resolution No. 2018-0014*. Available at: www.waterboards.ca.gov/board_decisions/adopted_orders/resolutions/2018/rs2018_0014_with_regs.pdf (2018). (Accessed: 4 November 2023).

Tang, J. *et al.* (2019) 'Municipal wastewater treatment plants coupled with electrochemical, biological and bio-electrochemical technologies: Opportunities and challenge toward energy self-sufficiency', *Journal of Environmental Management*, 234, pp. 396–403.

Tillman, A.-M., Svingby, M. and Lundström, H. (1998) 'Life cycle assessment of municipal waste water systems', *International Journal of Life Cycle Assessment*, 3, pp. 145–157.

Tortajada, C. (2020) 'Contributions of recycled wastewater to clean water and sanitation sustainable development goals', *NPJ Clean Water*, 3(1), p. 22.

Tsekouras, G.J. *et al.* (2022) Microbial fuel cell for wastewater treatment as power plant in smart grids: Utopia or reality?. *Frontiers in Energy Research*, 10, p.843768.

Turon, V. *et al.* (2016) 'Potentialities of dark fermentation effluents as substrates for microalgae growth: A review', *Process Biochemistry*, 51(11), pp. 1843–1854.

Van der Hoek, J.P., de Fooij, H. and Struker, A. (2016) 'Wastewater as a resource: Strategies to recover resources from Amsterdam's wastewater', *Resources, Conservation and Recycling*, 113, pp. 53–64.

Van Houtte, E. and Verbauwhede, J. (2012) 'Sustainable groundwater management using reclaimed water: The Torreele/St-André case in Flanders, Belgium', *Journal of Water Supply: Research and Technology—AQUA*, 61(8), pp. 473–483.

Verstraete, W. and Vlaeminck, S.E. (2011) 'ZeroWasteWater: Short-cycling of wastewater resources for sustainable cities of the future', *International Journal of Sustainable Development & World Ecology*, 18(3), pp. 253–264.

Vu, M.T. *et al.* (2022) 'Wastewater to R3–Resource recovery, recycling, and reuse efficiency in urban wastewater treatment plants', in *Clean Energy and Resource Recovery*. Elsevier, pp. 3–16.

Wang, Y.-H. *et al.* (2013) 'Electricity production from a bio-electrochemical cell for silver recovery in alkaline media', *Applied Energy*, 112, pp. 1337–1341.

Wang, Y.-P. *et al.* (2014) 'Improving electricity generation and substrate removal of a MFC–SBR system through optimization of COD loading distribution', *Biochemical Engineering Journal*, 85, pp. 15–20.

Wett, B., Buchauer, K. and Fimml, C. (2007) 'Energy self-sufficiency as a feasible concept for wastewater treatment systems', in. *IWA Leading Edge Technology Conference* , Singapore: Asian Water, pp. 21–24.

WHO. (2017) 'Potable reuse: Guidance for producing safe drinking-water'. World Health Organization (WHO). Available at: http://apps.who.int/iris/bitstream/handle/10665/258715/9789241512770-eng.pdf (Accessed: 4 December 2023).

Wilfert, P. *et al.* (2016) 'Vivianite as an important iron phosphate precipitate in sewage treatment plants', *Water Research*, 104, pp. 449–460.

Wilfert, P. *et al.* (2018) 'Vivianite as the main phosphate mineral in digested sewage sludge and its role for phosphate recovery', *Water research*, 144, pp. 312–321.

Wilsenach, J., Schuurbiers, C. and Van Loosdrecht, M. (2007) 'Phosphate and potassium recovery from source separated urine through struvite precipitation', *Water Research*, 41(2), pp. 458–466.

Winkler, M.-K. *et al.* (2013) 'The biodrying concept: An innovative technology creating energy from sewage sludge', *Bioresource Technology*, 147, pp. 124–129.

World Bank. (2019) 'World Bank's data on commodity markets: Fertilizers.' Available at: Retrieved from www.worldbank.org/en/ research/commodity–market (Accessed: 20 April 2023).

Wu, X. and Modin, O. (2013) 'Ammonium recovery from reject water combined with hydrogen production in a bioelectrochemical reactor', *Bioresource Technology*, 146, pp. 530–536.

Xie, M., Shon, H.K., Gray, S.R. and Elimelech, M. (2016) 'Membrane-based processes for wastewater nutrient recovery: Technology, challenges, and future direction', *Water Research*, 89, pp. 210–221.

Xu, K. *et al.* (2011) 'Simultaneous removal of phosphorus and potassium from synthetic urine through the precipitation of magnesium potassium phosphate hexahydrate', *Chemosphere*, 84(2), pp. 207–212.

Xu, K. *et al.* (2017) 'Removal and recovery of N, P and K from urine via ammonia stripping and precipitations of struvite and struvite-K', *Water Science and Technology*, 75(1), pp. 155–164.

Ye, Z.-L. *et al.* (2017) 'Adsorption behavior of tetracyclines by struvite particles in the process of phosphorus recovery from synthetic swine wastewater', *Chemical Engineering Journal*, 313, pp. 1633–1638.

Yuan, Q. *et al.* (2021) 'Biosynthesis of vivianite from microbial extracellular electron transfer and environmental application', *Science of The Total Environment*, 762, p. 143076.

Zang, G.-L. *et al.* (2012) 'Nutrient removal and energy production in a urine treatment process using magnesium ammonium phosphate precipitation and a microbial fuel cell technique', *Physical Chemistry Chemical Physics*, 14(6), pp. 1978–1984.

Zarei, M. (2020) 'Wastewater resources management for energy recovery from circular economy perspective', *Water-Energy Nexus*, 3, pp. 170–185.

Zhang, T. and Huang, C.H. (2020). Modeling the kinetics of UV/peracetic acid advanced oxidation process. *Environmental Science & Technology*, *54*(12), pp. 7579–7590.

Zhang, Y. *et al.* (2013). Phosphate separation and recovery from wastewater by novel electrodialysis. *Environmental Science & Technology*, 47(11), pp.5888–5895.

Zhao, X. *et al.* (2010) 'Study of the performance of an urban original source heat pump system', *Energy Conversion and Management*, 51(4), pp. 765–770.

Zhen, G. *et al.* (2017) 'Overview of pretreatment strategies for enhancing sewage sludge disintegration and subsequent anaerobic digestion: Current advances, full-scale application and future perspectives', *Renewable and Sustainable Energy Reviews*, 69, pp. 559–577.

Zheng, Y. *et al.* (2022) 'Recovery of phosphorus from wastewater: A review based on current phosphorous removal technologies', *Critical Reviews in Environmental Science and Technology*, pp. 1–25.

Zhou, Y. *et al.* (2018) 'Development of a novel membrane-less microbial fuel cell (ML-MFC) with a Sandwiched Nitrifying chamber for efficient wastewater treatment,' *Electroanalysis*, 30(9), pp. 2145–2152.

Zhou, Y. *et al.* (2023) 'A comprehensive review on wastewater nitrogen removal and its recovery processes', *International Journal of Environmental Research and Public Health*, 20(4), p. 3429.

Zittelli, G.C. *et al.* (2013) 'Photobioreactors for mass production of microalgae', In *Handbook of Microalgal Culture: Applied Phycology And Biotechnology*, pp. 225–266.

13 New Trends in Wastewater Recycling, Treatment, and Management

Helen Uchenna Modekwe, Olusola Olaitan Ayeleru, Peter Apata Olubambi, and Ishmael Matala Ramatsa

13.1 INTRODUCTION

The Earth's crust is covered with water, mainly salt water found in oceans and seas. However, only a minute portion, about 0.3% of the freshwater available in lakes and rivers is suitable for human consumption (Manasa & Mehta, 2020). In recent years, accessibility to fresh water has been limited due to the increasing human population, expansive globalization, and industrialization which give rise to competitive demand by many users (industry, agriculture, domestic, energy, etc.) and a surge in water pollution, all of which have resulted in a scarcity of clean water and the emergence of different water-related diseases. According to a report published by the United Nations (UN), going by current rates, billions of people will lack access to safe water, sanitation, and hygiene by 2030, unless progressive efforts (increased fourfold) are put in place to address the growing water challenges (United Nations, 2022). A significant percentage of people living around the globe in water-stressed regions, especially in developing countries, are without access to clean water. The reuse of water for nonpotable and potable applications is studied as a viable solution to water scarcity and water security. Wastewater treatment is, therefore, a restorative measure by which the quality of contaminated or already used water is restored to a desired quality for intended use (Lyu et al., 2016). Recycling and treatment of wastewater is of utmost importance to attain sustainable survival and growth of humans and every living organism. Recycled water could be used in irrigation for agricultural purposes and other areas such as landscaping (parks, lawns, etc.), as process water in power plants, factories, and refineries, in construction for concrete mixing, in surface cleaning and car washing, toilet flushing, etc. (Ngo et al., 2007).

Several countries have adopted regulations and action plans to regulate and oversee water reuse and treatment which are majorly aimed at minimizing the adverse effects

DOI: 10.1201/9781003441069-13

of contaminated water on public health and the environment. It is also established that to protect and ensure water quality and security, it is necessary to invest in proper developed and advanced wastewater treatment technologies, to incorporate water reuse into existing programs/policies, and to invest in new wastewater management approaches, for example, in the development of high-performance membrane technologies, cost-effective and low contaminant technologies, etc. (Englande Jr & Krenkel, 2003; Lyu et al., 2016).

Wastewater quality is characterized in terms of its physical, biological, and chemical properties and compositions (Manasa & Mehta, 2020; Kumar & Verma, 2023). The physical parameters used to define the quality of wastewater are temperature, color, odor, the presence of solids comprising the total dissolved solids and suspended solids, turbidity, etc. the effect of which may impact the extinction of flora and fauna (Manasa & Mehta, 2020). In addition, the chemical properties of wastewater are categorized under organic (total organic carbon (TOC), biochemical oxygen demand (BOD), total oxygen demand (TOD), chemical oxygen demand COD), inorganic (heavy metals, hardness, pH, acidity, basicity, salinity, sulfides, sulfates, and phosphorus, etc.) (Madhav et al., 2020; Manasa & Mehta, 2020), and biological properties (pathogens, bacteria, viruses, and coliforms, etc.). The occurrence of these in wastewater may cause various health problems, illnesses, and in most cases, death of organisms living in the water (Manasa & Mehta, 2020).

13.2 SOURCES OF WASTEWATER AND WATER POLLUTANTS

Wastewater comes from different sources and activities such as domestic, storm water/surface runoffs, industrial, agricultural, aquaculture, and horticultural activities, sewer infiltration/inflows, commercial activities, etc. Water pollution sources can be studied based on point: which describes the direct supply of polluted water and nonpoint contaminants which designates multiple supplies of polluted wastewater, as seen in the case of rain run-offs passing through several areas (Madhav et al., 2020). Another categorization of wastewater sources is based on anthropogenic (man-made) or natural activities (Madhav et al., 2020; Singh et al., 2020).

13.2.1 Natural Sources/Activities

The major natural sources of pollutants in waterbodies are through rainwater, underground rocks, volcanoes, natural/stormwater runoffs, surrounding vegetation, and the atmosphere (storms, dust). Rainwater dissolves surrounding pollutants such as the oxides of sulfur and nitrogen, and brings them down during downpours, this scenario results in the occurrence of acid rain (Singh et al., 2020). Rainfall also runs across lands and farms collecting contaminants such as toxic materials and agrochemicals, which are eventually deposited on the surface and groundwater (Singh et al., 2020). In addition, the presence of falling leaves, trunks, branches, or other vegetation enhances the organic content of water resulting in the increased concentration of organic and inorganic contents. Similarly, subsurface rocks and volcanoes enrich the salt and metal ion contents of water (Singh et al., 2020).

13.2.2 Anthropogenic Activities

The sources listed under anthropogenic activities are outlined below.

13.2.2.1 Domestic Wastewater

Domestic wastewater is contaminated water arising from day-to-day human and household activities such as laundry, kitchen, bathing, toilet flushing, etc. (Manasa & Mehta, 2020). Pollutants in domestic wastewater are usually from detergents, food waste, human feces, etc. (Singh et al., 2020). Based on the components present in domestic wastewater, they are further classified as blackwater which is wastewater that emanates from fecal discharge from sewage, and contains high concentrations of pathogens, inorganic and organic matter; yellow water is wastewater obtained in the form of urine. In addition, grey water is the category of wastewater that contains few pathogens unlike black water, but can be treated and reused for other purposes such as crop irrigation, flushing, etc. Grey water refers to all water from domestic households, businesses, and offices other than from toilets (Manasa & Mehta, 2020).

13.2.2.2 Industrial Wastewater

Enormous quantities of wastewater are produced from various production activities/steps in manufacturing industries and the class of pollutants generated from each industry differs from one to another (Manasa & Mehta, 2020; Kumar & Verma, 2023). The common industries and their prevalent effluents of interest are discussed below.

13.2.2.3 Food Industry

Food industries use water as the main ingredient in their production activities, in addition, a large portion of water is used in washing, cleaning (materials and machines), heating, disinfection, and cooling processes. Industries that fall under the food industry are dairy, confectionary, beverage, starch, brewery, and vegetable oil industries (Manasa & Mehta, 2020). Wastewater produced from food industries contains the majority of organic materials (such as sugar, fatty acids, flavoring agents, coloring additives, soluble starch, nitrogen, phosphorus, BOD, COD, suspended solids, ethanol, detergents, etc.). Effluents from these industries could be reused for other purposes since they are high in biodegradable organic matter and do not contain hazardous chemicals and pathogens. The methods for treating wastewater from food industries are usually biological treatment methods via aerobic and anaerobic means (Manasa & Mehta, 2020).

13.2.2.4 Metal Plating, Iron, and Steel Industries

Water is involved in various steps in the steel and iron plating process, such as rinsing in a bank of counter-current rinse baths (Hosseini et al., 2016), the spraying of water on a blast furnace to remove dust (Manasa & Mehta, 2020). Effluents from metal plating, iron, and steel industries contain dust, metal ions, sulfur-laden compounds, slags, ore particles, cyanides, phenol, and ash. Typical treatment technologies usually applied in the treatment of liquid waste influents from these industries are filtration, chemical neutralization, coagulation-settling, precipitation, aerobic biological methods, and air flotation (Manasa & Mehta, 2020).

13.2.2.5 Textile Industry

The textile industry is one of the major sources of wastewater among the diverse industries as a result of the involvement of water in the various stages (bleaching, scouring, dyeing, printing, mercerizing, finishing, etc.) of the process (Madhav et al., 2020). Wastewater effluents from this industry are usually treated by reverse osmosis, microfiltration, nanofiltration, and ultrafiltration methods. Other methods such as the aerobic biological technique, wet oxidation, and ozonation prior to ion exchange could be used to eliminate BOD, COD, and some chemical pollutants present (Manasa & Mehta, 2020).

13.2.2.6 Paper and Pulp Industry

This sector is one of the highest users of water and by extension the highest generator of contaminated wastewater (Manasa & Mehta, 2020; Kumar & Verma, 2023). Wastewater effluents from this industry are polluted with chlorinated organic compounds, absorbable organic halides, suspended solids, and sediments. Treatment of paper and pulp wastewater involves the combination of primary, secondary, and advanced treatment processes (Manasa & Mehta, 2020).

13.2.2.7 Nuclear and Thermal Power Industries

Wastewater from nuclear and thermal plants originates from wet scrubbing flue gases, fly ash, and bottom ash from thermal power plants using coal combustion for electricity generation, which are contaminated with total dissolved solids (TDS) saturated with oxides of nitrogen and sulfur, heavy metals, total suspended solids (TSS), selenium, arsenic, BOD, boron, etc. (Singh et al., 2020). The conventional treatment technologies for wastewater from thermal power plants are usually through multistep treatment methods while evaporation, precipitation/flocculation, ion exchange, and solid-phase separation are usually applied in the treatment of liquid radioactive wastewater from nuclear power plants before their release into the environment or before being recycled (Manasa & Mehta, 2020).

13.2.2.8 Pharmaceutical and Personal Care Industry

Pharmaceutical industries require vast amounts of water as active ingredients, raw materials, and solvents in the formulation, processing, and manufacturing of pharmaceuticals and personal care products (Manasa & Mehta, 2020). Wastewater produced from the pharmaceutical industry contains BOD, TOC, COD, high acidic and basic pH, nitrogen-containing compounds, organic compounds, etc. Due to the persistent nature of pharmaceutical wastewater, the common treatment method of wastewater from pharmaceutical industries is usually through membrane technology together with hybrid treatment processes or conventional treatment methods such as adsorption, etc. (Dey et al., 2019; Manasa & Mehta, 2020).

13.2.2.9 Mines and Quarries

Wastewater produced from mines is treated using flotation techniques, chemical precipitation, and biochemical and electrolyte methods. Water is mostly used for cooling and extraction purposes with common pollutants such as suspended solids from precipitated rocks, dust water, and some metallic ions, sulfides, sulfates, etc. (Manasa & Mehta, 2020).

13.2.2.10 Petroleum and Petrochemical Industries

Most wastewater generated from the petroleum industry is continually recycled and reused within the refinery. Free oil, emulsified oil, sulfur containing compounds, BOD, suspended solids, and phenolic and benzene, toluene, and xylene (BTX) compounds are the most frequent pollutants released from these industries (Madhav et al., 2020). Treatment technologies used for effluents from petroleum industries include biological treatment methods, oil-water separation, groundwater remediation, residual solids, and dewatering techniques (Manasa & Mehta, 2020). Other sources of industrial wastewater include confectionary, chemical, laundry, beverage, dairy, brewery and distillery industries.

13.2.2.11 Agricultural Wastewater

Agricultural wastewater from runoffs from farmlands, fish farming, and animal/poultry farms contains high concentrations of organic pollutants, pathogens, inorganic nutrients, nitrates, phosphates, suspended solids, and sediments due to contaminations with agrochemicals (pesticides, herbicides, fertilizers), antibiotics (pharmaceuticals), organic matter, animal droppings, and wastes, etc. (Singh et al., 2020). Effluents from these industries are treated by biological (aerobic digestion) and chemical treatment methods before being discharged (Madhav et al., 2020).

13.2.3 Types of Water Pollutants

13.2.3.1 Inorganic Pollutants

Contamination of waterbodies by inorganic compounds and chemicals of heavy metals, nitrogen (nitrites, nitrates, and ammonium compounds), and phosphorus-containing compounds from farmlands, sewage, and industrial sites are of great concern because most of them could accumulate in the soft tissues, are mostly carcinogenic and may result in numerous serious health-related problems (Singh et al., 2020).

13.2.3.2 Organic Pollutants

These pollutants comprise compounds of carbon, oxygen, hydrogen, sulfur, etc. mainly originating from personal care products, nanomaterials, medications/drugs, pesticides (diuron), paints, methyl tetrabutyl ether (MTBE), and so on, which are typical in effluents from municipal/industrial wastewater, sewage, and agricultural waste (Madhav et al., 2020; Singh et al., 2020). Some organic pollutants can be degraded by bacterial actions while some synthetic organic pollutants cannot be degraded by bacterial actions, and they also use up the oxygen content in water (Madhav et al., 2020; Singh et al., 2020).

13.2.3.2.1 Nutrients and Agricultural Pollutants

Chemical and inorganic fertilizers used in agriculture contain several concentrations of toxic metals (such as mercury, arsenic, cadmium, lead, copper, nickel, etc.), a few radionuclides, and some phosphates, ammonium, potassium, and nitrate-containing salts (Singh et al., 2020).

13.2.3.2.2 Suspended Solids

Suspended solids may be sand and other organic and inorganic matter that are less than 62 mm in diameter, whose presence in water could hinder the penetration of sunlight (Madhav et al., 2020). They are mostly found in sewage, municipal, and industrial wastewater treatment plants.

13.2.3.3 Pathogens

Pathogenic organisms that contaminate water include some species of parasites, viruses, bacteria, helminths, and protozoa. The presence of these microbes in water/wastewater results in various illnesses and water-related diseases (Singh et al., 2020).

13.2.3.4 Thermal Pollutants

Effluents from thermal power plants alter the temperature of water and waterbodies which affects the aquatic system, disrupting organisms living in the water due to global warming and the raised water temperature by the thermal effect (Singh et al., 2020).

13.2.3.5 Radioactive and Halides' Pollutants

Naturally occurring radioactive (such as uranium, radon, radium) and halogen (chloride, fluoride, and bromide) pollutants, usually found in dissolved form in surface and underground water, mostly originate from the Earth's crust during rock weathering and erosion (Madhav et al., 2020). Some anthropogenic activities also result in the release of these radioactive materials such as nuclear power plants during the manufacturing, testing, and usage of nuclear weapons. These pollutants have serious cancer-causing effects on some delicate human organs (Singh et al., 2020). Some of these radionuclides have very high half-lives and emit low and high radiation energy even at a trace amount (Madhav et al., 2020). Examples of radioactive elements found in wastewater are uranium, potassium-40 (^{40}K), radium-226 (^{226}Ra), tritium (^{3}H), radon-220 (^{220}Rn), polonium-210 (^{210}Po), carbon-14 (^{14}C), etc. (Madhav et al., 2020).

13.3 EFFECTIVE WASTEWATER MANAGEMENT AND CHALLENGES

Due to the potential toxicity of certain pollutants on public health, the environment, and aquatic life, stringent water quality requirements and limits have been adopted by governments and regulatory bodies. More attention has been focused on wastewater management, which embraces prevention in place of treatment to encourage clean technologies with more emphasis on ways to minimize waste generation and extend product (resource) life cycle analysis to lessen raw material and energy usage, in addition to environmental discharge and saving on waste treatment cost (Englande

Jr & Krenkel, 2003). Therefore, the management of wastewater involves effective collection, treatment, and discharge to receiving waterbodies with little or no environmental impact. Challenges to wastewater management could be classified as technical and social challenges (Villarín & Merel, 2020). As new chemical contaminants are regularly authorized and withdrawn, especially for pharmaceuticals and personal care products, detectable pollutants in wastewater vary from year to year and from country to country (Villarín & Merel, 2020), therefore, compound libraries used with high-resolution mass spectrometry need to be regularly updated to make provisions for new chemicals to assess their fate and occurrences in wastewater (Villarín & Merel, 2020). Most local wastewater treatment plants may not be designed to ensure the removal of some compounds such as engineered pollutants (nanomaterials and nanoparticles) and microplastics as such discharged treated water is considered a source of secondary contamination to the watercourses and environment. The occurrence of such chemicals and pollutants in wastewater could pose a major challenge to wastewater management especially when implementing water reuse.

13.4 WASTEWATER RECYCLING

13.4.1 Aims and Applications of Recycled Wastewater

Wastewater recycling and reuse are the major contributors to water resource management all over the world. In this case, water is employed for applications more than once and this helps to save freshwater from being over-exploited (Tortajada, 2020). Water has become crucial to industrial processes both locally and internationally; the needs of the industrial sector differ primarily from the municipal sector and differ visibly between industries and locations (DECHEMA, 2017). Water is a basic natural resource that several sectors of the economy rely on and despite its significance to human survival on Earth, it has been affected by climate change and this has resulted in ambiguities on the quantity needed for the sustenance of the environment (Akpor et al., 2015). Currently, many nations are confronted with water shortages since the supply of freshwater resources is unable to meet the needs of the hour. Therefore, water reuse or wastewater treatment become one of the viable choices for water security and reduction in environmental degradation arising from improper discharge of wastewater into the environment (Shahid et al., 2022).

Continuous urbanization, rapid population growth, and industrialization have also contributed immensely to the rapid wastewater generation originating from domestic and industrial usages (Armah et al., 2020). Improper wastewater management systems could lead to damage to aquatic and terrestrial biodiversity, oxygen diminution, beach closings, and contamination of groundwater and water bodies (Ahmed et al., 2022; Edokpayi et al., 2017). Hence, adequate management of wastewater is an important approach to safeguarding water systems and obtaining safe and quality drinking water. Wastewater management is usually referred to as the collection, treatment, and recycling of wastewater for secondary applications (Bhaskar, 2022; Fahad et al., 2019). With wastewater recycling and reuse in place, wastewater is now being viewed as a resource rather than waste since several resources such as potable water, nonpotable water, and energy are obtained from it (Libralato et al., 2012).

Wastewater management or wastewater recycling aims at closing the loop/cycles, achieving zero wastewater via obtaining diverse resources from wastewater or transiting from the conventional linear model of material usage to a circular one (Jhansi & Mishra, 2013; Lam et al., 2020; Pikaar et al., 2020). Wastewater recycling also aims at achieving Sustainable Development Goal 6 (SDG 6); ensuring access to water and sanitation services by all (World Business Council for Sustainable Development, 2020). Therefore, the overall aim of water recycling and management is to manage water resources in such a way that the present and future uses of these resources are not diminished and also to protect public health and the environment. Wastewater is utilized in irrigation farming, maintaining environmental flow, drinking water, etc. (Navarro et al., 2020).

13.4.2 Regulations/Policies on Water Reuse

Regulations/policies are major keys to facilitating the development of wastewater recycling projects. One of these key instruments is the Brazil National Water Resources Policy which outlines the commitment of the government toward fast-tracking wastewater treatment plants (Rodriguez et al., 2020). The goal of wastewater policy is to raise the quantity of treated wastewater via wastewater infrastructure or facilities (Peasey et al., 2000). Regulations backing the release of wastewater into the environment often address acceptable limits of the quality of treated effluents and can also inhibit the volume of effluents that can be released into the environment. Besides the restrictions on the intensity of pollutants, the wastewater discharge authorizations could comprise of restrictions on the total mass of a pollutant released into the environment. Therefore, permits must comply with a minimum standard (United States Environmental Protection Agency, 2012). To confirm safe water reuse, the application of water quality standards to a particular usage becomes crucial, ensuring functional monitoring implementation. Some of the standards that have been developed by European Countries (EU) include, Law 106 (I) 2002 Water and Soil Pollution Control and associated regulations (Cyprus), The Legal Framework for the Reuse of Treated Wastewater (Spain), etc. Additionally, some of the standards already developed for developing nations include those of the World Health Organization (WHO), International Standards Organization (ISO), etc. (EU Water Directors, 2016).

13.5 WASTEWATER TREATMENT TECHNOLOGIES AND PROCESSES

The main stages in wastewater treatment processes are preliminary, primary, secondary, and tertiary or advanced treatment processes.

13.5.1 Pretreatment/Preliminary Treatment Processes

The first phase in the wastewater treatment process is the preliminary phase. The pretreatment of the raw wastewater involves both mechanical, physical, and chemical processes. Here, influent is pumped into the treatment plant where floating suspended large particle-sized materials like leaves, stones, sticks, gravel, and sand are removed using screens (Englande Jr & Krenkel, 2003; Manasa & Mehta, 2020). Various screen bars (fine screens, mechanical, vibrating and micro, revolving drum, and coarse mesh screens) can be employed in the preliminary treatment phase (Manasa & Mehta, 2020). Furthermore, grit removal is carried out in a grit chamber where mineral materials, small inorganic solids, sand, silt, gravel, etc. are eliminated (Manasa & Mehta, 2020).

Depending on the type of wastewater to be treated, chemicals are used to regulate the growth and existence of microorganisms in water. Coagulation involves the neutralization of the charges of suspended solids resulting in the formation of sticky materials called "flocs," coagulation also assists in oil and grease removal (Englande Jr & Krenkel, 2003; Kamsonlian et al., 2021). Chemical coagulants such as alum and other salts of Fe such as ferric sulfate, ferric chloride, sodium aluminate, lime, sodium carbonate, etc. are utilized to break oil emulsions, enhance flocculation of the unstable particles into bulky flocs, which stabilize, attach, and are removed by flotation, sedimentation, and during filtration/sludge (Kamsonlian et al., 2021).

13.5.2 Primary Wastewater Treatment Processes

In primary wastewater treatment, a significant percentage of pathogens (bacteria) and microorganisms are removed, in addition, a substantial portion of suspended particles are removed by primary sedimentation and disinfection methods (Al-Rekabi et al., 2007).

13.5.2.1 Sedimentation

Here, treated water from the flocculation step is allowed to flow very slowly allowing larger-sized flocs to settle by gravity at the bottom of the sedimentation basin (Kamsonlian et al., 2021). The small-sized flocs at the base of the sedimentation basin stick together forming bigger flocs referred to as "sludge" which are removed at regular intervals (Manasa & Mehta, 2020).

13.5.2.2 Disinfection

Disinfection assists in removing more microorganisms present during primary treatment and ensures that bacteria and other harmful scum deposits from water, especially for drinking purposes, are removed. Disinfectants commonly used are fluorine and/or chlorine (calcium hypochlorite, sodium hypochlorite) for drinking water (Kamsonlian et al., 2021). Chlorination is a disinfection process using chlorine which reduces the concentration of disease-causing pathogens in water and wastewater. Fluoridation is also applied in the treatment of drinking water, and it entails the use of fluorine to destroy disease-causing microorganisms; it also assists in reducing tooth decay (Kamsonlian et al., 2021).

13.5.2.3 Ozonation

Ozonation of wastewater is more powerful and effective in removing bacteria causing water-borne diseases than chlorination (Kamsonlian et al., 2021). However, ozonation is one of the advanced chemical oxidation treatment methods which will be discussed later in this chapter. Ozonation is based on the initiation of ozone self-decomposition and oxidation in water at an elevated pH into hydroxyl radicals (Zhou & Smith, 2002). However, the problem associated with ozonation is the possible generation of secondary pollutants when bromide ions in water are oxidized into harmful brominated organic byproduct compounds such as polybrominated biphenyls, etc. (Zhou & Smith, 2002).

13.5.3 Secondary Wastewater Treatment

Secondary wastewater treatment assists in lessening the residual concentration of organic and suspended solids in effluents from primary treatment (Al-Rekabi et al., 2007). This involves the use of a biological process where aerobic microbes are used to decompose and remove biodegradable dissolved and colloidal organic materials in wastewater (Al-Rekabi et al., 2007). Secondary treatment processes are outlined below.

13.5.3.1 Activated Sludge Process

This consists of a continuous flow treatment process where sludge containing aerobic microorganisms such as bacteria and protozoa assists in the metabolic breakdown of organic matter in the influent wastewater (Manasa & Mehta, 2020). Active biological flocs are furnished with oxygen (typically, air in the form of bubbles or by turbulent agitation of the wastewater in the tank) in such a way that reasonable contact is maintained between the influent wastewater (within 30–90 mins) with the microorganism in the floc (Englande Jr & Krenkel, 2003). The concentration of microorganisms in the flocs is maintained in the tank by returning a portion of the sludge through the tank and allowing it to settle in a secondary sedimentation basin. Some portion of settled particles are discarded while a portion is fed back into the tank as active microorganisms for subsequent contact with the incoming raw wastewater (Englande Jr & Krenkel, 2003).

13.5.3.2 Extended Aeration

Extended aeration is more of a modified form of activated sludge system which is designed to give longer detention times for the aeration step, thereby resulting in endogenous respiration/oxidation of the sludge where BOD pollutants could be reduced to less than 10 mg/L (Englande Jr & Krenkel, 2003). This process is used in small-scale units such as motels, schools, resorts, and parks (Manasa & Mehta, 2020). Most industrial plants employ extended aeration mainly to facilitate the reduction of toxicity and removal to a minimum (µg/L) of some recalcitrant pollutants such as endocrine disruptive compounds (EDCs) and personal care products (PCPs) (Englande Jr & Krenkel, 2003).

13.5.3.3 Oxidation Ponds and Aerated Lagoons

Otherwise known as stabilization ponds, these are among the oldest practiced wastewater treatment processes and are still in use in many regions. They are a cost-effective treatment technique whose applicability depends on several factors such as land availability, weather, climate conditions, intended purpose/use of the treated wastewater, and location (Englande Jr & Krenkel, 2003). The detention time depends on the climatic conditions and the type of pond, which ranges between 7–180 weeks (Englande Jr & Krenkel, 2003). Ponds are classified as aerobic (maturation), anaerobic, aerated, and facultative ponds. They are applied in the treatment of wastewater from industrial facilities and small communities.

13.5.3.4 Trickling Filters

The trickling filter is a binding growth biological process (e.g., aerobic microbes) forming a biological film on the media surface (Englande Jr & Krenkel, 2003). Trickling filters are designed to treat about 50% of BOD. Here, effluent water coming from the sedimentation process flows from the top of the bed and penetrates through the film filters consisting of layers of sand, charcoal, and gravel where the organic matter, iron, manganese, silts, and very small flocs are absorbed onto the film as they flow through (Englande Jr & Krenkel, 2003). The quality of effluent water can be improved by subsequently directing the effluent to a biological contactor.

13.5.3.5 Rotating Biological Contactor

This is also an attached-growth biological process used in the removal of BOD. Contactors are made up of horizontal shafts which rotate through the wastewater (Manasa & Mehta, 2020). Microbes are usually attached over the shafts like slime layers which then remove organic materials as they rotate through the wastewater (Manasa & Mehta, 2020).

13.5.4 Advanced Wastewater Treatment Processes

Primary and secondary treatment processes are highly efficient in the removal of suspended solids, organic waste matter (BOD), and bacteria found in wastewater. However, other contaminants such as nutrients (phosphorus, nitrogen, etc.), heavy metals, refractory organics, recalcitrant pathogens, toxic organics, and volatile organic compounds (VOCs) are usually not removed as expected using these conventional methods (Zhou & Smith, 2002). Hence, additional treatment steps using advanced treatment technologies have been shown to be effective in removing these toxic materials and nutrients to provide reusable water for domestic and industrial processes (Englande Jr & Krenkel, 2003).

Advanced water treatment is classified based on the process flow steps utilized in the treatment of wastewater:

i. Biological processes for tertiary treatment of effluent for nutrient removal. This includes systems for nitrogen and phosphorus removal.
ii. Physicochemical processes using deep bed filtration and membrane separation.
iii. Hybrid processes consisting of a combination of physicochemical and biological processes as found in membrane bioreactors.

13.5.4.1 Biological Nutrient Removal Technologies

These approaches are not intended for the direct reuse of wastewater, rather, they provide a means for additional treatment resulting in reusable quality municipal wastewater.

i. Intermittently decanted aeration lagoon (IDAL) system: This system is generally used in domestic wastewater (sewage) treatment plants where nitrogen is biologically removed from wastewater following a two-step process (Ngo et al., 2007). In the first step, ammonia-nitrogen is converted into nitrate-nitrogen by nitrifying bacteria using nitrosomonas and nitrobacter, while in the second step, denitrification occurs in the absence of dissolved oxygen and nitrate-nitrogen is converted into nitrogen gas (Ngo et al., 2007). The IDAL system is carried out under different operations of aeration, settling, and decantation at different times but in the same tank. Nitrification operates during the aeration phase while denitrification occurs during the nonaeration duration of the cycle (Ngo et al., 2007)
ii. Biological enhanced phosphorus removal (BEPR) system: This system is based on the ability of most bacteria from the activated sludge to be able to break down the polyphosphates under anaerobic conditions to obtain energy for continued metabolic growth (Ngo et al., 2007). When activated sludge is exposed to anaerobic conditions, and afterward aerated, the polyphosphate-storing bacteria out-perform other microorganisms; they multiply and then phosphates are absorbed from the wastewater resulting in the formation of polyphosphate-supplies for new bacterial growth (Ngo et al., 2007).

13.5.4.1.1 Biodegradation

Microorganisms such as bacteria and fungi naturally degrade pollutants into their constituent materials which are not harmful. Microorganisms utilize these pollutant substrates as food, resulting in their conversion into harmless organic matter as a byproduct (Kamsonlian et al., 2021). Some of the bacterial strains that can break down contaminants in sewage wastewater are *Pseudomonas*, *Acinetobacter*, and *Proteus* (Mathews, 2014). The efficiency of the process is enhanced in the presence of oxygen (for aerobic respiration), moisture, and other nutrients (Kamsonlian et al., 2021). Biodegradation of polychlorinated biphenyls from sewage wastewater treatment plants was studied using bacterial strain MD2 (*Pseudomonas aeruginosa*), and it was found that this bacterial strain completely degraded into chlorobenzoate derivatives after 96 hours (Mathews, 2014).

13.5.4.1.2 Bioremediation

This involves the process of eliminating contaminants by biological means using microorganisms (Kamsonlian et al., 2021). Factors such as pH, soil type, oxygen content, temperature, and quantity of microbes present at the remediation site control the efficiency and effectiveness of the bioremediation process (Chan et al., 2022). Bioremediation could be undertaken at the site or off-site of the contamination. Some bioremediation processes include bioventing, biostimulation, bioaugmentation, composting, land-farming, and rhizorfiltration (Kamsonlian et al., 2021).

The selection of a bioremediator is based on its ability to decompose specific organic contaminants (Adhikari et al., 2019). For example, *Alcaligenes*, *Mycobacterium*, *Pseudomonas*, *Spingomonas*, and *Rhodococcus* are responsible for the degradation of hydrocarbons, pesticides, polyaromatic compounds, etc. while anaerobic bacteria are responsible for the decomposition of polychlorinated biphenyls (Kamsonlian et al., 2021). This process is efficient for the degradation of wastewater laden with dyes (Kamsonlian et al., 2021).

13.5.4.2 Physicochemical Processes

13.5.4.2.1 Deep Bed Filtration

i. Deep bed (granular bed) sand filters are used as the final clarification step in potable water treatment (Ngo et al., 2007). Some deep bed filtration processes include direct filtration, contact-flocculation filtration, mobile bed filters, and floating media filtration (Ngo et al., 2007).
ii. Direct filtration (DF) refers to the direct use of the flocculation process for a short time followed by the filtration step. DF is an advantageous process due to its ability to produce filterable flocs, and the availability of choosing from a wide range of filter media (Ngo et al., 2007). The efficiency of the DF process requires that information concerning the flocculation condition and flocculation type are provided so that flocs that are of filterable size are obtained in order not to clog the filter bed (Englande Jr & Krenkel, 2003).
iii. Contact-flocculation filtration (CFF) is where coagulation is carried out just before the filtration process (Ngo et al., 2007). Despite the low capital cost of CFF and DF operation, the major limitation of these technologies is fast filter clogging which requires constant backwashing.

13.5.4.2.2 Membrane Separation

Membranes are the semipermeable medium through which the feed stream is separated into permeate and retentate when a driving force is applied. Membrane separation can be classified based on: the size range of permeating species; the membrane rejection mechanism; the type of driving force utilized; the chemical structure and material used; the geometry of construction of the membrane; and the configuration of the membrane (Zhou & Smith, 2002). Membrane separation is employed in the removal of natural and synthetic organic materials, pathogens, toxic heavy metals, bacteria, and minute solid materials which can impart odor, color, and taste in water (Al-Rekabi et al., 2007). Figure 13.1 depicts the pore size, transmembrane pressure (TMP), and various particles removed using different membrane separation techniques. The major membrane-based separation processes are reverse osmosis, microfiltration, ultrafiltration, and nanofiltration. The common membrane configurations are spiral wound, hollow fiber, plate and frame, tubular, and rotating disc (Zhou & Smith, 2002). Successful membrane separation depends on the proper membrane material selection, which should be

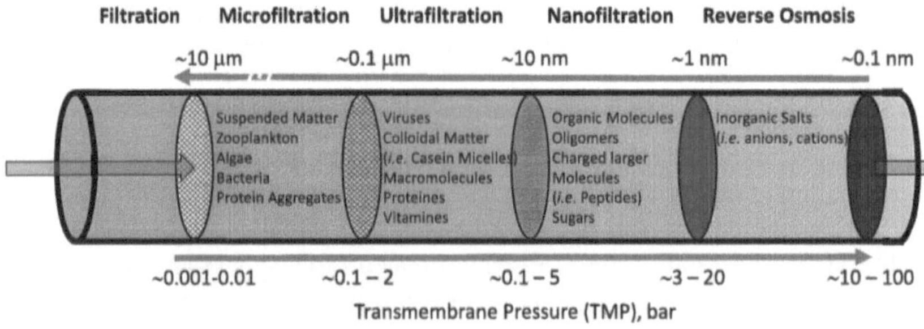

FIGURE 13.1 Pore size, TMP, and various particles removed using different membrane separation techniques.

Source: Abetz et al., (2021). Distributed under Open Access under the Creative Common Attribution 4.0 International Public License (CC BY 4.0) (http://creativecommons.org/licenses/by/4.0/))

based on excellent permeate flux, high rejection rate, good chemical resistance, cost-effectiveness, and durability of the membrane material employed (Zhou & Smith, 2002). In addition, the membrane performance also depends on the configuration/arrangement of the membrane, and the properties of the membrane such as pore size and molecular weight cut-off (MWCO) (Kurniawan et al., 2006; Zhou & Smith, 2002). However, the major drawbacks in the use of membrane technology is high fouling, particle cake deposition, and concentration polarization (Anjum et al., 2016; Kamsonlian et al., 2021). New advances in membrane technology research that incorporate nanomaterials including both inorganic or organic materials such as biochar, activated carbon, carbon nanotubes, graphene oxides, carbon nanofibers, metal and oxide nanoparticles, and nanometals, etc. into polymeric membrane materials resulting in the formation of mixed matrix membranes (MMM) have resurfaced (Anjum et al., 2016). The incorporated nanomaterials possess a high surface area, thermal stability, and strength which modifies the membrane by adding new functionalities that result in a reduction in membrane operational cost, improved fouling resistance, high permeate flux (permeability), high contaminate rejection rate, and high removal efficiency which benefits quality water treatment (Anjum et al., 2016).

i. Reverse osmosis (RO)

RO has the least pore size and can effectively retain nearly all ions under a high-pressure difference across the membrane with rather low permeability flux. RO is used in the removal of bacteria, viruses, pesticides, radium, and some naturally occurring organic materials (Kamsonlian et al., 2021). The RO system requires multiple filtration units and more pretreatment (disinfection) for heavily contaminated solutions. The major drawbacks to the use of this system are a high cost requirement for installation and maintenance and the inability of the system to remove chemicals like chlorine used in disinfection treatment (Kamsonlian et al., 2021).

ii. Microfiltration (MF)
An MF membrane is employed in the separation of microorganisms (algae), suspended solids (clay, sand, etc.), and colloidal micrometer particles from water. This type of membrane is permeable to species whose diameter is up to 0.5 μm (Zhou & Smith, 2002). The average pore size for MF is between 0.1–10 μm with a MWCO > 1,000,000 Da (Kamsonlian et al., 2021). MF membranes possess relatively high permeate flux under low transmembrane pressure. Daels et al. (2011) investigated the performance of WSCP (Poly[dimethylimino)(2-hydroxy-1,3-propanedily) Chloride] functionalized electrospun nanofiber microfiltration membranes in the separation and disinfection of hospital wastewater. They found that gram-positive bacteria such as *Staphylococcus aureus* and *Escherichia coli* were successfully removed from hospital wastewater using microfiltration membranes (Daels et al., 2011).

iii. Ultrafiltration (UF)
Ultrafiltration membranes are employed in the separation of wastewater for the removal of microorganisms, humic matter, and pathogens. UF membranes have a pore size in the range of 0.005–0.1 μm with a MWCO of about 10,000–100,000 Da. The operating transmembrane pressure of UF membranes is around 2–7 bar and for effective treatment by UF, disinfection treatment is employed (Kamsonlian et al., 2021). A study by Zhang et al. (2009) utilized ultrafiltration membranes consisting of polyvinyl chloride (H3-1060-V) in treating secondary water effluent at Scottside Water Campus as pretreatment for a RO system. Their results showed that PVC-UF membranes were effective in removing suspended solids, turbidity, and color from secondary water effluent with mostly organic compounds and few inorganic metals. The UF membrane was a successful pretreatment method in improving the average permeate flux, with recovery by the RO system up to 34% and 21%, respectively (Zhang et al., 2009).

iv. Nanofiltration
NF is a low-pressure driven membrane-based treatment process that can efficiently remove most organic molecules, nearly all viruses, and natural organic matter (NOM) with high permeate flux. In terms of separation by pore size, NF lies between UF and MF, with a pore size in the range of 0.2–2 nm and possesses an MWCO of 200–1000 Da (Kamsonlian et al., 2021). The main disadvantage of this technology is its high operational energy requirement (around 0.3–1 kWh/m^3) when compared to UF and MF systems (Kamsonlian et al., 2021). Again, NF possesses high fouling capabilities and has limited retention for salts and univalent ions. Agboola et al. (2016) studied the fouling properties of two different nanofiltration membranes, NF90-2450 and NF90-2540, in removing magnesium and manganese ions at three different pH values. Their results showed that pH affects the porosity and rejection of the metal ions, that is, a low pH value results in lower metal ion rejection. Similarly, titania ceramic nanofiltration membranes were tested in removing over 20 cations, anions, and compounds at a commercial pilot unit in a Canadian oil sand mine (Cabrera et al., 2021). According to the study by Cabrera et al. (2021), low

specific flux favors rejection of ions and solutes with about 100% rejection for TSS and 75% TOC. The studied nanofiltration membrane showed a high preference for both cation and anion rejection with high charge density (Cabrera et al., 2021). The potential of newly emerging 1D carbon nanotubes and 2D graphene, graphene oxide nanochannel membranes for filtration and separation applications have been explored. Han et al. (2013) tested the performance of fabricated ultrathin graphene nanofiltration membranes in treating organic dyes and salts. Their findings showed the membrane's excellent performance for organic charged dye retention based on the mechanism of physical sieving and electrostatic interactions (Han et al., 2013).

13.5.4.2.3 Adsorption

Adsorption is a mass transfer operation in which materials are transferred from one phase to another, that is from the liquid phase to the solid phase by either physical or chemical interaction (Adhikari et al., 2019). Adsorption is one of the recently applied treatment technologies in water and wastewater treatment due to its low cost, high efficiency, easy operation, and simple design (Ahmed et al., 2015). Adsorption has been applied in the treatment of industrial and municipal wastewater for the removal of emerging persistent organic pollutants of concern that cannot be removed by wastewater treatment plants (WWTPs) such as pharmaceuticals, surfactants, flame retardants, inorganic nutrients, heavy metals, pesticides, herbicides, preservatives, food additives, hormones, dyes, etc. from water (Sophia & Lima, 2018). Several adsorbents have been explored in the removal and treatment of water such as biochars, activated carbon, carbon nanotubes and carbon nanotube-based composites, graphene, graphene oxides, graphites and their composites, metallics and their oxide nanoparticles, etc. (Ahmed et al., 2015). Adsorbent materials utilized in adsorption processes can be obtained from naturally occurring materials and bipolymers including agricultural wastes and industrial byproducts (Adhikari et al., 2019). Several studies utilized graphene-based materials for the adsorption of organic pollutants and heavy metals from wastewater (Balasubramani et al., 2020; Cai et al., 2016; Guerra et al., 2021; Tene et al., 2022). Wang et al. (2018) employed GO loaded with TiO_2 (GO-TiO_2) in the absorptive removal of tetracycline. Both activities of GO nanosheets and granular activated carbon (GAC) were tested and compared in the absorption of methylene blue dyes and pharmaceutical compounds from water. The absorption capacity of the GO nanosheet for methylene blue was found to be three times higher than with GAC (Allgayer et al., 2020). Similarly, various researchers have used carbon nanotubes (CNTs) and CNT-based materials for the absorption of heavy metals from wastewater (Bhanjana et al., 2017; Mubarak et al., 2015). Modification of these adsorbents improves their adsorption capacity and efficiency for contaminant removal (Anjum et al., 2016).

13.5.4.2.4 Ion Exchange

Ion-exchange processes involve the exchange of toxic ions with nontoxic ones from a solid support material. The ion exchanger could be an anion or cation exchanger which are capable of exchanging anion and cation contaminants, respectively,

(Adhikari et al., 2019). Ion-exchange technology is efficient in the removal of inorganic and organic contaminants at low concentrations (within 250 mg/L) from industrial wastewater such as pharmaceutical wastewater, etc. (Adhikari et al., 2019). It can also be used in the production and purification of potable and softened water. Ion exchangers consist of active sites on their surfaces, with a high surface area and pore size, they can be either natural or synthetic resins. Some of the common exchangers applied in advanced wastewater treatment technology are zeolites, clinoptiloite, acrylic resins, metal-acrylic resins, e.g., Amberlite XAD resins, Amberlite IRC748, Amberlite IR-120, Amberjet 1200H, sodium silicate, polystyrene sulfonic acid, etc. (Adhikari et al., 2019; Kurniawan et al., 2006; Woodberry et al., 2007). In a field trial by Woodberry et al. (2007), Amberlite IR748, a chelating cation-exchange resin, was identified as an effective ion-exchange column, which can efficiently and effectively remove dissolved metal contaminants (Cd, Zn, Ni, Cu, and Fe) from surface waters discharged from an abandoned coastal disposal site. Brown macro-algae were also utilized as natural cation-exchangers in the removal of transition metal ions (Cu, Zn, and Ni) from petrochemical wastewater (Cechinel et al., 2016). Some of the challenges of ion-exchange technology are the inability of the system to handle highly concentrated contaminant solution and its high sensitivity to the solution pH; hence, in many cases, pretreatment techniques are always needed (Kurniawan et al., 2006).

13.5.4.2.5 Advanced Oxidative Processes (AOP)

The advanced oxidation process is a chemical oxidation process employed in the efficient removal of persistent and recalcitrant pollutants from wastewater, which are perhaps not degraded by other wastewater treatment techniques (Zhou & Smith, 2002). The principle of AOP treatment technology is based on highly reactive radicals of hydroxyl (OH^-) as the primary oxidant (Zhou & Smith, 2002). The hydroxyl radical is a strong oxidizing species with the potential to speed up the oxidation of contaminants in water and wastewater treatment. The generation of OH^- radicals is accelerated by hydrogen peroxide, ozone, heterogenous photocatalysts, titanium oxide, or by a combination of oxidants using UV radiation, ultrasound, and high electron beam irradiation (Zhou & Smith, 2002). AOP can degrade pollutants between the range of 100 ppm to 5 ppb and once OH^- radicals are generated, can breakdown nearly all organic complexes while larger organic chemicals are simultaneously disintegrated into water and minute gaseous products (Adhikari et al., 2019). Widely applied technologies are O_3-H_2O_2, H_2O_2-UV, O_3-UV, and heterogenous photocatalytic processes.

The Fenton process is frequently used in generating strong reactive hydroxyl radicals in the presence of hydrogen peroxide (H_2O_2) and metal ions (Fe^{2+}) (Adhikari et al., 2019).

Semiconductor-based photocatalysis is also utilized in the treatment of industrial and municipal wastewater. Here, semiconductive material is illuminated with light energy, which is greater than the bandgap energy of the material generating the electron-hole pair in the conductor and valence band of the semiconductor (Adhikari et al., 2019). The charge carriers then move up to the surface of the semiconductor for redox reaction producing reactive hydroxyl radicals (Adhikari et al., 2019). Commonly used semiconductors for the oxidation of contaminants are nanomaterials such as carbon nanotubes, graphene, graphites, graphene oxide, TiO_2, ZnS, CeO_2, ZnO, CdS, etc. (Adhikari et al., 2019; Anjum et al., 2016). The benefits of AOP technology include a fast rate of reaction, the ability to treat multiple refractory organic contaminants all at the same time, their ability to generate less harmful byproducts, and the potential to completely mineralize organic pollutants into carbon dioxide (Adhikari et al., 2019).

13.5.4.3 Hybrid Treatment Technologies

The combination of both biological and physicochemical processes in the treatment of water and wastewater endeavors to gain all the benefits of both processes in one step (Englande Jr & Krenkel, 2003; Kamsonlian et al., 2021; Mack et al., 2004). Among the hybrid treatment methods, membrane bioreactors, membrane filtration-powdered activated carbon (PAC), or granulated activated carbon have gained great interest in water and wastewater treatment areas (Englande Jr & Krenkel, 2003; Kamsonlian et al., 2021). One study developed a hybrid treatment process which combined both flotation/adsorption and membrane filtration (microfiltration) in the treatment and removal of heavy metals (up to 500 mg/L) from highly contaminated wastewater (Mavrov et al., 2003). Figure 13.2 illustrates a hybrid treatment technique combining membrane microfiltration, adsorption, and flotation processes in treating highly contaminated wastewater.

13.5.4.3.1 Membrane Bioreactors (MBRs)

MBRs are employed in the treatment of industrial and municipal wastewater. They consist of a combination of biological reactor units along with microfiltration and ultrafiltration membranes (Englande Jr & Krenkel, 2003). MBRs use biomass to decompose contaminants and membrane filtration (MF and UF) to separate the

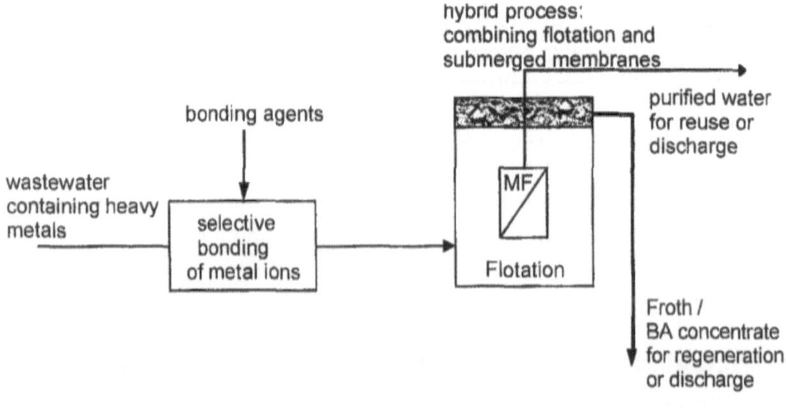

FIGURE 13.2 Hybrid (membrane microfiltration and floatation) wastewater treatment processes.

Source: Mavrov et al. (2003) with copyright permission from Elsevier Inc.

biomass from water (Zhou & Smith, 2002). The benefits of MBR systems is the generation of relatively small amounts of sludge with good discharge potentials; they are also operated at longer retention time which results in the complete oxidation of organic and nutrient pollutants, hence, they are very effective for nitrogen and BOD removal (Kamsonlian et al., 2021). However, the challenges of the MBR system are its low permeate flux due to fouling as a result of foulant deposition on the membrane surface, high energy demand and cost of operation, high capital cost, and the need for frequent membrane replacement (Kamsonlian et al., 2021).

13.6 SUMMARY

Over the years, stringent environmental regulations have facilitated improvement in the quality of treated water around the globe. Water recycling has been recognized as a feasible approach to alleviating water shortage. The recent advanced trend in the recycling and treatment of wastewater proves that every system is unique in its features and therefore, each technology has its merits and demerits in areas of its economic (capital investment, cost of operation, and cost of maintenance, etc.) and environmental (type of contaminant, and contaminant concentration) impacts. However, the major parameters considered in the selection of suitable wastewater treatment technology are the cost-effectiveness and simplicity of the system. The advanced technology in membrane adsorption and advanced oxidation technologies has been instrumental in achieving water recycling. It is expected that, in the future, high-performance hybrid-based plants that continuously treat wastewater will be designed. Finally, no matter the treatment technology employed, it is important that thorough and meticulous management strategies through regulations, acts, policies, and frameworks are implemented to ensure an effective, environmentally friendly management approach which encourages wastewater treatment at its source (point of use treatment) and by waste generators; and development of sustainable strategies for water resource management which also ensure adequate disposal of polluted water in such a way that waterbodies receiving discharged effluent water are well protected.

ACKNOWLEDGMENT

The authors would like to thank the University of Johannesburg and the Faculty of Engineering and the Built Environment of the University of Johannesburg for providing the financial support for this research under the platforms of Global Excellence Stature 4.0 (GES 4.0) and University Research Council (URC).

REFERENCES

Abetz, V., Brinkmann, T., and Mustafa, S. (2021). Fabrication and function of polymer membranes. *Chemistry Teacher International*, *3*(2), 141–154.

Adhikari, S., Mandal, S., Kim, D., and Mishra, A. K. (2019). An overview of treatment technologies for the removal of emerging and nanomaterials contaminants from municipal and industrial wastewater. In M. A. Kumar, A. H., and D. Nadjib (Eds.), *Emerging and Nanomaterial Contaminants in Wastewater* (pp. 3–40). Elsevier Inc.

Agboola, O., Mokrani, T., and Sadiku, R. (2016). Porous and fractal analysis on the permeability of nanofiltration membranes for the removal of metal ions. *Journal of Material Science, 51*, 2499–2511.

Ahmed, M. B., Zhou, J. L., Ngo, H. H., and Guo, W. (2015). Adsorptive removal of antibiotics from water and wastewater: Progress and challenges. *Science of the Total Environment, 532*, 112–126.

Ahmed, S. F., Mehejabin, F., Momtahin, A., Tasannum, N., Faria, N. T., Mofijur, M., …Mahlia, T. M. I. (2022). Strategies to improve membrane performance in wastewater treatment. *Chemosphere, 306*, 135527.

Akpor, O. B., Kayode-Oni, I., and Babalola, O. O. (2015). Wastewater management, recycling and discharge. *Hydrology, 3*(3), 33.

Al-Rekabi, W. S., Qiang, H., and Qiang, W. W. (2007). Improvements in wastewater treatment technology. *Pakistan Journal of Nutrition, 6*(2), 104–110.

Allgayer, R., Yousefi, N., and Tufenkji, N. (2020). Graphene oxide sponge as adsorbent for organic contaminants: Comparison with granular activated carbon and influence of water chemistry. *Environmental Science: Nano, 7*, 2669–2680.

Anjum, M., Miandad, R., Waqas, M., Gehany, F., and Barakat, M. A. (2016). Remediation of wastewater using various nano-materials. *Arabian Journal of Chemistry, 12,* 4897–4919.

Armah, E. K., Chetty, M., Adedeji, J. A., Kukwa, D. T., Mutsvene, B., Shabangu, K. P., and Bakare, B. F. (2020). Emerging trends in wastewater treatment technologies: The current perspective. In *Promising Techniques for Wastewater Treatment and Water Quality Assessment* (pp. 1–27): IntechOpen.

Balasubramani, K., Sivarajasekar, N., and Naushad, M. (2020). Effective adsorption of antidiabetic pharmaceutical (metformin) from aqueous medium using graphene oxide nanoparticles: Equilibrium and statistical modelling. *Journal of Molecular Liquids, 301*, 112426.

Bhanjana, G., Dilbaghi, N., Kim, K., and Kumar, S. (2017). Carbon nanotubes as sorbent material for removal of cadmium. *Journal of Molecular Liquids, 242*, 966–970.

Bhaskar, A. (2022, 25 October 2022). New technology trends in the wastewater management industry. Retrieved from www.forbes.com/sites/theyec/2022/10/25/new-technology-trends-in-the-wastewater-management-industry/?sh=50a097f14199

Cabrera, S. M., Winnubst, L., Richter, H., Voigt, I., and Nijmeijer, A. (2021). Industrial application of ceramic nanofiltration membranes for water treatment in oil sands mines. *Separation and Purification Technology, 256*, 117821.

Cai, M. Q., Su, J., Hu, J. Q., Wang, Q., Dong, C. Y., Pan, S. D., and Jin, M. C. (2016). Planar graphene oxide-based magnetic ionic liquid nanomaterial for extraction of chlorophenols from environmental water samples coupled with liquid chromatography–tandem mass spectrometry. *Journal of Chromatography A, 1459*, 38–46.

Cechinel, M. A. P., Mayer, D. A., Pozdniakova, T. A., Mazur, L. P., Boaventura, R. A. R., de Souza, A. A. U., de Souza, S. M. A. G. U., and Vilar, V. J. P. (2016). Removal of metal ions from a petrochemical wastewater using brown macro-algae as natural cation-exchangers. *Chemical Engineering Journal, 286*, 1–15.

Chan, S. S., Khoo, K. S., Chew, K. W., Ling, T. C., and Show, P. L. (2022). Recent advances biodegradation and biosorption of organic compounds from wastewater: Microalgae-bacteria consortium – A review. *Bioresource Technology, 344*, 126159.

Chandra, R. and Kumar, V. (2017). Detection of androgenic-mutagenic compounds and potential autochthonous bacterial communities during in-situ bioremediation of post methanated distillery sludge. *Frontiers in Microbiology, 8*, 87. https://doi.org/10.3389/fmicb.2017.00887

Daels, N., De Vrieze, S., Sampers, I., Decostere, B., Westbroek, P., Dumoulin, A., Dejans, P., De Clerck, K., and Van Hulle, S. W. H. (2011). Potential of a functionalised nanofibre microfiltration membrane as an antibacterial water filter. *Desalination, 275*, 285–290.

DECHEMA. (2017). Trends and perspectives in industrial water treatment. *Raw Water – Process – Waste Water Position Paper by the ProcessNet Subject Division Production-Integrated Water/Waste Water Technology.* Retrieved from Frankfurt: https://dechema.de/dechema_media/Downloads/Positionspapiere/Industrial_Watertechnologies_Positionpaper_ProcessNet2017.pdf

Dey, S., Bano, F., and Malik, A. (2019). Pharmaceuticals and personal care product (PPCP) contamination-a global discharge inventory. In *Pharmaceuticals and Personal Care Products: Waste Management and Treatment Technology* (pp. 1–26). Elsevier Inc.

Edokpayi, J. N., Odiyo, J. O., and Durowoju, O. S. (2017). Impact of wastewater on surface water quality in developing countries: A case study of South Africa. *Water Quality, 10*, 66561.

Englande Jr, A. J., and Krenkel, P. A. (2003). Wastewater treatment and water reclamation. In *Encyclopedia of Physical Science and Technology* (pp. 639–670).

EU Water Directors. (2016). Common implementation strategy for the water framework directive and the floods directive. *Guidelines on Integrating Water Reuse into Water Planning and Management in the Context of the WFD*, 95. Retrieved from https://ec.europa.eu/environment/water/pdf/Guidelines_on_water_reuse.pdf

Fahad, A., Mohamed, R. M. S., Radhi, B., and Al-Sahari, M. (2019). Wastewater and its treatment techniques: An ample. *Indian Journal of Science and Technology, 12*(25), 13.

Guerra, A. C. S., de Andrade, M. B., dos Santos, T. R. T., and Bergamasco, R. (2021). Adsorption of sodium diclofenac in aqueous medium using graphene oxide nanosheets. *Environmental Technology, 42*(16), 2599–2609.

Han, Y., Xu, Z., and Gao, C. (2013). Ultrathin graphene nanofiltration membrane for water purification. *Advanced Functional Materials, 23*, 3693–3700.

Hosseini, S. S., Bringas, E., Tan, N. R., Ortiz, I., Ghahramani, M., Shahmirzadi, M. A. A., and Alaei Shahmirzadi, M. A. (2016). Recent progress in development of high performance polymeric membranes and materials for metal plating wastewater treatment: A review. *Journal of Water Process Engineering, 9*, 78–110.

Jhansi, S. C., and Mishra, S. K. (2013). Wastewater treatment and reuse: sustainability options. *Consilience* (10), 1–15.

Kamsonlian, S., Yadav, S., Wasewar, K. L., Gaur, A., and Kumar, S. (2021). Treatment of contaminated water: Membrane separation and biological processes. In *Contamination of Water: Health Risk Assessment and Treatment Strategies* (pp. 339–350). Elsevier Inc.

Kurniawan, T. A., Chan, G. Y. S., Lo, W. H., and Babel, S. (2006). Physico-chemical treatment techniques for wastewater laden with heavy metals. *Chemical Engineering Journal, 118*, 83–98.

Kumar, V., Kumar, S., Pinê Américo-Pinheiro, J.H., Vinthange, M., Sher, F. (2023). Editorial: Emerging approaches for sustainable management for wastewater. *Frontiers in Environmental Science, 10*:1122659. https://doi.org/10.3389/fenvs.2022.1122659

Kumar, V., Verma, P. (2023). A critical review on environmental risk and toxic hazards of refractory pollutants discharged in chlorolignin waste of pulp and paper mills and their remediation approaches for environmental safety. *Environmental Research*, 236, 116728. https://doi.org/10.1016/j.envres.2023.116728

Lam, K. L., Zlatanović, L., and van der Hoek, J. P. (2020). Life cycle assessment of nutrient recycling from wastewater: A critical review. *Water Research, 173*, 115519.

Libralato, G., Ghirardini, A. V., and Avezzù, F. (2012). To centralise or to decentralise: An overview of the most recent trends in wastewater treatment management. *Journal of Environmental Management, 94*(1), 61–68.

Lyu, S., Chen, W., Zhang, W., Fan, Y., and Jiao, W. (2016). Wastewater reclamation and reuse in China: Opportunities and challenges. *Journal of Environmental Sciences*, *39*, 86–96.

Mack, C., Burgess, J. E., and Duncan, J. R. (2004). Membrane bioreactors for metal recovery from wastewater: A review. *Water SA*, *30*(4), 521–532.

Madhav, S., Ahamad, A., Singh, A. K., Kushawaha, J., Chauhan, J. S., Sharma, S., and Singh, P. (2020). Water pollutants: Sources and impact on the environment and human health. In D. Pooja, P. Kumar, P. Singh, and S. Patil (Eds.), *Sensors in Water Pollutants Monitoring: Role of Material, Advanced Functional Materials and Sensors* (pp. 43–62). Springer, Singapore.

Manasa, R. L., and Mehta, A. (2020). Wastewater: Sources of pollutants and its remediation. *Environmental Biotechnology*, 2, 197–219.

Mathews, S. (2014). Biodegradation of Polychlorinated Biphenyls (PCBs), Aroclor 1260, in wastewater by Isolate MD2 (*Pseudomonas aeruginosa*) from wastewater from Notwane sewage treatment plant in Gaborone, Botswana. *Bioremediation and Biodegradation*, *5*(7), 266.

Mavrov, V., Erwe, T., Blocher, C., and Chmiel, H. (2003). Study of new integrated processes combining adsorption, membrane separation and flotation for heavy metal removal from wastewater. *Desalination*, *157*, 97–104.

Mubarak, N. M., Sahu, J. N., Abdullah, E. C., and Jayakumar, N. S. (2015). Rapid adsorption of toxic Pb(II) ions from aqueous solution using multiwall carbon nanotubes synthesized by microwave chemical vapor deposition technique. *Journal of Environmental Sciences (China)*, *45*, 143–155.

Navarro, D., Cantero, R., Valls, E., and Puig, R. (2020). Circular economy: The case of a shared wastewater treatment plant and its adaptation to changes of the industrial zone over time. *Journal of Cleaner Production, 261*, 121242.

Ngo, H. H., Vigneswaran, S., and Sundaravadivel, M. (2007). Advanced treatment technologies for recycle/reuse of domestic wastewater. *Wastewater Recycle Reuse and Reclamation*, *I*, 77–98.

Peasey, A., Blumenthal, U., Mara, D., and Ruiz-Palacios, G. (2000). A review of policy and standards for wastewater reuse in agriculture: A Latin American perspective. *WELL Study, Task, 68*.

Pikaar, I., Huang, X., Fatone, F., and Guest, J. S. (2020). Resource recovery from water: From concept to standard practice. *Water Research* , 178, 115856.

Rodriguez, D. J., Serrano, H. A., Delgado, A., Nolasco, D., and Gustavo, S. (2020). *From Waste to Resource-Shifting Paradigms for Smarter Wastewater Interventions in Latin America and the Caribbean: Background Paper II. Showcasing the River Basin Planning Process through a Concrete Example-The Río Bogotá Cleanup Project*. International Bank for Reconstruction and Development / The World Bank, Washington.

Shahid, M. K., Kashif, A., Pathak, P., Choi, Y., and Rout, P. R. (2022). Water reclamation, recycle, and reuse. In A. An, V. Tyagi, M. Kumar, and Z. Cetecioglu (Eds.), *Clean Energy and Resource Recovery* (pp. 39–50): Elsevier, Amsterdam, Netherlands.

Singh, J., Yadav, P., Pal, A. K., and Mishra, V. (2020). Water Pollutants: Origin and Status. In D. Pooja, P. Kumar, P. Singh, and S. Patil (Eds.), *Sensors in Water Pollutants Monitoring: Role of Material, Advanced Functional Materials and Sensors* (pp. 5–20). Springer, Singapore.

Sophia A., C., and Lima, E. C. (2018). Removal of emerging contaminants from the environment by adsorption. *Ecotoxicology and Environmental Safety*, *150*, 1–17.

Tene, T., Bellucci, S., Guevara, M., Viteri, E., Polanco, M. A., Salguero, O., Vera-Guzmán, E., Valladares, S., Scarcello, A., Alessandro, F., Caputi, L. S., and Gomez, C. V. (2022). Cationic pollutant removal from aqueous solution using reduced graphene oxide. *Nanomaterials*, *12*(3), 309.

Tortajada, C. (2020). Contributions of recycled wastewater to clean water and sanitation Sustainable Development Goals. *NPJ Clean Water*, 3(1), 1–6.

United Nations. (2022). *Water and Sanitation – United Nations Sustainable Development*. The Sustainable Development Goals Report. www.un.org/sustainabledevelopment/water-and-sanitation/

United States. Environmental Protection Agency. (2012). *Guidelines for Water Reuse*. Retrieved from Washington, DC: www.epa.gov/sites/default/files/2019-08/documents/2012-guidelines-water-reuse.pdf

Villarín, M. C. and Merel, S. (2020). Paradigm shifts and current challenges in wastewater management. *Journal of Hazardous Materials*, *390*, 122139.

Wang, J., Liu, R., and Yin, X. (2018). Adsorptive removal of tetracycline on graphene oxide loaded with titanium dioxide composites and photocatalytic regeneration of the adsorbents. *Journal of Chemical and Engineering Data*, *63*, 409–416.

Woodberry, P., Stevens, G., Northcott, K., Snape, I., and Stark, S. (2007). Field trial of ion-exchange resin columns for removal of metal contaminants, Thala Valley Tip, Casey Station, Antarctica. *Cold Regions Science and Technology*, *48*, 105–117.

World Business Council for Sustainable Development. (2020). *Wastewater Zero. A Call to Action for Business to Raise Ambition for SDG 6.3*. Retrieved from Switzerland: file:///C:/Users/user/Downloads/Wastewater%20Zero_SGD%206.3.pdf

Zhang, X., Chen, Y., Konsowa, A. H., Zhu, X., and Crittenden, J. C. (2009). Evaluation of an innovative polyvinyl chloride (PVC) ultrafiltration membrane for wastewater treatment. *Separation and Purification Technology*, *70*(1), 71–78.

Zhou, H., and Smith, D. W. (2002). Advanced technologies in water and wastewater treatment. *Journal of Environmental Engineering Science*, *1*, 247–264.

14 Decreasing the Contaminant Load of Compounds Present in Whey by Using Microorganisms Capable of Generating Microbial Cellulose

Martha Irene Salazar-Manzanares, Julia Mariana Márquez-Reyes, Beatriz Adriana Rodríguez-Romero, Gerardo Méndez-Zamora, and Alejandro Isabel Luna-Maldonado

14.1 THE DAIRY INDUSTRY

Globally, dairy industry production in 2016 was 796 million tons. However, it has increased by about 59%, from 530 million tons in 1988 to 843 million tons in 2018 (FAO, 2018), increasing every four years by over 10% (FAOSTAT, 2017). Similarly, the average milk production in Mexico is more than 12 billion liters per year (Robledo-Padilla, , 2018). While it is characterized by not using water as a final component in producing its products, this does not mean water is not consumed. This sector requires more water for its auxiliary operations' processes, where approximately 20–40% of the total water is consumed, particularly in washing, cleaning, and disinfection processes. Therefore, it is classified as an industry that generates large volumes of wastewater with high organic matter content (Lucas & García, 2018). It requires high energy consumption in the start-up of machines, production, and waste management, directly impacting the increase of greenhouse gases. On the other hand, the most important environmental problem is the generation of wastewater, as it produces between 3739 and 11,217 million m^3 of waste per year; this is one to three times the volume of processed milk (Armijo et al., 2021). Due to their complexity, the dairy sector's processes are very varied. Therefore, the contamination generated by

 DOI: 10.1201/9781003441069-14

the cheese, cream, and butter production processes causes environmental problems due to the large amounts of liquid waste, mainly whey, diluted milk, separated milk and cream (Prieto-García et al., 2012).

One of the characteristics of dairy industries is the amount of organic matter present, the high biodegradability, the content of oils and fats, phosphorus, nitrates, suspended solids (especially cheese producers), and a significant variability of pH. According to OCDE/FAO projections (2015), world milk production by 2024 could increase to 175 million tons, with cheese being the most significant dairy product, with resulting higher whey production, which in turn represents a severe pollution problem due to its high chemical and biological oxygen demand (Santos et al., 2017).

14.1.1 Wastewater in the Dairy Industry

By definition, wastewater is water from processes used in different industrial management systems that require treatment prior to their discharge. In the dairy industry, effluents are characterized by their organic nature. According to Indigoyen-Ramirez (2019), liquid effluent temperature, pH variations, total solids, COD, high phosphorus levels, fats, and oils indicate dairy wastewater contamination. Additionally, wastewater's rapid acidification is due to the fermentation of the milk sugar. Therefore, this variation is a crucial point to consider when treating these industrial effluents. The volume of wastewater discharge produced is abundant, so it is estimated to be between 1.5–5 L of wastewater for each liter of milk processed. This differs by the technology used in the production processes, thus resulting in wastewater having variable characteristics (Chen et al., 2018). The stages of pasteurization, homogenization, and processing of dairy products, such as cheese, butter, and milk powder, produce a greater volume of wastewater (Tirado-Armesto et al., 2016). For example, for the manufacture of butter, between 1–3 L of wastewater is generated per liter of milk processed (L/L), and for the manufacture of cheese, it amounts to 2– 4 L/L (Centro de Actividad Regional de la producción, 2002).

According to Verduga-Vera (2020), effluents from dairy products are characterized by high COD and BOD_5, ranging from 2000 to 4000 mg/L and 2000 to 3000 mg/L, respectively. According to Ghaly et al. (2003), COD and BOD_5 organic matter are mainly contributed by lactose, amino acids, proteins, and cheese whey. They also contain other elements, such as suspended solids with values of 0.1–22 g/L, and inorganic elements, such as concentrations of 0.05 g/L phosphorus and 1.7 g/L nitrogen. On the other hand, they present elements such as calcium, magnesium, and iron. Therefore, the concentration in the effluent will depend on the final product processed and the management system. Chen et al. (2018) point out that the effluent from this sector is biodegradable, which means that the COD/BOD_5 ratio can fluctuate between 0.5–0.8, meaning that by using biological treatments, a removal of part of the organic matter present in the effluent is enabled. Within the dairy industry, the potentially polluting effluent comes from obtaining butter and cheese, especially whey, due to its high lactose and protein content (Gaibor, 2014). It has been identified that the effluents from the dairy industry, specifically whey, have an abundant amount of vitamins such as biotin, pantothenic acid, riboflavin, thiamine, minerals such as sodium, potassium, magnesium, and calcium, among others, which make it particularly suitable for biological treatment (Ghaly et al., 2003).

14.1.1.1 Whey

Whey is a byproduct characterized as a translucent yellow liquid, sometimes greenish; this can vary according to the quality and type of milk used to obtain it (Naula-Saez, 2017). On the other hand, the Codex Alimentarius defines it as a fluid that separates the curd from the coagulation of milk in cheese production. It can be obtained by acid, enzymatic, or bacterial coagulation; however, it is not a milk substitute (WHO-FAO, 2011).

According to Agualongo et al. (2022), the chemical composition of whey depends on the cheese-making conditions. It is composed of varied organic matter, mainly lactose (4–5% w/v), soluble proteins (0.4–0.5%p/v), and lipids (0.6–0.8% w/v), contains 50% total milk solids, and about 47% minerals (García-Casas et al., 2018; Kaur et al., 2020).

It contains proteins such as beta-lactoglobulin (β-Lg) and alpha-lactalbumin (α-LA), representing 10% and 4% of the milk protein content, while peptone-protease, serum albumin, immunoglobulins, and transferrin complement the protein composition of this byproduct. To a lesser extent, it contains proteins such as lactoferrase and lactoperoxidase, which are of commercial importance (Cisneros-Salazar, 2022). Other components found in whey with a high degree of availability are mineral salts, mainly potassium, followed by calcium, phosphorus, magnesium, zinc, iron, and copper, which can be used as an enriched culture medium for the growth of microorganisms (Parra-Huerta, 2009). These are primarily present in acid whey due to the solution of colloidal salts of the acidified milk. It also contains small but significant amounts of vitamins A, C, D, E, and B complex. Outstanding are those of the B group (thiamine, pantothenic acid, pyridoxine, nicotinic acid, riboflavin, and ascorbic acid). It is rich in essential amino acids, which allows its versatile use in the food industry (Agualongo et al., 2022; Parra-Huerta, 2009).

Of total milk production, more than 40% is processed as cheese, which means that it is demanding since, for every 10 liters of milk, approximately 1 kg of cheese is obtained, generating 9 liters of whey (Aider et al., 2009). Consequently, 90% of the milk used in the cheese industry is discarded as whey, which retains approximately 55% of the total milk substances, including soluble proteins, lactose, lipids, and mineral salts. Therefore, whey production has been considered for many years as waste, where 47% is dumped as waste by milk processing industries.

The worldwide production of whey per year is estimated to be around 160 million tons, of which 6 million tons are lactose (Sharma et al., 2018) with an increased rate of 1–2%; however, less than 50% of the total whey produced is processed with different technologies (Panghal et al., 2017).

14.1.1.2 Uses of Whey

Despite the nutritional composition of whey, one of the most abundant byproducts of the dairy industry, it is considered one of the most representative wastes because it generates high volumes of water. Due to the lack of effective technologies for its processing, its valorization remains a challenge today (Osorio-González, 2018). Some existing alternatives include fractionation, stabilization, transformation, and recombination (Muset & Castells, 2017). However, these strategies have limitations such as the use of accessible technologies, costs related to treatments, and factors such as quality and quantity (Palmieri et al., 2017). Traditionally, whey has been used as animal feed and fertilizer for crops, being discarded into the environment generating a negative environmental impact; the latter because of the absence of methods that allow its use and that are economically viable (Torres-Martínez, 2019).

Whey's high nutritional value makes it feasible for valorization through processing technologies. However, technological advances have allowed approximately 45% of this byproduct to be used, mainly transformed into whey powder, bioethanol, biopolymers, whey protein isolates, and probiotics (Yadav et al., 2015). Biotechnological alternatives, such as the use of fermentation to obtain organic acids, fermented beverages, protein isolates, utilization as a substrate, concentration, and recovery of lactose proteins and mineral salts are other utilization options (Chanfrau et al., 2017; Kolesovs & Semjonovs, 2020; Quille et al., 2021).

Fermentative processes are promising in research areas for the dairy industry since they seek to use dairy residues as energy sources using microorganisms capable of producing metabolites and biomass (Huaraca-Espinoza, 2019). Furthermore, these systems allow the valorization of this byproduct that provides various forms of transformation, given its composition of carbohydrates, lipids, proteins, and vitamins. There are reports of the use of whey in fermentation processes for the production of kefir beverages and byproducts such as lactic acid, ethanol, and acetic acid (Magalhães et al., 2010; Cury et al., 2017; Nastar-Marcillo, 2022), in addition to kombucha for obtaining bacterial cellulose (Nguyen et al., 2022).

14.1.2 Bacterial Cellulose (BC)

During kombucha fermentation, two distinct components are visualized: bacterial cellulose (BC) and culture medium. BC is an insoluble extracellular polymer synthesized by microorganisms corresponding to the genera *Acetobacter*, *Rhizobium*, *Agrobacterium*, and *Sarcina*, paying particular attention to the strains of *Gluconacetobacter* that have been considered as a study model (Jiménez-Sánchez, 2021). It is chemically identical to plant cellulose but differs in macroscopic organization (Laavanya et al., 2021).

In recent years, the production of BC has been the subject of study in different branches of science. Its unique properties (high degree of crystallinity, durability, high water retention capacity, malleability, strength, and biocompatibility) have made this material an excellent resource in medicine, the food industry, the paper industry, the textile industry, and cosmetics. Although many biopolymer-producing strains exist, the most efficient species is *Acetobacter xylinum*. However, some reports indicate that the genera *Kluyveromyces marxianus* and *Komagataeibacter* can ferment whey (Quille et al., 2021).

Therefore, this chapter aims to discuss reducing the load of organic and inorganic contaminants in whey and microbial cellulose production using a microbial fermenting consortium.

14.2 MATERIALS AND METHODS

14.2.1 Consortium

The working consortium (COR) was provided by the Microbial Biotechnology Laboratory, Faculty of Agronomy, University Autonomous Nuevo León. For COR maintenance, an infusion of green tea (3.6 g/L) and dextrose (50 g/L) was prepared at 25 °C to generate a symbiotic culture of bacteria and yeast (SCOBY) hotel (Fig. 14.1).

14.2.2 Whey

The whey was obtained from the Centro de Investigación y Desarrollo en Industrias Alimentarias dairy processing plant, exclusively from the mozzarella cheese production process. It was collected in 3 L plastic containers and kept refrigerated for no more than 48 h before use.

14.2.2.1 Variation of Whey Concentration

The nomenclature of the acid whey treatments used in this work is as follows: TC = control treatment; T1 = 100:0; T2 = 75:25; T3 = 50:50; T4 = 25:75; and T5 = 0:100 as a ratio of whey volume to water (% Whey: H_2O).

For the establishment of the batch fermentation, 200 mL of a mixture of whey: water (v/v) in different proportions (100:0, 75:25, 50:50, 25:75, and 0:100), and 15% (v/v) of the COR inoculum were added in plastic containers (500 mL). Green tea with dextrose (50 g/L) was used as a control for fermentation. Subsequently, the containers were covered with gauze and stored in the dark at 25±2 °C for 12 days, with five replicates per treatment.

FIGURE 14.1 a) SCOBY hotel (COR), b) green tea infusion.

14.2.2.2 Different Inoculum Concentrations

The establishment of the serum batch fermentation varying the inoculum concentration (5%, 10%, and 15% v/v) was in plastic containers (500 mL) containing 200 mL of serum that was covered with gauze, stored in the dark at 25 ±2° C for 12 days; five replicates per treatment were performed. Green tea with dextrose (50 g/L) with the same inoculum concentration (5%, 10%, and 15% v/v) was used as a control for fermentation.

The nomenclature of the acid whey treatments at different inoculum concentrations was as follows: control treatment 15% (TC1); control treatment 10% (TC2); control treatment 5% (TC3); whey 15% (T1i); whey 10% (T2i); whey 5% (T3i).

14.2.2.3 Characterization of Whey

For each experiment, the whey was characterized at the beginning and end, in which the chemical oxygen demand, total solids, dissolved solids, proteins, total sugars, and pH were determined.

14.2.2.3.1 COD

In HACH® tubes, 2.5 mL of sample, 1.5 mL of digester solution, and 3.5 mL of silver sulfate solution were added; they were digested for 2 h at 150 ºC to determine the concentration by UV-vis spectrophotometry at 620 nm (APHA, 2005).

14.2.2.3.2 Total Solids

A 5 mL homogenized sample was placed in a crucible and weighed, then heated at 105 °C for 24 hours, after which it was allowed to cool and weighed again. The concentration was calculated using Equation 1 (APHA, 2005):

$$\text{ST mg/L} = ([[(A-B)]])/(\text{Sample mL}) * 1000 \qquad \text{Equation (1)}$$

Where: A = Initial weight and B = Filter weight, both expressed in mg.

14.2.2.3.3 Volatile Total Solids

The final sample obtained from the total solids was placed in a muffle at 550 °C for 20 minutes, and Equation 2 used to determine the concentration (APHA, 2005):

$$\text{STV mg/L} = ([[(A-B)]])/(\text{Sample mL}) * 1000 \qquad \text{Equation (2)}$$

Where: A = weight of the dry sample at 105 °C and B = weight of the dry sample at 550 °C, expressed in mg.

14.2.2.3.4 Determination of Acidity

Two drops of 0.1% phenolphthalein were added to 10 mL of the sample and titrated with NaOH solution (1N) until a pink coloration was obtained (Vitas et al., 2020). The percentage acidity was calculated with Equation 3:

% Acidity = ([[(mL NaOH)(N NaOH)(0.06)])/(mL Sample) x100 Equation (3)

14.2.2.3.5 Determination of pH

The pH value was obtained using a multiparametric pH meter (Thermo Scientific™ Orion Star™ A211 benchtop pH meter).

14.2.2.3.6 Determination of Total Sugars

The procedure described by Dubois et al. (1956) was performed with some modifications to determine total sugars. First, 1 mL of the sample plus 1 mL of phenol (5%) and 5 mL of concentrated sulfuric acid was taken, allowed to react for 5 min, and quantified by UV-vis spectrophotometry at 490 nm.

14.2.2.3.7 Protein Determination

Using the method of Bradford (1976), 200 μL of the sample plus 800 μL of Bradford solution were taken and quantified at 595 nm by UV-vis spectrophotometry.

14.2.2.3.8 Microbial Cellulose Yield

The BC formed in the experiment was collected manually, washed with running water to remove solids, and the excess water was removed and finally weighed to calculate the yield according to Equation 4:

Yield (%) = (Weight of BC (g/))/Volume (L))/Concentration of volatile solids at start-up (g/L) × 100 Equation (4)

14.2.3 Experimental Design

The results of different acid whey concentrations and different inoculum percentages were subjected to analysis of variance and comparison of means with Tukey's test ($p \leq 0.05$) using the statistical package Minitab 17.

14.3 RESULTS AND DISCUSSION

14.3.1 Physicochemical Characterization of Whey

The characterization of whey is shown in Table 14.1. The whey presented similarities to those reported by other authors, in terms of pH value (5.3), protein concentration (16.12 ± 0.0 g/L), total sugar (33 ± 2 g/L), and acidity (1.1± 0.1%) (Pires et al., 2021; Panesar et al., 2007; Lione et al., 2021). However, its composition variability depended on which product was previously elaborated (Bernal-Aldama, 2022).

14.3.2 Effect of Whey Concentration

The behavior of the parameters evaluated in the fermentation of whey at different concentrations is shown in Figure 14.2. The statistical analysis indicated significant differences between the experimental times ($p < 0.05$). At the beginning of

TABLE 14.1
Characterization of whey

Parameter	pH	At (%)	COD (g/L)	P (g/L)	TS (g/L)	TVS (g/L)	TA (g/L)
Whey	5.3±0.01	1.1±0.1	61.6±39.1	16.12±0.0	54.46±3.3	48.45±3	33 ±2

Note: Mean ± standard deviation. pH = potential of hydrogen; At = total acidity; COD = chemical oxygen demand; P = protein; TS = total solids; TVS = total volatile solids; TA = total sugars.

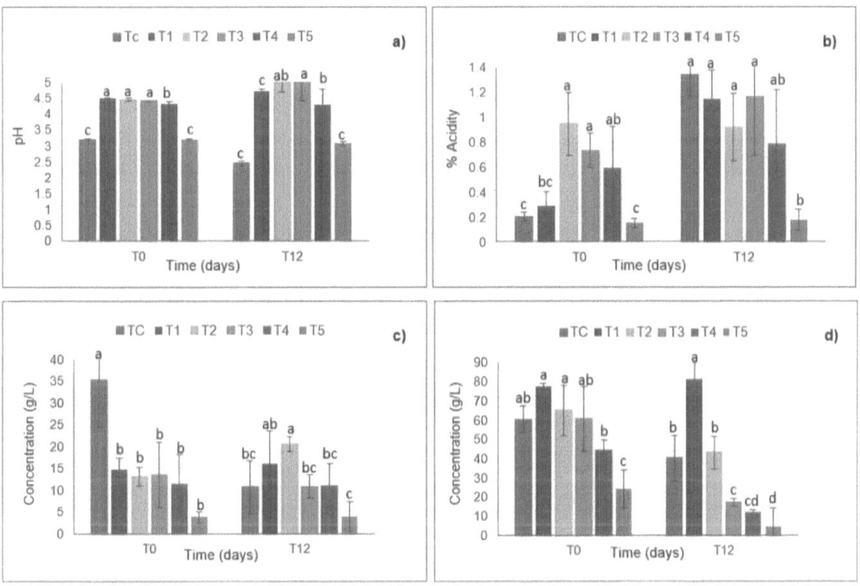

FIGURE 14.2 (a) pH values, (b) total sugar, (c) total acid, and (d) COD in the fermentation of whey at different concentrations. ± = Standard deviation. TC: control; T1 (100% whey); T2 (75 Whey:25 H_2O); T3 (50 Whey:50 H_2O); T4 (25 Whey:75 H_2O); and T5 (0 Whey:100 H_2O).

fermentation, the pH ranged from 3.1 to 4.4; however, on day 12, the values increased in almost all treatments except for TC and T5, which decreased to 2.5 and 3, respectively (Fig. 14.2a). The increase in pH was related to the origin of the whey (Primiani et al., 2018), and the decrease in this parameter depended on the microbial activity that produced organic acids.

The behavior of acidity in acid whey is observed in Figure 14.2b. The initial values ranged from 0.1 to 0.9%, with T2 having the lowest acidity. At the end of fermentation, T4 and T5 maintained similar values to the initial ones, which was related to the low concentration of whey in the fermentation mixture. TC, T1, and T3 increased, which may be related to the metabolic processes of the lactic and acetic acid bacteria that form the microbial consortium that increased the acidity in the medium. It was also observed that the concentration of whey in the mixture increased the total acidity of the fermented product. It was found that there were statistical differences between treatments at each fermentation time.

The results in determining total sugar content are shown in Figure 14.2c. The highest concentration of total sugars was found in the control (35 g/L), and the rest of the treatments were below 15 g/L. At the end of fermentation, a decrease ($p < 0.05$) in the concentration was observed for treatments TC, T3, and T4, while for treatments T1, T2, and T5, an increase ($p < 0.05$) in the concentration of total sugars was observed with final values of 15.8, 20.5, and 3.91 g/L, respectively. The decrease in sugar concentration was directly related to substrate consumption by the microorganisms of the consortium, the increase may be due to the production of mono, di, and polysaccharides during fermentation, which were also quantified (Amarasinghe et al., 2018).

Figure 14.2d shows the behavior of COD during fermentation. The concentrations of the treatments at the beginning are significantly different ($p < 0.05$), with values ranging from 24 to 77 g/L and not exceeding the maximum permissible limits (150 mg/L) in the Mexican regulations for discharge into water bodies or drains (NOM-001-SEMARNAT, 2021). At the end of fermentation, the values decreased for all treatments ($p < 0.05$), with concentrations ranging from 4 to 43 g/L. However, for T1, the concentration increased (81 g/L) due to the production of organic compounds by biological activity and to the same biomass that increased during fermentation. It was possible to remove 55.5% of the chemical oxygen demand, which was directly attributed to avoiding contamination problems in the effluent water.

The total solids' concentration at the experiment's beginning ranged from 2 to 45 g/L. It was observed that the concentration of whey determined the weight of total solids in the mixture (Fig. 14.3a). The higher the whey level, the higher the concentration of solids in the fermentation. The concentrations were different among treatments but similar among fermentation times. This behavior was due to cell biomass production (Yadav et al., 2015) and microbial cellulose generation. However, the decrease in initial total solids' concentration was independent of the above, but with the analysis methodology proposed here, it could not be easily differentiated.

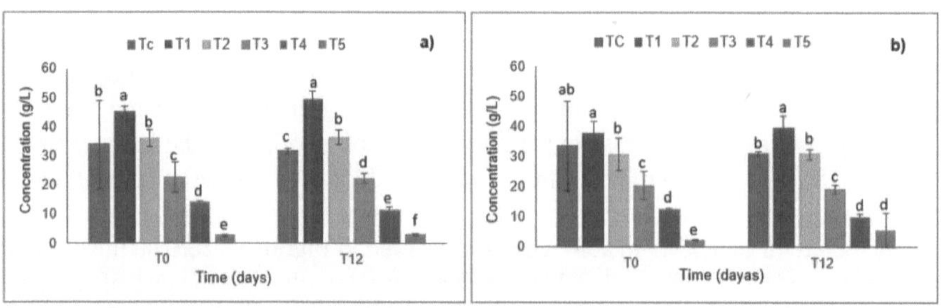

FIGURE 14.3 (a) Total solids and (b) total volatile solids in the fermentation of whey at different concentrations ± = Standard deviation. TC: control; T1 (100% whey); T2 (75 whey:25 H_2O); T3 (50 whey:50 H_2O); T4 (25 whey:75 H_2O); and T5 (100% H_2O).

Similar behavior was observed with total volatile solids at TC, T1, and T2 where the concentration remained almost constant at the end of fermentation (Fig. 14.3b). However, for T3 and T4 with 50% and 25% whey, respectively, a decrease in the concentration of total volatile solids was observed, which was related to lower microbial cellulose production and cell generation.

14.3.3 Effect of Inoculum Concentration on Whey

The pH values at the beginning of the experiment for TC1, TC2, and TC3 ranged from 3 to 3.5; all of them did not contain whey. In comparison, the rest that only contained whey (T1i, T2i, and T3i) maintained a pH of 4.5 (Fig. 14.4a). It was observed that the initial pH values did not depend on the inoculum concentration and depended directly on the fermentation medium. At the end of fermentation, the pH values decreased in treatments TC1, TC2, and TC3, which is the expected behavior for green tea-based fermentation (Treviño-Garza et al., 2020). In TC1 and TC3, pH values also decreased due to the production of organic acids generated by the microbial activity, but in TC2 (10%), there was a slight increase, probably due to the presence of NH_4 due to the degradation of proteins and intracellular components from dead cells (Reyes-Alvarado, 2012).

The lowest initial values of acidity percentage were obtained in TC with percentages from 0.2% to 0.4%, while for serum, it ranged from 1.3% to 1.7% (p

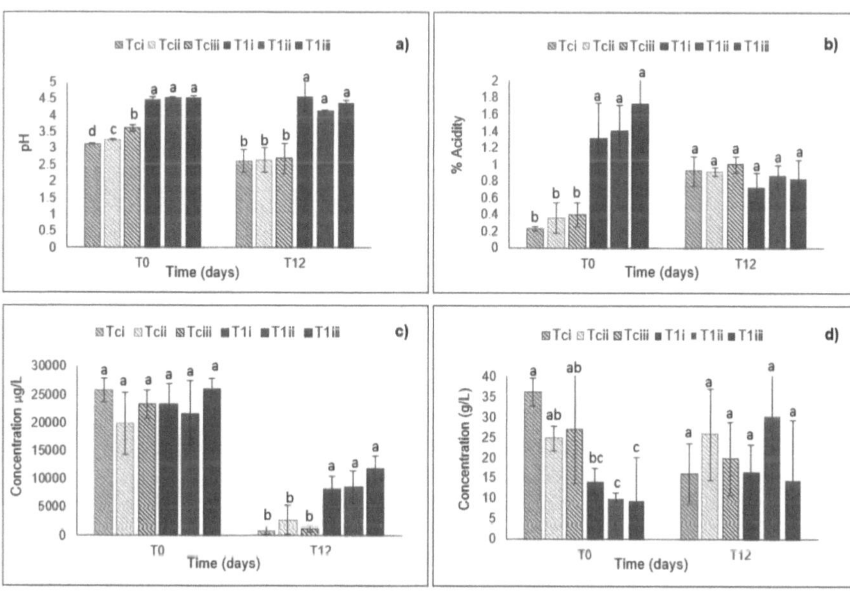

FIGURE 14.4 pH values (a), total acidity (b), protein (c), and total sugar (d) in the fermentation of whey with different concentrations of inoculum ± = Standard deviation. TCi: control (15%); TCii: control (10%); TCiii: control (5%); T1i: whey (15%); T1ii: whey (10%); T1iii: whey (5%).

< 0.05). The percentage of acidity increased in relation to fermentation time in the control treatments with values from 0.91% to 0.99% (Fig. 14.4b), the highest in the TCiii, which corresponded to the highest concentration of inoculum in the green tea. In the serum, the percentages were between 0.7–0.8%, with T1i having the lowest value of 0.68%. However, no significant differences were found between treatments ($p > 0.05$). Acidity in fermentation depends on pH, temperature, and acid generation.

The pH, temperature, and generation of organic acids by the metabolism of sugar consumption influence the growth parameters of microorganisms (Tu et al., 2019). The initial protein concentration ranged from 18 to 23 g/L in the green tea treatments, and the whey from 21 to 16 g/L (Fig. 14.4c). However, the protein concentration decreased significantly at the end of fermentation ($p < 0.05$). The decrease in protein was associated with using nitrogen to promote microbial cell development (Khosravi et al., 2019).

At the beginning of the experiment, the treatments presented total sugar concentrations ranging from 9 g/L to 36 g/L ($p < 0.05$). Treatments TCi and TCiii (Fig. 14.4d) presented carbohydrate degradation for the elapsed time of fermentation, with final concentrations of 16 and 19 g/L, while all serum treatments and TCii increased at the end of fermentation (14 to 30 g/L), being nonsignificant ($p > 0.05$). The same behavior occurred previously due to the production of mono-, di- and polysaccharides resulting from the activity of acidifying bacteria.

The initial concentrations of chemical oxygen demand were between 52–62 g/L· and nonsignificant among the treatments evaluated ($p > 0.05$). At the end of

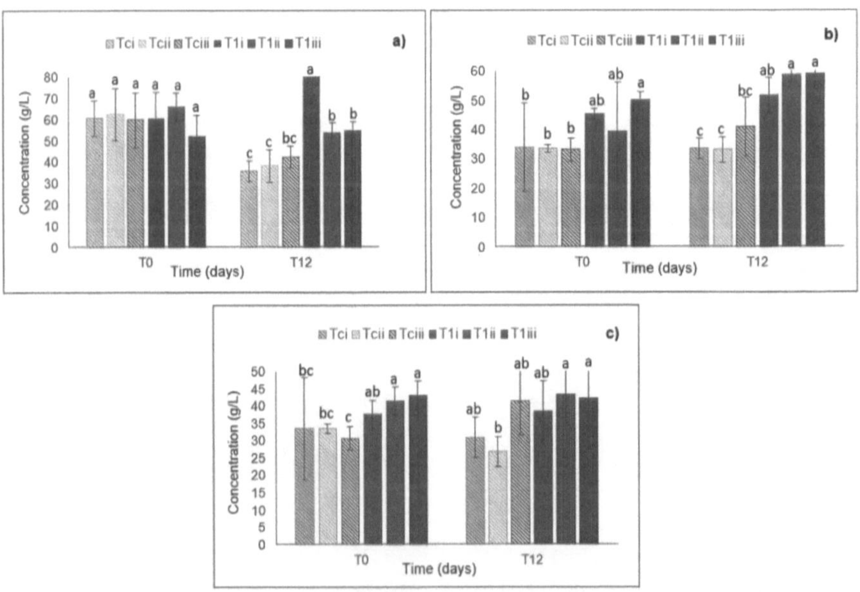

FIGURE 14.5 COD values (a), TS (b), and TVS (c) in the fermentation of whey acid with different percentages of inoculum ± = Standard deviation. TCi: control (15%); TCii: control (10%); TCiii: control (5%); T1i: whey (15%); T1ii: whey (10%); T1iii: whey (5%).

fermentation, the concentration increased in treatments T1i and T1iii, probably because small BC fragments were trapped at sampling (Fig. 14.5a). For the rest of the treatments, the concentrations decreased. The response of COD during fermentation is influenced by the cell concentration in the medium and directly by the percentage of inoculation used.

Figure 14.5b shows that as the glucose concentration increased at the end of fermentation, the total solids' content increased, related to the concentration used to inoculate the treatments. At the end of the experiment, the concentrations remained between 51–58 g/L, which was attributed to the metabolism of the microorganisms when consuming the carbon source of the fermentation (Jakubczyk et al., 2020). The control treatments showed 33–40 g/L, decreasing in the treatment inoculated at 15% (TCi) and 10% (TCii) v/v. This effect was similar to that obtained by Hernández (2006), who evaluated the removal of solids with different serum concentrations. The values of total volatile solids (Fig. 14.5c) in the treatments presented initial values ranging from 30 to 43 g/L, increasing slightly at the end of fermentation ($p < 0.05$). In treatments TCiii, T1i, and T1ii, the concentrations were 41, 38, and 43 g/L, respectively.

14.3.4 Production of BC

The highest quantified mass of BC was in T3 (51 g/L) and T2 (46.9 g/L) and the lowest in T5 (3.7 g/L). It should be noted that the latter treatment only contained water and inoculum, making clear the importance of the carbon source for BC production (Fig. 14.6). The BC production is directly related to substrate concentration (Goh et al., 2012). Treviño-Garza et al., (2020) obtained a higher yield of BC (60–160 g/L) from two microbial consortia using dextrose, glucose, sucrose, and fructose as carbon sources. Davalos-Cerron (2022) obtained higher BC yields in fermented whey media than in ethylene glycol H-S media. The same behavior was observed in the yields ranging from 139% to 152%.

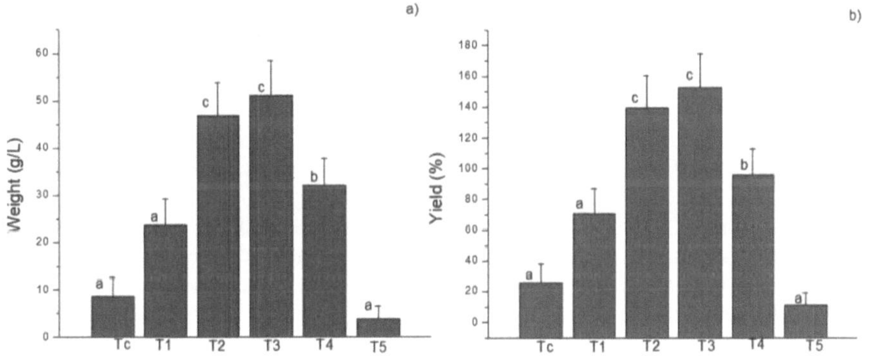

FIGURE 14.6 Weight and yield of BC with different concentrations of whey acid inoculated at 10%. TC: control; T1 (100% whey); T2 (75 whey:25 H_2O); T3 (50 whey:50 H_2O); T4 (25 whey:75 H_2O); and T5 (100% H_2O).

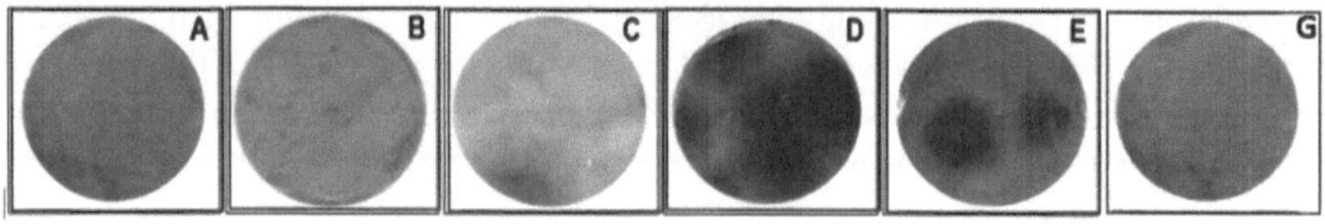

FIGURE 14.7 Cellulose disc produced at different whey concentrations. (A) TC: control; (B) T1 (100% whey); (C) T2 (75 whey:25 H_2O); (D) T3 (50 whey:50 H_2O); (E) T4 (25 whey:75 H_2O); and (G) T5 (100% H_2O).

The biopolymer formation presented white and yellow shades (Fig. 14.7), similar to those reported by Nguyen et al., (2022). However, regardless of their concentration, a lighter coloration was observed in the serum treatments concerning the BC generated in green tea. The hue was associated with microbial growth in the acidic medium (Jayabalan et al., 2014).

14.4 CONCLUSIONS

The microbial community capable of fermenting whey is an excellent option to decrease the concentration of total solids, volatile total solids, chemical oxygen demand, and improve essential water quality parameters. The concentration of the inoculum does not modify the production of microbial cellulose, but increasing its concentration promotes the reduction of proteins, total solids, volatile total solids, and chemical oxygen demand, generating an effluent with a lower pollutant load.

On the other hand, whey is an ideal fermentation medium for producing microbial cellulose, with production being favored when it is in low proportions. Furthermore, microbial cellulose can be used as an environmentally friendly material in different applications in the area of knowledge.

The treatment of effluents from the dairy industry through microorganisms capable of fermenting and producing microbial cellulose is possible, making its implementation attractive for industrial scale-up.

REFERENCES

Agualongo, L., Aucatoma, D., Sagnay, D., Santillan, N., and Jácome, C. 2022. El suero de leche, subproducto de la industria de queso: Composición, recuperación de proteínas y aplicaciones. *Journal of Agro-Industry Sciences* 4, no. 1: 13–22. htpps://doi.org/10.17268/jais.2022.002.

Aider, M., de Halleux, D., and Melnikova, I. 2009. Skim acidic milk whey cryoconcentration and assessment of its functional properties: Impact of processing conditions. *Innovative Food Science and Emerging Technologies* 10, no. 3: 334–341. htpps://doi.org/10.1016/j.ifset.2009.01.005.

Amarasinghe, H., Weerakkody, N. S., and Waisundara, V. Y. 2018. Evaluation of physicochemical properties and antioxidant activities of kombucha "Tea fungus" during extended periods of fermentation. *Food Science & Nutrition*, 6, no. 3: 659–665. htpps://doi.org/doi.org/10.1002/fsn3.605.

APHA, AWWA, and WEF. 2017. Method 5220 C, chemical oxygen demand. Standard methods for the examination water and wastewater. www.revistatyca.org.mx/public/journals/1/documentos/2021/proximos_numeros/enero_febrero_2021/2151_final.pdf.

Armijo, J., Azogue, H., Barragán, S., and Freire, A. 2021. Biotratamientos de aguas residuales en la industria láctea. *Journal of Agro-Industry Sciences* 3, no. 1: 21–26. htpps://doi.org/10.17268/jais.2021.003.

Bernal-Aldana, A. S. 2022. Aplicaciones y tecnologías utilizadas para el aprovechamiento del suero lácteo, la producción del suero en polvo, derivados y sus aplicaciones en la industria en general de alimentos. https://repository.unad.edu.co/handle/10596/51553.

Bradford, Marion M. 1976. A rapid and sensitive method for the quantitation of microgram quantities of protein utilizing the principle of protein-dye binding. *Analytical Biochemistry* 72, no. 1-2: 248–54. htpps://doi.org/10.1016/0003-2697(76)90527-3.

Centro de Actividad Regional de la producción. 2002. Prevención de la contaminación en la industria láctea. Universidad de Salamanca. http://coli.usal.es/web/demo_appcc/demo_ej ercicio/lac_es.pdf.

Chanfrau, J. M. P., Pérez, J. N., Fiallos, M. V. L., Intriago, L. M. R., Toledo, L. E. T., and Guerrero, M. J. C. 2017. Valorización del suero de leche: Una visión desde la biotecnología. *Bionatura* 2, no.4: 468–76. htpps://doi.org/10.21931/rb/2017.02.04.11.

Chen, Z., Luo, J., Hang, X., and Wan, Y. 2018. Physicochemical characterization of tight nanofiltration membranes for dairy wastewater treatment. *Journal of Membrane Science* 547: 51–63. htpps://doi.org/10.1016/j.memsci.2017.10.037.

Cisneros-Salazar, A. A. 2022. Beneficios de la utilización del suero de leche en la elaboración de suplementos proteicos en la industria láctea. Thesis, Universidad Central del Ecuador. www.dspace.uce.edu.ec/bitstream/25000/28180/1/FCQ-CQA-CISNEROS%20ALISSON.pdf.

Cury R, K., Aguas, Y., Martínez, A., Olivero, R., and Ch, L. C. 2017. Residuos agroindustriales su impacto, manejo y aprovechamiento. *Revista Colombiana de Ciencia Animal-RECIA* 9, S: 122–132. htpps://doi.org/ 10.24188/recia.v9.nS.2017.530.

Dávalos-Cerrón, P. Y. 2022. Aislamiento y caracterización de cepas nativas de *Komagataeibacter xylinus* y comparación de su crecimiento en diferentes sustratos. Thesis, Universidad Técnica de Ambato. https://repositorio.uta.edu.ec/handle/123456789/34967.

Dubois, M., Gilles, K.A., Hamilton, J.K., Robers and P.A., Smith, F. 1956. Colorimetric method for the determination of sugars and related substances. *Analytical Cemistry* 28, no.3: 350–356. htpps://doi.org/10.1021/ac60111a017.

FAO. 2018. Producción lechera. www.fao.org/dairy-production-products/production/es/ (accessed 5-March- 2023).

FAOSTAT. 2017. Tendencias y desafíos del futuro de la agricultura y la alimentación. www.fao.org/3/i6583e/i6583e.pdf (accessed 5-March- 2023).

Gaibor, J. 2014. Caracterización del agua residual generada en la planta de lácteos el Salinerito–Parroquia Salinas–cantón Guaranda para el diseño de una planta de tratamiento. *Revista de Investigación Talentos* 1, no. 1(I):107–112.

García-Casas, V., Sánchez, Companioni, R., and Ramón, Ramón, T. 2018. *Suero de Leche. La ciencia detrás de su rescate.* Ecuador: Cámara Ecuatoriana de Libro.

Ghaly, A. E., Tango, M. A., and Adams, M. A. 2003. Enhanced lactic acid production from cheese whey with nutrient supplement addition. *International Commission of Agricultural Engineering CIGR Journal* no. 5: 20. https://ecommons.cornell.edu/bitstream/handle/1813/10321/FP%2002%20009a%20Ghaly.pdf?sequence=1&isAllowed=y.

Goh, W. N., Rosma, A., Kaur, B., Fazilah, A., Karim, Abd. A., and Bhat, R. 2012. Fermentation of black tea broth (kombucha): I. effects of sucrose concentration and fermentation time on the yield of microbial cellulose. *International Food Research Journal* 19, no. 1: 109–117.

Hernández-Preciado, B. 2006. *Filtración de lactosuero a diferentes concentraciones como pretratamiento*. Thesis, Universidad Autónoma del Estado de Hidalgo.

Huaraca-Espinoza, M. 2019. *Evaluación de tres Sistemas de Fermentación para producir Celulasa a partir de Suero de Leche* y *Aspergillus niger*. Thesis, Universidad Nacional Micaela Bastidas de Apurímac.

Indigoyen-Ramirez, D. 2019. *Remoción de materia orgánica de aguas residuales de la industria láctea por el método de electrocoagulación, utilizando energía eólica.* Thesis, Universidad Nacional del Centro del Perú.

Jakubczyk, K., Ladunska, J., Kochman., and J., Janda, K. 2020. Chemical profile and antioxidant activity of the kombucha beverage derived from white, green, black and red tea. *Antioxidants* 9, no. 5: 447. htpps://doi.org/10.3390/antiox9050447.

Jayabalan, R., Malbaša, R. V., Lončar, E. S., Vitas, J. S., and Sathishkumar, M. 2014. A review on kombucha tea—microbiology, composition, fermentation, beneficial effects, toxicity, and tea fungus. *Comprehensive Reviews in Food Science and Food Safety* 13, no.4: 538–550. https://doi.org/10.1111/1541-4337.12073.

Jiménez-Sánchez, Y. Y. 2021. Aplicación de la celulosa bacteriana en el diseño de productos: un camino a la sustentabilidad. *Diseño arte y arquitectura*, no. 11: 41–57. https://doi.org/doi:10.33324/daya.vi11.458.

Kaur, N., Sharma, P., Jaimni, S., Kehinde, B. A., and Kaur, S. 2020. Recent developments in purification techniques and industrial applications for whey valorization: A review. *Chemical Engineering Communications*, no.1: 123–138. https://doi.org/ doi:10.1080/00986445.2019.1573169.

Khosravi, S., Safari, M., Emam-Djomeh, Z., and Golmakani, M. T. 2019. Development of fermented date syrup using Kombucha starter culture. *Journal of Food Processing and Preservation* 43, n.° 2: e13872. https://doi.org/ doi: 10.1111/jfpp.13872.

Kolesovs, S., and Semjonovs, P. 2020. Production of bacterial cellulose from whey current state and prospects. *Applied Microbiology and Biotechnology* 104, no.18: 7723–7730. https://doi.org/ doi:10.1007/s00253-020-10803-9.

Laavanya, D., Shirkole, S., and Balasubramanian, P. 2021. Current challenges, applications and future perspectives of SCOBY cellulose of Kombucha fermentation. *Journal of Cleaner Production* 295: 126454. https://doi.org/doi:10.1016/j.jclepro.2021.126454.

Lione, D., Chenevier, D. T., and Fideleff, S. (2021). Tratamiento de lactosuero acoplado a la producción de biomasa: estudio de factibilidad a escala laboratorio para el cultivo de cianobacterias con fines biotecnológicos. https://repositorio.uca.edu.ar/bitstream/123456789/13101/1/tratamiento-lactosuero-acoplado.pdf (accessed 1-march-2023).

Lucas, S. M. and García, R. S. 2018. El agua en la industria alimentaria. *Boletín de la Sociedad Española de Hidrología Médica,* 33(2): 157–171. https://doi.org/doi: 10.23853/bsehm.2018.0571.

Magalhães, K. T., Pereira, M. A., Nicolau, A., Dragone, G., Domingues, L., Teixeira, J. A., and Schwan, R. F. 2010. Production of fermented cheese whey-based beverage using kefir grains as starter culture: Evaluation of morphological and microbial variations. *Bioresource Technology* 101(22): 8843–50. https://doi.org/10.1016/j.biortech.2010.06.083.

Muset, G. B. and Castells, M. L. 2017. *Valorización del lactosuero*. Brasil: Instituto Nacional de Tecnología Industrial. www.embrapa.br/busca-de-publicacoes/-/publicacao/1104688/valorizacion-del-lactosuero.

Naula-Sáez, J. R. 2017. *Diseño de un proceso para la obtención de proteína del lactosuero mediante la operación unitaria de secado por atomización*. Thesis. Escuela Superior Politécnica de Chimborazo.

Nastar-Marcillo, D. A. 2022. *Producción de una bebida funcional a partir de mezclas de suero de leche de cabra y jugos de frutos andinos, fermentadas con gránulos de kéfir.* Thesis. Universidad Técnica del Norte.

Nguyen, H. T., Ngwabebhoh, F. A., Saha, N., Zandraa, O., Saha, T., and Saha, P. 2022. Development of novel biocomposites based on the clean production of microbial cellulose from dairy waste (sour whey). *Journal of Applied Polymer Science* 139, no.1: 51433. . https://doi.org/10.1002/app.51433.

NOM-001-SEMARNAT. 2021. Norma Oficial Mexicana. 2021. Que establece los límites permisibles de contaminantes en las descargas de aguas residuales en cuerpos receptores propiedad de la nación. www.dof.gob.mx/nota_detalle.php?codigo=5645374&fecha=11/03/2022#gsc.tab=0 (accessed 5-march-2023).

OCDE-FAO. 2015. Perspectivas Agrícolas 2015-2024. Organización para la Cooperación y Desarrollo Económico (OCDE)-Organización de las Naciones Unidas para la Alimentación y la Agrícultura (FAO), Roma.

Osorio-González, C. S., Sandoval-Salas, F., Hernández-Rosas, F., Hidalgo-Contreras, J. V., Gómez-Merino, F. C., and Ávalos de la Cruz, D. A. 2018. Potencial de aprovechamiento del suero de queso en México. *Agro Productividad* 11, n.° 7 (Julio): 1–6. https://revista-agroproductividad.org/index.php/agroproductividad/article/view/922 (accessed 7-march-2023).

Palmieri, N., Forleo, M. B., and Salimei, E. 2017. Environmental impacts of a dairy cheese chain including whey feeding: An Italian case study. *Journal of Cleaner Production* 140: 881–889. https://doi.org/doi: 10.1016/j.jclepro.2016.06.185.

Panesar, P. S., Kennedy, J. F., Gandhi, D. N., and Bunko, K. 2007. Bioutilisation of whey for lactic acid production. *Food Chemistry* 105 (1):1–14). https://doi.org/doi:10.1016/j.foodchem.2007.03.035.

Panghal, A., Kumar, V., Dhull, S. B., Gat, Y., and Chhikara, N. 2017. Utilization of dairy industry waste-whey in formulation of papaya RTS beverage. *Current Research in Nutrition and Food Science Journal* 5, n.° 2: 168–174. https://doi.org/doi: 10.12944/crnfsj.5.2.14.

Parra-Huertas, R. A. 2009. Lactosuero: importancia en la industria de alimentos. *Revista facultad nacional de agronomía Medellín* 62(1): 4967–82. https://revistas.unal.edu.co/index.php/refame/article/view/24892.

Pires, A. F., Marnotes, N. G., Rubio, O. D., Garcia, A. C., and Pereira, C. D. 2021. Dairy by-products: A review on the valorization of whey and second cheese whey. *Foods* 10, n.° 5: 1067. https://doi.org/doi:10.3390/foods10051067.

Prieto-García, F., Callejas Hernández, J., Reyes Cruz, V. E., and Marmolejo Santillán, Y. 2012. Electrocoagulación: una alternativa para depuración de lactosuero residual. *AIDIS de Ingeniería y Ciencias Ambientales.* Investigación, desarrollo y práctica 5(3): 51–77. https://revistas.unam.mx/index.php/aidis/article/view/34727.

Primiani, C. N., Mumtahanah, M., and Ardhi, W. 2018. Kombucha fermentation test used for various types of herbal teas. *Journal of Physics: Conference Series* 1025: 012073. https://doi.org/doi:10.1088/1742-6596/1025/1/012073.

Quille, L. Q., Vilca, O. M. L., and Ordoñez, F. P. A. 2021. Potencialidades del lactosuero generado por la industria quesera y su valorización. *Innovación tecnológica y valoración económica* 1, n.° 2: 16–23. https://doi.org/doi:10.51392/rcidas.v1i2.10.

Reyes-Alvarado, L. C. 2012. Producción de etanol por fermentación de suero de quesería con levaduras del género Kluyveromyces y Saccharomyces. Thesis, Universidad Veracruzana. https://cdigital.uv.mx/bitstream/handle/123456789/46948/ReyesAlvaradoLuisCarlos.pdf?sequence=1&isAllowed=y.

Robledo-Padilla, R. 2018. *Producción de leche en México y el impacto de las importaciones de leche en polvo.* México. http://ru.iiec.unam.mx/4223/1/2-Vol1_Parte2_Eje2_Cap2-192-Robledo.pdf.

Santos, L. F. D., Gonçalves, C. M., Ishii, P. L., and Suguimoto, H. H. 2017. Deproteinization: an integrated-solution approach to increase efficiency in β-galactosidase production using cheese whey powder (CWP) solution. *Revista Ambiente y Agua, Interdisciplinary Journal of Applied Science* 12, no. 4: 643–651. https://doi.org/doi:10.4136/ambi-agua.1936.

Sharma, D., Manzoor, M., Yadav, P., Sohal, J. S., Aseri, G. K., and Khare, N. 2018. Bio-valorization of dairy whey for bioethanol by stress-tolerant yeast. *Fungi and their Role in Sustainable Development: Current Perspectives*, 349–66. Singapore: Springer Singapore. https://doi.org/doi:10.1007/978-981-13-0393-7_20.

Tirado-Armesto, D. F., Gallo García, L. A., Acevedo Correa, D., and Mouthon Bello, J. A. 2016. Biotratamientos de aguas residuales en la industria láctea. *Producción más limpia* 11(1): 171–84. https://doi.org/doi:10.22507/pml. v11n1a16.

Torres-Martínez, Q. 2019. Evaluación de estrategias para el manejo de residuos de lactosuero en la localidad de Miahuatlán, Veracruz. Thesis, Universidad Veracruzana.

Treviño-Garza, M. Z., Romero, R., Vidales Contreras, J. A., Báez González, J. G., and Márquez Reyes, J. M. 2020. Efecto de la concentración de dextrosa en la producción de películas de celulosa microbiana a partir de Kombucha de té verde. I*nvestigación y Desarrollo en Ciencia y Tecnología de Alimentos* 5:118–123. http://eprints.uanl.mx/23519/1/20.pdf.

Tu, C., Tang, S., Azi, F., Hu, W., and Dong, M. 2019. Use of kombucha consortium to transform soy whey into a novel functional beverage. *Journal of Functional Foods* 52: 81–89. https://doi.org/doi:10.1016/j.jff.2018.10.024.

Vitas, J., Vukmanović, S., Čakarević, J., Popović, L., and Malbaša, R. 2020. Kombucha fermentation of six medicinal herbs: Chemical profile and biological activity. *Chemical Industry and Chemical Engineering Quarterly* 26 (2): 157–170. https://doi.org/doi:10.2298/ciceq190708034v.

Verduga-Vera, M. E. 2020. Cultivo en batch de Scenedesmus spp., en aguas residuales de industrias lácteas: Crecimiento, Productividad y Composición bioquímica. Thesis, Universidad de Guayaquil. http://repositorio.ug.edu.ec/bitstream/redug/48682/1/CD_TESIS_ELOIZA.pdf.

WHO-FAO. 2011. Leche y productos lácteos.fao.org. Disponible en:www.fao.org/3/a-i2085s.pdf.

Yadav, J. S. S., Yan, S., Pilli, S., Kumar, L., Tyagi, R. D., and Surampalli, R. Y. 2015. Cheese whey: A potential resource to transform into bioprotein, functional/nutritional proteins and bioactive peptides. *Biotechnology Advances* 33, n.° 6: 756–774. https://doi.org/doi: 10.1016/j.biotechadv.2015.07.002.

15 Unveiling the Prospects of High-throughput Microbial Biotechnology in the Effective Management of Wastewater

Munawar Sultana, Anamica Hossain, Aksa Hossain Nizhum, and M. Anwar Hossain

15.1 BACKGROUND

Waste management is the process of safely and efficiently processing and disposing of waste products to minimize the harmful effects on the surrounding environment and human health. There is evidence of waste management since ancient civilizations when waste was dumped in adjacent landfills, open pits, or burned. As cities developed and waste quantities increased in the early twentieth century, there was a demand for more structured waste management procedures. In the 1920s and 1930s, the first incinerators were erected, and landfills became more controlled to minimize pollution and health problems (Worrell & Vesilind, 2015). Concerns over pollution and the environment prompted the foundation of the United States Environmental Protection Agency (EPA) in 1970 and the enactment of the Clean Air Act, Clean Water Act, and Resource Conservation and Recovery Act (RCRA) in the 1960s (www.epa.gov/). These rules established a framework for controlling waste disposal techniques while also safeguarding human health and the environment. Waste management has evolved since then, with the introduction of new technologies and practices such as recycling, composting, and waste-to-energy systems. The emphasis has switched from waste disposal to waste reduction, sustainable methods, and limiting the environmental effect of waste management operations. Hydrocarbons, pesticides, heavy metals, and other harmful organic substances have been widely treated using bioremediation technology. This technology relies on various organisms such as bacteria, fungi, and algae that can break down contaminants into less dangerous chemicals or immobilize them. Furthermore, advanced technologies employ high-throughput methodologies

DOI: 10.1201/9781003441069-15

such as metagenomic analysis, functional gene analysis, synthetic biology, and many others which are needed for improving efficiency while managing a continuously rising volume of wastes cost effectively as well as providing greater outlooks for recycling and recovery for considerably better results that will be described later in the chapter. The nexus approach for integrated management and planning of utilization of natural resources in waste treatment is also one of the important aspects to be explored further.

15.2 WASTE TREATMENT PROCESSES AT A GLANCE

Treatment of waste can be done through applying different types of techniques. The employment of physical techniques to remove or separate impurities from water is referred to as physical water treatment. Sedimentation, filtration, disinfection, nanofiltration, reverse osmosis, and electrodialysis are a few examples (Bahuguna et al., 2021). Sedimentation is the process of allowing suspended materials to settle down at the bottom, whereas filtration is the removal of particles from water through a physical barrier. Disinfection is the process of destroying dangerous bacteria in water by using physical or chemical treatments.

There are biological techniques available where microorganisms are used in biological water treatment to break down and eliminate pollutants from water. Activated sludge treatment, bioreactors, and engineered wetlands are all examples (Shankar et al., 2021). Microorganisms break down organic materials in wastewater via aeration and mixing in activated sludge treatment. Microorganisms are utilized in bioreactors to digest pollutants in water, whereas plants and microorganisms work together to eliminate toxins in built wetlands.

Chemical treatment is the process of employing chemicals to remove pollutants from water. Coagulation, flocculation, and disinfection are a few examples (Suhardi et al., 2009). Coagulation is the process of introducing a chemical that causes pollutants to clump together, making them simpler to remove from water. Flocculation is the gentle mixing of pollutants, whereas disinfection is the use of chemicals or UV rays to destroy hazardous microbes in water.

Sludge treatment is the process of treating wastewater sludge to eliminate impurities and reduce the volume of the sludge. Anaerobic digestion, composting, and incineration are some examples (Bahuguna et al., 2021). Microorganisms digest organic materials in sludge anaerobically, generating biogas that may be utilized for electricity. In composting, sludge is combined with other organic materials to generate a soil amendment, whereas in incineration, sludge is burned to reduce its volume and remove germs.

15.3 TECHNOLOGICAL APPROACHES TO TREATING WASTEWATER

15.3.1 Classical or Traditional Approaches

There are different types of approaches employed to treat wastewater depending on its source, composition, physicochemical properties, and availability of resources. Some of the most common approaches are briefly discussed below:

Microbial fuel cells (MFCs): The principle behind MFC devices includes the implication of the metabolic processes of microorganisms that can convert chemical energy into electrical energy. In this case, electricity is derived from a variety of organic substrates, including wastewater, and applies to the treatment of wastewater to remove pollutants while generating electricity. Microbes can oxidize organic/inorganic molecules followed by sequential processes that can transform chemical energy into adenosine triphosphate (ATP). In this process, electrons are ferried to the terminal electron acceptor, and an electrical current is generated (Torres et al., 2009). There are reports on the usage of organic or inorganic materials from various types of wastewater as substrates for energy production.

Anaerobic digestion (AD): In the absence of oxygen, organic matter is biologically broken down during AD. In this process, organic waste and wastewater are commonly treated upon conversion of the organic matter into biogas, which can be used for energy production (Ugwu et al., 2022; Cruza et al., 2023). An example of AD application is in the treatment of animal manure and food waste. Available or developing AD systems using sewage sludge for operation exceed 1300 units around the world (IEA Bioenergy, 2001). It is worth mentioning that, wastewater treatment plants (WWTPs) are the prime sources of raw materials for biogas produced in AD plants, indicating that AD has become an essential and necessary aspect of recent WWTPs (Hanum et al., 2019).

Aerobic treatment systems: These systems utilize aerobic microorganisms to break down and treat wastewater. This process requires the addition of oxygen, and it can be used for both domestic and industrial wastewater treatment purposes. An example of an aerobic treatment system is the activated sludge process. Aerobic treatment units oxidize soluble organic and nitrogenous chemicals at a rapid rate (Seabloom & Buchanan, 2004). Because of its ease of installation, simplicity of operation, minimum maintenance, and better energy efficiency, the sequencing batch reactor, which is a type of aerobic treatment unit, is preferable and is also ideal for small communities (U.S EPA, 1992).

Constructed wetlands (CWs): These are systems that have been engineered to use natural processes including wetland plants, microbes, and soil for wastewater treatment. Their classical use is the treatment of household wastewater or stormwater runoff. A specific example of CWs' application is the treatment of septic tank effluent. Other examples include the treatment of wastewater including household wastewater (Cooper et al., 1997), acid mine drainage (Kleinmann et al., 1987), and industrial wastewater.

Membrane bioreactors (MBRs): MBRs represent a fusion of biological treatment along with membrane filtration that can produce high-quality treated water. Municipal wastewater is commonly treated using this method along with industrial wastewater. One notable example of MBR application is the treatment of textile wastewater. Prior to the 1990s, the majority of MBRs installed were utilized for industrial water treatment. The introduction of submerged membranes (Yamamoto et al.,1989) boosted the number of MBRs handling municipal wastewater, while the MBR market is now witnessing rapid expansion (Judd, 2006).

Bioelectrochemical systems (BESs): BESs use the metabolic activity of microorganisms to generate electricity and/or treat wastewater. This system has the potential to treat a variety of wastewaters, including domestic, agricultural, and industrial. An example of a BES application is the treatment of metal in wastewater. As is commonly known, metal pollution is a worry for the ecosystem since these chemicals are not biodegradable. Since half-cell redox is advantageous compared to organic matter, the metal reduction can be spontaneous on the cathode in BES (Cheng et al., 2013; Huang et al., 2015).

Algae-based treatment systems: Algae-based treatment systems utilize algae to remove pollutants from wastewater. Both domestic and industrial wastewater, as well as agricultural runoff, can be treated by this system. Microalgae biomass from wastewater streams has a high potential for long-term bioproducts such as proteins, fatty acids, pigments, biofertilizers/biochar, and animal feed (Wollmann et al., 2019).

Biological adsorption: This refers to an inherent mechanism of biological systems that can be utilized to remove pollutants from wastewater. Such adsorption relies on using different natural materials such as plant fibers, algae, and bacteria. An example of the application of biological adsorption is in the elimination/reduction of heavy metals from industrial wastewater. The use of nonliving biomass from *Pseudomonas* species has been experimentally employed for biosorption tests to Cr (VI), Cu (II), Cd (II), and Ni (II) (Hussein et al., 2004).

Fungal-based treatment systems: Fungal-based treatment systems utilize fungi to remove pollutants from wastewater. These are generally used for industrial wastewater treatment. There are recent reports on fungal-based treatment systems where white-rot fungi were used to remove different phenolics from wastewater (Martinez & Ng, 2018).

Biorefineries: Biorefineries are facilities that convert biomass into biotechnological products, including fuels, chemicals, and materials. These facilities may use a wide range of feedstocks, including municipal solid waste and agricultural residues. A noteworthy example of a biorefinery application is the bioconversion of corn stover into ethanol fuel (Martinez Hernandez & Ng, 2018; Kumar et al., 2023).

Trickling filters: Trickling filters are biological treatment systems that use a bed of rocks, plastic, or other media to support a biofilm of microorganisms that can degrade pollutants in wastewater. A common usage of these is household wastewater treatment. The textile wastewater decolorization efficacy of *Marasmius* sp. has previously been studied in immersion and trickling filter systems (Suhardi et al., 2009).

15.3.2 High-throughput Approaches Aiding Advances in Wastewater Treatment

With advances in science, a wide array of technologies is available to analyze, study, screen, and understand the wastewater composition and community more precisely. This not only helps in more efficient treatment but also simplifies the strategy designing and optimizing stage. Some of the high-throughput approaches are outlined in the next sections.

15.3.2.1 Metagenomics

Metagenomics refers to the analysis of total genetic material contained in an environmental sample using sequencing technology. This unveils the microbial community of that sample and has the potential to provide valuable information about the composition, diversity, and functional capabilities of these microbial communities (Kumar et al., 2020; Kumar et al., 2022; Wani et al., 2022). Talking about the biological waste management process includes discussing the microbial consortium suitable for waste degradation, the metabolic pathways and metabolites specifically responsible for the process, protection against infiltration of pathogenic microbes, and so on. As compared to traditional culture-based isolation, which misses out on the non-culturable microbes and is time-consuming, metagenomic analysis can achieve all the above-mentioned goals within a short period.

The role of metagenomics as a high-throughput approach is briefly discussed in the following sectors:

15.3.2.1.1 Characterization of the Microbial Community in the Waste stream

A very important parameter to be considered before subjecting any waste stream to treatment is its microbial community. For example, a waste stream from industry incorporates bacteria which might be able to reduce heavy metals and toxic chemicals whereas waste from hospitals harbors bacteria with extensive antibiotic resistance. Since biological treatment frequently employs indigenous microbial communities, it is essential to know the exact composition present in waste streams so that they can be filtered of pathogens and enriched with bioremediating isolates (Kumar & Chandra 2020). Investigation of municipal activated sludge through high-density microarrays detected a bacterial core that was dominated by *Proteobacteria*, followed by other phyla such as *Firmicutes*, *Bacteroidetes*, and *Actinobacteria* (Piceno et al., 2010). A similar report was indicated by several amplicon sequencing studies (Zhang et al., 2011). Interestingly, independent of reactor size, layout, or operating mode, this core seemed to be dispersed in comparable proportions throughout municipal wastewater treatment plants (WWTPs) and was found to be constant across continents (Piceno et al., 2010). In contrast, the bacterial composition of industrial-activated sludge differed significantly from municipal activated sludge, indicating that different parameters of influent wastewater have a considerable effect on bacterial community diversity. Moreover, the treatment process and its effectiveness are also dependent on the compositions of microbial diversity and the metabolic potential of microbial consortia (Flowers et al., 2013), which can be rapidly studied by none other than the metagenomic approach.

15.3.2.1.2 Efficiency of Treatment Evaluation

High-throughput sequencing can rapidly provide an overview of pathogenic bacteria present in sewage samples before and after treatment is applied. This can not only evaluate the efficiency of a particular treatment strategy but also provide solid evidence to compare several treatment processes. In fact, the particular stage at which maximum reduction of pathogens occurs can be shortlisted, and those steps optimized, altered, or redesigned as per specific requirements. For example, raw sewage dominated by different bacteria such as *Arcobacter butzleri*, *Aeromonas hydrophila*, and *Klebsiella pneumonia* had most eliminated by employing the oxidation ditch step (Lu et al., 2015). Furthermore, whole metagenome sequencing was done to profile the biofilm-layer-microbial population residing on plant root surfaces and was compared to the microbial population within wastewater. This technology focused on elucidating whether plant-microbe interactions can promote the removal of micropollutants in wastewater treatment plants (WWTPs), just like the remediation of recalcitrant xenobiotics. The report suggests a steep decrease in the concentration of PPCPs in the anaerobic treatment tank dominated by bacilli within the biofilm population (Balcom et al., 2016). The most prevalent xenobiotic molecule discovered across the system was benzoate degradation. The wastewater microbial communities were more diversified in terms of taxonomy and metabolism than the submerged plant root biofilm. The root biofilm, on the other hand, exhibited metabolic genes for xenobiotic metabolism at an increased abundance and variation (Balcom et al., 2016).

15.3.2.1.3 Identify Novel Enzymes and Metabolic Pathways for Waste Degradation

Several enzymes such as oxygenases, peroxidases, polyphenol oxidase, tyrosinase, and laccase can treat inorganic pollutants and heterocyclic aromatic compounds whereas protease, amylase, lipase, cellulase, urease, and xylanases work specifically on organic pollutants. Frequently, pathogenic bacteria can degrade a wide range of xenobiotics and often the degradation ability is acquired from the environment. A metagenomic approach can be used to delineate the specific enzymes produced by the bacteria, pinpoint the metabolic pathways involved, distinguish acquired and inherent genes associated through pathway reconstruction and functional analysis and thus, provide a clear overview to work with. For example, a metagenomic study identified unique biosurfactant generating genes in marine bacteria, which aided in the discovery of numerous routes and techniques for better biosurfactant synthesis. Metagenomic studies revealed Palmitoyl putrescine and N-acyl amino acids as two new biosurfactants (Jackson et al., 2015). The bacterial population retrieved from heavy metal and radionucleotide-contaminated areas through metagenomic analysis showed the emergence of novel bacteria capable of heavy metal as well as radionucleotide remediation. Furthermore, heavy metal resistance genes have also been identified in metagenomic investigation of heavy metal-polluted locations, indicating the presence of denitrifying bacteria in heavy metal-contaminated areas. Additionally, from uranium-contaminated sites, *Geobacter* species have been identified in metagenomic research as the dominant bacteria that are difficult to isolate through culture assay (Datta et al., 2020).

15.3.2.1.4 Development of Microbial-based Bioreactors for Waste Treatment

Metagenomics may be used to find microbes capable of decomposing certain waste chemicals and then using them to construct microbial-based waste treatment bioreactors. These bioreactors may be constructed to optimize circumstances for the development and activity of these microorganisms as well as the waste management system's efficiency. In a study, a 16S rRNA gene-based study was performed to identify bacterial populations residing within dye effluent in an anoxic-oxic bioreactor. The anoxic-oxic bioreactor was set up on a lab scale from where azo-dye decolorizing and degrading bacteria were enriched in two stages. Separate 16S rDNA metagenomic libraries of enriched populations were created and sequenced, and phylogenetic analysis was done to evaluate their potential as dye degraders (Dafale et al., 2010). To cite another example, the gold ore processing mechanism involves the use of cyanide (CN), which leads to the generation of thiocyanate- (SCN) polluted effluent and it is necessary to treat it. Although it is reported that certain microbes can degrade SCN and CN, detailed information on their composition and metabolic capability is still lacking. What is known so far is their potential to break SCN and CN into carbon, sulfur, and nitrogen which can be an option to utilize in microbial bioremediation of such toxic elements. Such a species is reported as specific *Thiobacillus* spp. through metagenomic investigation and its whole genome was found to possess a novel SCN degrading operon. The study also revealed a bigger autotrophic population and a smaller heterotrophic population. These findings point to reactor design optimization options such as enhanced aerobic/anaerobic partitioning and the removal of organic carbon from the reactor feed (Kantor et al., 2015).

15.3.2.1.5 Pollution Biomarkers

Metagenomic analyses are potent for identifying diverse microbial presence as well as inheritance of their specific genes that can function as pollution indicators. For example, the prediction of hexane compound contamination can be done by detecting the presence of cyclo-alkane degrading genes within an environment (Techtmann & Hazen, 2016).

15.3.2.2 Analysis of Microbial Population

The technique of ultra-high-throughput microbial diversity analysis is used to explore the diversity and community composition of microbes at such a high resolution and throughput that their inherent metabolic pathways can be determined for broader benefit. It entails the use of sophisticated sequencing technologies, such as Illumina and nanopore sequencing, to create massive volumes of sequence data from microbial communities' DNA or RNA. Sequencing of rDNA clone libraries, terminal restriction fragment length polymorphism, and microarrays are three very important techniques employed (Guttman et al., 2016).

This method enables researchers to analyze microbial communities in a thorough and unbiased manner, revealing the identities and roles of the microorganisms present. It also allows for the finding of previously undiscovered and uncommon species within the community, which may have major ecological or biological consequences.

Aside from sequencing, ultra-high-throughput microbial community study frequently employs bioinformatics' tools and statistical analyses to analyze and interpret the data. Taxonomic categorization, phylogenetic analysis, and functional annotation are examples of these procedures. In a previous study, the microbial diversity of residential wastewater discharged from toilets, known as blackwater, was analyzed before resource recovery and treatment (Mogili & Mohapatra, 2021). High-throughput pyrosequencing of the 16S rRNA gene was done to organize and study the leachate communities along with their composition profile. *Firmicutes* dominance, together with *Bacteroidetes*, *Proteobacteria*, and *Spirochaetes*, revealed a microbial community that destroys organic materials in raw leachate before the methanogenic phase. The discovered genera correlated well with the traditional paths of AD processes (Köchling et al., 2015). Another study discovered that waste-associated ecosystems had a distinct bacterial variety, with *Proteobacteria* and *Bacteroidetes* predominating. *Paracaedibacteraceae* (uncultured EF667926), *Ralstonia*, *Chroococcidiopsis*, *Chitinophagaceae* (uncultured FN428761), *Sphingobium*, and *Heliimonas* were among the abundant genera (> 1%). Approximately one-third of the species in these genera were uncultured (Pan et al., 2021). According to an investigation of the microbiome of effluent exiting typical WWTPs between 2015–2016 and entering the aquatic environment, the bacterial populations were investigated using Illumina MiSeq 16S rRNA gene amplicon sequencing and analyzed using Calypso software. The WWTP microbiomes shared several similarities with the human fecal microbiome and included potential sources of human infections. This information is useful in assessing the danger presented by WWTP effluent to the environment and human health (Do et al., 2019).

Overall, ultra-high-throughput microbial community analysis has transformed our understanding of microbial communities and their involvement in a variety of ecosystems ranging from soil and water to the human microbiome.

15.3.2.3 Use of Microbial Culture or Consortia in Wastewater Treatment

Microbial culturing can be molded to be a high-throughput waste management technology. This is because microorganisms like bacteria and fungi may degrade a wide spectrum of organic chemicals and contaminants, lowering their concentration in waste.

In this technique, waste is collected and then injected into a bioreactor or a fermenter containing microorganisms or their consortia. The microbes then degrade the waste's organic molecules into simpler chemicals like carbon dioxide, water, and biomass. The biomass produced may subsequently be collected and processed into a variety of valuable products, including biofuels, biobased chemicals, and animal feed. Furthermore, the carbon dioxide created during the degradation process may be recovered and utilized for a variety of purposes, including greenhouse gas mitigation and carbon sequestration.

A study used a "Tecan Freedom Evo 200 pipetting robot" to cultivate *Synechocystis sp.* PCC6803 in microtiter plates which were placed within a cultivation chamber of CO_2 atmosphere associated with programmed shaking, temperature, and illumination conditions. Several indicators, including growth, whole absorption spectrum, chlorophyll concentration, MALDI-TOF-MS, and a unique vitality assessment methodology, were already automated and could be monitored during culturing. At present, the platform should be suitable for numerous phototrophic bacteria and extendable to even more (Tillich et al., 2014). Through biofilm development and the synthesis of extracellular polymeric molecules, bacteria isolated from marine settings are expected to be better exploited in the bioremediation of heavy metals, hydrocarbons, and many other refractory chemicals and xenobiotics. Many marine microorganisms have been shown to have bioremediation capabilities (Dash & Mangwani, 2013). High-throughput methods for microorganism cultivation in low-nutrient media yield *Proteobacteria* and other new marine isolates (Connon et al., 2002). Droplet-based microfluidics, microbial microdroplet culture systems, microcapsules, etc. are a few of the various techniques for high-throughput microbial cultivation.

Overall, microbial cultivation offers a promising approach for high-throughput waste management, as it is relatively low cost, environmentally friendly, and can generate useful products from waste materials that would otherwise be disposed of in a landfill or incinerator.

15.3.2.4 Microbial Functional Gene Analysis

Identification of specific microbial genes and pathways involved in the degradation of different types of waste materials can be used to optimize the conditions within the bioreactor or fermenter, such as adjusting the pH, temperature, and nutrient availability, to improve the performance of the microbial community in the context of high-throughput waste treatment processes. Furthermore, functional gene analysis may be utilized to track changes in the microbial population across time. This can aid in identifying any changes in the community structure or function that may signal difficulties with the waste treatment process or the need for operational modifications. In one study, both pyrosequencing and Illumina high-throughput sequencing were employed along with molecular technology to investigate the composition and functions of nitrifying and denitrifying microbial communities residing in the aerobic and anaerobic sludge of two full-scale tannery WWTPs. A dominance of *Nitrosomonas europaea* in the WWTPs' aerobic and anaerobic sludge was indicated through *amo*A gene cloning of ammonia-oxidizing bacteria with a coexistence of *Proteobacteria*. There was the presence of *Thauera*, *Paracoccus*, *Hyphomicrobium*, *Comamonas*, and *Azoarcus*, as denitrifiers along with functional genes of nitrifiers and denitrifiers such as *amo*A, *nir*K, *nir*S, and *nos*Z. A synergistic correlation between the qPCR and metagenomic analysis was observable and therefore, can be a potential method for a comprehensive investigation of the abundance of functional genes in the environment (Wang et al., 2014). In another study, several activated sludge samples, collected from selected WWTPs in Beijing, were subjected to a comprehensive functional gene array study. The purpose was to determine the microbial functional genes for different biogeochemical processes such as carbon, nitrogen, phosphorous, and sulfur cycles, metal resistance, antibiotic resistance, and organic contaminant degradation. Several common functional genes for diverse bacteria were reported from the samples, which included gene categories such as *cgl*, *amyA*, *lip*, *nirS*, *nirK*, *nosZ*, *ureC*, *ppx*, *ppk*, *aprA*, *dsrA*, *sox*, *and benAB*. Canonical correspondence analysis (CCA) revealed that microbial functional patterns were substantially associated with water temperature, dissolved oxygen (DO), ammonia concentrations, and COD loading rate. This study offered an overall view of the functional architecture of activated sludge microbial communities in WWTPs, as well as the relationships between microbial communities and environmental factors in WWTPs (Wang et al., 2014). Other research found that the acute toxicity of sludge in a 4-chlorophenol acclimated sequencing batch bioreactor (SBR) was considerably greater than in a control SBR that did not use 4-CP. In addition, the functional gene expression of *Pseudomonas* in the acclimated SBR was down-regulated, and the methods of their expression need additional investigation. This research provides theoretical evidence for a thorough understanding of sludge performance in industrial wastewater treatment (Zhao et al., 2018). As a whole, microbial functional gene analysis provides many significant insights.

15.3.2.5 Microbial Synthetic Biology in Accelerating Waste Treatment Technologies

Microbial synthetic biology is a branch of biology that uses genetic engineering and synthetic biology techniques to create novel microbial systems with specified purposes. In the context of wastewater management, microbial synthetic biology may be utilized to develop high-throughput techniques for removing toxins from wastewater.

One possible solution is to design microbes to produce enzymes or other compounds capable of breaking down certain contaminants. Bacteria, for example, can be genetically modified to generate enzymes that degrade certain types of organic molecules, such as those present in industrial wastes. This can improve treatment efficiency and minimize the amount of time necessary to treat the effluent. Another strategy is to create microbial populations to collaborate in the removal of pollutants from wastewater. Microorganisms can be genetically modified to emit signals or communicate with one another in ways that favor the growth of certain species or the elimination of certain contaminants. It may be able to construct more efficient and effective wastewater treatment procedures by engineering microbial communities in this manner.

High-throughput approaches can also be utilized to speed up the development of microbial wastewater treatment systems. Researchers, for example, might utilize high-throughput screening approaches to rapidly identify microorganisms with desirable features, such as the capacity to manufacture certain enzymes or collaborate in specified ways. This can drastically save the time and money needed to build successful wastewater treatment methods. In 2016, researchers announced the discovery of a unique phenomenon for enhanced degradation of Direct Red 5B azo dye by *Irpex lacteus* CD2 using lignin as a co-substrate (Sun et al., 2015). Under DR5B and lignin treatments, a wide range of lignin-degrading peroxidases, oxidases, radical-generating enzymes, and other related components were upregulated. To supply necessary enzymes and redox conditions for aromatic compound breakdown, lignin-induced genes were supplemented DR5B-induced genes (Sun et al., 2015). The study found that adding phosphate, nitrate, or a combination of nitrate and phosphate to the sludge increased the abundance of microorganisms such as *Bacillus*, *Coprothermobacter*, *Rhodobacter*, *Pseudomonas*, *Achromobacter*, *Desulfitobacter*, *Desulfosporosinus*, *T78*, *Methanobacterium*, and *Methanosaeta*, which are known to degrade hydrocarbons. This led to a 46–55% reduction in total petroleum hydrocarbon (TPH) levels. Additionally, when indigenous *Bacillus* strains were used to generate biosurfactants and produce hydrocarbons, along with proper nutrition, the TPH levels were reduced even further, reaching a reduction of 57–75% (Roy et al., 2018). Reineke (1998) discussed the use of a patchwork assembly to engineer recombinant strains for the complete degradation of chloroaromatics. This involved combining pathways obtained from various bacteria via conjugation into a single recombinant host, allowing the mineralization of a specific compound. For instance, Hrywna et al. (1999) transformed *Comamonas testosteroni* strain VP44 with plasmids bearing the *ohb* and *fcb* operons, resulting in a strain capable of completely mineralizing monochlorobiphenyls. *Deinococcus radiodurans* is a useful host for genetic engineering approaches involving mixed waste. It has been engineered with genes encoding toluene dioxygenase from *Pseudomonas putida* F1, enabling effective oxidation of toluene, chlorobenzene, 3,4-dichloro-1-butene, and TCE in a highly irradiating environment. *E. coli* has also been used as a host for genetic engineering to enhance heavy metal bioremediation. For example, Crameri et al. (1997) assembled an arsenate detoxification pathway, resulting in increased arsenate resistance and the ability to detoxify arsenate by reduction. Qin et al. (2006) engineered *E. coli* to detoxify mercury by expressing the Hg(II) binding domain of the MerR protein from *Shigella flexneri* on the cell surface. Additionally, Ohtsubo et al. (2003) improved the PCB and biphenyl degrading ability of *Pseudomonas* sp. strain KKS102, while Wang et al. (2002) engineered *E. coli* to produce fusions of OPH with domains allowing the fusion proteins to be expressed on the cell surface (Urgun-demirtas & Stark, 2006). Therefore, microbial synthetic biology holds great promise for the development of high-throughput wastewater treatment processes that are more efficient, effective, and sustainable.

15.3.2.6 Microbial Bioreactor Design

Microbial bioreactors are commonly used in bioremediation to efficiently and effectively degrade various pollutants. These bioreactors provide a controlled environment for microorganisms to degrade pollutants, including industrial chemicals, pesticides, and petroleum hydrocarbons. High-throughput bioremediation is achieved through the use of bioreactors, which can process large volumes of contaminated water or soil. Engineered bioreactors that provide optimal conditions for microbial growth and biodegradation have been created for use in bioremediation procedures to meet various remediation goals. Bioreactors in use range in operation mode from batch to continuous to fed-batch and are designed to optimize microbial processes with respect to contaminated media and pollutant nature. Bioreactors designed for bioremediation include packed, stirred tanks, airlift, slurry phase, and partitioning phase reactors, among others (Tekere, 2019). Examples include the use of slurry bioreactors (SB) as an effective technology for the bioremediation of problematic sites with high contents of clay and organic matter and pollutants that are toxic, recalcitrant, or display hysteretic behavior. SB technology allows for convenient control and manipulation of various environmental parameters for faster and enhanced treatment of polluted soils. SB with simultaneous electron acceptors is an emerging area and has proved useful for bioremediation of soils polluted with hydrocarbons and some organochlorinated compounds. Although the significance of microbial community characterization for reactor operation and design optimization is undeniable, studies in this area are still in their infancy (Robles-gonzález et al., 2008). Overall, microbial bioreactors offer a promising approach to high-throughput bioremediation, and ongoing research continues to improve the efficiency and effectiveness of these systems.

15.3.3 Combined Approaches of Traditional and High-throughput Techniques in Bioremediation

So far, we have discussed both long-used traditional procedures and recently found high-throughput techniques as shown in Figure 15.1. The classical approaches are constructed wetlands, microbial fuel cells, trickling filters and such. Metagenomic analysis, microfluidics, and synthetic biology ushered in the era of high-throughput analysis. Without a doubt, the greatest approach to managing waste efficiently is to combine high-throughput screening findings with long-established optimized conventional processes (Fig. 15.1). Traditional culture techniques, for example, may be employed to isolate bacteria capable of degrading a specific contaminant from a polluted location. The whole microbial population present at the location,

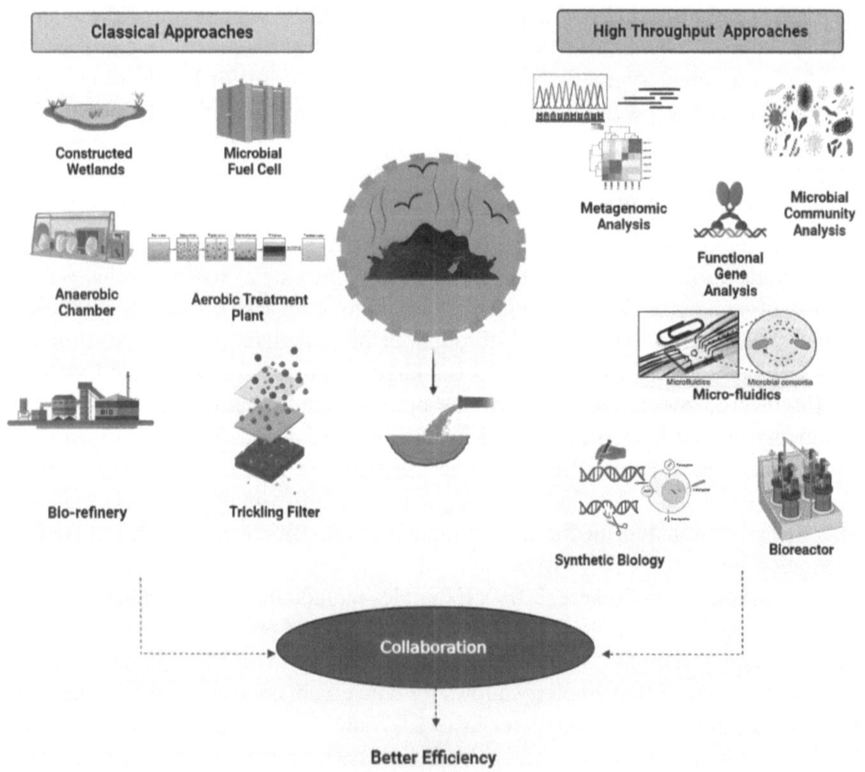

FIGURE 15.1 Overview of classical and high-throughput approaches for waste water management (This image is generated using BioRender).

including additional possible biodegrades, may be subsequently identified using high-throughput sequencing. These data may then be utilized to develop a bioremediation plan that not only targets the individual contaminant but also considers the larger microbial population and its interconnections. Furthermore, high-throughput approaches may be utilized to evaluate the efficacy of the bioremediation process in real-time, allowing modifications to be made as needed to optimize efficiency. Again, synthetic biology may adopt appropriate microorganisms for maximum activity, and using them in standard bioreactors will allow for faster action and higher production. The metagenome now allows us to examine the entire microbial population and learn all of the specifics, making it easy to determine whether an aerobic, anaerobic, indigenous, or engineered mode or mix of tactics would produce the greatest results.

15.4 NEXUS APPROACH

Stringent discharge regulations and the introduction of new contaminants pose ongoing challenges for wastewater treatment. Examples include micropollutants like pesticides, medications, personal care products, endocrine-disrupting chemicals (EDCs), and nanoparticles, which are continuously emerging in wastewater treatment plant (WWTP) effluents and residual activated sludge. These newly identified substances often occur at low concentrations, making it difficult for conventional wastewater treatment systems to effectively remove them. However, viewed from an evolutionary standpoint, certain bacteria have the potential to adapt to the presence of these novel chemical compounds, which could aid in their elimination (Wu & Yin, 2020). Consequently, addressing current and future wastewater treatment issues would greatly benefit from understanding how to optimize microbial habitats to enhance the removal of various types of contaminants.

In recent times, the concept of a nexus has gained prominence in the field of sustainable development. The nexus approach holds great significance in environmental management as it recognizes the interconnectedness of natural resources and emphasizes the need for integrated planning and management to promote a more sustainable and efficient utilization of these resources. Within the environmental context, the nexus approach proves valuable in identifying and tackling issues concerning water, energy, and food systems, which can have substantial repercussions on ecosystems and biodiversity (Ayres et al., 2021).

For instance, excessive water usage in irrigation can deplete underground water sources and adversely affect aquatic ecosystems. Similarly, energy production processes can lead to air and water pollution while contributing to climate change, thereby exacerbating the impact on natural systems. By employing the nexus approach, environmental managers can effectively identify and address these interconnected challenges comprehensively and cohesively. This approach takes into account the interdependencies between water, energy, and food systems, as well as their implications for natural systems (Paul et al., 2016). The application of the nexus approach can also play a significant role in advancing sustainable practices and technologies. For example, it can support the adoption of renewable energy sources and the implementation of water-efficient irrigation systems, which contribute to a more sustainable and resilient environment. Overall, the nexus approach offers a valuable framework for environmental management, facilitating integrated and sustainable approaches to resource utilization, and aiding in the preservation and safeguarding of ecosystems and biodiversity.

The concept of the microbial nexus pertains to the interconnections and interdependencies among microorganisms in the environment. It encompasses their crucial roles in processes such as nutrient cycling and carbon sequestration, which are vital for ecosystem functioning. The microbial nexus is influenced by various factors, including:

Microbial diversity: Microbial diversity is a critical factor in the microbial nexus, as different microorganisms play different roles in ecosystem processes. The diversity of microorganisms can be influenced by various factors, including environmental conditions such as temperature, pH, and moisture.

Nutrient availability: Nutrient availability is another critical factor in the microbial nexus, as microorganisms require nutrients to carry out their metabolic processes. Nutrient availability can be influenced by factors such as soil type, organic matter content, and nutrient inputs.

Carbon cycling: Microorganisms play a crucial role in carbon cycling, both as decomposers and as producers of greenhouse gases such as carbon dioxide and methane. The balance between carbon inputs and outputs can influence the microbial community composition and the rates of carbon cycling.

Water availability: Water availability is an essential factor in the microbial nexus, as microorganisms require water for their metabolic processes. Water availability can influence microbial diversity and activity, as well as nutrient cycling rates.

Climate change: Climate change can have significant impacts on the microbial nexus, as changing environmental conditions can affect the diversity and activity of microorganisms. For example, changes in temperature and precipitation patterns can alter microbial community composition and nutrient cycling rates.

Overall, the microbial nexus is a complex and dynamic system that is influenced by various biotic and abiotic factors. Understanding the factors involved in the microbial nexus can help us better manage and protect ecosystem processes and functions.

15.4.1 Elemental Cycles of Sulfur, Carbon, and Nitrogen

The removal of organic carbon through anaerobic treatment is a specialized process that involves the cooperative action of functional microorganisms to convert organic carbon into renewable bioenergy in the form of methane (CH_4). During anaerobic treatment, complex organic substances like proteins and sugars are initially broken down into simpler monomers by fermenting bacteria (Ijoma et al., 2022). These monomers are then utilized by acidogenic bacteria to produce acetate and hydrogen. Methanogens, in turn, use acetate, hydrogen, and carbon dioxide (CO_2) to produce CH_4 as the end product (Mackenzie et al., 1998). The stability of an anaerobic system relies on the ability to maintain a thriving microbial population and support the growth of these microorganisms. In contrast, aerobic biological wastewater treatment processes primarily involve the breakdown of organic carbon by heterotrophic microorganisms, resulting in the production of CO_2 and biomass.

By cultivating microbial communities through a sequential arrangement of anaerobic, anoxic, and aerobic reactors, it becomes possible to enrich specific groups of bacteria such as ammonia-oxidizing bacteria (AOB), nitrite-oxidizing bacteria (NOB), denitrifiers, and phosphorus-accumulating organisms (PAOs). This enrichment contributes to the efficient removal of organic carbon, nitrogen, and phosphorus from wastewater. PAOs, for instance, can store volatile fatty acids (VFAs) in the anaerobic reactor, utilizing energy from intracellularly stored polyphosphate. Denitrification occurs in the anoxic reactor, with organic carbon in the wastewater serving as the electron donor and recirculated oxidized nitrogen acting as the electron acceptor. In the aerobic reactor, PAOs take up phosphorus, nitrifiers perform ammonia nitrification, and heterotrophic microorganisms engage in activities related to organic carbon removal. Sulfate-reducing bacteria (SRB), which control the sulfidogenic bioprocess, will become essential microorganisms for sulfate removal when treating wastewater that contains sulfate. During sulfate reduction, SRB converts sulfate first into sulfite and subsequently into sulfide. There are two pathways for the conversion of sulfite to sulfide. The first is a direct pathway where sulfite is reduced to sulfide by accepting six electrons. The second is an indirect pathway involving intermediates such as trithionate and thiosulfate. In the metabolism of sulfate-reducing bacteria (SRB), carbon sources play a role as electron donors. For instance, some SRBs utilize Acetyl CoA or a modified tricarboxylic acid (TCA) cycle to break down organic compounds and serve as electron donors. SRB can also metabolize various intermediate products from anaerobic fermentation and hydrolysis, including amino acids, carbohydrates, long-chain fatty acids, and volatile fatty acids (VFAs). In such cases, sulfate can be effectively utilized to simultaneously remove organic carbon.

Simultaneous removal processes are well-established, including the concurrent removal of carbon and nitrogen through denitrification, the removal of sulfur and nitrogen through sulfur-based denitrification, and the removal of organic carbon, nitrogen, and phosphorus through denitrifying phosphorus-accumulating organisms (PAOs). In marine sediments, it has been observed that sulfate reduction (sulfammox) can occur alongside anaerobic denitrification and ammonium oxidation. This process results in the generation of sphalerite, elemental sulfur, and free sulfide through sulfate reduction and ammonium oxidation. To facilitate the growth of sulfammox, wastewater needs to contain certain amounts of sulfate and ammonium. However, linking specific microorganisms to these processes remains challenging.

15.4.2 The r/K Selection Theory

In both natural and engineered ecosystems, two common types of strategists can be characterized using the r/K selection theory. The K-strategists are nonopportunistic organisms that are well adapted to crowded and resource-limited environments, where they can survive near the carrying capacity of the ecosystem (Andrews & Harris, 1986. On the other hand, r-strategists thrive in uncrowded and resource-rich

TABLE 15.1
Features that define r- and K-strategists

Characteristics	r-strategist	K- strategist
Ribosomal RNA operon copy number	High	Low
Resistance to mortality	Low	Variable High
Tolerance to inhibitory chemicals	Variable Low	Variable High
Maximum growth rate	High	Low
Substrate affinity	Low	High
Efficacy of biomass production from food	Low	High
Ability to compete with substrate limitations	Low	High

Source: Yin et al. (2022).

environments, where competition is minimal and resources are abundant. These two strategist types exhibit distinct traits. R-strategists are copiotrophs and flourish in eutrophic conditions with high resource availability, as described by the Monod equation, showing fast growth rates and low resource affinity. In contrast, K-strategists are oligotrophs characterized by slow growth rates and high resource affinity (Yin et al., 2022). Therefore, when optimizing microbial habitats, it is crucial to consider both biochemical redox conditions and substrate availability.

For example, in a plug flow reactor, r-strategist microorganisms may become enriched in the front section where high nutrient concentrations allow for efficient contaminant removal. In the rear section of the reactor, where contaminants are primarily degraded and oligotrophic conditions prevail, K-strategist microorganisms may dominate. This concept suggests that systems with pronounced substrate gradients can simultaneously achieve high removal rates and efficiency for a wide range of contaminants by providing specialized niches for both r- and K-strategists. By considering this substrate-based approach, microbial habitats can be optimized to enhance the removal of various types of contaminants. R- and K-strategist characteristics (Andrews & Harris 1986) are shown in Table 15.1 and Figure 15.2.

15.4.3 Creation of a Microenvironment to Support a Variety of Microbial Activities

The promotion of a diverse microbial environment can be achieved by combining suspended flocs and biofilm in wastewater treatment systems. Microorganisms in these systems tend to form either suspended flocs or a biofilm, and different microorganisms exhibit preferences for living in one or the other. Microorganisms with fast growth rates generally thrive in suspended flocs, while slower-growing microorganisms may have better survival in biofilm. For instance, the enrichment of the slow-growing complete ammonia oxidizer in biofilm systems has been observed after several years of acclimation or enrichment. By coupling suspended flocs and

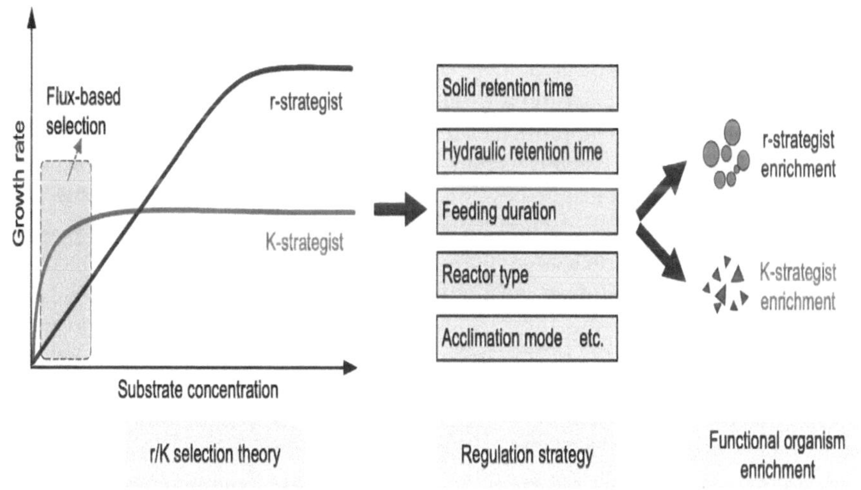

FIGURE 15.2 The r-/K-strategists' development characteristics (The image is adapted from the source).

Source: Yin et al. (2022).

biofilm, different microorganisms can access distinct microbial niches and facilitate interactions among different microbial species. This arrangement promotes diverse microbial functions and enhances the efficiency of wastewater treatment.

Suspended flocs provide a suitable environment for fast-growing microorganisms, while a biofilm offers a favorable habitat for slow-growing microorganisms. Achieving the coupling of suspended flocs and biofilm requires careful design of wastewater treatment systems, including appropriate reactor sizes, shapes, and configurations. Additionally, the system should provide adequate nutrients, pH, temperature, and oxygen availability to support the growth and activity of microorganisms. Controlling environmental factors within biofilm systems is crucial to ensure stable process operation and facilitate the connection between suspended flocs and biofilm. One critical aspect to manage is the gradient of dissolved oxygen (DO) within the biofilm. The biofilm can be divided into three layers based on the oxygen levels present: aerobic, anoxic, and anaerobic. Different microorganisms develop and metabolize within these distinct strata, contributing to the comprehensive degradation of various contaminants. It is essential to carefully manage the DO gradient to ensure all layers of the biofilm receive an appropriate amount of oxygen for microbial growth and function. Excessive oxygen supply can harm anaerobic niches within the biofilm and restrict certain microbial activities. Therefore, maintaining a well-managed DO gradient is crucial. This can be achieved by incorporating suitable reactor sizes, shapes, and configurations into the design of the wastewater treatment system. Furthermore, providing the right nutrients, pH, temperature, and oxygen availability is essential for the thriving of microorganisms in the system.

15.4.4 Outlooks for the Future

Despite the discoveries indicated, there are still gaps in our understanding of how to apply the r/K selection theory to biological wastewater treatment systems. There are additional (functional) creatures that cannot be easily grouped into these two classes, even if certain functional organisms may be classed as r- or K-strategists. Therefore, the r/K selection theory must be developed to broaden its application. Another obstacle to the use of the r/K selection theory in engineered bioprocesses is the development of r-/K-strategist detection tools. According to recent reports, the ribosomal RNA operon (rrn) copy number is an important feature for determining life strategy under various resource availability conditions. The amount of rrn copies is positively connected with high maximum growth rates, hence r-strategists should have more rrn copies than K-strategists. For instance, it was demonstrated that the dynamics of the average rrn copy number and its correlations with the concentrations of phosphate and ammonium represent the succession of the r/K-strategists and their susceptibility to nutritional perturbations. However, more work should be put into developing efficient techniques for the discovery and analysis of r-/K-strategists. Particularly, further approaches in molecular biology, quantitative modeling, and bioinformatics are required. Finally, since there is a mixed flora of microorganisms in wastewater biological treatment systems, it is important to clarify in future research if and how the interaction between microbial populations impacts r-/K-properties.

15.5 SOCIOECONOMIC IMPACTS OF NEXUS APPROACHES

The utilization of high-throughput and nexus techniques in bioremediation can yield significant socioeconomic impacts. High-throughput techniques applied to the cleanup of contaminated sites can lead to faster and more efficient remediation processes, resulting in reduced costs and minimized adverse effects on human health and the environment. However, the initial investment required for implementing high-throughput techniques may be prohibitively expensive, and there may be social and ethical considerations to account for, such as potential job displacement due to automation. The nexus approach, which acknowledges the interconnectedness of water, energy, and food systems, has significant potential for generating substantial socioeconomic impacts. For example, the utilization of artificial wetlands for bioremediation can provide additional benefits such as creating wildlife habitats and offering recreational opportunities while simultaneously addressing water pollution (Huang et al., 2020). However, the adoption of nexus approaches can be complex and require collaboration among diverse stakeholders, including government agencies, private industries, and local communities.

In a study conducted in Chile, the cost of bioremediation for chronically hydrocarbon-contaminated soils using different methods ranged from USD 50.7 to USD 310.4 per cubic meter of soil. This study was conducted in a country without an established environmental remediation industry or economic policies to guide treatment costs (Siddiqi & Anadon, 2011). The most cost-effective treatment identified was biostimulation with 10% compost (BE1), which was found to be less expensive than bioaugmentation-based approaches. This treatment resulted in significant reductions in direct material expenses, provision, and covering assets.

Another research article published in 2022 (Akinsete et al., 2022) proposed a framework for the integrated management of the water–energy–food nexus. The framework combined four independent models: Economic Growth (Stochastic Game Model), Environmental Policy (Legal Principles and Norms Model), Demographic, Cultural, and Social Development (Model of Systems Dynamics), and Water Governance Principles (Law/Policy Classification and Expectation Matrix). This integrated approach addressed the less-studied socio-anthropological components of the nexus. It highlighted the human element as a component of the larger ecosystem, considering sociocultural and economic activities, the laws and regulations governing these activities, and the potential socioeconomic implications and consequences. The article suggested that while environmental models are valuable decision-making tools, their integration with socioeconomic and policy-based models provides a more comprehensive understanding of the ecosystem. Overall, the socioeconomic impacts of high-throughput and nexus approaches in bioremediation are influenced by several factors, including the technologies used, the social and political context in which they are implemented, and the level of community engagement and participation in decision-making processes.

15.6 CONCLUSION

Therefore, we may conclude that a high-throughput approach to waste management refers to a system that can handle a huge volume of waste in a short period by leveraging sophisticated technologies such as metagenomics, synthetic biology, microfluidics, and many others. High-throughput waste management systems offer the advantage of minimizing the quantity of waste that is disposed of in landfills, which can assist in reducing the environmental and health concerns connected with waste disposal. The nexus method, on the other hand, believes in an integrated strategy that considers the interdependence of water, energy, and food systems. The nexus method for waste management entails recognizing the connections between waste creation, water consumption, energy use, and food production. A nexus approach to waste management allows for the identification of synergies and trade-offs that can improve the overall sustainability of the waste management system. For example, a high-throughput waste management system that generates electricity through incineration can assist in minimizing greenhouse gas emissions and reliance on fossil fuels. However, if not handled appropriately, this strategy may result in the generation of toxic ash, which might endanger water quality. It is feasible to discover solutions to reduce the negative effects of incineration on water quality while optimizing the advantages of energy generation by using a nexus approach. Overall, when paired with the nexus approach, the high-throughput approach to waste management can assist in developing a more sustainable and resilient waste management system that satisfies societal demands while reducing environmental consequences.

REFERENCES

Akinsete, Ebun, Phoebe Koundouri, Xanthi Kartala, Nikos Englezos, and Jonathan Lautze. 2022. "Sustainable WEF Nexus Management: A Conceptual Framework to Integrate Models of Social, Economic." *Policy, and Institutional Developments* 4: 1–13. https://doi.org/10.3389/frwa.2022.727772.

Andrews, John H., and Robin F. Harris.1986. "*r*-and *K*-Selection and Microbial Ecology." In: Marshall, K.C. (eds) Advances in Microbial Ecology. Advances in Microbial Ecology, vol 9. Springer, Boston, MA. https://doi.org/10.1007/978-1-4757-0611-6_3

Ayres Rebello, Thais, Ricardo Franci Gonçalves, and João Luiz Calmon. 2021. "Impacts of a NEXUS Wastewater Treatment Plant: A Comparison between Traditional and Resource Recovery Systems." www.researchgate.net/publication/348481066.

Bahuguna, Ayush, Sachin Sharma, S. K. Singh, Akshat Bahuguna, and Basant Kumar Dadarwal. 2021. "Physical Method of Wastewater Treatment-A Review." *Quest Journals Journal of Research in Environmental and Earth Sciences*. 7. www.questjournals.org.

Balcom, Ian N., Heather Driscoll, James Vincent, and Meagan Leduc. 2016. "Metagenomic Analysis of an Ecological Wastewater Treatment Plant s Microbial Communities and Their Potential to Metabolize Pharmaceuticals [Version 1; Referees: Awaiting Peer Review]," no. 0. https://doi.org/10.12688/f1000research.9157.1.

Cheng S.-A., Wang B.-S. and Wang Y.-H. 2013. "Increasing Efficiencies of Microbial Fuel Cells for Collaborative Treatment of Copper and Organic Wastewater by Designing Reactor and Selecting Operating Parameters." *Bioresourcing Technology* 147, 332–337. doi:10.1016/j.biortech.2013.08.040.

Connon, Stephanie A., Stephen J. Giovannoni, Stephanie A. Connon, and Stephen J. Giovannoni. 2002. "High-Throughput Methods for Culturing Microorganisms in Very-Low-Nutrient Media Yield Diverse New Marine Isolates High-Throughput Methods for Culturing Microorganisms in Very-Low-Nutrient Media Yield Diverse New Marine Isolates." *Applied and Environmental Microbiology* 68 (8), 3878–3885. https://doi.org/10.1128/AEM.68.8.3878.

Cooper, P., Smith, M., Maynard, H. 1997. "The Design and Performance of a Nitrifying Vertical-Low Bed Treatment System." *Water Science Technology* 6, 35: 215–21

Crameri, A., Dawes, G., Rodriguez, E., Jr, Silver, S., & Stemmer, W. P. 1997. Molecular evolution of an arsenate detoxification pathway by DNA shuffling. *Nature Biotechnology*, 15(5), 436–438. https://doi.org/10.1038/nbt0597-436

Cruza, I. A., Nascimento, V. R. S., Felisardoa, R. J. A., dos Santos, A .M. G., de Jesusc, A. A., de Vasconcelos, B. R., Kumar, V., Cavalcanti, E. B., de Souza, R. L., Ferreira, L. F. R. 2023. "Evaluation of Artificial Neural Network Models for Predictive Monitoring of Biogas Production from Cassava Wastewater: A Training Algorithms Approach." *Biomass and Bioenergy* 175, 106869. https://doi.org/10.1016/j.biombioe.2023.106869

Dafale, Nishant, Leena Agrawal, Atya Kapley, Sudhir Meshram, Hemant Purohit, and Satish Wate. 2010. "Bioresource Technology Selection of Indicator Bacteria Based on Screening of 16S RDNA Metagenomic Library from a Two-Stage Anoxic – Oxic Bioreactor System Degrading Azo Dyes." *Bioresource Technology* 101 (2): 476–84. https://doi.org/10.1016/j.biortech.2009.08.006.

Dash, Hirak R. and Neelam Mangwani. 2013. "Marine Bacteria: Potential Candidates for Enhanced Bioremediation," 561–71. https://doi.org/10.1007/s00253-012-4584-0.

Datta, Saptashwa, K. Narayanan Rajnish, Melvin S. Samuel, Arivalagan Pugazlendhi, and Ethiraj Selvarajan. 2020. "Metagenomic Applications in Microbial Diversity, Bioremediation, Pollution Monitoring, Enzyme and Drug Discovery . A Review Human

Virome Protein Cluster Database." *Environmental Chemistry Letters*, no. May. https://doi.org/10.1007/s10311-020-01010-z.

Do, Thi Thuy, Sarah Delaney, and Fiona Walsh. 2019. "16S RRNA Gene Based Bacterial Community Structure of Wastewater Treatment Plant Effluents," no. January: 1–10. https://doi.org/10.1093/femsle/fnz017.

Flowers, Jason J., Tracey A. Cadkin, and Katherine D. Mcmahon. 2013. "Seasonal Bacterial Community Dynamics in a Full-Scale Enhanced Biological Phosphorus Removal Plant." *Water Research*, 1–13. https://doi.org/10.1016/j.watres.2013.07.054.

Gregor, C. B., R. M. Garrels, F.T. Mackenzie, and Barry Maynard J. 1988. *Chemical Cycles in the Evolution of the Earth Publisher*. In C.B. Gregor, R.M. Garrels, F.T. Mackenzie and J.B. Maynard (Eds.), (276 pp). John Wiley & SonsEditors: New York, N.Y. ISBN 0-47i-08911-7

Guttman, Hadassa Shira, Susan Youngren, and Kimberly Pause Tucker. 2016. "Environmental Microbial Community Analysis Methods as Applied to Bioremediation." *Bios* 87(3): 85–95. http://www.jstor.org/stable/24878641.

Hanum, Farida, Lee Chang Yuan, Hirotsugu Kamahara, Hamidi Abdul Aziz, Yoichi Atsuta, Takeshi Yamada, and Hiroyuki Daimon. 2019. "Treatment of Sewage Sludge Using Anaerobic Digestion in Malaysia: Current State and Challenges." *Frontiers in Energy Research*. Frontiers Media S.A. https://doi.org/10.3389/fenrg.2019.00019.

Hrywna, Y., Tsoi, T. V., Maltseva, O. V., Quensen, J. F., 3rd, & Tiedje, J. M. 1999. Construction and Characterization of Two Recombinant Bacteria that Grow on Ortho- and Para-substituted Chlorobiphenyls. *Applied and Environmental Microbiology* 65: 163–2169.

Huang L., Wang Q., Jiang L., Zhou P., Quan X. and Logan B. E. 2015. "Adaptively Evolving Bacterial Communities for Complete and Selective Reduction of Cr(VI), Cu(II), and Cd(II) in Biocathode Bioelectrochemical Systems." *Environmental Science Technology* 49, 9914–9924. doi:10.1021/acs.est.5b00191

Huang, D., Li, G., Sun, C., and Liu, Q. (2020). "Exploring Interactions in the Local Water-Energy-Food Nexus (Wef-Nexus) Using a Simultaneous Equations Model." *Science Total Environmental*. 703, 135034. doi: 10.1016/j.scitotenv.2019.135034

Hussein, Hany, Soha Farag Ibrahim, Kamal Kandeel, and Hassan Moawad. 2004. "Biosorption of Heavy Metals from Waste Water Using Pseudomonas sp." *Electronic Journal of Biotechnology* 7 (1): 45–53. https://doi.org/10.2225/vol7-issue1-fulltext-2.

IEA Bioenergy. 2001. *Biogas and More! Systems and Markets Overview of Anaerobic Digestion*. Abingdon: AEA Technology Environment.

Ijoma, Grace N., Asheal Mutungwazi, Thulani Mannie, Weiz Nurmahomed, Tonderayi S. Matambo, and Diane Hildebrandt. 2022. "Addressing the Water-Energy Nexus: A Focus on the Barriers and Potentials of Harnessing Wastewater Treatment Processes for Biogas Production in Sub Saharan Africa." *Heliyon* 8(5). https://doi.org/10.1016/j.heliyon.2022.e09385.

Jackson, Stephen A, Erik Borchert, Fergal O. Gara, and Alan D. W. Dobson. 2015. "Science Direct Metagenomics for the Discovery of Novel Biosurfactants of Environmental Interest from Marine Ecosystems." *Current Opinion in Biotechnology* 33: 176–82. https://doi.org/10.1016/j.copbio.2015.03.004.

Ju, Feng, and Tong Zhang. 2015. "16S RRNA Gene High-Throughput Sequencing Data Mining of Microbial Diversity and Interactions." https://doi.org/10.1007/s00253-015-6536-y.

Judd S. 2006. *The MBR Book Principles and Applications of Membrane Bioreactors for Water and Wastewater Treatment*. Elsevier, Oxford, UK.

Kantor, Rose S., A. Wynand Van Zyl, Robert P. Van Hille, Brian C. Thomas, Susan T. L. Harrison, Jillian F. Banfield, Cape Town, Cape Town, and South Africa. 2015.

"Bioreactor Microbial Ecosystems for Thiocyanate and Cyanide Degradation Unravelled with Genome-Resolved Metagenomics." 17: 4929–41. https://doi.org/10.1111/1462-2920.12936.

Kleinmann R. L. P., Girts M. A., Reddy K. R., and Smith, W.H. 1987."Acid Mine Water Treatment: A Overview of An Emergent Technology", (pp. 255–261). In K.R. Reddy and W.H. Smith (Eds.) *Aquatic plants for water treatment and resource recovery.* Magnolia Publications, Orlando, FL.

Köchling, Thorsten, José Luis Sanz, and Sávia Gavazza. 2015. "Analysis of Microbial Community Structure and Composition in Leachates from a Young Landfill by 454 Pyrosequencing." https://doi.org/10.1007/s00253-015-6409-4.

Kumar, V., Thakur, I. S., Singh, A. K., and Shah, M. P., 2020. "Application of metagenomics in remediation of contaminated sites and environmental restoration." In: Shah M., Rodriguez-Couto S., and Sengor S. S. (Eds.), *Emerging Technologies in Environmental Bioremediation*. Elsevier. https://doi.org/10.1016/B978-0-12-819860-5.00008-0.

Kumar, V. and Chandra, R., 2020. Metagenomics analysis of rhizospheric bacterial communities of *Saccharum arundinaceum* growing on organometallic sludge of sugarcane molasses-based distillery. *3 Biotech* 10(7), 316. https://doi.org/10.1007/s13205-020-02310-5

Kumar, V., Garg, V. K., Kumar, S., and Biswas, J. K. (Eds.). (2022). *Omics for Environmental Engineering and Microbiology Systems*. CRC Press, Boca Raton, FL.

Kumar, V., Bhat, S. A., Kumar, S., Verma, P., Badruddin, I. A., Américo Pinheiro, J. H. P., Sathyamurthy, R., and Atabani. A. E., 2023. Tea byproducts biorefinery for sustainable energy production and value-added products development: A step toward environmental sustainability. *Fuel* 350, 128811. https://doi.org/10.1016/j.fuel.2023.128811

Lu, Xin, Xu-xiang Zhang, Zhu Wang, Kailong Huang, Yuan Wang, and Weigang Liang. 2015. "Bacterial Pathogens and Community Composition in Advanced Sewage Treatment Systems Revealed by Metagenomics Analysis Based on High-Throughput Sequencing." *PLoS One* 10(5), e0125549. https://doi.org/10.1371/journal.pone.0125549.

Martinez Hernandez, Elias, and Kok Siew Ng. 2018. "Design of Biorefinery Systems for Conversion of Corn Stover into Biofuels Using a Biorefinery Engineering Framework." *Clean Technologies and Environmental Policy* 20 (7): 1501–14. https://doi.org/10.1007/s10098-017-1477-z.

McKenzie Smith, K. M.S. 2013. Methanogens and Methane Oxidizing Bacteria in Forested, Urban Unrestored, and Urban Restored Streams. Directed by Dr. Parke A. Rublee and Dr. Anne E. Hershey. pp. 63.

Nitish Venkateswarlu Mogili and Sanjeeb Mohapatra. 2021. "Microbial Community Analysis of Wastewater During Blackwater Treatment." In M. P. Shah & S. Rodriguez-Couto (Eds.), *Wastewater Treatment Reactors*, (pp. 483–507). Elsevier. ISBN 9780128239919. DOI: 10.1016/B978-0-12-823991-9.00024-1

Ohtsubo, Yoshiyuki & Shimura, Minoru & Delawary, Mina & Kimbara, Kazuhide & Takagi, Masamichi & Kudo, Toshiaki & Ohta, Akinori & Nagata, Yuji. (2003). Novel Approach to the Improvement of Biphenyl and Polychlorinated Biphenyl Degradation Activity: Promoter Implantation by Homologous Recombination. *Applied and Environmental Microbiology* 69: 146–53. 10.1128/AEM.69.1.146-153.2003.

Orellana, Roberto, Andrés Cumsille, Paula Piña-Gangas, Claudia Rojas, Alejandra Arancibia, Salvador Donghi, Cristian Stuardo, et al. 2022. "Economic Evaluation of Bioremediation of Hydrocarbon-Contaminated Urban Soils in Chile." *Sustainability* 14 (19). https://doi.org/10.3390/su141911854.

Orellana, Roberto, Paula Piña-gangas, Claudia Rojas, Alejandra Arancibia, Salvador Donghi, Cristian Stuardo, and Patricio Cabrera. 2022. "Economic Evaluation of Bioremediation of Hydrocarbon- Contaminated Urban Soils in Chile." Sustainability. 14(19):11854. https://doi.org/10.3390/su141911854.

Pan, Yimin, Qiaoqiao Ren, Pei Chen, Jiguo Wu, Zhendong Wu, and Guoxia Zhang. 2021. "Insight Into Microbial Community Aerosols Associated With Electronic Waste Handling Facilities by Culture-Dependent and Culture-Independent Methods." 9: 1–9. https://doi.org/10.3389/fpubh.2021.657784.

Paul, Parneet, Ameena Al Tenaiji, and Nuhu Braimah. 2016. "A Review of the Water and Energy Sectors and the Use of a Nexus Approach in Abu Dhabi." *International Journal of Environmental Research and Public Health* 13(4) : 364. https://doi.org/10.3390/ijerph13040364.

Piceno, Yvette M., Gary L. Andersen, and Ivan Moreno-andrade. 2010. "Bacterial Community Structure in Geographically Distributed Biological Wastewater Treatment Reactors," *Environmental Science & Technology*, 44(19): 7391–7396. https://doi.org/10.1021/es101554m

Qin, Jie & Song, Lingyun & Brim, Hassan & Daly, Michael & Summers, Anne. 2006. Hg(II) sequestration and protection by the MerR metal-binding domain (MBD). *Microbiology* (Reading, England). 152. 709–19. 10.1099/mic.0.28474-0

Reineke W. 1998. Development of hybrid strains for the mineralization of chloroaromatics by patchwork assembly. *Annual Review of Microbiology*, 52, 287–331. https://doi.org/10.1146/annurev.micro.52.1.287

Robles-gonzález, Ireri V., Fabio Fava, and Héctor M. Poggi-varaldo. 2008. "Sediments" 16: 1–16. https://doi.org/10.1186/1475-2859-7-5.

Roy, Ajoy, Avishek Dutta, Siddhartha Pal, Abhishek Gupta, Jayeeta Sarkar, Ananya Chatterjee, Anumeha Saha, Poulomi Sarkar, Pinaki Sar, and Sufia K. Kazy. 2018. "Biostimulation and Bioaugmentation of Native Microbial Community Accelerated Bioremediation of Oil Refinery Sludge." *Bioresource Technology*. https://doi.org/10.1016/j.biortech.2018.01.004.

Seabloom, R. W. and J. R. Buchanan. 2005. Aerobic Treatment of Wastewater and Aerobic Treatment Units Text. in Gross M. A. and Deal N.E., (eds.) *University Curriculum Development for Decentralized Wastewater Management. National Decentralized Water Resources Capacity Development Project*. University of Arkansas, Fayetteville, AR.

Shankar, Ravi, Sanjay Kumar, Amarjeet Kumar Prasad, Prateek Khare, Anil Kumar Varma, and Vinod Kumar Yadav. 2021. "Biological Wastewater Treatment Plants (WWTPs) for Industrial Wastewater." In *Microbial Ecology of Wastewater Treatment Plants*, 193–216. Elsevier. https://doi.org/10.1016/B978-0-12-822503-5.00023-0.

Siddiqi, A., and Anadon, L. D. (2011). "The Water–Energy Nexus in Middle East and North Africa." *Energy Policy* 39, 4529–4540. doi: 10.1016/j.enpol.2011.04.023

Suhardi, Sri Harjati, Tjandra Setiadi, H. C. Hambali, E. Suwito, S. H. Suhardi, and T. Setiadi. 2009. "Textile Wastewater Decolorization Performance Using Marasmius sp. in Immersion and Trickling Systems Special Issue 'Frontier Research: Waste Management for Sustainable Development' (Impact Factor 3.889) View Project Syngas Fermentation for Bioethanol Production View Project Textile Wastewater Decolorization Performance Using *Marasmius* Sp. in Immersion and Trickling Systems." www.researchgate.net/publication/343099220.

Sun, Su, Shangxian Xie, Hu Chen, Yanbing Cheng, Yan Shi, Xing Qin, Susie Y. Dai, Xiaoyu Zhang, and Joshua S. Yuan. 2015. *Genomic and Molecular Mechanisms for Efficient Biodegradation of Aromatic Dye*. Elsevier B.V. https://doi.org/10.1016/j.jhazmat.2015.09.071.

Techtmann, Stephen & Hazen, Terry. 2016. "Metagenomic Applications in Environmental Monitoring and Bioremediation. *Journal of Industrial Microbiology & Biotechnology* 43. 10.1007/s10295-016-1809-8.

Tekere, Memory. 2019. 'Microbial Bioremediation and Different Bioreactors Designs Applied'. *Biotechnology and Bioengineering*. IntechOpen. doi:10.5772/intechopen.83661.

Tillich, Ulrich M., Nick Wolter, Katja Schulze, Dan Kramer, Oliver Brödel, and Marcus Frohme. 2014. "High-Throughput Cultivation and Screening Platform for Unicellular Phototrophs," *BMC Microbiology* 14(239), 1–13. https://doi.org/10.1186/s12866-014-0239-x

Torres C. I., Brown R. K., Parameswaran P., Marcus A. K., Wanger G., Gorby Y. A., and Rittmann B. E. 2009. "Selecting Anode-Respiring Bacteria Based on Anode Potential." *Environmental Science & Technology*, 43(24), 9519–9524. https://doi.org/10.1021/es902165y

Ugwu, E. I., Pinê Américo, J. H., Nwobia, L. I., Lam, S. S., Kumar, V., Ikechukwu, E. L., and Victor, E. C. 2022. Opimization of parameters in biomethanization process with co-digested poultry wastes and palm oil mill effluents. *Cleaner Chemical Engineering* 3, 100033. https://doi.org/10.1016/j.clce.2022.100033

Urgun-demirtas, Meltem and Benjamin Stark. 2006. "Use of Genetically Engineered Microorganisms (GEMs) for the Bioremediation of Contaminants," 145–64. https://doi.org/10.1080/07388550600842794.

U. S. EPA. 1992. *Summary Report: "Small Community Water and Wastewater Treatment.* Office of Water, Washington, DC, (EPA/625/R-92/010).

Wang, Aijun & Mulchandani, Ashok & Chen, Wilfred. (2002). Specific Adhesion to Cellulose and Hydrolysis of Organophosphate Nerve Agents by a Genetically Engineered *Escherichia coli* Strain with a Surface-Expressed Cellulose-Binding Domain and Organophosphorus Hydrolase. *Applied and Environmental Microbiology* 68, 1684–1689. 10.1128/AEM.68.4.1684-1689.2002.

Wang, Xiaohui, Yu Xia, Xianghua Wen, Yunfeng Yang, and Jizhong Zhou. 2014. "Microbial Community Functional Structures in Wastewater Treatment Plants as Characterized by GeoChip." *PLoS ONE* 9 (3), e93422. https://doi.org/10.1371/journal.pone.0093422.

Wang, Zhu, Xu-xiang Zhang, Xin Lu, Bo Liu, Yan Li, Chao Long, and Aimin Li. 2014. "Abundance and Diversity of Bacterial Nitrifiers and Denitrifiers and Their Functional Genes in Tannery Wastewater Treatment Plants Revealed by High- Throughput Sequencing." *PLoS ONE* 9(11): e113603. https://doi.org/10.1371/journal.pone.0113603.

Wani, A. K., Akhtar, N., Naqash, N., Chopra, C., Singh, R., Kumar, V., Kumar, S. Mulla, S. I., and Américo-Pinheiro, J. H. P. 2022. "Bioprospecting Culturable and Unculturable Microbial Consortia Through Metagenomics for Bioremediation." *Cleaner Chemical Engineering* 2, 100017. https://doi.org/10.1016/j.clce.2022.100017

William A. Worrell and P. Aarne Vesilind. 2016. *Solid Waste Engineering: A Global Perspective*. Cengage Learning, Boston, Massachusetts, United States. https://books.google.com.bd/books?id=UsgaCgAAQBAJ

Wollmann, Felix, Stefan Dietze, Jörg Uwe Ackermann, Thomas Bley, Thomas Walther, Juliane Steingroewer, and Felix Krujatz. 2019. "Microalgae Wastewater Treatment: Biological and Technological Approaches." In *Engineering in Life Sciences*. Wiley-VCH Verlag. https://doi.org/10.1002/elsc.201900071.

Wu, Guangxue, and Qidong Yin. 2020. "Microbial Niche Nexus Sustaining Biological Wastewater Treatment." *NPJ Clean Water* 3 (1). https://doi.org/10.1038/s41545-020-00080-4.

Yin, Qidong, Yuepeng Sun, Bo Li, Zhaolu Feng, and Guangxue Wu. 2022. "The r/K Selection Theory and Its Application in Biological Wastewater Treatment Processes." in *Science of the Total Environment*. Elsevier B.V. https://doi.org/10.1016/j.scitotenv.2022.153836

Yamamoto K., Hiasa M., Mahmood T., and Matsuo T. 1989. "Direct Solid-Liquid Separation Using Hollow Fiber Membrane in an Activated Sludge Aeration Tank. *Water Science Technology* 21:43. ISSN: 0273-1223

Zhang, Tong, Ming-fei Shao, and Lin Ye. 2011. "454 Pyrosequencing Reveals Bacterial Diversity of Activated Sludge from 14 Sewage Treatment Plants." *The ISME Journal* 6 (6): 1137–47. https://doi.org/10.1038/ismej.2011.188.

Zhao, Jianguo, Yahe Li, Yu Li, Zeya Yu, and Xiurong Chen. 2018. "Effects of carbon sources on sludge performance and microbial community for 4-chlorophenol wastewater treatment in sequencing batch reactors". *Bioresource Technology* 255: 22–28. https://doi.org/10.1016/j.biortech.2018.05.102.

16 Strategies for Emerging Compounds' Removal through Microbial Niche Tuning

Sakshi Sharma, Shiromi Shailesh, Preksha Palsania, and Garima Kaushik

16.1 INTRODUCTION

A broad phrase used to describe compounds and substances that are environmentally important and are the subject of current scientific or political concern is "emerging contaminants." Emerging contaminants (ECs) are classified as "compounds that are not presently covered by the current water quality regulatory frameworks, have not previously been investigated, and are thought to be imminent hazards to environmental ecosystems as well as to human health and safety" (La Farre et al., 2008) from the field of water analytics.

The terms "emerging toxins," "compounds of increasing concerns," "previous contaminants," and "re-emerging toxins" refer to classes of substances that are innately hazardous to the environment (though not essentially toxic) but lack standard specific regulations because scientific research has not yet clearly illustrated their effects (Daughton, 2004). Due to the absence of a precise description regarding terminology for each of these classes, many toxic compounds can be classified in more than one category. Thus, we shall generically refer to them as ECs for simplicity's sake. The presence, distribution, and toxicity testing of several emerging contaminants [or emerging pollutants (EPs)] terrestrially (Caliman & Gavrilescu, 2009) and on the seafloor (Bigus et al., 2014) in addition to sewage and waterways (Deblonde et al., 2011; Stuart et al., 2012; Ozcan et al., 2013) have all been the subject of countless research investigations over the past few decades. Concerning their dispersal, it was verified that residues of artificial pollution are now detectable everywhere, including the stratosphere, deep oceans, poles, many species of wildlife, modern communities, the food chain, and most importantly newborns (Naidu & Wong, 2013).

The hazardous metals that combine chemical sensitivity and nuclear radiation are the most well-known of them. Natural or depleted uranium is an excellent showcase for "standard" EPs. The World Health Organization has provided numerous drinking water regulations. Regarding uranium toxicity (Briner, 2010), its health-based

DOI: 10.1201/9781003441069-16

guideline increased from 2 g/L in 1998 (the year it was introduced) to 30 g/L in the 2011 guidance document. However, many scientists have implored the United Nations World Health Organization (WHO) to re-evaluate its judgment given the increased environmental contamination caused by generations of nuclear activities (Frisbie et al., 2013). ECs in the form of pharmaceuticals, personal care products (PPCPs), and pesticides enter into the environment and pose severe consequences for aquatic species, animals, and human beings due to their toxic and bioactive properties. They have adverse effects due to phenomena like biomagnification, bioaccumulation, etc. The regulation of PPCPs and pesticides has been determined in African water systems (Okoye et al., 2022).

A new class of toxins, i.e., "endocrine-disrupting chemicals" (EDCs), have been identified as harmful for human health and the environment. These endocrine disruptors are associated with cancer, cardiovascular risk, behavioral abnormalities, autoimmune defects, and reproductive issues. The interaction effects of EDCs are drawing more attention because these substances can be found in well-known sources like medicines and plasticizers as well as nonpoint sources, including agricultural runoff and stormwater infiltration. Pesticides, heavy metals, detergents, plasticizers, industrial chemicals, and medications are examples of natural and synthetic chemical compounds that also fall within the category of EDCs (Hadibarata et al., 2007; Liu et al., 2009; Hadibarata et al., 2012; Tapia-Orozco et al., 2016; Sathishkumar et al., 2020; Ho & Adnan, 2021; Lai, 2021; Maharjan et al., 2021; Sathishkumar et al., 2021; Tang, 2021; Wang et al., 2021). Natural EDCs are derived from organisms like animals and plants that include phytoestrogens, genistein, and coumestrol, as well as secondary metabolites (Kabir et al., 2015; Tapia-Orozco et al., 2016). These substances act as a reservoir for pollution (contaminants) through natural sources like rivers, streams, and oceans (Tang et al., 2021; Ng & Elshikh, 2021). EDCs enter the water bodies in several ways including point and nonpoint sources. This chapter is a comprehensive effort to summarize the presence of such EPary and human medications, hygiene items, perfluorinated compounds, polycyclic aromatic hydrocarbons (PAHs), heavy metals, detergents, pesticides, fire retardants, algal toxins, and engineered nanomaterials as shown in Figure 16.1.

Water is frequently used for activities associated with farming, businesses, and public use. At the same time, wastewater discharges after use are impregnated with various chemical contaminants. Industrial discharges and farming practices substantially cause the problem of contaminants in wastewater. These activities have created different contaminants and impacted the hydrological cycle, generating concerns around the globe about their possible implications for fauna, flora, and human health. Multiple studies have documented the emergence of novel substances, also ascribed as "emerging contaminants," in wastewater and aquatic habitats throughout the past 20 years (Pham & Proulx, 1997; Vogelsang et al., 2006; Rosal et al., 2010). Phthalates, pesticides, metals, polycyclic aromatic hydrocarbons, and estrogenic compounds include the 33 priority chemicals or categories of substances listed in Annex X of the EU framework regulation 2000/06/CE. Additionally, the Registration, Evaluation, Authorisation and Restriction of Chemicals (REACH) legislation was developed in Europe in 2007 and tried to identify toxic substances and less toxic substitutes. Three

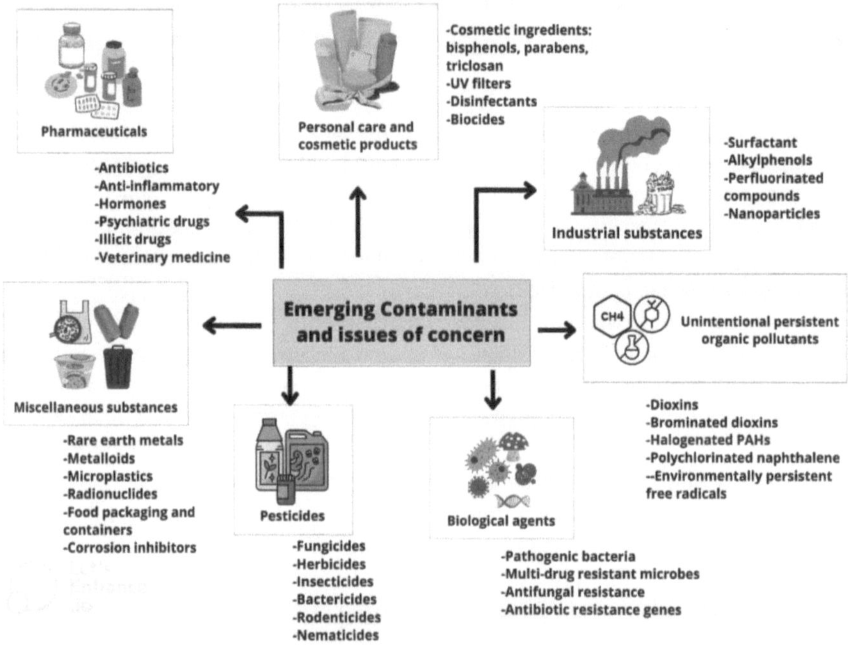

FIGURE 16.1 Classification of emerging contaminants and issues of concern in ecosystems.

phthalates [di (2- ethylhexyl) phthalate (DEHP), Di-butyl phthalate, and Benzyl butyl phthalate (BBP)], are grouped as carcinogenic, harmful for reproduction, and declared pervasive in the environment, and must be treated as compounds of immediate concern.

Metals, bacteria, hydrocarbons, and other ions like nitrates and ammonia have all been found in water for several decades. These contaminants are subject to regulation and management as they have been identified as affecting human health and the environment. However, information on the presence and consequences of phthalates, medicinal chemicals, PAHs, polychlorinated biphenyls (PCBs), and Bisphenol A (BPA) is still lacking. Drugs such as analgesics, antibiotics, beta-blockers, anticonvulsants, lipid-regulators, contrast media, anti-cancer media, hormones, and disinfectants are among the common pharmaceutical compounds that are found in the environment (Halling-Sørensen et al., 1998; Garric & Ferrari, 2005).

In this chapter, we present a detailed account of the occurrence, fate, impacts, and remediation measures to deal with ECs. The effects of these contaminants on our ecosystem as well as public health are presented and discussed to formulate a sustainable approach and help produce a policy document for consideration.

16.2 OCCURRENCE OF EMERGING CONTAMINANTS

a. Rivers: Extensive research on the occurrence and distribution of EPs in rivers worldwide has been documented. All 28 sampling locations in the Yangtze River Delta region of China continuously recorded 44 analytes, the preponderance of which were industrial compounds such as Benzotriazole and flame retardants made of organophosphates (Peng et al., 2018). All 55 sampling sites of the Danube River were monitored for 235 selected EPs, of which 125 EPs were discovered in at least 50% of the stations (Ginebreda et al., 2018). In the freshwater and sediment of Pira Creek and Jundia River, Brazil, seven drugs for medical use, three steroid hormones, and PPCPs were recovered (de Sousa et al., 2018). The South African Hartbeespoort Dam water was tested for various pharmaceuticals including efavirenz, nevirapine, and carbamazepine (Rimayi et al., 2018). The rivers in Serbia's Vojvodina Region were mostly contaminated by residential wastewater. A thorough investigation revealed that almost all examined samples frequently quantified sterols. With EPs in Ukrainian urban streams reportedly exceeding EU criteria, the maximum loading from Ukrainian cities into EU rivers can be over 0.7 t/ac for diclofenac and 0.4 t/ac for nonylphenols (Vystavna et al., 2018).

b. Estuary-Area: Estuary regions are highly urbanized and industrialized since they have historically been established as economic zones. Since anthropogenically produced organic contaminants are easily adsorbed on the water's surface, the lower flow rate of rivers and streams causes these suspended particles to settle out in estuaries. Anti-inflammatory medications, hypertensive medications, stimulants, and artificial sweeteners were the most common substances in the Basque Country's estuaries of Spain (Mijangos et al., 2018a; Mijangos et al., 2018b). Multiclass EPs such as diazinon, Bisphenol A, amoxicillin, progesterone, and estrone (E1) occurred in tropical coastal sediments of the Klang River Estuary across all sampling sites investigated, while other molecules such as primidone, diclofenac, testosterone, 17-estradiol (E2), and 17-ethynyl estradiol (EE2) were remarkably consistent in sediment samples, with percentages of detection ranging from 89% to 98% (Omar et al., 2018). Nonylphenol constituted the most common compound among the examined EDCs in seafloor sediments from the proximity of subsurface wastewater discharges (SSOs) along the So Paulo State Coast (Brazil), and the variability of EDCs grew with an increase in inhabitants supported by SSOs (Santos et al., 2018).

c. Surface Water and Drinking Water: It has been demonstrated that toxic compounds are being traced, especially in public supply drinking water systems. In Shanghai's water source, 65 emerging trace organic contaminants were found (Sun et al., 2018; Sun et al., 2020). The surface waters of the Yangtze River and the Han River, which are drinking water sources for the nearby communities, included 33 pharmaceuticals and their metabolites. Low concentrations of EPs were discovered in potable water treatment facilities in north-western Italy (Magi et al., 2018).

d. Other Environment Sources: In rainwater collected in Moscow, almost 700 organic molecules from diverse chemical families were found; among these are trichloronitromethane, organophosphates, and pyridines that may be regarded as EPs (Polyakova et al., 2018).

In Poland, EPs were widely recognized in groundwater and wastewater. It was determined that a significant source of EPs in surface and groundwater was the drainage from wastewater treatment facilities, which only largely removed EPs. In contrast, the primary source of EPs in groundwater was the penetration of landfill leachate (Kapelewska et al., 2018). In the Dongjiang River basin in South China, artificial sweeteners were extensively found at significant concentrations in surface water and groundwater (Yang et al., 2018).

16.3 IMPACTS OF EMERGING CONTAMINANTS

a. Ecosystem

Although ECs have biological effects, there are insufficient data to estimate their potential ecotoxicological implications. Perfluorooctane sulfonate (PFOS) and perfluorooctanoate (PFOA) are two of the most extensively researched chemicals in this class of polyfluoroalkyl compounds (PFC) (Apelberg et al., 2007). These substances, known as surfactants, are employed in various applications, from paper coatings and carpets that repel water and oil to specialized chemical uses, including pesticides and firefighting foams. PFOS causes structural deformities, growth delay, lower birth mass, shortened gestational duration, and elevated neonatal mortality in rats and mice, according to studies by Lau et al. (2003) and Fuentes et al. (2007). Fox (2001) examined the impact of ECs on the Canadian Great Lakes. Alterations in metabolic abnormalities, sex ratios, thyroid dysfunction, feminization/demasculinization, decreased reproductive success, and altered brain development and behavior in water birds, turtles, and fish are just a few organisms reported for the toxic effects of these ECs (Tillitt et al., 1992; Henny et al., 1996; Bishop et al., 1998).

Diclofenac, butylated hydroxyanisole, hexabromocyclododecane, ibuprofen, organochlorine pesticides, polychlorinated dibenzofurans, 17a-ethynylestradiol, and 17b-trenbolone are some of the ECs that have a detrimental effect on aquatic organisms. These compounds have adverse consequences for aquatic organisms' genetic makeup, aberrant development, cancer, hormonal imbalance, and the prevalence of castrating effects in men. Research has shown that ECs are hazardous to land animals such as cats, dogs, poultry, hogs, rats, alligators, and other rodents. Such ECs can be transferred by surface water, aquifers, coastal regions, food, air, soil, sediment, and biota through the food chain. Ingestion, inhalation, and cutaneous exposure of biota through polluted media are prominent exposure pathways. Water runoff and sewage significantly impact downstream aquifers, surface water bodies, and drinking water supplies for ECs. A significant source of exposure is the consumption of ECs through potable water, tainted food, and biota.

b. Public Health

Trihalomethanes (THMs), such as chloroform ($CHCl_3$), bromodichloromethane ($CHBrCl_2$), dibromochloromethane ($CHBr_2Cl$), and bromoform ($CHBr_3$), are disinfection byproducts that are present in some indoor swimming pools

(Aggazzotti et al., 1998). Humans are prone to cancer in the presence of chemicals like $CHCl_3$ and $CHBrCl_2$ (IARC, 1991; WHO, 1993). Male humans exposed to such substances may have lower sperm counts (Kavlock et al., 1996; Sheiner et al., 2003; Bonde, 2010). Certain ECs can potentially be biomagnified in the system and negatively affect health. The USEPA (2003) estimated that Americans' blood contains 5 ppb of PFOA. According to Kosaka et al. (2007), percolates affect surface water quality to a reasonable extent. They can affect thyroid gland iodide, which may influence metabolism, growth, and development (Lamm et al., 1999).

Among the most hazardous ECs for public health, perchlorates, Bisphenol A, PFOS, disinfectants, and PCBs have been considered significant in terms of the risks associated with them. Numerous businesses employ these substances to manufacture polymers, electric equipment, additives, and cleaning goods. Studies show these substances are blamed for human cancer, hormone disturbances, and embryo growth restriction.

Research has recently focused on consumer goods potentially affecting the endocrine system (Rosner & Markowitz, 2013). Exogenous compounds known as endocrine disruptors impair the endocrine system's functioning and harm organisms' growth, reproduction, and progeny, even at concentrations as low as a few micrograms or nanograms per liter (Bila & Dezotti, 2003). Through changing the production, metabolism, expression, transport, or activity of endogenous hormones, this class of substances can mimic, oppose, or change the levels of such hormones.

16.4 SCREENING AND DETECTION OF EMERGING CONTAMINANTS

The recurrent uncovering of EPs has caused great alarm about the environmental challenges they may pose. However, because EPs are present in the environment at trace concentrations, challenges with water sample enrichment, interference from the environmental experimental parameters, and complicated analysis processes continue to stand in the way of future research on EPs. Many investigations were conducted in 2018 to create sensitive approaches and techniques for the preconcentration, introduction, and diagnosis of EPs.

Suspected screening is a valuable method to identify unidentified substances using data about individual compounds gathered from different places such as academic papers and web databases. Solid phase extraction (SPE) and liquid chromatography-high resolution mass spectrometry (LC-HRMS) were used to establish a systematic technique by Asghar et al. (2018) to identify compounds and measure the human drug residues in surface water. Using a planned study to quantify 154 chemicals, 84 precursors, and 70 active metabolites, Hermes et al. (2018) devised a direct injection and multiresidue methodological approach separated into two chromatographic runs.

This effectively yielded samples with liquid chromatography-tandem mass spectrometry to discover EPs in waterways at minimal concentrations. Although polar organic chemical integrative samplers (POCIS) used in inactive sampling may be bought from professionals, a sampler's storage lifespan should be considered. According to Magi et al. (2018), commercial POCIS samplers held for more than nine years exhibited meager sampling rates, suggesting that the samplers' adsorption capacities may have changed. Gabbana et al. (2018) developed a novel cell arrangement for in situ extraction, preconcentration (based on liquid-liquid microextraction), to identify EPs in water at the trace level subsequent electroanalysis. Abujaber et al. (2018) suggested the very first magnetic solid phase extraction technique for the identification of EPs (diclofenac, ibuprofen, naproxen, and paracetamol) in water sources using magnetic cellulose nanoparticles (MCNPs) coated with 1 butyl-3 methylimidazolium hexafluorophosphate [C4MIM] [PF_6] ionic liquid through electrostatic interactions. The procedure is straightforward, efficient, and eco-friendly, as the extraction yields ranged from 85% to 116%.

Using polyethersulfone (PES) microextraction and liquid chromatography-tandem mass spectrometry (LC-MS/MS) analysis, Mijangos et al. (2018a) and Mijangos et al. (2018a) developed an innovative method to simultaneously recognize 41 multiclass EPs in saltwater, wastewater treatment plant (WWTP) effluents, and estuarine samples. The PES technique, as compared to the SPE methodology, enabled the expense extraction of complicated aqueous samples with less matrix effect using 120 milliliters of the water sample. Rocío-Bautista et al. (2018) used an innovative green process that included water and urea as co-reactants to effectively synthesize the CIM-80 material (aluminum (III) mesaconate) in higher production. With a small ecological footprint, the CIM-80 substance was tested in a membrane separation methodology for tracking up to 22 contaminants in water. With two EPs, the typical extraction efficiency may be as high as 70% (triclosan and carbamazepine).

16.5 MICROBIAL ELECTROCHEMICAL TECHNOLOGY/ BIOELECTRIFICATION

Microbes catalyze an electrochemical process that is used by microbial electrochemical technology (MET). This technique relies on the interactions among biological microorganisms and active electrodes (Mohan et al., 2019). The anodic chamber of a conventional microbial fuel cell (MFC) is injected with anaerobic sludge, and the cathode chamber is isolated from the anode chamber by a proton exchange membrane (PEM). In the case of a MFC, electrical energy is captured by connecting the load externally, yet external power is supplied in the case of microbial electrochemical (MEC). Bioelectrochemical systems can be categorized as built wetland-microbial fuel cells (CW-MFC), plant-microbial fuel cells (plant-MFC), soil-microbial fuel cells (soil-MFC), and sediment-microbial fuel cells (sediment-MFC) based on their applicability in various situations.

MET and a few in situ bioremediation methods, such as CW-MFCs, soil-MFCs, sediment-MFCs, and plant-MFCs are discussed here. The degradation of several xenobiotic chemicals and ECs, including dyes, insecticides, herbicides, PAHs, and heavy metals has been discussed. By fusing natural systems with METs, these hybrid microbial electrochemical technologies (HMETs) enable the advantages of both approaches.

Bioelectrification, or the use of electrically active microbes to remediate contaminants, is an emerging field of study with great potential for addressing the issue of ECs. (Tan et al., 2021) The process typically involves using electrodes, which serve as electron donors or acceptors. The microbes can interact with the electrodes and transfer electrons, which can drive reactions that break down the contaminants (Fig. 16.2). This process is known as microbial electrochemistry.

The use of bioelectrification and microbial niche tuning is done to remove ECs. For example, a study by Yang et al. (2020) investigated the use of bioelectrochemical systems to treat wastewater containing the emerging contaminant triclosan. Triclosan is a joint antibacterial agent found in PPCPs that has been shown to affect aquatic organisms negatively. The study found that by manipulating the operating conditions of the bioelectrochemical system, it was possible to create a microbial community that could effectively remove triclosan from the wastewater. Another study by Li et al. (2020) explored microbial niche tuning to remove the emerging contaminant Bisphenol A (BPA) from wastewater. Bisphenol A (BPA) is a chemical commonly found in plastics and has been linked to various health problems. The study found that by adjusting the pH of the wastewater, it was possible to create conditions that favored the growth of a specific type of microbe that could remove Bisphenol A (BPA) effectively (Ghangrekar et al., 2020). Another study by Zhang et al. (2021) investigated using bioelectrochemical systems to remove the emerging contaminant sulfamethoxazole from wastewater. Sulfamethoxazole is a common antibiotic that has been shown to affect aquatic organisms negatively. The study found that a bioelectrochemical system incorporating a specific type of microbe could achieve a removal efficiency of over 90%.

ECs are a class of pollutants that pose significant risks to human and environmental health due to their persistent, toxic, and nondegradable nature. Bioelectrification is an innovative technology that uses electrical current to enhance the biodegradation of ECs by microbes in a treatment system. The approach is based on the principle

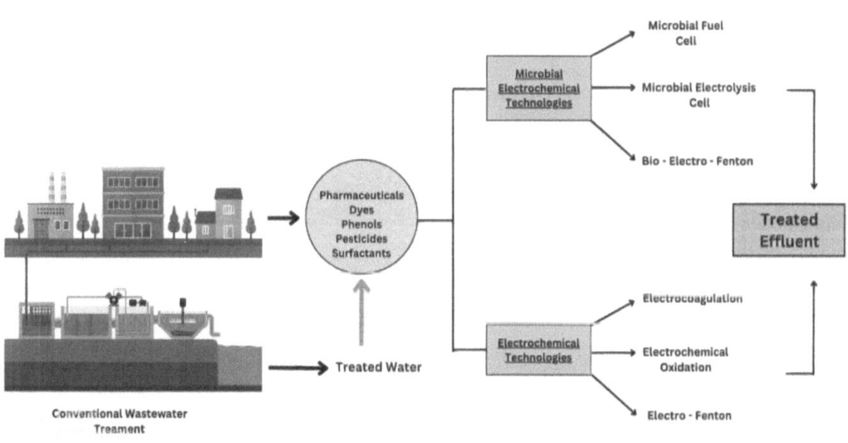

FIGURE 16.2 Microbial electrochemical technology process.

that electrical current can influence microbial metabolism, increasing biodegradation rates and efficiency. The following are some strategies for emerging compound removal through microbial niche tuning-bioelectrification:

- Microbial Community Engineering: The microbial community in a treatment system can be selectively engineered to enhance the degradation of ECs. This can be achieved by adding microbial strains that can degrade the target contaminants or by manipulating the microbial community to favor the growth of degraders.
- Electrostimulation of Microbes: Electrical current can be used to directly stimulate the metabolism of microbes, thereby increasing their degradation rate. This can be achieved by applying direct electrical current or using electrodes to generate electrons that microbes can take up.
- Electro-Fenton Process: The electro-Fenton process involves using electrical current to generate hydroxyl radicals, which are highly reactive and can degrade a wide range of contaminants. The process can enhance the degradation of ECs, particularly those that are difficult to degrade through conventional biological processes.
- Combined Technologies: Bioelectrification can be used with other treatment technologies, such as advanced oxidation processes or bioreactors, to enhance the degradation of ECs. For example, bioelectrification can be used to enhance the degradation of contaminants in bioreactors or to provide additional electron donors to enhance the performance of advanced oxidation processes.

16.6 MICROBIAL ENZYMES

Protein-based substances, also known as enzymes, serve as specialized catalysts for chemical processes. They have become increasingly valuable in industrial operations thanks to their capacity to carry out extremely particular chemical changes (Li et al., 2012). Due to their stability, catalytic activity, simplicity of manufacture, and optimization, microbial enzymes are more often used in industries and medicine than plant and animal enzymes. Enzymes have been widely used in various sectors because of their quick processing times, low energy requirements, economic effectiveness, nontoxicity, and environmental friendliness (including food, agriculture, chemicals, and pharmaceuticals). Since enzymes are more specialized, effective, and environmentally friendly catalysts, they have a solid potential to replace chemical processes. They can also degrade toxic chemical compounds found in industrial and household wastes (phenolic compounds, nitriles, amines, and so on) through either degradation or conversion mechanisms (Singh et al., 2016). Enzymatic reactions, compared to chemical ones, are more catalytically effective, specific, and free of side reactions. They are also biodegradable, simple to remove from contaminated streams, simple to standardize in commercial preparations, as well as operate at mild temperatures, pressures, and pH levels.

Enzymes among oxidoreductase groups catalyze the transfer of electrons from a donor to an acceptor (a different substrate). These enzymes have numerous uses in the chemical, textile, paper, pulp, pharmaceutical, agrochemical, polymer, biofuel, amino acid, nutraceutical, and cosmetic industries because they enable water use as a solvent and facilitate regio-, stereo-, and enantioselective reactions. Oxygen serves as the final electron acceptor for one class of oxidoreductase enzymes. The molecule of oxygen is highly reactive and packed with energy. Complex enzymes have been created by nature that efficiently use the energy contained in the oxygen molecule while preserving the creation of reactive oxygen species with the least amount of energy.

Laccase is an enzyme that has been researched since the nineteenth century. It was initially isolated by Yoshida from the exudates of the Japanese lacquer tree *Rhus vernicifera* in 1883, giving rise to laccase. Bertrand and Laborde were the first to demonstrate the presence of laccase in fungi in 1896 (Desai & Nityanand, 2011). Laccases (p-diphenol: dioxygen oxidoreductase) are blue multicopper oxidases that can generate reactive radical molecules. With oxygen as the ultimate electron acceptor, they may catalyze the one-electron oxidation of various organic and inorganic substrates. Enzymes typically have a high level of substrate specificity, whereas oxidases—like laccases—have a lower level of specificity. Laccases oxidize phenolic and non-phenolic compounds, making them helpful in bioremediating natural pollutants (Singh et al., 2016).

In nature, laccases are often found in plants, fungi, certain bacteria, and insects. They may be extracellular or intracellular, and while they perform physiologically distinct roles in various animals, they all share the ability to catalyze either polymerization or depolymerization (Patel et al., 2020). Numerous plants have been found to contain laccases, but detailed studies of these enzymes are challenging because crude extracts often contain a variety of oxidative enzymes with broad substrate specificities (Ranocha et al., 1999). As a result, little is known about plant laccase in detail. Laccase isolated from *R. vernicifera* has been utilized frequently to investigate the mechanism of action. Plant laccases are found in the xylem, where they synthesize radical-based lignin polymer and polymerize monolignols into dimers and trimers during the early phases of lignification (Gavnholt & Larsen, 2002).

Few bacterial laccases have been identified; the first one, *Azospirillum lipoferum*, was found in plant root bacteria and was responsible for melanin synthesis. The thermostable CotA laccase produced by *Bacillus subtilis* creates the color in the endospore's coat. Laccases are also present in *Streptomyces cyaneus* and *Streptomyces lavendulae*, in prokaryotes they are not often clustered together, whereas in eukaryotes, they are only found in white-rot fungus (Patel et al., 2020). To manufacture biotechnologically significant laccases, more than 60 fungal strains (*Basidiomycetes*, *Ascomycetes*, and *Deuteromycetes*) with high redox potential are required (Upadhyay et al., 2016). Numerous *Ascomycetes* and *Basidiomycetes* fungal species, including *Magnaporthe grisea* (Iyer & Chattoo, 2003), *Cerrena unicolor* (Bittner et al., 2012), *Trametes versicolour* (Minussi et al., 2007), *Trichoderma reesei* (Levasseur et al., 2010), and others, have had laccase activity which was further improved and purified. *Aspergillus*, *Curvularia*, and *Penicillium*, three species of terrestrial *Ascomycete*, and several aquatic *Ascomycetes* have shown the ability to produce laccase. In fungi, there are many laccase-encoding genes; however, physiological functions are still unknown (Patel et al., 2020).

16.7 REDUCTIVE CHLORINATION

Among the methods being explored for ECs' removal, reductive chlorination through microbial niches has been gaining attention due to its efficacy and environmental compatibility. Reductive chlorination is a process that involves the transformation of a pollutant into a less harmful or nontoxic compound by the addition of chlorine and the removal of electrons. This process has been widely used in treating contaminated soil and water, especially for removing halogenated organic compounds such as chlorinated solvents, pesticides, and PCBs. In the case of ECs, reductive chlorination can be applied through microbial niches. Microorganisms use the reductive power of chlorine to break down the pollutant into simpler and less toxic compounds.

The mechanism of reductive chlorination in microbial niches involves the interaction between the target compound and the microorganisms present in the environment. The microorganisms use enzymes and other metabolic pathways to transform the target compound into a chlorinated intermediate. Chlorine is then added to the intermediate, resulting in the formation of a less toxic compound. This process is facilitated by electron donors such as organic matter or hydrogen gas, which provide the necessary reducing power for the reaction. Reductive chlorination is a chemical reaction that involves the substitution of a chlorine atom with a hydrogen atom on an organic molecule. This reaction effectively removes various contaminants, including persistent organic pollutants, PCB, and ECs, such as pharmaceuticals, PPCPs, and pesticides, from the environment. The effectiveness of reductive chlorination is due to its ability to convert these contaminants into less toxic and less persistent compounds that are more easily degraded by microorganisms.

The use of microbial niches to remove contaminants has gained increasing attention in recent years due to their ability to transform and degrade a wide range of organic compounds. Microbial niches are locations within the environment where specific microbial communities can carry out a particular metabolic process. Reductive dechlorination is a key process in microbial niches to degrade chlorinated organic compounds.

Several studies have investigated the role of reductive chlorination in removing ECs through microbial niches. For example, a study conducted by Dong et al. (2020) investigated the removal of Bisphenol A (BPA) from wastewater by a mixed microbial culture. The study found that reductive dechlorination was essential in removing Bisphenol A (BPA), producing nontoxic intermediates. Another study by Zhang et al. (2021) investigated the removal of triclosan, an antimicrobial agent commonly found in PPCPs, from wastewater by a mixed microbial culture. The study found that reductive dechlorination was the main pathway for removing triclosan, producing fewer toxic intermediates.

Reductive chlorination is effective in removing ECs through microbial niches, but several challenges still need to be addressed to optimize its use. One challenge is the identification of suitable microbial cultures that can efficiently degrade specific contaminants. This can be achieved through molecular biology techniques such as deoxyribonucleic acid (DNA) sequencing and metagenomics. Another challenge is optimizing environmental conditions to promote the growth and activity of the microbial culture. This can include optimizing pH, temperature, and nutrient availability and providing appropriate electron donors and acceptors for reductive dechlorination. Advantages of Reductive Chlorination:

- One of the advantages of using microbial niches to remove contaminants is their ability to carry out biotransformation reactions that can lead to the formation of less toxic and more easily degradable compounds. For example, reductive chlorination can lead to the formation of compounds that are more easily degraded by microorganisms, such as alkanes and alkenes.
- Another advantage of using microbial niches to remove contaminants is their ability to adapt to changing environmental conditions. This can include changes in temperature, pH, and nutrient availability, as well as the presence of other contaminants. Microbial communities can also adapt to changes in the type and concentration of electron donors and acceptors available for the reductive dechlorination process.
- Reductive chlorination can also be necessary for removing contaminants in natural systems, such as groundwater and soils. Microbial communities in these environments can carry out reductive dechlorination reactions that can lead to the removal of contaminants over time. However, the efficiency of these processes can be affected by factors such as the availability of electron donors and acceptors, as well as the presence of competing microbial communities.
- In addition to removing ECs, reductive chlorination has also been used to treat other contaminants, such as chlorinated solvents and pesticides. In these cases, the reductive dechlorination process can more easily lead to the formation of nontoxic compounds.

16.8 BIOAUGMENTATION

Bioaugmentation is a bioremediation technique that involves the addition of exogenous microorganisms to enhance the natural degradation of contaminants. In recent years, there has been growing concern over the presence of ECs in the environment, which are compounds that are not traditionally monitored in routine environmental monitoring programs but have the potential to cause adverse effects on human health and the environment. ECs are often resistant to conventional treatment processes and can persist in the environment for long periods. Bioaugmentation offers a promising approach for removing ECs by tuning the microbial niche to enhance the degradation of these compounds. Microbial niche tuning involves optimizing environmental conditions, such as temperature, pH, and nutrient availability, to support the growth and activity of the added microorganisms. In this way, bioaugmentation can enhance the natural degradation of ECs and improve the overall efficiency of bioremediation.

Bioaugmentation can be done using different microorganisms, such as bacteria, fungi, and algae. Bacteria are the most used microorganisms in bioaugmentation due to their ability to degrade various pollutants. For instance, *Pseudomonas putida* (discovered by Ananda Mohan Chakrabarty) is a bacterium that has been extensively studied for its ability to degrade organic pollutants, such as PAHs and PCBs (Herrero & Stuckey, 2020; Kanaly & Harayama, 2000).

Fungi are also used in bioaugmentation due to their ability to degrade recalcitrant pollutants, such as lignin and cellulose. For instance, white-rot fungi have been shown to degrade various pollutants, including chlorinated solvents and dioxins (Pointing, 2001). The success of white-rot fungi in degrading pollutants is attributed to their ability to produce lignin-degrading enzymes, such as laccases and peroxidases that can break down recalcitrant organic compounds.

Algae are also used in bioaugmentation due to their ability to remove pollutants through photosynthesis and bioaccumulation. Algae can remove pollutants, such as heavy metals and nutrients, through adsorption and precipitation. Moreover, algae can accumulate pollutants within their cells, leading to their removal from the environment. For instance, *Chlorella vulgaris* is an alga that has been shown to remove heavy metals, such as cadmium and lead, from water bodies (Cheng et al., 2016). The success of *C. vulgaris* in removing pollutants is attributed to its ability to accumulate heavy metals within its cells, leading to their removal from the environment.

16.8.1 Selection of Microorganisms for Bioaugmentation

The success of bioaugmentation depends on selecting appropriate microorganisms that can degrade the target ECs. The selection of microorganisms can be based on various factors, such as their ability to degrade the target contaminants, their tolerance of environmental conditions, and their compatibility with the indigenous microbial community. The choice of microorganisms also depends on the specific application, such as wastewater treatment, soil remediation, or bioreactor systems. One approach for selecting microorganisms is to screen them from contaminated environments rich in the target contaminants. This approach can identify the natural microbial communities already adapted to the contaminants and potentially be used for bioaugmentation. For example, a study by Quan et al. (2008) screened a soil sample contaminated with the herbicide 2,4-dichloro phenoxy acetic acid (2,4-D) for bacteria capable of degrading the compound. The study identified several bacterial strains that were able to degrade 2,4-D and could potentially be used for bioaugmentation. Another approach is using genetically engineered microorganisms designed to degrade the target contaminants. Genetic engineering can introduce new metabolic pathways into the microorganisms or enhance their existing degradation pathways. For example, a study by Sahota & Sharma (2022) used a genetically modified strain of *Pseudomonas putida* engineered to degrade the antibiotic sulfamethoxazole. The study found that the genetically modified strain could degrade sulfamethoxazole efficiently and potentially be used for bioaugmentation.

16.8.2 Optimization of Environmental Conditions for Microbial Niche Tuning

The bioaugmentation process success rate largely depends on optimizing the environmental conditions to support the growth and activity of the added microorganisms. The environmental conditions can be optimized by adjusting temperature, pH, and nutrient availability parameters. Optimizing environmental conditions can improve the degradation efficiency of the microorganisms and enhance the overall bioremediation process.

1. Temperature is an essential factor affecting microorganisms' growth and activity. Different microorganisms have different temperature optima, ranging from 5 °C to 70 °C. Therefore, it is crucial to optimize the temperature conditions to support the growth and activity of the selected microorganisms. For example, a study by Li et al. (2018) investigated the effect of temperature on the degradation of the antibiotic norfloxacin by a bacterial consortium. The study found that the degradation efficiency of norfloxacin was highest at 30 °C, which was the temperature that supported the highest microbial activity.
2. pH is another crucial factor affecting microorganisms' growth and activity. Different microorganisms have different pH optima, ranging from acidic to alkaline conditions. Therefore, it is essential to optimize the pH conditions to support the growth and activity of the selected microorganisms. For example, a study by Sun et al. (2020) investigated the effect of pH on the degradation of the pesticide chlorpyrifos by a bacterial consortium. The study found that the degradation efficiency of chlorpyrifos was highest at pH 7.0.
3. Nutrient availability also plays a crucial role in the growth and activity of microorganisms. Nutrient limitation can lead to reduced microbial activity and degradation efficiency. Therefore, it is essential to optimize the nutrient conditions to support the growth and activity of the selected microorganisms. For example, a study by (Chen et al., 2017) investigated the effect of nutrient availability on the degradation of the pharmaceutical ibuprofen by a bacterial consortium. The study found that adding carbon and nitrogen sources significantly enhanced the degradation efficiency of ibuprofen, indicating the importance of nutrient availability in bioaugmentation.

16.8.3 Implementation of Bioaugmentation in Different Environments

Bioaugmentation can be implemented in different environments, such as wastewater treatment plants, soil remediation sites, and bioreactor systems. Implementing bioaugmentation can be challenging due to the complex nature of the environments and the potential interactions between the added microorganisms and the indigenous microbial community.

1. Wastewater treatment plants are one of the most common bioaugmentation applications. They receive various contaminants, including ECs such as

pharmaceuticals and PPCPs. Bioaugmentation can enhance the degradation of these contaminants and improve the overall efficiency of wastewater treatment. For example, a study by Feng et al. (2019) investigated using a bacterial consortium to remove the antibiotic ciprofloxacin in a wastewater treatment plant. The study found that bioaugmentation significantly enhanced the removal of ciprofloxacin and supported the improvement of the overall efficiency of wastewater treatment.

2. Soil remediation sites are another potential application of bioaugmentation. Soil remediation sites often receive contaminants from industrial activities, such as pesticides and PAHs. Bioaugmentation can enhance the degradation of these contaminants and improve the overall efficiency of the soil remediation process. For example, a study by Sarker et al. (2021) investigated the use of a bacterial consortium for the remediation of soil contaminated with the pesticide chlorpyrifos, and found an effective role of bioaugmentation in soil remediation.
3. Bioreactor systems are also a potential application of bioaugmentation. They are commonly used to treat industrial wastewater and can receive diverse contaminants. Bioaugmentation can enhance the degradation of these contaminants and improve the overall efficiency of bioreactor systems. For example, a study by Zur et al. (2020) investigated using a bacterial consortium to remove the pharmaceutical ibuprofen in a sequencing batch reactor. The study found that bioaugmentation significantly enhanced the removal of ibuprofen and improved the overall efficiency of the bioreactor system.
4. Bioaugmentation is a promising approach for removing ECs through microbial niche tuning. However, challenges still need to be addressed to implement bioaugmentation successfully. One challenge is the potential interactions between the added microorganisms and the indigenous microbial community. The added microorganisms can compete with the indigenous microbial community for resources, potentially disrupting the ecosystem balance. Therefore, it is essential to carefully consider the choice of microorganisms and the environmental conditions to minimize the potential impact on the indigenous microbial community.

Another challenge is the potential for microorganisms to lose their activity over time. Microorganisms can lose their activity due to various factors, such as changes in environmental conditions and competition from other microorganisms. Therefore, it is crucial to monitor the activity of the added microorganisms and adjust the environmental conditions as necessary to maintain their activity. The microbial community can be tuned to degrade specific contaminants by selecting and adding specific microorganisms. However, successful implementation of bioaugmentation requires careful consideration of the choice of microorganisms, environmental conditions, and potential interactions with the indigenous microbial community. Future research in bioaugmentation can lead to the development of more efficient and robust microbial consortia and the optimization of environmental conditions for the growth and activity of the selected microorganisms. Overall, bioaugmentation can potentially be a valuable tool for the remediation of contaminated environments and protection of public health.

16.9 NANOTECHNOLOGY

Nanotechnology has been recognized as a promising tool for removing ECs from water, soil, and air. Removal of ECs through traditional methods has been expensive, inefficient, or environmentally harmful. Nanotechnology offers a new approach that can potentially overcome some of these challenges. In this context, microbial niche tuning has been proposed as a strategy that combines the use of nanoparticles with the manipulation of microbial communities to enhance the removal of ECs from contaminated environments (Shahi et al., 2021).

Nanotechnology has received much interest lately because of the unique physical properties that substances exhibit at the nanoscale. Compared to their bulkier equivalents, nanomaterials have higher surface-to-volume ratios, increasing reactivity and, consequently, their efficacy. Nanomaterials can also be functionalized or grafted with functional groups that can target specific molecules of interest (pollutants) for efficient remediation thanks to different surface chemistry, which is impossible with traditional approaches (Rimayi et al., 2018). In addition, purposeful modifications to the physical characteristics of the nanoparticles (such as size, shape, porosity, and chemical composition) may result in the inclusion of new advantageous properties that directly affect the efficiency of the substance for pollutant cleaning. The nanomaterial's flexible physical characteristics and wide surface chemistry provide significant advantages over conventional methods for removing environmental pollutants. It follows that methods developed around a single nano platform may be less efficient, selective, and stable than those developed by mixing numerous unique materials (hybrids/composites) and pulling certain desirable qualities from each component. The stability of the material may be improved by adding nanoparticles to a framework rather than utilizing them alone (Del Prado-Audelo et al., 2021; Mukhopadhyay et al., 2022).

Functionalising a material with specific chemicals targeting contaminant molecules can increase the selectivity and effectiveness of nanoparticles. The materials used to clean up pollutants must not turn into new sources of pollution and should be a target for any technology applied in environmental cleanup. Therefore, using biodegradable polymers in this application is quite exciting. Using biodegradable materials can boost consumer acceptability of a specific technology while also offering a cleaner and greener option for cleaning contaminants in the environment by reducing the need to dispose of material waste after treatment. For example, microplastics are a growing concern due to their small size and persistence in the environment. Nanoparticles can be used to break down microplastics into smaller particles, which microbes or other treatment methods can then degrade. This approach offers a potential solution to the growing problem of environmental microplastic pollution (Wang et al., 2021).

It is now possible to safely remove toxins due to the development of nanomaterials that are acceptable to the environment and that use extracts from microorganisms and other living organisms, e.g., Fe nanoparticles are one of the green nanoparticles used in remediation because of their nontoxic nature, magnetic susceptibility, and redox potential when interacting with water (Anam et al., 2019). A summarized view of all these technologies is presented in Table 16.1 with their advantages and limitations.

TABLE 16.1
Advantages and limitations of removal techniques

Sr. No.	Techniques	Remediation mechanism	Advantages	Limitations
1	Microbial electrochemical technology	This approach encourages microbial transformation of pollutants into nontoxic or less dangerous forms by acting as nonexhaustible electron donors and acceptors	Recovery of nutrients Low-cost treatment Fast biosensing of wastewater quality in real-time Energy conversion from waste is an advantage	Low COD removal rate Little power production Expensive catalyst
2	Microbial reductive dechlorination	Used to remove harmful pollutants. To comprehend the mechanisms underlying these phenomena, it is crucial to have a complete mechanistic understanding of electron competition and electron selectivity in natural hydrochemical environments	There are no expenses for regular upkeep and operation Affordable	High contaminant and metal concentrations can hinder reductive dechlorination Using the technology results in higher treatment costs, which calls for carefully examining the polluted area
3	Microbial enzymes	With the aid of these enzymes, microorganisms can break chemical bonds and transfer electrons from one chemical molecule (the acceptor) to another (the donor) in energy-producing biochemical processes	Eliminates the emission of secondary pollutants Lessens invasiveness and makes it simpler to fix environmental issues without compromising delicate ecosystems Budget-friendly Recyclable cleanup method	Arduous procedure Microbial enzymes can only break down biodegradable materials Since bioremediation treatments have no predetermined endpoint, evaluating their effectiveness is challenging

4	Bioaugmentation	Adding exogenous or native microbial biomass speeds up the degradation of the contaminant at the pollutant site	Enhancing solid settling efficiency Once the microbes are present, the work is less labor-intensive because they take over Virtually minimal environmental deterioration	Proper humidity, nutrients, and temperatures must be present for microbes to develop Not every waste can be metabolized by bacteria
5	Nanotechnology	Adsorption, nanoparticle agent, and the target pollutant must contact for a detoxifying or immobilizing reaction	Reduces costs Improves fuel efficiency It has the potential to save energy It may efficiently remove pollutants due to its small surface area, high adsorption capacity, and potential for selectivity	Harmful impacts on the environment's beneficial bacteria and consequences for agricultural plant growth and productivity After the wastewater is treated, recovering the nanocatalysts can be challenging It could unintentionally generate new dangerous chemicals

16.10 NANOTECHNOLOGY AND BIOAUGMENTATION COLLECTIVELY IN THE REMOVAL OF EMERGING CONTAMINANTS

The combined use of nanotechnology and bioaugmentation can lead to synergistic effects, resulting in improved removal efficiency of ECs. Nanotechnology can enhance the adsorption and uptake of pollutants by microorganisms, leading to improved biodegradation and bioaccumulation. Moreover, nanotechnology can enhance the delivery of microorganisms to the contaminated site, leading to improved colonization and survival of microorganisms.

The adsorption and uptake of pollutants by microorganisms can be enhanced through the functionalization of nanomaterials with microbial adhesion molecules, such as peptides and antibodies.

Moreover, nanomaterials can enhance the delivery of microorganisms to the contaminated site. For instance, magnetic nanoparticles (MNPs) can be used to deliver microorganisms to the contaminated site, where they can be immobilized and used to degrade pollutants (Shahi et al., 2021). Magnetic nanoparticles (MNPs) can be functionalized with microbial adhesion molecules, such as antibodies, which can bind to the microorganisms, leading to their targeted delivery to the contaminated site.

Bioaugmentation can also be combined with nanotechnology to remove ECs from the environment. For instance, the combined use of bioaugmentation with graphene oxide (GO) has been shown to enhance the removal of ECs, such as pharmaceuticals and PPCPs, from water bodies (Wang et al., 2020). GO can be used as a carrier for microorganisms, leading to their targeted delivery to the contaminated site, where they can degrade the pollutants. Moreover, GO can enhance microorganisms' adsorption and uptake of ECs, leading to improved removal efficiency.

16.11 CONCLUSION

Large numbers and volumes of ECs have been released into the surroundings since World War II, with detrimental impacts on aquatic/terrestrial animals and human health, including population loss, a rise in malignancies, and decreased reproductive function in organisms. Traditional treatment techniques are employed to eliminate nutrients, sediments, particle pollutants, and dissolved biodegrading organic mass; however, they are insufficient to eliminate the ECs that are complex and synthetic. Currently used wastewater treatment procedures are inadequate in removing trace contaminants. Activated carbon adsorption, nanofiltration, and reverse osmosis are among the most successful advanced technologies; however, they are more expensive than any other advanced treatment techniques. HMETs look promising and can be scaled up for field applications. Enzymatic systems play a very significant role in the bioremediation of xenobiotics and pollutants created as a result of the extensive use of chemicals in several industrial processes. The purifying procedure raises the overall cost of enzyme preparations, which becomes a main obstacle to its commercialization. The application of bioaugmentation coupled with nanotechnology has not been harnessed in remediation technologies, even though nanomaterials are increasingly utilized in the remediation of polluted surroundings. It is essential to create more efficient and effective processes for developing such technologies at commercial scales since doing so would make it possible to provide "greener" solutions for a "clean" environment.

REFERENCES

Abujaber, F., Zougagh, M., Jodeh, S., Ríos, Á., Bernardo, F. J. G., & Martín-Doimeadios, R. C. R. (2018). Magnetic cellulose nanoparticles coated with ionic liquid as a new material for the simple and fast monitoring of emerging pollutants in waters by magnetic solid phase extraction. *Microchemical Journal*, *137*, 490–495.

Aggazzotti, G., Fantuzzi, G., Righi, E., & Predieri, G. (1998). Blood and breath analyses as biological indicators of exposure to trihalomethanes in indoor swimming pools. *Science of the Total Environment*, *217*(1-2), 155–163.

Anam, G. B., Choi, J., & Ahn, Y. (2019). Reductive dechlorination of perchloroethene (PCE) and bacterial community changes in a continuous-flow, two-stage anaerobic column. *International Biodeterioration & Biodegradation*, *138*, 41–49.

Apelberg, B. J., Witter, F. R., Herbstman, J. B., Calafat, A. M., Halden, R. U., Needham, L. L., & Goldman, L. R. (2007). Cord serum concentrations of perfluorooctane sulfonate (PFOS) and perfluorooctanoate (PFOA) about weight and size at birth. *Environmental Health Perspectives*, *115*(11), 1670–1676.

Asghar, M. A., Zhu, Q., Sun, S., & Shuai, Q. (2018). Suspect screening and target quantification of human pharmaceutical residues in the surface water of Wuhan, China, using UHPLC-Q-Orbitrap HRMS. *Science of the Total Environment*, *635*, 828–837.

Bigus, P., Tobiszewski, M., & Namieśnik, J. (2014). Historical records of organic pollutants in sediment cores. *Marine Pollution Bulletin*, *78*(1-2), 26–42.

Bila, D. M., & Dezotti, M. (2003). Fármacos no meio ambiente. *Química Nova*, *26*, 523–530.

Bishop, C. A., Ng, P., Pettit, K. E., Kennedy, S. W., Stegeman, J. J., Norstrom, R. J., & Brooks, R. J. (1998). Environmental contamination and developmental abnormalities in eggs and hatchlings of the common snapping turtle (Chelydra serpentina serpentina) from the Great Lakes—St Lawrence River basin (1989–1991). *Environmental Pollution*, *101*(1), 143–156.

Bittner, B., Kellner, H., Jehmlich, N., Ullrich, R., Pecyna, M. J., Nousiainen, P., & Liers, C. (2012). The wood-rot Ascomycetes, *Xylaria polymorpha* produces a novel GH 78 glycoside hydrolase that exhibits α-L-rhamnosidase and feruloyl esterase activity and releases hydroxycinnamic acids from lignocelluloses. *Applied and Environmental Microbiology*, *78*(14), 4893–4901.

Bonde, J. P. (2010). Male reproductive organs are at risk from environmental hazards. *Asian Journal of Andrology*, *12*(2), 152.

Briner, W. (2010). The toxicity of depleted uranium. *International Journal of Environmental Research and Public Health*, *7*(1), 303–313.

Caliman, F. A., & Gavrilescu, M. (2009). Pharmaceuticals, personal care products and endocrine disrupting agents in the environment–A review. *CLEAN–Soil, Air, Water*, *37*(4-5), 277–303.

Chen, Y., Lan, S., Wang, L., Dong, S., Zhou, H., Tan, Z., & Li, X. (2017). A review: driving factors and regulation strategies of microbial community structure and dynamics in wastewater treatment systems. *Chemosphere*, *174*, 173–182.

Cheng, J., Qiu, H., Chang, Z., Jiang, Z., & Yin, W. (2016). The effect of cadmium on the growth and antioxidant response for freshwater algae Chlorella vulgaris. *SpringerPlus*, *5*, 1–8.

Daughton, C. G. (2004). Non-regulated water contaminants: Emerging research. *Environmental Impact Assessment Review*, *24*(7-8), 711–732.

de Sousa, D. N. R., Mozeto, A. A., Carneiro, R. L., & Fadini, P. S. (2018). Spatio-temporal evaluation of emerging contaminants and their partitioning along a Brazilian watershed. *Environmental Science and Pollution Research*, *25*, 4607–4620.

Deblonde, T., Cossu-Leguille, C., & Hartemann, P. (2011). Emerging pollutants in wastewater: A review of the literature. *International Journal of Hygiene and Environmental Health*, *214*(6), 442–448.

Desai, S. S., Tennali, G. B., Channur, N., Anup, A. C., Deshpande, G., & Murtuza, B. A. (2011). Isolation of laccase producing fungi and partial characterization of laccase. *Biotechnol. Bioinf. Bioeng*, *1*(4), 543–549.

Del Prado-Audelo, M. L., García Kerdan, I., Escutia-Guadarrama, L., Reyna-González, J. M., Magaña, J. J., & Leyva-Gómez, G. (2021). Nanoremediation: nanomaterials and nanotechnologies for environmental cleanup. *Frontiers in Environmental Science*, *9*, 793765.

Dong, F., Li, C., Ma, X., Lin, Q., He, G., & Chu, S. (2020). Degradation of estriol by chlorination in a pilot-scale water distribution system: Kinetics, pathway and DFT studies. *Chemical Engineering Journal*, *383*, 123187.

Dos Santos, D. M., Buruaem, L., Gonçalves, R. M., Williams, M., Abessa, D. M., Kookana, R., & de Marchi, M. R. R. (2018). Multiresidue determination and predicted risk assessment of contaminants of emerging concern in marine sediments from the vicinities of submarine sewage outfalls. *Marine Pollution Bulletin*, *129*(1), 299–307.

Feng, N. X., Yu, J., Xiang, L., Yu, L. Y., Zhao, H. M., Mo, C. H., ... & Li, Q. X. (2019). Co-metabolic degradation of the antibiotic ciprofloxacin by the enriched bacterial consortium XG and its bacterial community composition. *Science of the Total Environment*, *665*, 41–51.

Fox, G. A. (2001). Effects of endocrine disrupting chemicals on wildlife in Canada: Past, present and future. *Water Quality Research Journal*, *36*(2), 233–251.

Frisbie, S. H., Mitchell, E. J., & Sarkar, B. (2013). World Health Organization increases its drinking-water guideline for uranium. *Environmental Science: Processes & Impacts*, *15*(10), 1817–1823.

Fuentes, S., Colomina, M. T., Vicens, P., Franco-Pons, N., & Domingo, J. L. (2007). Concurrent exposure to perfluorooctane sulfonate and restraint stress during pregnancy in mice: Effects on postnatal development and behavior of the offspring. *Toxicological Sciences*, *98*(2), 589–598.

Gabbana, J. V., de Oliveira, L. H., Paveglio, G. C., & Trindade, M. A. G. (2018). Narrowing the interface between sample preparation and electrochemistry: Trace-level determination of emerging pollutant in water samples after in situ microextraction and electroanalysis using a new cell configuration. *Electrochimica Acta*, *275*, 67–75.

Garric, J., & Ferrari, B. (2005). Les substances pharmaceutiques dans les milieux aquatiques. Niveaux d'exposition et effet biologique: Que savons nous?. *Revue des Sciences de l'eau*, *18*(3), 307–330.

Gavnholt, B., & Larsen, K. (2002). Molecular biology of plant laccases in relation to lignin formation: Minireview (pp. 273–280). Physiologia Plantarum (Denmark).

Ghangrekar, M. M., Sathe, S. M., & Chakraborty, I. (2020). In situ bioremediation techniques for the removal of emerging contaminants and heavy metals using hybrid microbial electrochemical technologies. In *Emerging Technologies in Environmental Bioremediation* (pp. 233–255). Elsevier.

Ginebreda, A., Sabater-Liesa, L., Rico, A., Focks, A., & Barceló, D. (2018). Reconciling monitoring and modeling: An appraisal of river monitoring networks based on a spatial autocorrelation approach-emerging pollutants in the Danube River as a case study. *Science of the Total Environment*, *618*, 323–335.

Hadibarata, T., Abdullah, F., Yusoff, A. R. M., Ismail, R., Azman, S., & Adnan, N. (2012). Correlation study between land use, water quality, and heavy metals (Cd, Pb, and Zn) content in water and green lipped mussels *Perna viridis* (Linnaeus.) at the Johor Strait. *Water, Air, & Soil Pollution*, *223*, 3125–3136.

Hadibarata, T., Tachibana, S., & Itoh, K. (2007). Biodegradation of phenanthrene by fungi screened from nature. *Pakistan Journal of Biological Sciences*, *10*(15), 2535–2543.

Halling-Sørensen, B. N. N. S., Nielsen, S. N., Lanzky, P. F., Ingerslev, F., Lützhøft, H. H., & Jørgensen, S. E. (1998). Occurrence, fate and effects of pharmaceutical substances in the environment-A review. *Chemosphere*, *36*(2), 357–393.

Henny, C. J., Grove, R. A., & Hedstrom, O. R. (1996). *A Field Evaluation of Mink and River Otter on the Lower Columbia River and the Influence of Environmental Contaminants*. National Biological Service, Forest and Rangeland Ecosystem Science Center, Northwest Research Station.

Hermes, N., Jewell, K. S., Wick, A., & Ternes, T. A. (2018). Quantification of more than 150 micropollutants including transformation products in aqueous samples by liquid chromatography-tandem mass spectrometry using scheduled multiple reaction monitoring. *Journal of Chromatography A*, *1531*, 64–73.

Herrero, M., & Stuckey, D. C. (2015). Bioaugmentation and its application in wastewater treatment: A review. *Chemosphere*, *140*, 119–128.

Ho, Z. H., & Adnan, L. A. (2021). Phenol removal from aqueous solution by adsorption technique using coconut shell activated carbon. *Tropical Aquatic and Soil Pollution,* 404.

IARC. (1991). Chlorinated drinking water. In: *Chlorinated Drinking Water; Chlorination by Products; Some Other Halogenated Compounds; Cobalt and Cobalt Compounds* (Vol. 52, pp. 45–141). International Agency for Research on Cancer, Lyon (IARC Monographs on the evaluation of the Carcinogenic Risk of Chemicals to Humans).

Iyer, G., & Chattoo, B. B. (2003). Purification and characterization of laccase from the rice blast fungus, Magnaporthe grisea. *FEMS Microbiology Letters, 227*(1), 121–126.

Kabir, E. R., Rahman, M. S., & Rahman, I. (2015). A review on endocrine disruptors and their possible impacts on human health. *Environmental Toxicology and Pharmacology*, *40*(1), 241–258.

Kanaly, R. A., & Harayama, S. (2000). Biodegradation of high-molecular-weight polycyclic aromatic hydrocarbons by bacteria. *Journal of Bacteriology*, *182*(8), 2059–2067.

Kapelewska, J., Kotowska, U., Karpińska, J., Kowalczuk, D., Arciszewska, A., & Świrydo, A. (2018). Occurrence, removal, mass loading and environmental risk assessment of emerging organic contaminants in leachates, groundwaters and wastewaters. *Microchemical Journal*, *137*, 292–301.

Kavlock, R. J., Daston, G. P., DeRosa, C., Fenner-Crisp, P., Gray, L. E., Kaattari, S., ... & Tilson, H. A. (1996). Research needs for the risk assessment of health and environmental effects of endocrine disruptors: A report of the US EPA-sponsored workshop. *Environmental Health Perspectives*, *104*(suppl 4), 715–740.

Kosaka, K., Asami, M., Matsuoka, Y., Kamoshita, M., & Kunikane, S. (2007). Occurrence of perchlorate in drinking water sources of metropolitan area in Japan. *Water Research*, *41*(15), 3474–3482.

Lai, H. J. (2021). Adsorption of remazol brilliant violet 5R (RBV-5R) and remazol brilliant blue R (RBBR) from aqueous solution by using agriculture waste. *Tropical Aquatic and Soil Pollution*, *1*(1), 11–23.

Lamm, S. H., Braverman, L. E., Li, F. X., Richman, K., Pino, S., & Howearth, G. (1999). Thyroid health status of ammonium perchlorate workers: A cross-sectional occupational health study. *Journal of Occupational and Environmental Medicine*, 41(4), 248–260.

Lau, C., Thibodeaux, J. R., Hanson, R. G., Rogers, J. M., Grey, B. E., Stanton, M. E., ... & Stevenson, L. A. (2003). Exposure to perfluorooctane sulfonate during pregnancy in rat and mouse. II: postnatal evaluation. *Toxicological Sciences*, *74*(2), 382–392.

Levasseur, A., Saloheimo, M., Navarro, D., Andberg, M., Pontarotti, P., Kruus, K., & Record, E. (2010). Exploring laccase-like multicopper oxidase genes from the ascomycete *Trichoderma reesei*: a functional, phylogenetic and evolutionary study. *BMC Biochemistry*, *11*, 1–10.

Li, S., Hua, T., Li, F., & Zhou, Q. (2020). Bio-electro-Fenton systems for sustainable wastewater treatment: mechanisms, novel configurations, recent advances, LCA and challenges. An updated review. *Journal of Chemical Technology & Biotechnology*, *95*(8), 2083–2097.

Li, S., Yang, X., Yang, S., Zhu, M., & Wang, X. (2012). Technology prospecting on enzymes: Application, marketing and engineering. *Computational and Structural Biotechnology Journal, 2*(3), e201209017.

Li, X., Chen, S., Angelidaki, I., & Zhang, Y. (2018). Bio-electro-Fenton processes for wastewater treatment: Advances and prospects. *Chemical Engineering Journal*, *354*, 492–506.

Liu, Z. H., Kanjo, Y., & Mizutani, S. (2009). Removal mechanisms for endocrine disrupting compounds (EDCs) in wastewater treatment—physical means, biodegradation, and chemical advanced oxidation: A review. *Science of the Total Environment*, *407*(2), 731–748.

Magi, E., Di Carro, M., Mirasole, C., & Benedetti, B. (2018). Combining passive sampling and tandem mass spectrometry for the determination of pharmaceuticals and other emerging pollutants in drinking water. *Microchemical Journal*, *136*, 56–60.

Maharjan, A. K., Wong, D. R. E., & Rubiyatno, R. (2021). Level and distribution of heavy metals in Miri River, Malaysia. *Tropical Aquatic and Soil Pollution*, *1*(2), 74–86.

Mijangos, L., Ziarrusta, H., Olivares, M., Zuloaga, O., Möder, M., Etxebarria, N., & Prieto, A. (2018b). Simultaneous determination of 41 multiclass organic pollutants in environmental waters by means of polyethersulfone microextraction followed by liquid chromatography–tandem mass spectrometry. *Analytical and Bioanalytical Chemistry*, *410*, 615–632.

Mijangos, L., Ziarrusta, H., Ros, O., Kortazar, L., Fernández, L. A., Olivares, M., ... & Etxebarria, N. (2018a). Occurrence of emerging pollutants in estuaries of the Basque Country: analysis of sources and distribution, and assessment of the environmental risk. *Water Research*, *147*, 152–163.

Minussi, R. C., Miranda, M. A., Silva, J. A., Ferreira, C. V., Aoyama, H., Marangoni, S., & Dura´n, N. (2007). Purification, characterization and application of laccase from Trametes versicolor for colour and phenolic removal of olive mill wastewater in the presence of 1-hydroxybenzotriazole. *African Journal of Biotechnology, 6*(10), 1248–1254.

Mohan, S. V., Sravan, J. S., Butti, S. K., Krishna, K. V., Modestra, J. A., Velvizhi, G., ... & Pandey, A. (2019). Microbial electrochemical technology: emerging and sustainable platform. In *Microbial electrochemical technology* (pp. 3–18). Elsevier.

Mukhopadhyay, R., Sarkar, B., Khan, E., Alessi, D. S., Biswas, J. K., Manjaiah, K. M., ... & Ok, Y. S. (2022). Nanomaterials for sustainable remediation of chemical contaminants in water and soil. *Critical Reviews in Environmental Science and Technology, 52*(15), 2611–2660.

Naidu, R., & Wong, M. H. (2013). Contaminants of emerging concern. Foreward. *The Science of the Total Environment*, *463*, 1077–1078.

Ng, M. H., & Elshikh, M. S. (2021). Utilization of Moringa oleifera as natural coagulant for water purification. *Industrial and Domestic Waste Management*, *1*(1), 1–11.

Okoye, C. O., Okeke, E. S., Okoye, K. C., Echude, D., Andong, F. A., Chukwudozie, K. I., ... & Ezeonyejiaku, C. D. (2022). Occurrence and fate of pharmaceuticals, personal care products (PPCPs) and pesticides in African water systems: A need for timely intervention. *Heliyon*, 8(3): e09143

Omar, T. F. T., Aris, A. Z., Yusoff, F. M., & Mustafa, S. (2018). Occurrence, distribution, and sources of emerging organic contaminants in tropical coastal sediments of anthropogenically impacted Klang River estuary, Malaysia. *Marine Pollution Bulletin*, *131*, 284–293.

Ozcan, S., Tor, A., & Aydin, M. E. (2013). Investigation on the levels of heavy metals, polycyclic aromatic hydrocarbons, and polychlorinated biphenyls in sewage sludge samples and ecotoxicological testing. *Clean–Soil, Air, Water*, *41*(4), 411–418.

Patel, A., Patel, V., Patel, R., Trivedi, U., & Patel, K. (2020). Fungal laccases: Versatile green catalyst for bioremediation of organic pollutants. In *Emerging Technologies in Environmental Bioremediation* (pp. 85–129). Elsevier.

Peng, Y., Fang, W., Krauss, M., Brack, W., Wang, Z., Li, F., & Zhang, X. (2018). Screening hundreds of emerging organic pollutants (EOPs) in surface water from the Yangtze River Delta (YRD): occurrence, distribution, ecological risk. *Environmental Pollution*, *241*, 484–493.

Pham, T. T., & Proulx, S. (1997). PCBs and PAHs in the Montreal Urban Community (Quebec, Canada) wastewater treatment plant and in the effluent plume in the St Lawrence River. *Water Research*, *31*(8), 1887–1896.

Pointing, S. (2001). Feasibility of bioremediation by white-rot fungi. *Applied Microbiology and Biotechnology*, *57*, 20–33.

Polyakova, O. V., Artaev, V. B., & Lebedev, A. T. (2018). Priority and emerging pollutants in the Moscow rain. *Science of the Total Environment*, *645*, 1126–1134.

Quan, X., Shi, H., Liu, H., Lv, P., & Qian, Y. (2004). Enhancement of 2,4-dichlorophenol degradation in conventional activated sludge systems bioaugmented with mixed special culture. *Water Research* 2004, 38, 245–253.

Ranocha, P., McDougall, G., Hawkins, S., Sterjiades, R., Borderies, G., Stewart, D., & Goffner, D. (1999). Biochemical characterization, molecular cloning and expression of laccases—A divergent gene family—in poplar. *European Journal of Biochemistry*, *259*(1-2), 485–495.

Rimayi, C., Odusanya, D., Weiss, J. M., de Boer, J., & Chimuka, L. (2018). Contaminants of emerging concern in the Hartbeespoort Dam catchment and the uMngeni River estuary 2016 pollution incident, South Africa. *Science of the Total Environment*, *627*, 1008–1017.

Rocío-Bautista, P., Pino, V., Ayala, J. H., Ruiz-Pérez, C., Vallcorba, O., Afonso, A. M., & Pasán, J. (2018). A green metal–organic framework to monitor water contaminants. *RSC Advances*, *8*(55), 31304–31310.

Rosal, R., Rodriguez, A., Perdigon-Melon, J.A., Petre, A., Garcia-Calvo, E., Gomez, M.J., Aguera, A., & Fernandez-Alba, A.R. (2010). Occurrence of emerging pollutants in urban wastewater and their removal through biological treatment followed by ozonation. *Water Research*, *44*, 578–588.

Rosner, D., & Markowitz, G. (2013). Persistent pollutants: A brief history of the discovery of the widespread toxicity of chlorinated hydrocarbons. *Environmental Research*, *120*, 126–133.

Sahota, N. K., & Sharma, R. (2022). Insight into pharmaceutical waste management by employing bioremediation techniques to restore environment. In *Handbook of Solid*

Waste Management: Sustainability through Circular Economy (pp. 1795–1826). Singapore: Springer Nature Singapore.

Santos, M. M., Ruivo, R., Capitão, A., Fonseca, E., & Castro, L. F. C. (2018). Identifying the gaps: Resources and perspectives on the use of nuclear receptor based-assays to improve hazard assessment of emerging contaminants. *Journal of Hazardous Materials, 358*, 508–511.

Sarker, A., Nandi, R., Kim, J. E., & Islam, T. (2021). Remediation of chemical pesticides from contaminated sites through potential microorganisms and their functional enzymes: Prospects and challenges. *Environmental Technology & Innovation, 23*, 101777.

Sathishkumar, P., Meena, R. A. A., Palanisami, T., Ashokkumar, V., Palvannan, T., & Gu, F. L. (2020). Occurrence, interactive effects and ecological risk of diclofenac in environmental compartments and biota-A review. *Science of the Total Environment, 698*, 134057.

Sathishkumar, P., Mohan, K., Ganesan, A. R., Govarthanan, M., Yusoff, A. R. M., & Gu, F. L. (2021). Persistence, toxicological effect and ecological issues of endosulfan–A review. *Journal of Hazardous Materials, 416*, 125779.

Shahi, M. P., Kumari, P., Mahobiya, D., & Shahi, S. K. (2021). Nano-bioremediation of environmental contaminants: applications, challenges, and future prospects. *Bioremediation for Environmental Sustainability*, 83–98.

Sheiner, E. K., Sheiner, E., Hammel, R. D., Potashnik, G., & Carel, R. (2003). Effect of occupational exposures on male fertility: Literature review. *Industrial Health, 41*(2), 55–62.

Singh, R., Kumar, M., Mittal, A., & Mehta, P. K. (2016). Microbial enzymes: industrial progress in 21st century. *3 Biotech, 6*, 1–15.

Stuart, M., Lapworth, D., Crane, E., & Hart, A. (2012). Review of risk from potential emerging contaminants in UK groundwater. *Science of the Total Environment, 416*, 1–21.

Sun, R., Luo, X., Li, Q. X., Wang, T., Zheng, X., Peng, P., & Mai, B. (2018). Legacy and emerging organohalogenated contaminants in wild edible aquatic organisms: Implications for bioaccumulation and human exposure. *Science of the Total Environment, 616*, 38–45.

Sun, X., Chen, L., Liu, C., Xu, Y., Ma, W., & Ni, H. (2020). Biodegradation of CP/TCP by a constructed microbial consortium after comparative bacterial community analysis of long-term CP domesticated activated sludge. *Journal of Environmental Science and Health, Part B, 55*(10), 898–908.

Tan, Q., Chen, J., Chu, Y., Liu, W., Yang, L., Ma, L., ... & He, F. (2021). Triclosan weakens the nitrification process of activated sludge and increases the risk of the spread of antibiotic resistance genes. *Journal of Hazardous Materials, 416*, 126085.

Tang, K. H. D. (2021). Interactions of microplastics with persistent organic pollutants and the ecotoxicological effects: A review. *Tropical Aquatic and Soil Pollution, 1*(1), 24–34.

Tang, Y. Y., Tang, K. H. D., Maharjan, A. K., Aziz, A. A., & Bunrith, S. (2021). Malaysia moving towards a sustainability municipal waste management. *Industrial and Domestic Waste Management, 1*(1), 26–40.

Tapia-Orozco, N., Ibarra-Cabrera, R., Tecante, A., Gimeno, M., Parra, R., & Garcia-Arrazola, R. (2016). Removal strategies for endocrine disrupting chemicals using cellulose-based materials as adsorbents: A review. *Journal of Environmental Chemical Engineering, 4*(3), 3122–3142.

Thibodeaux, J. R., Hanson, R. G., Rogers, J. M., Grey, B. E., Barbee, B. D., Richards, J. H., ... & Lau, C. (2003). Exposure to perfluorooctane sulfonate during pregnancy in rat and mouse. I: Maternal and prenatal evaluations. *Toxicological Sciences, 74*(2), 369–381.

Tillitt, D. E., Ankley, G. T., Giesy, J. P., Ludwig, J. P., Kurita-Matsuba, H., Weseloh, D. V., ... & Kubiak, T. J. (1992). Polychlorinated biphenyl residues and egg mortality in double-crested cormorants from the Great Lakes. *Environmental Toxicology and Chemistry: An International Journal, 11*(9), 1281–1288.

Upadhyay, P., Shrivastava, R., & Agrawal, P. K. (2016). Bioprospecting and biotechnological applications of fungal laccase. *3 Biotech, 6*(1), 15.

Vogelsang, C., Grung, M., Jantsch, T. G., Tollefsen, K. E., & Liltved, H. (2006). Occurrence and removal of selected organic micropollutants at mechanical, chemical and advanced wastewater treatment plants in Norway. *Water Research, 40*(19), 3559–3570.

Vystavna, Y., Frkova, Z., Celle-Jeanton, H., Diadin, D., Huneau, F., Steinmann, M., ... & Loup, C. (2018). Priority substances and emerging pollutants in urban rivers in Ukraine: Occurrence, fluxes and loading to transboundary European Union watersheds. *Science of the Total Environment, 637*, 1358–1362.

Wang, S., Li, L., Yu, S., Dong, B., Gao, N., & Wang, X. (2021). A review of advances in EDCs and PhACs removal by nanofiltration: Mechanisms, impact factors and the influence of organic matter. *Chemical Engineering Journal, 406*, 126722.

WHO. (1993). *Guidelines for Drinking Water Quality. Vol. 1. Recommendations*, 2nd edn. Word Health Organization, Geneva.

Yang, K., Ji, M., Liang, B., Zhao, Y., Zhai, S., Ma, Z., & Yang, Z. (2020). Bioelectrochemical degradation of monoaromatic compounds: Current advances and challenges. *Journal of Hazardous Materials, 398*, 122892.

Yang, Y. Y., Zhao, J. L., Liu, Y. S., Liu, W. R., Zhang, Q. Q., Yao, L., ... & Ying, G. G. (2018). Pharmaceuticals and personal care products (PPCPs) and artificial sweeteners (ASs) in surface and ground waters and their application as indication of wastewater contamination. *Science of the Total Environment, 616*, 816–823.

Zhang, L., Wu, D., Liang, J., Wang, L., & Zhou, Y. (2021). Triclosan transformation and impact on an elemental sulfur-driven sulfidogenic process. *Chemical Engineering Journal, 421*, 129634.

Żur, J., Michalska, J., Piński, A., Mrozik, A., & Nowak, A. (2020). Effects of low concentration of selected analgesics and successive bioaugmentation of the activated sludge on its activity and metabolic diversity. *Water, 12*(4), 1133.

17 Use of a Microbial Nexus System in High Andean Areas

Effectiveness and Feasibility for Wastewater Treatment

Edwin Hualpa-Cutipa, Richard Andi Solorzano Acosta, Milagros Estefani Alfaro Cancino, and Bernabe Luis-Alaya

17.1 INTRODUCTION

Environmental problems such as population growth, urbanization, and constant changes in the climate have generated the need to use alternative strategies to improve the management of water resources, particularly when it comes to wastewater (Pardo, 2022). Wastewater contains a considerable number of pollutants such as heavy metals, dyes, surfactants, pharmaceuticals, and metalloids (Chandra & Kumar 2017a, 2017b; Kumar et al., 2021, 2022), which have become one of the main environmental problems faced by humanity as water is a finite natural resource and its availability and quality is a global demand at all times (Rashid et al., 2021). Unfortunately, the conventional process of wastewater remediation cannot completely remove all the contaminants, or due to the nature of the process they accumulate or generate other contaminants (Elgarahy et al., 2021; Kumar & Verma, 2023).

Even when talking about wastewater treatment using microorganisms that have been put into practice for more than a century, there is still ongoing research and development to improve efficiency and sustainability in the process. Advances in technology and scientific understanding continue to drive innovation in this field (Wu & Yin, 2020). After such a long time and despite the development of infrastructure such as treatment plants (WWTPs) that ensure efficient wastewater management, innovations have emerged around the sustainable management of wastewater globally (Van Loosdrecht & Brdjanovic, 2014). For example, resource reuse and energy recovery from wastewater whose ultimate objective is to simultaneously achieve water purification and energy and/or nutrient recuperation (Mohan et al., 2016).

On the other hand, it is evident that wastewater treatment still faces challenges due to the appearance of new emerging pollutants of low concentrations that are not easy

DOI: 10.1201/9781003441069-17

to remove using traditional technologies, including recalcitrant organic compounds, nanoparticles, and even endocrine disruptors (Völker et al., 2016). Faced with this problem, environmental biotechnology is key to wastewater management (Wu & Yin, 2020).

Environmental biotechnology employs biological systems ranging from microorganisms to their components or byproducts such as enzymes (Cota-Ruiz et al., 2019). Nevertheless, the effectiveness of these biological systems has its limitations in certain regions of the planet such as the Andean region, where environmental conditions are limiting, for example, high radiation near the equatorial line, low temperatures at night that can reach freezing point, and low nutrient availability (Buytaert et al., 2007; Bader et al., 2007). The high Andean region, found from 3500 to 5000 m altitude above sea level, is composed of Latin American countries including southern Venezuela, Colombia, Ecuador, and northern Peru. This area is experiencing rapid population growth that together with urbanization have become the main factors deteriorating its river ecosystems due to the generation of wastewater which is more difficult to treat in this environment (Jiménez-Rivillas et al., 2018; Iñiguez-Armijos et al., 2022; Rascon et al., 2021).

The Central Andes region covers an altitude between 3100–4200 m.a.s.l and is characterized by the presence of closed basins, where you can find many salt flats, volcanoes, lakes with a high number of salts, hydrothermal springs, and desert ecosystems (Saona et al., 2020). Due to its altitudinal gradients and location, it is conditioned by the particular environmental characteristics of that area that make it one of the most extreme places on Earth. This is mainly due to low oxygen pressure, high UV radiation, persistent volcanic eruptions, high heat fluctuations, high drought, or nutrient-poor environments (Albarracín et al., 2015a). Consequently, these factors severely condition the life of those microorganisms that are part of the natural landscape (Fig. 17.1) of the Andean zones (Saona et al., 2020). All these

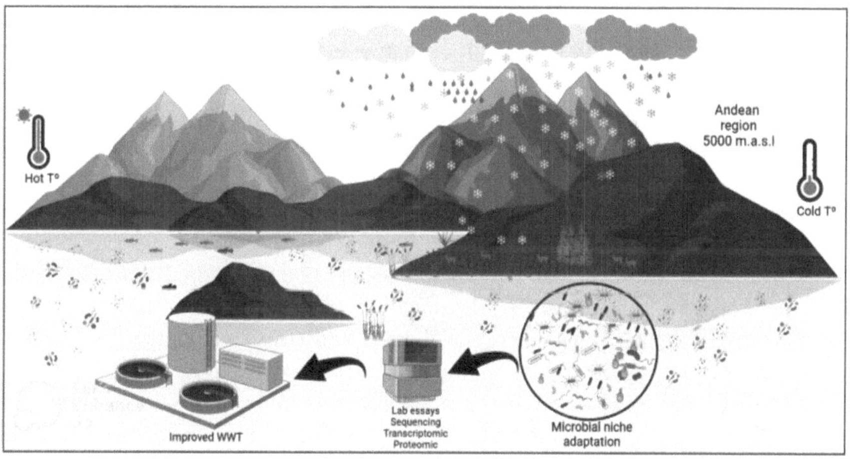

FIGURE 17.1 Wastewater treatment by the microbial nexus approach in a high Andean environment.

complex microorganisms are called microbial ecosystems and mainly consist of *Proteobacteria*, *Bacteroidetes*, and *Firmicutes* (Fernandez et al., 2016).

Thanks to the evolution of these autochthonous microbial communities, microbial niches have developed, which are characteristic spaces that present a variety of biotic (bacterial interactions) and abiotic (biotope) conditions that provide positive growth rates (Chandra & Kumar 2017b; Baquero et al., 2021). Among the various niches that make up these microbes are surface waters, sediments, microbial mats, microbialites, among others (Dorador et al., 2020). Microorganisms present in an environment with simultaneous extreme factors are designated as "polyelectrophilic" microorganisms (Kurth et al., 2015). Study strategies for this class of microbes are currently being applied using morpho- and phylogenetic methods through proteomic and metagenomic approaches. It is known that the extremoliths generated by these polyelectrophilic microorganisms have a high potential in biotechnology use for bioremediation and biomineralization, even in drug development (Albarracín et al., 2015b).

Currently, environmental issues are those that affect the population the most, and mining is one of the main causes of pollution, although it generates lucrative profits (Ortiz-Ojeda et al., 2013). One of the wastes that causes the most pollution is wastewater, which further increases its harmful action due to the appearance of new micropollutants. Since its elimination is complicated with the technology usually used, it is necessary to look for other solutions that can meet this objective. One of them is the development of an organic alternative to improve Andean crops, that is, the use of microbial inoculants with Plant Growth Promoting Rhizobacteria (PGPR) capacity isolated from high Andean areas, which are adapted to low-temperature conditions and have the potential to bioremediate the contaminants present in the soil (Chandra et al., 2017; Huasasasquiche et al., 2020).

Another alternative is the microbial niche nexus which is the adaptation of microbial niches to accommodate different microbial communities for efficient removal of pollutants (Wu & Yin, 2020). Therefore, wastewater treatment processes can be improved by elucidating biological metabolic mechanisms. Syntrophic bacteria and methanogens (see Table 17.1) can exchange electrons through conductive pili, or cytochromes, in their outer membranes to produce syntrophic methane (Wu & Yin, 2020). This demonstrates a faster and more energy-efficient route to methane production, while also being an important way to improve energy conversion from wastewater. To advance and strengthen the idea of a microbial niche nexus, an understanding of microbial ecology and the regulation and interactions of known and unknown microorganisms is essential. In addition, more environmental biology research based on the microbial niche nexus needs to be considered, considering the enrichment of microorganisms, the identification of microbial functions and metabolism, system design and operation control, and finally the development and application of new technologies.

Although according to research, the main problem in this field is the low coverage of domestic sewage treatment, this requires the development of inexpensive, reliable, and easy-to-use sewage treatment systems using endemic microorganisms that are extremely adapted to these environmental conditions (Peña & Mara, 2003). In the

TABLE 17.1
Some of the microorganisms used in wastewater treatment

Microorganisms	Pollutant and/or application	Reference
Bacillus subtilis and *Pseudomonas cepacea*	Nickel (Ni)	Abdel-Monem et al. (2010)
Synechocystis sp (Cyanobacteria)	Biosorption of chromium (VI) (Cr)	Ozturk et al. (2011)
Arthrobacter sp.	Cadmium (Cd)	Hasan and Srivastava (2011)
Bacillus subtilis	Chromium (Cr) (III)	Aravindhan et al. (2012)
Bacillus subtilis, *Pseudomonas fluoresens*, *Burkholderia cepacia*, and *Chryseomonas luteola*	Heavy metals as a nickel (Ni)	Al-Gheethi et al. (2014)
Bacillus amyloliquefaciens and *Pseudomonas aeruginosa*	Applications as pollutant biosurfactants	Ndlovu et al. (2017)
Pseudomonas aeruginosa	Alkaline lipase for biodegradation in food wastewater treatment	Hu et al. (2018)
Rhodopseudomonas and *Pseudomonas*	Pollutants in oil refinery	Sun et al. (2022))
Chlorococcum sp. and indigenous bacteria	Process of municipal wastewater treatment	Yang et al. (2022)
Chlorella vulgaris -bacteria co-cultures	Parabens	Sousa et al. (2023)

high Andean areas, the spatial separation of the mountains creates a unique set of habitats, forcing microbes to adapt by increasing biodiversity with unique potential for the restoration of degraded environments (Cortez et al., 2018).

In the context of the use of biological systems and wastewater reuse, the application of the "nexus approach," which initially emerged for the food–energy–water system, has been proposed as an option for water resource treatment, since it gives an integrated management strategy through which to assess the interactions between food, energy, and water system components. However, in the Latin American region, also located in the high Andes, studies and reports evaluating the nexus are limited compared to other regions of the Earth. The nexus approach assumes scarcity of natural resources and recognizes that water, energy, food, and other resources are interconnected in complex webs of relationships, and that resource use and availability are interdependent (Leck et al., 2015).

More recently, the nexus concept has been applied to the field of sustainable development (Mo & Zhang, 2013; Leck et al., 2015). By adopting a nexus concept with the wastewater industry, sustainable development of wastewater treatment should be achieved, considering other system components. Thus, the concept of microbial relationships can be proposed for biological wastewater treatment, like systems related to water relationships. A microbial nexus implies adaptation of microbial niches to suit different microbial communities for efficient removal of known and unknown pollutants (Pardo, 2022).

In this sense, optimizing microbial niches to remove more types of pollutants will help address wastewater treatment challenges by designing and operating biological treatment processes in microenvironments, as certain types of microbial migration can adapt to new emergent, sustainable substrate gradient compounds over time, and even favor removal of newly formed compounds (Wu & Yin, 2020; Daims et al., 2006). Through proper adjustment, functional microorganisms suitable for the process can be selected and enriched to achieve the removal of various pollutants in the nexus system except for energy recovery and resource recycling. On the other hand, the isolation of microorganisms from remote habitats with extreme climatic conditions has led to the discovery of many enzymes with attractive properties that may be used as valuable reagents in industrial energy-saving processes (Brück et al., 2022). Especially interesting for wastewater treatment and environmental bioremediation in low-temperature regions is the need for in situ applications without heating and stability under these conditions (Jiang et al., 2021). Additionally, the development of cold-adapted microbial consortia could potentially enhance bioremediation efficiency in these environments.

Therefore, the development and prospects are addressed in this text through aspects of the microbial nexus, microbial function, and the development and application of new technologies based on such an approach for wastewater treatment (Wu & Yin, 2020). Wastewater treatment has yet to resolve stringent discharge standards and the emergence of new pollutants, so this chapter puts forward the principles of the microbial nexus for biological wastewater treatment to achieve efficient removal of known and unknown pollutants by various microbial communities.

The microbial nexus can be used to solve new challenges emerging in wastewater treatment through the development and application of principles such as microbial enrichment, microbial function, specific microbial metabolism, system design and operational control, and new technologies that support the application of such principles as infrastructure construction and microenvironment establishment. The aim of this approach is to achieve simultaneous water purification and energy and/or nutrient recovery.

17.2 MICROBIAL ENRICHMENT

The enrichment process consists of supplying appropriate and sufficient substrate at the proper pH and temperature conditions so that microorganisms can rapidly proliferate during wastewater treatment (Chandra & Kumar, 2015; Agrawal et al., 2021; Liu et al., 2021). There are several organic compounds such as acetate, cellulose, and glucose that can be used as substrates in the enrichment process (Ebadinezhad et al., 2019). The presence of these substrates and conditions that ensure their utilization as an energy source in such a way that they can improve the efficiency of the remediating microorganisms is an additional strategy that aims to increase the initial microbial population until it reaches the level where there is no risk of losing the microbial load and having adapted to the conditions of the wastewater to be treated, they can thrive in less ideal conditions (Rizzo et al., 2021). When the substrate is finished, the initial microbial population will be replaced by another one according to the substrate that is now in greater quantity or has been generated as a result of the degradation of the previous substrate. In this sense, we also speak of microbial enrichment when, depending on the availability of the substrate, the wastewater can be inoculated at a certain time with specific species according to the need for treatment of the pollutant (Yin et al., 2022). It is usual, for example, to first remove nitrogen-rich organic matter with nitrifying or photosynthetic microorganisms that also remove phosphates and potassium and then treat other pollutants by enriching the water with molasses to treat some metals (Akgul et al., 2021).

17.3 MICROBIAL FUNCTION

Various elements, mainly metals, can be accumulated in the microbial biomass (Medfu Tarekegn et al., 2020). Therefore, microorganisms are capable of mediating the precipitation of metals and minerals through the production of metabolites (Fig. 17.2), modifying the physicochemical conditions of their cellular environment, and releasing substances of a diverse nature capable of absorbing, binding, or trapping soluble or insoluble metal species (Kanamarlapudi et al., 2018). These capabilities are recognized as microbial functions because they involve maintaining balance or fulfilling a role within nutrient cycling or mediating between organic and inorganic phases of elements that are also present in wastewater (Nealson & Stahl, 2018). This implies that there are specific niches in the transformation of each element, including the specificity of microbial strains, species, or ecotypes adapted to adverse conditions such as low temperatures in high Andean areas (Jing & Kjellerup, 2018).

17.4 SPECIFIC MICROBIAL METABOLISM

With proper management, these functional microorganisms can perform continuous biological metabolic processes in time or space that can be used to successfully purify wastewater (Wu & Yin, 2020). Furthermore, studying the metabolic dynamics, diversity, and microbial interactions of microorganisms is crucial for the development of novel wastewater treatment technologies based on microbial niche improvement (Lawson et al., 2019).

Heterotrophs, for example, are very common microorganisms in wastewater treatment and usually responsible for hydrolysis under both aerobic and anaerobic conditions (Fig. 17.3). Under aerobic conditions, only the soluble substrate is considered to be involved in the growth of heterotrophs. The previously suspended substrate must undergo hydrolysis and be utilized by the microorganisms for their growth. This process contributes the most to COD removal, production of new biomass, and oxygen demand. Ammonia nitrogen is consumed in the growth process by its incorporation into the cells while also changing the alkalinity. In the absence of

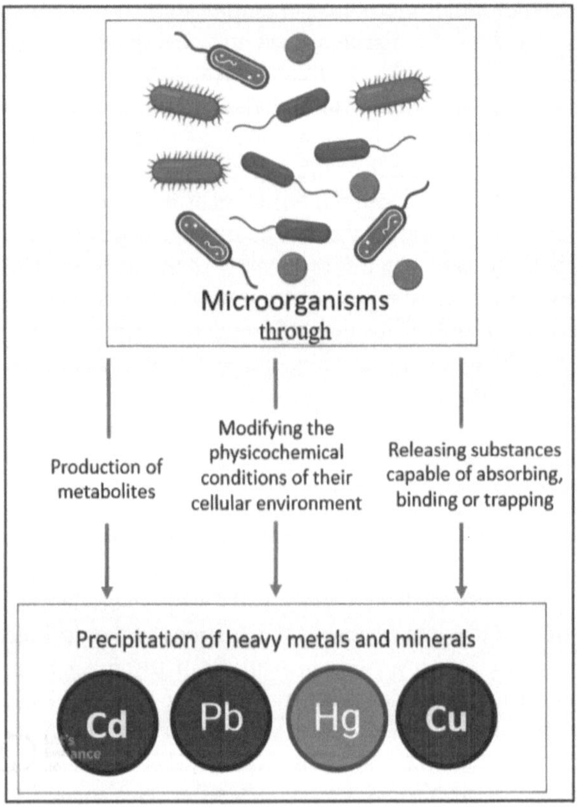

FIGURE 17.2 Diagram of the forms of metal accumulation in the biomass of microorganisms.

oxygen, heterotrophic organisms can utilize nitrates as a terminal electron acceptor with soluble organic substrate as the carbon and energy source The process results in a supplementary production of heterotrophic biomass and gaseous nitrogen (denitrification). The latter results from the reduction of nitrates with its subsequent change in alkalinity (Duque-Sarango et al., 2018).

17.5 MICROBIAL ECOLOGY-BASED INFRASTRUCTURE DESIGN AND OPERATION

Processes that consume more energy are negative in the approach to climate change and there is a lot of information comparing aerobic and anaerobic processes without taking into account the quality of discharge output, nutrient removal, or establishing comparisons in specific contexts that are not generalizable to all environments (Rahman et al., 2018). In this sense, it is necessary to look for treatments that solve problems of pollutants in wastewater such as the use of biologically based treatment modules in the communities of rural villages of the Andes, due to their ease of use and construction with materials from the area (Stefanakis, 2020). For the design of infrastructure, the biological principle that determines the performance of bacterial

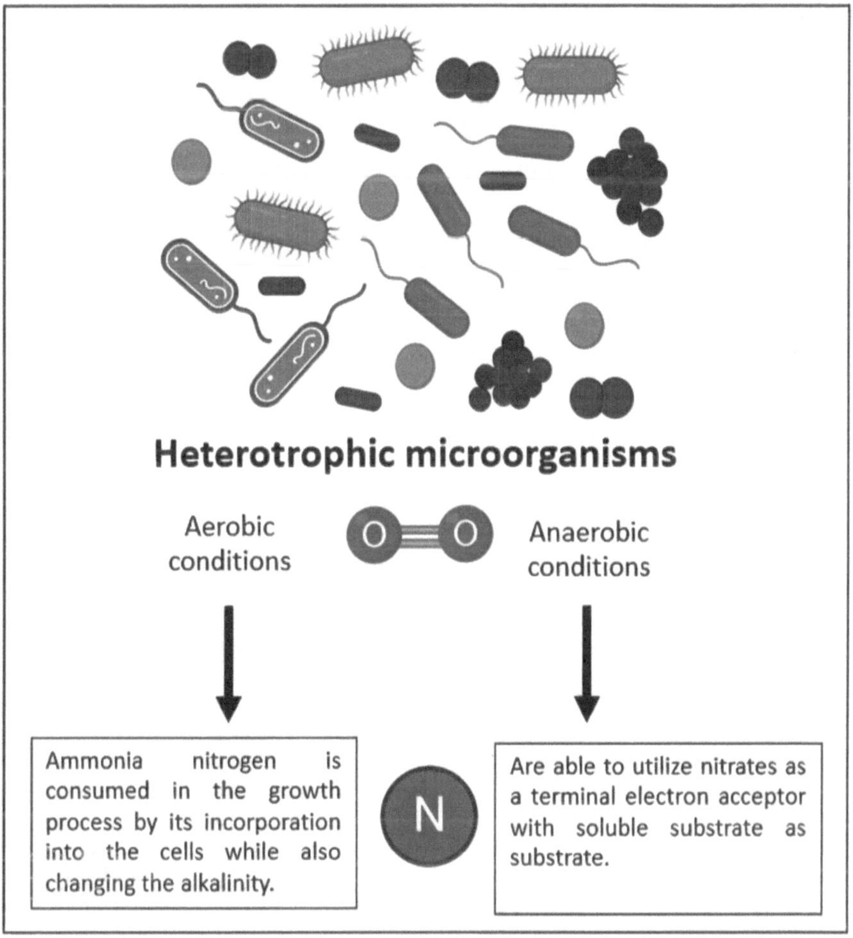

FIGURE 17.3 Diagram of the role of heterotrophic microorganisms in aerobic and anaerobic conditions.

consortia due to their antioxidant capacity must be respected. In addition, when these microbes come into contact with organic matter, they excrete beneficial substances such as vitamins, organic acids, and minerals. Similarly, they thrive by competitive exclusion, both in contaminated and decomposing niches, and then die when conditions are clean, so there is no risk of secondary contamination, but there is an alternation of species and generation of biomass (Kataki et al., 2021).

17.6 DEVELOPMENT OF NEW TECHNOLOGIES FOR THE USE OF MICROORGANISMS

Currently, the use of microorganisms is indisputable given the evidence; however, some limitations go beyond their ecological requirements (Hutchins et al., 2019). Limitations are also found at the technological level, so more efficient methods should be explored that allow working in nonaxenic conditions, allowing the absorption of contaminants without modifying the rheology of the wastewater, such as immobilization in matrices or the use of consortia without the need to alternate structures or reactors of different types to work on each specific contaminant (Jain et al., 2022). The use of live and/or complete microorganisms presents a low efficiency concerning the use of their enzymes or purified metabolites, so that work on the improvement of these organisms in the face of the problem of genetic manipulation and what it represents in ethical and evolutionary terms is also a latent problem (Patel et al., 2023).

REFERENCES

Abdel-Monem M. O., Al-Zubeiry A. H., and Al-Gheethi A. A. (2010). Biosorption of nickel by Pseudomonas cepacia 120S and *Bacillus subtilis* 117S. *Water Science Technology, 61*, 2994–3007

Agrawal, N., Kumar, V., and Shahi, S. K. (2021). Biodegradation and detoxification of phenanthrene in in-vitro and in-vivo conditions by a newly isolated ligninolytic fungus *Coriolopsis byrsina* strain APC5 and characterization of their metabolites for environmental safety. *Environmental Science and Pollution Research.* https://doi.org/10.1007/s11356-021-15271-w

Akgul, V., Cirik, K., Duyar, A., and Koroglu, E. O. (2021). The effect of olive oil mill and molasses wastewater as a co-substrate during simultaneous textile wastewater treatment and energy generation. *International Journal of Global Warming*, *23*(1), 30–42.

Albarracín, V. H., Gärtner, W., and Farias, M. E. (2015b). Forged under the sun: Life and art of extremophiles from Andean Lakes. *Photochemistry and Photobiology*, *92*(1), 14–28. Available at: https://pubmed.ncbi.nlm.nih.gov/26647770/ [Accessed 21 Mar. 2023].

Albarracín, V. H., Kurth, D., Ordoñez, O. F., Belfiore, C., Luccini, E., Salum, G. M., Piacentini, R. D., and Farías, M. E. (2015a). High-up: A remote reservoir of microbial extremophiles in Central Andean Wetlands. *Frontiers in Microbiology*, 6. Available at: www.frontiersin.org/articles/10.3389/fmicb.2015.01404/full [Accessed 21 Mar. 2023].

Al-Gheethi A. A. S., Norli I., Lalung J., Megat-Azlan A., Nur-Farehah Z. A., and Ab. Kadir M. O. (2014). Biosorption of heavy metals and cephalexin from secondary effluents by tolerant bacteria. *Clean Technology Environmental Policy 16*, 137–148. doi:10.1007/s10098-013- 0611-9

Aravindhan R., Fathima A., Selvamurugan M., Rao J. R., and Balachandran U. N. (2012). Adsorption, desorption, and kinetic study on Cr(III) removal from aqueous solution using *Bacillus subtilis* biomass. *Clean Technology Environmental Policy, 14*, 727–735. doi:10.1007/s10098- 011-0440-7

Bader, M. Y., M. Rietkerk, and A. K. Bregt. (2007). Vegetation structure and temperature regimes of tropical alpine treelines. *Arctic, Antarctic, and Alpine Research*, 39:353–364.

Baquero, F., Coque, T. M., Galán, J. C., and Martinez, J. L. (2021). The origin of niches and species in the bacterial world. *Frontiers in Microbiology*, [online] 12. Available at: www.frontiersin.org/articles/10.3389/fmicb.2021.657986/full [Accessed 21 Mar. 2023].

Brück, S. A., Contato, A. G., Gamboa-Trujillo, P., de Oliveira, T. B., Cereia, M., and de Moraes Polizeli, M. de L. T. (2022). Prospection of psychrotrophic filamentous fungi isolated from the high Andean Paramo Region of Northern Ecuador: Enzymatic activity and molecular identification. *Microorganisms*, *10*(2), Article 2. https://doi.org/10.3390/microorganisms10020282

Buytaert, W., Deckers, J., and Wyseure, G. (2007). Regional variability of volcanic ash soils in south Ecuador: The relation with parent material, climate and land use. *CATENA*, *70*(2), 143–154. https://doi.org/10.1016/j.catena.2006.08.003

Chandra, R. and Kumar, V. (2017a). Detection of *Bacillus* and *Stenotrophomonas* species growing in an organic acid and endocrine-disrupting chemicals rich environment of distillery spent wash and its phytotoxicity. *Environmental Monitoring and Assessment, 189*, 26. https://doi.org/10.1007/s10661-016-5746-9

Chandra, R. and Kumar, V. (2017b). Detection of androgenic-mutagenic compounds and potential autochthonous bacterial communities during in-situ bioremediation of post methanated distillery sludge. *Frontiers in Microbiology, 8*, 87. https://doi.org/10.3389/fmicb.2017.00887

Chandra R., Kumar V., and Yadav S., 2017. Extremophilic ligninolytic enzymes. In: Sani R., Krishnaraj R. (eds.) *Extremophilic Enzymatic Processing of Lignocellulosic Feedstocks to Bioenergy*. Springer, Cham. doi:10.1007/978-3-319-54684-1_8

Chandra, R. and Kumar, V. (2015). Biotransformation and biodegradation of organophosphates and organohalides. In: Chandra, R. (Ed.) *Environmental Waste Management*. Boca Raton: CRC Press. doi:10.1201/b19243-17.

Cortez, A., Orlando Olivares, B., Mayela Parra, R., Lobo, D., Rey, J. C., and Rodríguez, M. F. (2018). Descripción de los eventos de sequía meteorológica en localidades de la Cordillera Central, Venezuela. *Cienc., Ing. Apl.* 1, 23–45. doi:10.22206/cyap.2018.v1i1.pp23–45

Cota-Ruiz, K., Nuñez-Gastelúm, J. A., Delgado-Rios, M., and Martinez-Martinez, A. (2019). Biorremediación: Actualidad de conceptos y aplicaciones. *Biotecnia*, *21*(1), 37–44. Recuperado 8 de abril de 2023, de https://biotecnia.unison.mx/index.php/biotecnia/article/view/811

Daims, H., Taylor, M.W., Wagner, M. (2006). Wastewater treatment: a model system for microbial ecology. *Trends in Biotechnology*, 24, 483–489. https://doi.org/10.1016/j.tibtech.2006.09.002

Dorador, C., Molina, V., Hengst, M., Eissler, Y., Cornejo, M., Fernández, C., and Pérez, V. (2020). Microbial communities composition, activity, and dynamics at Salar de Huasco: A polyextreme environment in the Chilean Altiplano. *Microbial Ecosystems in Central Andes Extreme Environments*, [online] 123–139. Available at: https://link.springer.com/chapter/10.1007/978-3-030-36192-1_9 [Accessed 21 Mar. 2023].

Duque-Sarango, P., Heras-Naranjo, C., Lojano-Criollo, D., and Viloria, T. (2018). Modelamiento del tratamiento biológico de aguas residuales; estudio en planta piloto de contactores biológicos rotatorios. *Revista ciencia UNEMI*, *11*(28), 88–96.

Ebadinezhad, B., Ebrahimi, S., and Shokrkar, H. (2019). Evaluation of microbial fuel cell performance utilizing sequential batch feeding of different substrates. *Journal of Electroanalytical Chemistry*, *836*, 149–157.

Elgarahy, A. M., Elwakeel, K. Z., Mohammad, S. H., and Elshoubaky, G. A. (2021). A critical review of biosorption of dyes, heavy metals and metalloids from wastewater as an efficient and green process. *Cleaner Engineering and Technology*, *4*, 100209. https://doi.org/10.1016/j.clet.2021.100209

Fernandez, A.B., Rasuk, M.C., Visscher, P.T., Contreras, M., Novoa, F., Poire, D.G., Patterson, M.M., Ventosa, A., and Farias, M.E. (2016). Microbial diversity in sediment ecosystems (Evaporites Domes, microbial mats, and crusts) of Hypersaline Laguna Tebenquiche, Salar de Atacama, Chile. *Frontiers in Microbiology*, [online] 7. Available at: https://pubmed.ncbi.nlm.nih.gov/27597845/ [Accessed 21 Mar. 2023].

Hasan S. H. and Srivastava P. (2011). Biosorptive abatement of Cd^{2+} by water using immobilized biomass of *Arthrobacter* sp. response surface methodological approach. *Industrial and Engineering Chemical Research* 50(1):247–258

Hu, J., Cai, W., Wang, C., Du, X., Lin, J., and Cai, J. (2018). Purification and characterization of alkaline lipase production by Pseudomonas aeruginosa HFE733 and application for biodegradation in food wastewater treatment. *Biotechnology and Biotechnological Equipment*, *32*(3), 583–590.

Huasasquiche Sarmiento, L., Moreno Díaz, P., Jiménez Dávalos, J. (2020). Caracterización y evaluación del potencial PGPR de la microflora asociada al cultivo de Tarwi (Lupinus mutabilis Sweet). *Ecología Aplicada*, 19, 65–76. https://doi.org/10.21704/rea.v19i2.1557

Hutchins, D. A., Jansson, J. K., Remais, J. V., Rich, V. I., Singh, B. K., and Trivedi, P. (2019). Climate change microbiology—problems and perspectives. *Nature Reviews Microbiology*, *17*(6), 391–396.

Iñiguez-Armijos, C., Tapia-Armijos, M. F., Wilhelm, F., and Breuer, L. (2022). Urbanisation process generates more independently acting stressors and ecosystem functioning impairment in tropical Andean streams. *Journal of Environmental Management*, *304*, 114211. https://doi.org/10.1016/j.jenvman.2021.114211

Jain, M., Khan, S. A., Sharma, K., Jadhao, P. R., Pant, K. K., Ziora, Z. M., and Blaskovich, M. A. (2022). Current perspective of innovative strategies for bioremediation of organic pollutants from wastewater. *Bioresource Technology*, *344*, 126305.

Jiang, G., Chen, P., Bao, Y., Wang, X., Yang, T., Mei, X., Banerjee, S., Wei, Z., Xu, Y., and Shen, Q. (2021). Isolation of a novel psychrotrophic fungus for efficient low-temperature composting. *Bioresource Technology*, *331*, 125049. https://doi.org/10.1016/j.biortech.2021.125049

Jiménez-Rivillas, C., García, J. J., Quijano-Abril, M. A., Daza, J. M., Morrone, J. J., Jiménez-Rivillas, C., García, J. J., Quijano-Abril, M. A., Daza, J. M., and Morrone, J. J. (2018). A new biogeographical regionalisation of the Páramo biogeographic province. *Australian Systematic Botany*, *31*(4), 296–310. https://doi.org/10.1071/SB18008

Jing, R. and Kjellerup, B. V. (2018). Biogeochemical cycling of metals impacting by microbial mobilization and immobilization. *Journal of Environmental Sciences*, *66*, 146–154.

Kanamarlapudi, S. L. R. K., Chintalpudi, V. K., and Muddada, S. (2018). Application of biosorption for removal of heavy metals from wastewater. *Biosorption*, *18*(69), 70–116.

Kataki, S., Chatterjee, S., Vairale, M.G., Dwivedi, S.K., and Gupta, D.K. (2021). Constructed wetland, an eco-technology for wastewater treatment: A review on types of wastewater treated and components of the technology (macrophyte, biolfilm and substrate). *Journal of Environmental Management,* 283, 111986. https://doi.org/10.1016/j.jenvman.2021.111986

Kurth, D., Belfiore, C., Gorriti, M. F., Cortez, N., Farias, M. E., and Albarracín, V.H. (2015). Genomic and proteomic evidences unravel the UV-resistome of the poly-extremophile *Acinetobacter* sp. Ver3. *Frontiers in Microbiology*, [online] 06. Available at: www.frontiersin.org/articles/10.3389/fmicb.2015.00328/full [Accessed 21 Mar. 2023].

Kumar, V., Ameen, F., Islam, M. A., Agrawal, S., Motghare, A., Dey, A., Shah, M. P., Américo-Pinheiro, J. H. P., Singh, S., and Ramamurthy, P. C. (2022). Evaluation of cytotoxicity and genotoxicity effects of refractory pollutants of untreated and biomethanated distillery effluent using *Allium cepa. Environmental Pollution, 300*, 118975. https://doi.org/10.1016/j.envpol.2022.118975

Kumar, V., Shahi, S. K., Ferreira, L. F. R., Bilal, M., Biswas, J. K., and Bulgariu, L. (2021). Detection and characterization of refractory organic and inorganic pollutants discharged in biomethanated distillery effluent and their phytotoxicity, cytotoxicity, and genotoxicity assessment using *Phaseolus aureus* L. and *Allium cepa* L. *Environmental Research, 201*, 111551. https://doi.org/10.1016/j.envres.2021.111551

Kumar, V. and Verma, P. (2023). A critical review on environmental risk and toxic hazards of refractory pollutants discharged in chlorolignin waste of pulp and paper mills and their remediation approaches for environmental safety. *Environmental Research, 236*, 116728. https://doi.org/10.1016/j.envres.2023.116728

Lawson, C. E., Harcombe, W. R., Hatzenpichler, R., Lindemann, S. R., Löffler, F. E., O'Malley, M. A., and McMahon, K. D. (2019). Common principles and best practices for engineering microbiomes. *Nature Reviews Microbiology, 17*(12), 725–741.

Leck, H., Conway, D., Bradshaw, M., and Rees, J. (2015). Tracing the water–energy–food nexus: Description, theory and practice. *Geography Compass*, *9*(8), 445–460. https://doi.org/10.1111/gec3.12222

Liu, W., Wang, Q., Shen, Y., and Yang, D. (2021). Enhancing the in-situ enrichment of anammox bacteria in aerobic granules to achieve high-rate CANON at low temperatures. *Chemosphere*, *278*, 130395.

Medfu Tarekegn, M., Zewdu Salilih, F., and Ishetu, A. I. (2020). Microbes used as a tool for bioremediation of heavy metal from the environment. *Cogent Food and Agriculture*, *6*(1), 1783174.

Mo, W., and Zhang, Q. (2013). Energy–nutrients–water nexus: Integrated resource recovery in municipal wastewater treatment plants. *Journal of Environmental Management*, *127*, 255–267. https://doi.org/10.1016/j.jenvman.2013.05.007

Mohan, S. V., Butti, S. K., Amulya, K., Dahiya, S., and Modestra, J. A. (2016). Waste biorefinery: A new paradigm for a sustainable bioelectro economy. *Trends in Biotechnology*, *34*(11), 852–855. https://doi.org/10.1016/j.tibtech.2016.06.006

Ndlovu, T., Rautenbach, M., Vosloo, J.A., et al. (2017). Characterisation and antimicrobial activity of biosurfactant extracts produced by *Bacillus amyloliquefaciens* and *Pseudomonas aeruginosa* isolated from a wastewater treatment plant. *AMB Expr, 7*, 108. https://doi.org/10.1186/s13568-017-0363-8

Nealson, K. H., and Stahl, D. A. (2018). Microorganisms and biogeochemical cycles: what can we learn from layered microbial communities?. In *Geomicrobiology* (pp. 5–34). De Gruyter.

Ortiz-Ojeda, P., Ogata-Gutiérrez, K., Zúñiga-Dávila, D., Ortiz-Ojeda, P., Ogata-Gutiérrez, K., Zúñiga-Dávila, D. (2017). Evaluation of plant growth promoting activity and heavy metal tolerance of psychrotrophic bacteria associated with maca (<em>Lepidium meyenii</em> Walp.) rhizosphere. AIMSMICRO 3, 279–292. https://doi.org/10.3934/microbiol.2017.2.279

Ozturk S, Aslim B, and Tunceli A. (2011). Biosorption of chromium (VI) ions from aqueous system by free and immobilized biomass of wild *Synechocystis* sp a comparative study. *Fresenius Environmental Bulletin 20*:2412–2418,

Pardo, C. T. (2022). *Application of a Food-Energy-Water (FEW) Nexus Approach to Water Resources Management in the Colombian Andean Region* [Thesis, Purdue University Graduate School]. https://doi.org/10.25394/PGS.19678401.v1

Patel, A. K., Dong, C. D., Chen, C. W., Pandey, A., and Singhania, R. R. (2023). Production, purification, and application of microbial enzymes. In *Biotechnology of microbial enzymes* (pp. 25–57). Academic Press.

Peña, M. and Mara, D. (2003). High-rate anaerobic pond concept for domestic wastewater treatment: Results from pilot scale experience. In: Water and environmental management series (WEMS). IWA Publishing, London. www.bvsde.paho.org/bvsacd/agua2003/pilot.pdf.

Rahman, S. M., Eckelman, M. J., Onnis-Hayden, A., and Gu, A. Z. (2018). Comparative life cycle assessment of advanced wastewater treatment processes for removal of chemicals of emerging concern. *Environmental Science and Technology*, *52*(19), 11346–11358.

Rascón, J., Corroto, F., Leiva-Tafur, D., and Gamarra Torres, O. A. (2021). Spatial and temporal dynamics of the trophic state in a tropical high Andean Lake in northern. *Ecologia Austral*, *31*(2), 343–356. Scopus. https://doi.org/10.25260/EA.21.31.2.0.1200

Rashid, R., Shafiq, I., Akhter, P., Iqbal, M. J., and Hussain, M. (2021). A state-of-the-art review on wastewater treatment techniques: The effectiveness of adsorption method. *Environmental Science and Pollution Research*, *28*(8), 9050–9066. https://doi.org/10.1007/s11356-021-12395-x

Rizzo, C., Caldarone, B., De Luca, M., De Domenico, E., and Giudice, A. L. (2021). Native bilge water bacteria as biosurfactant producers and implications in hydrocarbon-enriched wastewater treatment. *Journal of Water Process Engineering*, *43*, 102271.

Saona, L. A., Soria, M., Villafañe, P. G., Lencina, A. I., Stepanenko, T. and Farías, M. E. (2020). Andean microbial ecosystems: Traces in hypersaline lakes about life origin. *Astrobiology and Cuatro Ciénegas Basin as an Analog of Early Earth*, [online] 167–181. Available at: https://link.springer.com/chapter/10.1007/978-3-030-46087-7_8 [Accessed 21 Mar. 2023].

Sousa H., Sousa C. A., Vale F., Santos L., and Simões M. (2023). Removal of parabens from wastewater by Chlorella vulgaris-bacteria co-cultures. *Science Total EnvironMental*, *884*, 163746. doi:10.1016/j.scitotenv.2023.163746.

Stefanakis, A. I. (2020). Constructed wetlands: Description and benefits of an eco-tech water treatment system. In *Waste Management: Concepts, Methodologies, Tools, and Applications* (pp. 503–525). IGI Global.

Sun Y, Li X, and Liu G. (2022). Enhanced pollutants removal and high-value cell inclusions accumulation with Fe^{2+} in heavy oil refinery treatment system using Rhodopseudomonas and Pseudomonas. *Chemosphere*. doi:10.1016/j.chemosphere.2022.133520.

Van Loosdrecht, M. C. M., and Brdjanovic, D. (2014). Anticipating the next century of wastewater treatment. *Science*, *344*(6191), 1452–1453. https://doi.org/10.1126/science.1255183

Völker, J., Castronovo, S., Wick, A., Ternes, T. A., Joss, A., Oehlmann, J., and Wagner, M. (2016). Advancing biological wastewater treatment: Extended anaerobic conditions enhance the removal of endocrine and dioxin-like activities. *Environmental Science and Technology*, *50*(19), 10606–10615. https://doi.org/10.1021/acs.est.5b05732

Wu, G., and Yin, Q. (2020). Microbial niche nexus sustaining biological wastewater treatment. *NPJ Clean Water*, *3*(1), 33.

Yang, X., Liu, X., Xie, S., Feng, J., and Lv, J. (2022). The interaction between Chlorococcum sp. GD and indigenous bacteria in the process of municipal wastewater treatment. *Journal of Cleaner Production, 362*, 132472.

Yin, Q., Sun, Y., Li, B., Feng, Z., and Wu, G. (2022). The r/K selection theory and its application in biological wastewater treatment processes. *Science of the Total Environment*, 824, 153836. https://doi.org/10.1016/j.scitotenv.2022.153836

Index

Note: Page numbers in *italics* and **bold** refer to figures and tables, respectively.

N

O

P

T

X

Z